Advances in Geographical and Environmental Sciences

Advances in Geographical and Environmental Sciences synthesizes series diagnostigation and prognostication of earth environment, incorporating challenging interactive areas within ecological envelope of geosphere, biosphere, hydrosphere, atmosphere and cryosphere. It deals with land use land cover change (LUCC), urbanization, energy flux, land-ocean fluxes, climate, food security, ecohydrology, biodiversity, natural hazards and disasters, human health and their mutual interaction and feedback mechanism in order to contribute towards sustainable future. The geosciences methods range from traditional field techniques and conventional data collection, use of remote sensing and geographical information system, computer aided technique to advance geostatistical and dynamic modeling.

The series integrate past, present and future of geospheric attributes incorporating biophysical and human dimensions in spatio-temporal perspectives. The geosciences, encompassing land-ocean-atmosphere interaction is considered as a vital component in the context of environmental issues, especially in observation and prediction of air and water pollution, global warming and urban heat islands. It is important to communicate the advances in geosciences to increase resilience of society through capacity building for mitigating the impact of natural hazards and disasters. Sustainability of human society depends strongly on the earth environment, and thus the development of geosciences is critical for a better understanding of our living environment, and its sustainable development.

Geoscience also has the responsibility to not confine itself to addressing current problems but it is also developing a framework to address future issues. In order to build a 'Future Earth Model' for understanding and predicting the functioning of the whole climatic system, collaboration of experts in the traditional earth disciplines as well as in ecology, information technology, instrumentation and complex system is essential, through initiatives from human geoscientists. Thus human geoscience is emerging as key policy science for contributing towards sustainability/survivality science together with future earth initiative.

Advances in Geographical and Environmental Sciences series publishes books that contain novel approaches in tackling issues of human geoscience in its broadest sense—books in the series should focus on true progress in a particular area or region. The series includes monographs and edited volumes without any limitations in the page numbers.

Kusum Pandey · N. L. Kushwaha ·
Chaitanya B. Pande · K. G. Singh
Editors

Artificial Intelligence and Smart Agriculture

Technology and Applications

Springer

Editors
Kusum Pandey
Garhwal Regional Centre
G. B. Pant National Institute of Himalayan Environment
Srinagar, Uttarakhand, India

Chaitanya B. Pande
Indian Institute of Tropical Meteorology (IITM)
Pune, Maharashtra, India

N. L. Kushwaha
ICAR-Indian Agricultural Research Institute
New Delhi, Delhi, India

K. G. Singh
Punjab Agricultural University
Ludhiana, Punjab, India

ISSN 2198-3542 ISSN 2198-3550 (electronic)
Advances in Geographical and Environmental Sciences
ISBN 978-981-97-0340-1 ISBN 978-981-97-0341-8 (eBook)
https://doi.org/10.1007/978-981-97-0341-8

This Springer imprint is published by the registered company Springer Nature Singapore Pte Ltd.
The registered company address is: 152 Beach Road, #21-01/04 Gateway East, Singapore 189721, Singapore

Contents

Part I Fundamentals of Smart Agriculture

1 **Artificial Intelligence and Smart Farming: An Overview** 3
Varsha Kanojia, A. Subeesh, and N. L. Kushwaha

2 **Remote Sensing in Precision Agriculture: Current Status and Applications** 23
Dilip Kumar, Asit Kumar Pradhan, Rajni Jain, Vikas Kumar, Sneha Murmu, Ipsita Samal, and Himanshu Shekhar Chaurasia

3 **Revolutionizing Agriculture: Geospatial Technologies and Precision Farming in India** 43
Wasim Ayub Bagwan

4 **Assessment of Water Use Efficiency for Sustainable Water Management by Integration of Remote Sensing and Machine Learning** 61
K. Krupavathi, M. Raghu Babu, A. Mani, P. R. K. Parasad, and L. Edukondalu

5 **Pathways for Intensified Agriculture Through Sustainable Water Management: Need of an Hour** 79
Pooja Purushotham, G. Sahana, Gaurav Rakhonde, Namburi Karunakar Reddy, Anusha M. Nayak, and K. M. Pravalika

6 **Crop Yield Prediction Using Artificial Intelligence and Remote Sensing Methods** 103
Rahul Banerjee, Bharti, Pankaj Das, and Sadaf Khan

Part II Technology, Tools and Techniques

7 Role of Artificial Intelligence in Revolutionizing Agricultural Technology 121
Ali Raza, Yongguang Hu, Siham Acharki, Noman Ali Buttar, Muhammad Zubair, Neyha Rubab Syed, Dinesh Kumar Vishwakarma, and Ahmed Elbeltagi

8 Weather Intelligence for Climate-Resilient Agriculture 147
V. A. Blessy, K. K. Shaheemath Suhara, Anu Varughese, M. R. Namitha, P. R. Anjitha Krishna, Ardra Wilson, and K. Sreeja

9 Big Data Analytics for Improved Weather Forecasting and Disaster Management 175
Gaurav Y. Rakhonde, Shalaka Ahale, Namburi Karunakar Reddy, Pooja Purushotham, and Ananya Deshkar

10 Crop Monitoring System Integrating with Internet of Things and Artificial Intelligence 193
Asit Kumar Pradhan, Dilip Kumar, Himanshushekhar Chaurasia, Sneha Murmu, Ipsita Samal, Tanmaya Kumar Bhoi, G. A. K. Kumar, and Biswajit Mondal

11 Soil Moisture and Temperature Management Using IoT for Sustainable Farming 209
P. Kannan, M. Mohamed Roshan Abu Firnass, Jay Bose, S. Ponmani, and C. Swaminathan

12 Internet of Things-Enabled Irrigation Management System for Precision Agriculture 231
Siddharam, L. Aiswarya, Venkatesh Gaddikeri, G. M. Rajesh, Malkhan Singh Jatav, and Jitendra Rajput

13 Unmanned Aerial Vehicles (UAV) for Smart Agriculture 251
R. Pungavi and C. Praveenkumar

14 IoT-Enabled Unmanned Aerial Vehicle: An Emerging Trend in Precision Farming 271
Manish Yadav, B. B. Vashisht, Niharika Vullaganti, S. K. Jalota, S. L. Yadav, Gurjeet Singh, Arun Kumar, and Shivani Kothiyal

15 Smart Irrigation Management Through Unmanned Aerial Vehicles (UAVs) 293
L. Aiswarya, Siddharam, G. M. Rajesh, Venkatesh Gaddikeri, Malkhan Singh Jatav, Jitendra Rajput, and K. R. Asha

16 Biotic Stress Management in Field Crops Using Artificial Intelligence Technologies 315
Shubham Anand and Sarabjot Kaur Sandhu

17 **Soilless Cultivation: A Distinct Vision for Sustainable Agriculture** 337
Alka Chaudhary and Shubham Anand

18 **Nutrient Management Strategies for Water and Nutrient Saving in Substrate Soilless Culture Under Protected Cultivation** 369
Kusum Pandey

19 **Integration of AI and IoT in Soilless Cultivation to Power Sustainable Agricultural Revolution** 387
Ajay N. Satpute, Kishor P. Gavhane, Simranpreet Kaur, Ayushi Jha, Nrusingh Charan Pradhan, and Manojit Chowdhury

20 **Plant Disease Detection, Diagnosis, and Management: Recent Advances and Future Perspectives** 413
Priyanka Negi and Shubham Anand

21 **Disease and Pest Control Through Advance Technology** 437
Shalaka Ahale, Gaurav Rakhonde, Sheenam Bhateja, J. Kuppuraj, and Soumya Mishra

22 **A Critical Analysis of Convolutional Neural Networks for Leaf Disease Detection in Plants** 461
Gurdit Singh

23 **Robotic Seeding or Sowing System in Smart Agriculture** 495
Nrusingh Charan Pradhan, Mude Arjun Naik, Manojit Chowdhury, Ajay Kushwah, K. R. Asha, Tushar Dhar, Kishor P. Gavhane, Sumit B. Urhe, and Ajay N. Satpute

24 **Revolutionising Crop Yield Prediction: The Synergy of Remote Sensing and Artificial Intelligence Technologies** 521
Sarabjot Kaur Sandhu and Shubham Anand

25 **Harvesting the Future: Navigating the Challenges and Limitations of AI in Agriculture** 547
Pooja Purushotham, Gaurav Rakhonde, Namburi Karunakar Reddy, Anusha M. Nayak, K. M. Pravalika, and G. Sahana

Index 565

Part I
Fundamentals of Smart Agriculture

Chapter 1
Artificial Intelligence and Smart Farming: An Overview

Varsha Kanojia, A. Subeesh, and N. L. Kushwaha

Abstract The integration of artificial intelligence (AI) into the field of agriculture has given rise to the concept of smart farming. AI has significantly transformed the way agriculture is approached through precision agriculture. This approach leverages AI algorithms and data analytics to optimize farming practices at a granular level. By utilizing technologies like remote sensing, drones, and satellite imagery, farmers can collect extensive data on crop conditions, soil quality, and environmental factors. AI algorithms then process this data to generate valuable insights, enabling farmers to make informed decisions regarding resource allocation such as water, fertilizers, and pesticides. The result is improved crop yields, reduced resource waste, and minimized environmental impact. This chapter provides an overview of the advancements and applications of AI in smart farming, highlighting its potential to revolutionize traditional agricultural practices. The chapter covers various aspects, including precision agriculture, crop monitoring and management, livestock management, and agricultural robotics. Additionally, it explores the benefits and challenges associated with AI in smart farming, emphasizing the need for further research and development in this rapidly evolving field.

Keywords Machine learning · IoT · Precision farming · Sensors · Smart agriculture

V. Kanojia (✉) · N. L. Kushwaha
Punjab Agricultural University, Ludhiana, Punjab 141004, India
e-mail: varsha.cae023@gmail.com

N. L. Kushwaha
e-mail: nand.kushwaha@icar.gov.in

A. Subeesh
ICAR—Central Institute of Agricultural Engineering, Bhopal, Madhya Pradesh 462038, India

N. L. Kushwaha
ICAR—Indian Agricultural Research Institute, New Delhi 110012, India

K. Pandey et al. (eds.), *Artificial Intelligence and Smart Agriculture*,
Advances in Geographical and Environmental Sciences,
https://doi.org/10.1007/978-981-97-0341-8_1

1.1 Introduction

Smart farming, also known as precision agriculture, involves the use of cutting-edge technologies such as artificial intelligence (AI), Internet of Things (IoT), and big data analytics to optimize agricultural processes (Patel et al. 2023). The integration of AI into various fields has revolutionized industries, and agriculture is no exception. Smart farming, empowered by AI technologies, has the potential to optimize crop production, resource management, and decision-making processes. Artificial intelligence (AI) has emerged as a transformative force across various industries, and one area where its potential is being increasingly harnessed is smart farming. By combining AI algorithms with advanced sensing technologies, smart farming enables more efficient and sustainable agricultural practices (Dhanya et al. 2022; Elbeltagi et al. 2022).

Artificial intelligence has revolutionized the agricultural landscape, empowering farmers with data-driven insights and automation capabilities (Pandey et al. 2020; Kushwaha et al. 2021). In smart farming, AI plays a crucial role in optimizing crop management, resource allocation, and decision-making processes. By leveraging machine learning algorithms, AI systems can analyze vast amounts of data collected from sensors, drones, and satellites, providing valuable insights into soil quality, crop health, weather patterns, and pest infestations (Patrício and Rieder 2018). These insights enable farmers to make informed decisions about irrigation, fertilization, and pest control, leading to increased yields, reduced costs, and minimized environmental impact.

One key application of AI in smart farming is precision agriculture, which involves the use of AI algorithms to precisely tailor farming practices to individual plants or sections of fields. For instance, AI-powered robots equipped with computer vision can identify and remove weeds, reducing the need for herbicides and enhancing crop quality (Wu et al. 2021; Dhanya et al. 2022). Similarly, AI-based systems can monitor plant health in real time, detecting diseases and nutrient deficiencies early on, thus enabling timely interventions (Venkatesh et al. 2022). By optimizing resource utilization and minimizing waste, precision agriculture can significantly improve productivity and sustainability in farming operations.

Furthermore, AI-driven predictive analytics are enhancing agricultural planning and risk management. By analyzing historical data, AI algorithms can forecast crop yields, market demand, and price fluctuations, enabling farmers to make strategic decisions about crop selection, timing, and marketing (Ait Issad et al. 2019). Such insights help farmers mitigate risks and maximize profitability. Additionally, AI-powered systems can provide early warnings about extreme weather events, allowing farmers to take preventive measures and protect their crops (Kamilaris et al. 2017).

However, the adoption of AI in smart farming is not without challenges. Data collection and privacy concerns, as well as the digital divide in rural areas, pose hurdles to widespread implementation (Akkem et al. 2023). Moreover, there is a need for robust AI models that can handle the variability and complexity of agricultural systems, accounting for factors such as soil types, crop varieties, and local conditions (Lu 2019). Addressing these challenges requires collaborative efforts among researchers, policymakers, and industry stakeholders. The integration of artificial intelligence and smart farming holds immense promise for the agricultural sector. By leveraging AI algorithms, farmers can optimize resource allocation, improve crop management, and enhance decision-making processes. Precision agriculture, predictive analytics, and automation are just a few of the areas where AI is transforming farming practices. However, challenges related to data privacy, access, and model robustness must be addressed for the widespread adoption of AI in smart farming. Nevertheless, the potential benefits of AI in revolutionizing agriculture are undeniable. Therefore, this chapter aims to shed light on the transformative role of AI in smart farming and its potential to address the challenges faced by the agriculture industry. Also, it provides the intersection of artificial intelligence and smart farming, discussing its applications, benefits, and future prospects. Automation of agricultural operations using data-driven precision agriculture is shown in Fig. 1.1.

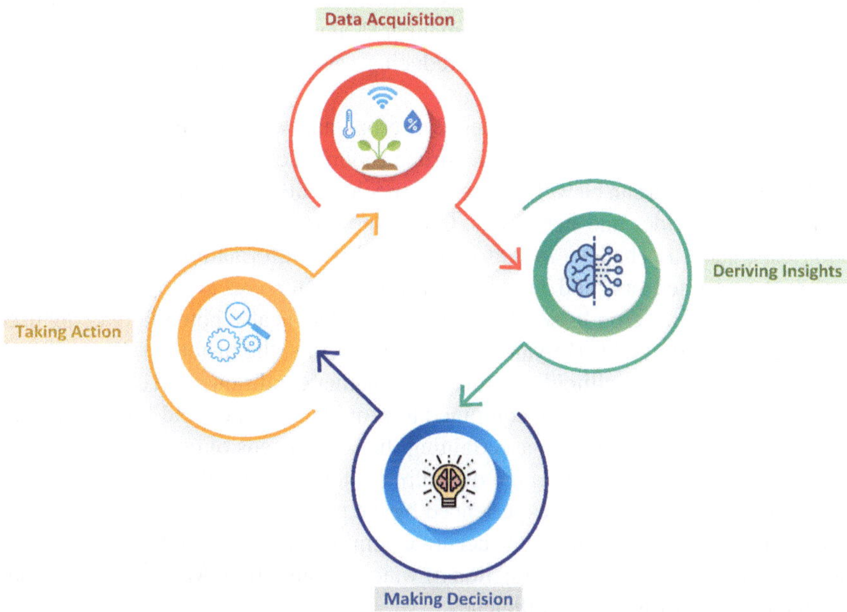

Fig. 1.1 Stages in automation of agricultural operations using data-driven precision agriculture

1.2 Precision Agriculture: Enhancing Crop Production

1.2.1 Remote Sensing and Imaging Techniques

AI algorithms, coupled with remote sensing technologies, enable farmers to monitor crop health, identify pest infestations, and detect nutrient deficiencies. This section explores the applications of AI in remote sensing and imaging techniques for precision agriculture. Remote sensing and computer vision, combined with artificial intelligence, have made significant strides in smart agriculture. These technologies facilitate more efficient and precise control on large farms by gathering important crop information using satellite imagery and aerial drones (Jung et al. 2021). This imagery, when coupled with deep learning and machine learning models, provides insights into crop health status, growth information, and nutrient status. Additionally, information on soil parameters such as soil moisture and temperature can also be collected through the sensors on satellites and drones, and this can be further processed using AI models to create maps that can assist farmers in applying pesticides, fertilizers, and irrigation in a much more efficient manner (Kalambukattu et al. 2018). With this, resource utilization becomes efficient, reducing wastage and improving agricultural yields.

AI algorithms may distinguish between crops and weeds or pests by analyzing high-resolution images captured using drone-mounted cameras (Genze et al. 2022), and this information can be well utilized for mechanical removal or targeted spraying to use less pesticides and herbicides. Crops show unique signs in case of disease outbreaks, such as changes in leaf colour, and this information can be precisely captured using sensors, and farmers can quickly intervene to stop the spread of diseases and reduce crop losses. Remote sensing data is also potentially being used for predicting crop yields (Iniyan et al. 2023). For this, the historical data and real-time data, including satellite/drone imagery, weather conditions, soil characteristics, irrigation information, etc., are combined to provide accurate yield forecasts. Intelligent AI models also assist in ascertaining the water needs of crops by taking into account various factors such as the moisture content of the soil, rates of evapotranspiration, and levels of stress experienced by the plants. The utilization of precision irrigation techniques empowers the accurate administration of water at specific locations and times, resulting in a reduction of water consumption (Maia et al. 2022) and an enhancement of water use efficiency.

Integrating remote sensing with computer vision applications in smart agriculture relies on a range of sensors, including multispectral, hyperspectral, thermal, etc. (Gevaert et al. 2015; Deng et al. 2018). Multispectral sensors can capture data across multiple electromagnetic spectrum bands, providing useful insights into different crop attributes. These sensors measure the reflectance characteristics of the crops, which helps in the identification of plant stress when modelled using computer vision algorithms. Additionally, it can also be used in vegetation indices monitoring, potential disease outbreaks, etc., for assessing crop health. In comparison to multispectral sensors, hyperspectral sensors work at much finer spectral resolutions, collecting data over hundreds of narrow, continuous bands. With the use of this finer and more extensive spectrum data, crop conditions may be analyzed more precisely, allowing for the detection of certain diseases, nutritional deficits, and pest attacks (Adão et al. 2017). Precision farming techniques and focused interventions are made possible by AI algorithms that can properly classify crops and identify minute changes in vegetation after being trained on hyperspectral data (Khan et al. 2022). For tracking changes in heat distribution and crop temperature, thermal sensors are crucial. These sensors track the thermal radiation that crops release and provide useful data on heat sensitivity, plant stress, and water availability. Farmers may improve irrigation practices, spot locations with uneven water distribution, and lessen the effect of severe temperatures on crop growth by combining thermal data with AI algorithms.

1.2.2 Decision Support Systems

Artificial intelligence (AI)-based decision support systems play a crucial role in smart agriculture by providing farmers with valuable insights, predictions, and recommendations to optimize their decision-making processes (Zhai et al. 2020; Pandey et al. 2023). These intelligent systems require massive amounts of data captured from diverse sources such as weather stations, soil sensors, agricultural machinery, and satellite images. These systems gather data and process data from several sources to create a comprehensive picture of the environment and growing conditions. The typical dataset for a decision support system comprises crop health, weather information, soil moisture, nutrient levels, pest-disease information, environmental factors, etc. The data for an intelligent decision support system can be structured, semi-structured, or unstructured, covering historical records, textual information, images, or videos. AI algorithms such as classification, regression, clustering, and anomaly detection are commonly employed to identify the patterns, correlations, and trends in the data collected. Information from textual data is extracted using natural language processing algorithms. To improve decision-making, the system incorporates domain-specific knowledge, principles, and heuristics. Expert systems, which simulate human expertise in particular domains, employ knowledge representation for reasoning and generate recommendations based on predefined rules and logical inferences (Puppe 2012). In decision modelling, the system constructs intelligent models based on the data processed and knowledge representations. For example, a simple decision tree may be used to predict outcomes (class or continuous values) and generate decision rules. Further, the system generates predictions and recommendations based on the modelling results and the results may be presented as interactive dashboards containing charts and reports. An architecture of decision support system is presented in Fig. 1.2.

The integration of AI-powered decision support systems in the context of smart agriculture presents a multitude of advantages; however, it is not without its inherent obstacles. AI-powered decision support systems enable farmers to make data-driven

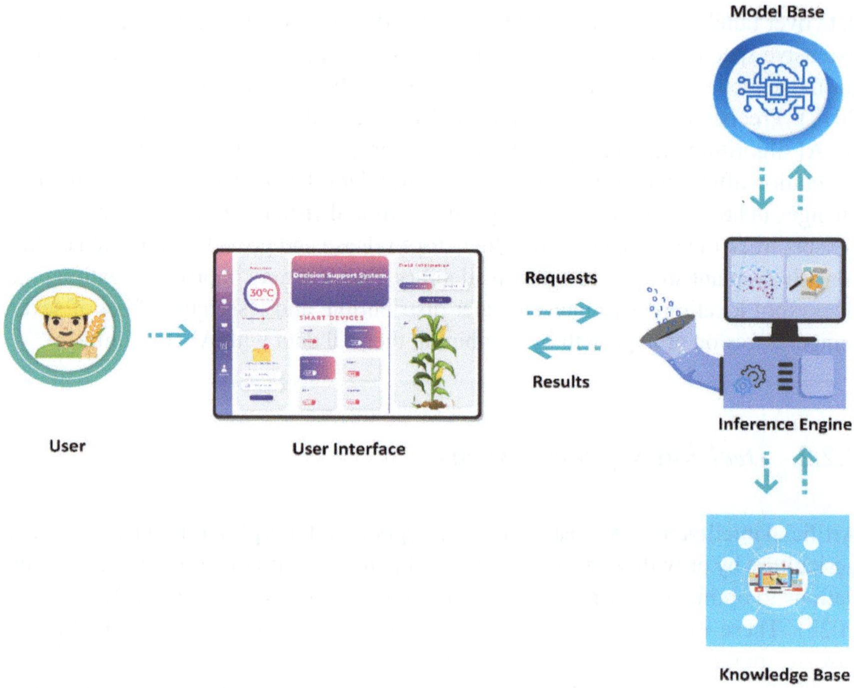

Fig. 1.2 Architecture of decision support system

decision-making, resulting in enhanced productivity, decreased expenses, and optimized allocation of resources (Lynch et al. 2000; Yousaf et al. 2023). Decision support systems powered by artificial intelligence have the capability to promptly detect and distinguish diseases, pests, and weeds. Through the integration of computer vision and machine learning approaches, these systems possess the capability to scrutinize images or sensor data in order to detect indications of diseases, thereby facilitating timely intervention to avert crop impairment. Irrigation scheduling, yield estimation, and crop–soil health information can help farmers to optimize and do strategic planning on harvesting, logistics, and marketing strategies. However, the success of a data-driven decision support system depends on the quality and accessibility of the data. The adoption of these systems by farmers may be impeded by factors such as lack of awareness, unfamiliarity, or resistance to change. Overcoming these obstacles necessitates specialized training, demonstration, and illustration of the advantages and practicality of these systems. For widespread adoption of the DSS, it is imperative to address various factors such as privacy, technical expertise, cost, integration, and ethical considerations.

1.3 Crop Monitoring and Management

1.3.1 Automated Monitoring Systems

AI-based monitoring systems equipped with cameras, sensors, and drones allow farmers to collect vast amounts of data on crop growth, moisture levels, and pest presence. This section explores the role of AI in automating crop monitoring and management processes. Crop monitoring systems based on AI employ advanced prediction modelling and computer vision models to monitor and manage crops in agricultural environments (Subeesh and Mehta 2021) (Fig. 1.3). These systems can analyze data, make predictions, and provide actionable to agricultural practitioners. Crop monitoring systems that utilize artificial intelligence can evaluate the condition of crops through the examination of multiple factors. Sensor data and imagery can be analyzed to detect indications of stress (Chandel et al. 2022), nutrient deficiencies (Yi et al. 2020), diseases (Rangarajan et al. 2018), or pest infestations (Liu and Wang 2021). Historical crop data and present weather conditions can also be employed for predictive analysis regarding agricultural productivity, disease and pest outbreaks, and identifying the optimal time for harvesting.

The distinguishing characteristics between healthy and sick plants may be determined by looking at their colour and texture. For image-based crop monitoring, CNNs are a prevalent choice for building models related to image classification and object detection (Jena et al. 2022). These models can be trained on labelled images and possess the capability to recognize particular crop varieties, detect the presence of diseases or pests, evaluate indicators of plant health, and make estimations regarding potential crop yield. The AI systems have the capability to analyze the data obtained from diverse sensors situated in the field to monitor crop physiological parameters, soil parameters, and environmental parameters such as temperature and humidity. The machine learning models can further process this data to identify nutrient deficiencies, optimize irrigation schedules, detect anomalies, or predict crop stress. Various machine learning models such as decision trees, support vector machines (SVMs), and random forests can be effectively utilized for the analysis of sensor data and predict crop stress. Unmanned aerial vehicles (UAVs) that are outfitted with cameras or multispectral sensors are also utilized to obtain high-resolution imagery of agricultural fields. The images offer comprehensive data regarding the crop in the field and the plausible occurrences of pest invasions. These aerial images can be further processed using deep learning models for the classification or identification of potential problems in the agricultural field (Bouguettaya et al. 2023). The monitoring of crop health has become increasingly convenient with the availability of AI-powered mobile and web applications, affording farmers the flexibility to monitor and analyze their crops' health. The purpose of these applications is to streamline the procedure of gathering, presenting, and analyzing crop health information, thereby enabling farmers to make informed decisions, choices, and make strategic plans.

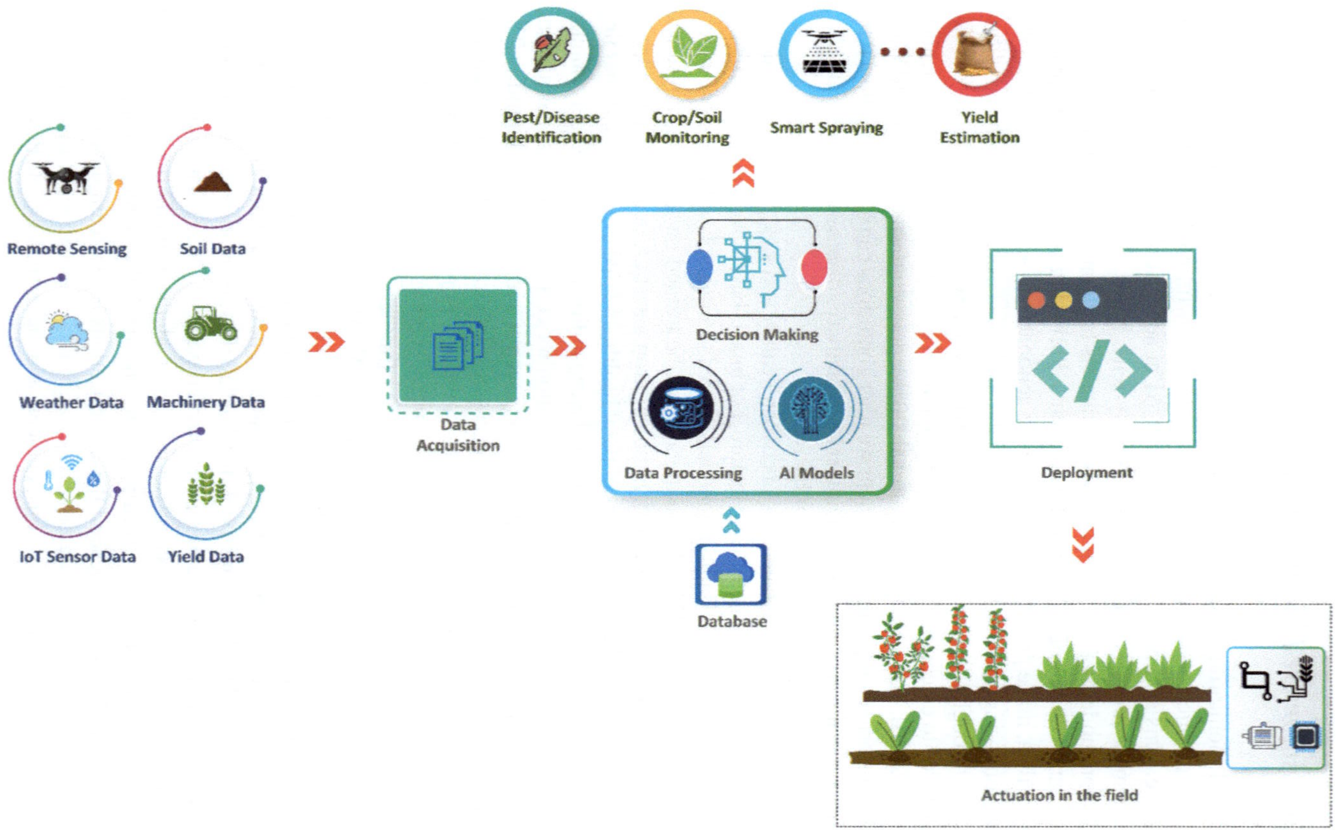

Fig. 1.3 Schematic of data-driven smart agriculture system

1.3.2 Predictive Analytics for Yield Optimization

Utilizing machine learning algorithms, AI can analyze historical and real-time data to predict crop yields accurately. This section discusses the potential of predictive analytics in optimizing crop yields and reducing resource wastage. Artificial intelligence algorithms can be effectively used for yield estimation by analyzing complicated patterns, correlations, and interactions among diverse agricultural yield-influencing parameters (Chlingaryan et al. 2018; Babaee et al. 2021). The dataset comprises information pertaining to crop productivity, encompassing past performance metrics, meteorological trends, soil quality metrics, irrigation details, fertilization methodologies, and other relevant factors. Machine learning techniques facilitate the detection of patterns, trends, and correlations in the data, thereby aiding in the comprehension of the interrelationships among various variables and their influence on crop yields. Machine learning algorithms for yield estimation identify the most pertinent features or variables for yield estimation using the feature selection approach (Sharma et al. 2021; Pant et al. 2021). These features can be weather information (temperature, humidity, precipitation, number of sunshine hours, etc.), soil characteristics (nutrient information, pH levels, soil moisture, etc.), crop genetics, and agronomic practices. This feature selection approach is vital, allowing the algorithms to focus only on the major factors that are significantly influencing crop yields. For predictive modelling, machine learning models such as decision trees, random forest regression, and gradient boosting models are well utilized. They generate predictions on the basis of learned relationships and provide a quantitative understanding of how changes in features/variables affect the yield attributes. Deep learning models, ANN, and RNN models have also been utilized for yield estimation by capturing complex relationships among multiple variables and revealing latent insights that may not be discernible through conventional regression analysis. Satellite and UAV imagery can also provide significant insights into crop health and growth stages and further develop models for yield estimation. Deep learning-based computer vision models can estimate the growth stages and maturity level of orchard crops. By analyzing the visual cues pertaining to the flowering and fruiting (mature/immature fruits), the object detection models can provide accurate estimates of the yield. Estimating the canopy cover and total biomass can also be achieved using computer vision models. The quantification of pixel intensity or greenness present in the images allows the models to easily estimate the biomass and canopy coverage of different crops. This approach provides insights into plant density, vigour, and plant health, which are some crucial factors for yield estimation. These data-driven yield estimation approaches enable growers and stakeholders to optimize agricultural operations, enhance resource allocation, and make informed decisions to improve crop productivity.

1.4 Livestock Management

1.4.1 Health Monitoring and Disease

Detection AI-based systems enable real-time monitoring of livestock health and early detection of diseases. This section explores the applications of AI in livestock management, including wearable devices, computer vision, and data analytics. Artificial intelligence (AI) has the potential to be utilized for the purpose of monitoring the health status of livestock and detecting diseases or abnormalities in real time. These algorithms require vast amounts of data to generate precise predictions by identifying patterns and correlations. For this reason, data from several sensors deployed on livestock farms is gathered to get information on animal behaviour, physiological parameters, and environmental conditions. Wearable devices have the capability to monitor essential physiological parameters, including but not limited to body temperature, heart rate, respiratory rate, and activity levels (Fuentes et al. 2022; Wu et al. 2022). Furthermore, the utilization of cameras and microphones can facilitate the observation of the behaviour and identification of indications of illness or discomfort. Below are some commonly used sensors in livestock monitoring.

- **Wearable Sensors**: these sensors are attached to the body of the animals, commonly in the form of collars and ear tags, and are capable of gathering a range of data including but not limited to body temperature, heart rate, respiratory rate, activity levels, and geospatial coordinates of the position. These sensors provide real-time data that can be closely monitored to track changes in animal health.
- **Environmental Sensors**: these sensors are deployed in the environment where the animal is residing, and they capture the environmental parameters that affect the well-being of the animals, including atmospheric temperature, humidity, air quality, lighting conditions, etc. These parameters can also significantly influence the well-being of the animals.
- **Imaging Sensors**: imaging sensors like RGB cameras, thermal imaging, etc., are helpful in creating real-time intelligent monitoring systems. These images can be effectively utilized to evaluate animal behaviour, detection of physical symptoms and diseases, etc.

Multiple sensors used in smart livestock management are often connected to a centralized system, where it can collect and store real-time information, and further, it can be processed using advanced ML or DL algorithms. Once the machine learning or deep learning model is trained on the data captured, the fine-tuned model can be deployed to monitor the livestock in real time. The input from the farms can be directly fed into the model, which analyzes and provides insights into the health status of individual animals or the entire herd. In the case any anomalies or indications of illness are identified by the models, notifications can be dispatched to farmers or veterinarians for prompt intervention. AI-based livestock health monitoring system comes with numerous advantages. Primarily, it facilitates prompt identification of health issues and ailments, thereby enabling timely intervention and medical attention. This helps in mitigating the transmission of illnesses and diminishes the likelihood of severe issues and mortality. The AI-based system facilitates continuous monitoring of livestock activities and health, found to enhance animal welfare by facilitating proactive care and intervention (Riekert et al. 2020). The automation of livestock monitoring gives enormous time saving for farmers and enables them to allocate time and resources more effectively and enhance productivity.

1.4.2 Automated Feeding and Tracking Systems

AI-driven feeding and tracking systems help farmers optimize feed distribution, monitor animal behaviour, and improve overall livestock productivity. This section highlights the benefits and challenges associated with automated livestock management. Automated feeding systems that utilize artificial intelligence have the capability to distribute animal feed according to pre-established schedules or personalized feeding plans. These automated systems can dispense precise amounts of food, thereby guaranteeing that every animal receives the necessary nutritional intake. AI algorithms can modify feeding schedules by considering various factors such as the weight of the animal, its health status, and growth phases (Bao and Xie 2022). The animal housing area can be equipped with cameras capable of capturing images or video. These visual data are subsequently given to image recognition algorithms for processing. This functionality facilitates the monitoring of individual animals, quantification of their population, and examination of their behaviour. DL model can utilize image inputs to count the animals and their posture to determine the feed dispensing time and quantity. Through the examination of visual data, AI algorithms possess the capability to identify and monitor diverse behavioural patterns exhibited by animals. This may encompass the recognition of indications of emotional or psychological discomfort, abnormal movements, or alterations in dietary patterns (Wu et al. 2023). The aforementioned observations may aid in the identification of potential health concerns, facilitating prompt intervention and mitigating the occurrence of disease outbreaks. Through the examination of patterns and correlations, it is possible to construct predictive models that can forecast animal behaviour, optimize feeding regimens, and anticipate health conditions. The utilization of these models can facilitate the process of making well-informed decisions pertaining to the management and welfare of animals. It is also possible to couple remote monitoring solutions with AI-based animal feeding and tracking systems. This facilitates the monitoring of animals in real time through mobile applications. In the event of detecting anomalies or critical incidents, such as abrupt changes in feeding patterns or abnormal movements, notifications can be sent to the farm owners or veterinary professionals to initiate timely intervention.

In general, the implementation of automated tracking and tracing systems for livestock offers numerous benefits. These include the elimination of subjective human measurements, improved capacity to provide optimal care for the animals throughout their lifespan on the farm, the ability to anticipate events and facilitate timely interventions, etc. (Alanezi et al. 2022; Yousefi et al. 2022).

1.5 Agricultural Robotics

1.5.1 Robotic Applications in Smart Farming

AI-powered agricultural robots are transforming labour-intensive tasks such as planting, harvesting, and weeding. This section provides an overview of the robotic applications in smart farming and their impact on productivity and efficiency. Agribots, also known as agricultural robots, are automated systems that are designed to execute several agricultural operations (Gollakota and Srinivas 2011). The utilization of cutting-edge technologies, including artificial intelligence, computer vision, and robotics, enables these robots to effectively automate and enhance agricultural tasks. The most common harvesting robots are used to automate the most labor-intensive and time-consuming harvesting operations in agriculture (Zhang et al. 2018; Onishi et al. 2019). They are designed to pick fruits, vegetables, and other objects. These robots rely heavily on advanced computer vision models, mainly object detection models, to detect objects, locate them, and identify their ripeness level. Through object detection models, the robots can be trained on massive fruit/vegetable image datasets to recognize the visual characteristics regarding the ripeness level and thus ensure precise harvesting. AI algorithms can also assist in precise control of the robotic arms utilized by the agribots. The integration of machine learning and sensor feedback enables the robots to acquire knowledge of the target and adjust to diverse crop structures, dimensions, and orientations. This also enables individuals to delicately grasp and handle crops, thereby reducing the likelihood of damage during the harvesting process. Similar to harvesting robots, weeding robots can also be equipped with advanced computer vision and artificial intelligence algorithms that enable them to accurately detect and selectively eliminate weeds without causing any damage to the crops. Mechanical methods like cutting or uprooting, as well as precise spraying techniques that target specific areas, can be utilized to apply herbicides in a targeted manner. Autonomous agricultural machinery, including autonomous tractors, has the capability to execute a range of tasks in the field, such as ploughing, tilling, and crop spraying. They have the capability to function independently without the need for human intervention by adhering to predetermined routes or GPS coordinates. It is possible to optimize fuel consumption, minimize human error, and operate continuously for prolonged durations.

Agribots have been found to offer several advantages in the field of agriculture, such as enhanced productivity, increased efficiency, decreased labour expenses, optimized resource utilization, and improved environmental sustainability (Oliveira et al. 2021). It also minimizes the occurrence of human error and fatigue through their tireless and consistent performance. Agribots have the capability to function in challenging and perilous environments, thereby enabling farmers to concentrate on higher-level duties and strategic decision-making. It is worth noting that agribots are not designed to completely substitute human labour. The purpose of these devices is to provide support and enhance the performance of human farmers and labourers, especially in tasks that are monotonous or require significant physical exertion. The management and maintenance of robots still require human supervision and intervention, along with the interpretation of the data and insights they generate.

1.5.2 *Ethical and Social Implications of Agricultural Robotics*

As agricultural robots become more prevalent, ethical considerations regarding job displacement, environmental impact, and data privacy emerge. This section addresses these concerns and emphasizes the need for responsible deployment of AI-driven agricultural robotics. Although agricultural robots bring a plethora of opportunities in agriculture, they encounter several obstacles and several ethical concerns that need to be carefully addressed to facilitate their wider adoption (Dara et al. 2022; Ryan 2022). The increasing automation of agricultural tasks through the use of robots and AI technologies has raised concerns regarding the potential displacement of human labour. Sufficient measures should be taken to provide support to impacted employees via retraining programs and alternative job prospects. Data privacy and security are another major concern as agribots generate massive amounts of data; their protection and privacy are of utmost importance. The data generated by agribots mainly consists of farm-specific information, crop data, and data from sensors. Safeguarding data from unauthorized access, misuse, or breaches is crucial. It is imperative to establish suitable data governance measures and security protocols in order to safeguard the confidentiality and accuracy of sensitive agricultural data. Ensuring transparency and explainability of AI systems is also vital to enable comprehension among farmers and stakeholders regarding the decision-making process. AI models should be developed more responsibly that allows for interpretation and explanation of the reasoning behind their actions. Establishing clear and concrete guidelines and legal frameworks to address accountability and liability issues raised by agribots is crucial to support farmers, consumers, and other stakeholders in years to come. Additionally, it is essential to conduct a thorough assessment of the energy consumption and ecological impacts associated with the use of agricultural robots.

In order to address the ethical considerations, it is necessary to engage a diverse range of stakeholders, encompassing farmers, AI experts, policymakers, and community representatives (Deshpande and Sharp 2022). It is important to establish ethical protocols and benchmarks tailored to the implementation of AI in the agricultural sector, taking into account the unique circumstances and perspectives of the local, regional, and global contexts. The implementation of routine evaluations, inspections, and surveillance of AI systems can aid in the detection and reduction of ethical threats, as well as guarantee the conscientious and beneficial implementation of AI-driven robots within the agriculture sector (Fig. 1.4).

1.6 Data Management and Connectivity

An essential aspect of AI-enabled smart farming is efficient data management and connectivity. With the proliferation of sensors, drones, and other data-gathering devices, there is a massive influx of agricultural data. AI algorithms can process and analyze this vast amount of data to extract meaningful insights. However, data management, storage, and connectivity pose significant challenges. Farmers need reliable and secure systems to handle large datasets, ensure data privacy, and establish seamless connectivity across different devices and platforms. Addressing these challenges will facilitate the effective utilization of AI in smart farming.

Ensuring data security is vital as AI is purely data-driven, and data breaches can disrupt the solutions (Hu et al. 2021). Agricultural data comprises confidential information, including farm operations, financial records, and personal data, and unauthorized access to such information has the potential to result in serious issues. In order to address this, it is advisable to implement resilient access control mechanisms, encryption methodologies, and secure authentication protocols. Conducting routine security audits and vulnerability assessments and implementing intrusion detection systems can effectively detect and mitigate potential vulnerabilities, thereby ensuring the protection of sensitive data from unauthorized access and potential security breaches. The ownership and control of data are also some critical factors to be taken into account. Farmers may have concerns regarding the ownership and utilization of data gathered from their agricultural land. Hence, unambiguous agreements and policies are to be set for farmers to maintain ownership of their data and possess authority over its utilization. Ensuring transparency and securing explicit consent from farmers with regard to the collection and utilization of data can foster trust and safeguard privacy rights. Additionally, there can be concerns regarding exploitation of data or secondary objectives. The sharing or utilization of agricultural data collected for specific purposes without the knowledge and consent of the data owners for unrelated activities is deemed inappropriate. The implementation of data governance frameworks and privacy policies is essential to guarantee that data is utilized solely for its designated objectives.

The success of AI-based agricultural solutions also heavily relies on the connectivity. The establishment of uninterrupted connectivity plays a key role in facilitating the prompt collection, analysis, and decision-making based on the data. Notwithstanding, challenges may arise due to connectivity issues, such as inadequate network coverage in rural regions. In order to tackle this issue, it is imperative to enhance the network infrastructure and broaden the scope of internet connectivity in agricultural areas. Moreover, the utilization of edge computing, a technique that carries out data processing in proximity to the origin, may mitigate the consequences of unstable or sporadic connectivity.

1.7 Environmental Sustainability and Resource Optimization

One of the critical advantages of AI in smart farming is its potential to contribute to environmental sustainability. By precisely monitoring and managing crop health, water usage, and fertilizer application, AI can help to minimize resource wastage and reduce environmental impact. AI-powered systems can optimize irrigation schedules, reducing water usage while maintaining crop productivity. Furthermore, the precise application of fertilizers and pesticides based on real-time data can minimize chemical usage, promoting sustainable farming practices. AI's ability to balance resource optimization with yield maximization is crucial for ensuring the long-term viability and ecological resilience of agricultural systems. The implementation of artificial intelligence (AI) solutions in the field of agriculture is crucial for promoting environmental sustainability and optimizing resource utilization. Artificial intelligence algorithms conduct analysis of soil conditions, weather patterns, and crop health by leveraging real-time data obtained from sensors, satellites, and drones. The utilization of data-driven techniques empowers farmers to make well-informed decisions regarding the allocation of resources, including the optimization of irrigation schedules, the judicious application of fertilizers, and the selective use of pesticides. Artificial intelligence (AI) contributes to the preservation of soil health, prevention of water contamination, and mitigation of environmental impacts by minimizing resource waste and reducing the use of chemicals. The implementation of a site-specific pest and disease management approach diminishes the overuse of chemical pesticides, alleviates environmental contamination, and safeguards advantageous organisms that are crucial in preserving the equilibrium of the ecosystem. Also, through the implementation of AI-based solutions, farmers can make informed decisions and can take actions to mitigate soil erosion, augment biodiversity, and enhance the overall health of the ecosystem.

1.8 Adoption Challenges and Future Outlook

While AI holds great promise for smart farming, there are several challenges that need to be addressed for widespread adoption. One of the primary challenges is the digital divide, as not all farmers have access to the necessary infrastructure and technical expertise required for implementing AI solutions. Bridging this divide requires targeted efforts to provide training, support, and affordable access to technology in rural areas. Additionally, cost-effectiveness, scalability, and standardization of AI

Fig. 1.4 Ethical AI specifications (Dara et al. 2022)

technologies and platforms are areas that require further research and development. An additional obstacle to developing an efficient AI-based agriculture solution is the accessibility and quality of data. The optimal performance of AI is dependent on the availability of substantial amounts of high-quality data. But the data collection and management systems utilized in agriculture are frequently disconnected, and the quality of data can exhibit variability. To tackle this challenge, it is vital to foster collaborative endeavours aimed at setting up frameworks for data-sharing, advocating for data interoperability, and constructing resilient infrastructure for data collection. Ethical consideration and associated problem will definitely continue to arise in the smart agriculture solutions. The key ethical AI principles includes fairness (monitoring and removing bias, fair acquisition of information, and fair accessibility to digital assets), transparency, accountability, robustness, sustainability (sustainable usage and deployment), and privacy (Dara et al. 2022).

Looking ahead, the future of AI in smart farming looks promising. Continued advancements in AI algorithms, data analytics, and IoT technologies will enhance the capabilities and applications of AI in agriculture. Collaborative initiatives involving researchers, farmers, industry stakeholders, and policymakers are crucial for driving innovation, addressing challenges, and ensuring the responsible and sustainable integration of AI in the farming sector. Moreover, interdisciplinary research that combines AI with other emerging technologies, such as blockchain and edge computing, can unlock new possibilities for transforming agricultural practices and creating more resilient and efficient food systems.

1.9 Conclusion

In conclusion, the integration of AI into smart farming has the potential to revolutionize the agriculture industry. Artificial intelligence has revolutionized the agricultural landscape, empowering farmers with data-driven insights and automation capabilities. In smart farming, AI plays a crucial role in optimizing crop management, resource allocation, and decision-making processes. By leveraging machine learning algorithms, AI systems can analyze vast amounts of data collected from sensors, drones, and satellites, providing valuable insights into soil quality, crop health, weather patterns, and pest infestations. These insights enable farmers to make informed decisions about irrigation, fertilization, and pest control, leading to increased yields, reduced costs, and minimized environmental impact. This chapter has provided an in-depth overview of the applications of AI in precision agriculture, crop monitoring and management, livestock management, and agricultural robotics. While AI offers numerous benefits, including enhanced productivity, resource optimization, and environmental sustainability, challenges such as data management, connectivity, adoption barriers, and ethical considerations need to be addressed. By overcoming these challenges and fostering collaboration between stakeholders, the agriculture industry can harness the full potential of AI to create a more efficient, sustainable, and resilient future for farming. Continued research, innovation, and knowledge sharing will play a pivotal role in shaping the future of AI-enabled smart farming.

References

Adão T, Hruška J, Pádua L, Bessa J, Peres E, Morais R, Sousa JJ (2017) Hyperspectral imaging: a review on UAV-based sensors, data processing and applications for agriculture and forestry. Remote Sens 9(11):1110. https://doi.org/10.3390/rs9111110

Ait Issad H, Aoudjit R, Rodrigues JJPC (2019) A comprehensive review of data mining techniques in smart agriculture. Eng Agric Environ Food 12(4):511–525. https://doi.org/10.1016/j.eaef.2019.11.003

Akkem Y, Biswas SK, Varanasi A (2023) Smart farming using artificial intelligence: a review. Eng Appl Artif Intell 120:105899. https://doi.org/10.1016/j.engappai.2023.105899

Alanezi MA, Shahriar MS, Hasan MdB, Ahmed S, Sha'aban YA, Bouchekara HREH (2022) Livestock management with unmanned aerial vehicles: a review. IEEE Access 10:45001–45028. https://doi.org/10.1109/ACCESS.2022.3168295

Babaee M, Maroufpoor S, Jalali M, Zarei M, Elbeltagi A (2021) Artificial intelligence approach to estimating rice yield*. Irrig Drain 70(4):732–742. https://doi.org/10.1002/ird.2566

Bao J, Xie Q (2022) Artificial intelligence in animal farming: a systematic literature review. J Clean Prod 331:129956. https://doi.org/10.1016/j.jclepro.2021.129956

Bouguettaya A, Zarzour H, Kechida A, Taberkit AM (2023) A survey on deep learning-based identification of plant and crop diseases from UAV-based aerial images. Cluster Comput 26(2):1297–1317. https://doi.org/10.1007/s10586-022-03627-x

Chandel NS, Rajwade YA, Dubey K, Chandel AK, Subeesh A, Tiwari MK (2022) Water stress identification of winter wheat crop with state-of-the-art AI techniques and high-resolution thermal-RGB imagery. Plants 11(23):3344

Chlingaryan A, Sukkarieh S, Whelan B (2018) Machine learning approaches for crop yield prediction and nitrogen status estimation in precision agriculture: a review. Comput Electron Agric 151:61–69. https://doi.org/10.1016/j.compag.2018.05.012

Dara R, Hazrati Fard SM, Kaur J (2022) Recommendations for ethical and responsible use of artificial intelligence in digital agriculture. Front Artif Intell 5

Deng L, Mao Z, Li X, Hu Z, Duan F, Yan Y (2018) UAV-based multispectral remote sensing for precision agriculture: a comparison between different cameras. ISPRS J Photogramm Remote Sens 146:124–136. https://doi.org/10.1016/j.isprsjprs.2018.09.008

Deshpande A, Sharp H (2022) Responsible AI systems: who are the stakeholders? In: Proceedings of the 2022 AAAI/ACM conference on AI, ethics, and society. Association for Computing Machinery, New York, NY, pp 227–236

Dhanya VG, Subeesh A, Kushwaha NL, Vishwakarma DK, Nagesh Kumar T, Ritika G, Singh AN (2022) Deep learning based computer vision approaches for smart agricultural applications. Artif Intell Agric 6:211–229. https://doi.org/10.1016/j.aiia.2022.09.007

Elbeltagi A, Kushwaha NL, Srivastava A, Zoof AT (2022) Chapter 5—Artificial intelligent-based water and soil management. In: Poonia RC, Singh V, Nayak SR (eds) Deep learning for sustainable agriculture. Academic Press, pp 129–142

Fuentes S, Viejo CG, Tongson E, Dunshea FR (2022) The livestock farming digital transformation: implementation of new and emerging technologies using artificial intelligence. Anim Health Res Rev 23(1):59–71. https://doi.org/10.1017/S1466252321000177

Genze N, Ajekwe R, Güreli Z, Haselbeck F, Grieb M, Grimm DG (2022) Deep learning-based early weed segmentation using motion blurred UAV images of sorghum fields. Comput Electron Agric 202:107388. https://doi.org/10.1016/j.compag.2022.107388

Gevaert CM, Suomalainen J, Tang J, Kooistra L (2015) Generation of spectral-temporal response surfaces by combining multispectral satellite and hyperspectral UAV imagery for precision agriculture applications. IEEE J Selected Topics Appl Earth Observ Remote Sens 8(6):3140–3146. https://doi.org/10.1109/JSTARS.2015.2406339

Gollakota A, Srinivas MB (2011) Agribot—a multipurpose agricultural robot. In: 2011 annual IEEE India conference, pp 1–4

Hu Y, Kuang W, Qin Z, Li K, Zhang J, Gao Y, Li W, Li K (2021) Artificial intelligence security: threats and countermeasures. ACM Comput Surv 55(1):20:1–20:36. https://doi.org/10.1145/3487890

Iniyan S, Akhil Varma V, Teja Naidu C (2023) Crop yield prediction using machine learning techniques. Adv Eng Softw 175:103326. https://doi.org/10.1016/j.advengsoft.2022.103326

Jena B, Nayak GK, Saxena S (2022) Convolutional neural network and its pretrained models for image classification and object detection: a survey. Concurr Comput Pract Exp 34(6):e6767. https://doi.org/10.1002/cpe.6767

Jung J, Maeda M, Chang A, Bhandari M, Ashapure A, Landivar-Bowles J (2021) The potential of remote sensing and artificial intelligence as tools to improve the resilience of agriculture production systems. Curr Opin Biotechnol 70:15–22. https://doi.org/10.1016/j.copbio.2020.09.003

Kalambukattu JG, Kumar S, Arya Raj R (2018) Digital soil mapping in a Himalayan watershed using remote sensing and terrain parameters employing artificial neural network model. Environ Earth Sci 77(5):203. https://doi.org/10.1007/s12665-018-7367-9

Kamilaris A, Kartakoullis A, Prenafeta-Boldú FX (2017) A review on the practice of big data analysis in agriculture. Comput Electron Agric 143:23–37. https://doi.org/10.1016/j.compag.2017.09.037

Khan A, Vibhute AD, Mali S, Patil CH (2022) A systematic review on hyperspectral imaging technology with a machine and deep learning methodology for agricultural applications. Eco Inform 69:101678. https://doi.org/10.1016/j.ecoinf.2022.101678

Kushwaha NL, Rajput J, Elbeltagi A, Elnaggar AY, Sena DR, Vishwakarma DK, Mani I, Hussein EE (2021) Data intelligence model and meta-heuristic algorithms-based pan evaporation modelling

in two different agro-climatic zones: a case study from northern India. Atmosphere 12(12):1654. https://doi.org/10.3390/atmos12121654
Liu J, Wang X (2021) Plant diseases and pests detection based on deep learning: a review. Plant Methods 17(1):22. https://doi.org/10.1186/s13007-021-00722-9
Lu Y (2019) Artificial intelligence: a survey on evolution, models, applications and future trends. J Manage Anal 6(1):1–29. https://doi.org/10.1080/23270012.2019.1570365
Lynch T, Gregor S, Midmore D (2000) Intelligent support systems in agriculture: how can we do better? Aust J Exp Agric 40(4):609–620. https://doi.org/10.1071/ea99082
Maia RF, Lurbe CB, Hornbuckle J (2022) Machine learning approach to estimate soil matric potential in the plant root zone based on remote sensing data. Front Plant Sci 13
Oliveira LFP, Moreira AP, Silva MF (2021) Advances in agriculture robotics: a state-of-the-art review and challenges ahead. Robotics 10(2):52. https://doi.org/10.3390/robotics10020052
Onishi Y, Yoshida T, Kurita H, Fukao T, Arihara H, Iwai A (2019) An automated fruit harvesting robot by using deep learning. Robomech J 6(1):1–8
Pandey K, Kumar S, Malik A, Kuriqi A (2020) Artificial neural network optimized with a genetic algorithm for seasonal groundwater table depth prediction in Uttar Pradesh, India. Sustainability 12(21):8932
Pandey K, Singh K G, Singh A (2023) Multi-sensors based smart nutrient reuse management system for closed soilless culture under protected cultivation. Comput Electr Agric 204:107495
Pant J, Pant RP, Kumar Singh M, Pratap Singh D, Pant H (2021) Analysis of agricultural crop yield prediction using statistical techniques of machine learning. Mater Today Proc 46:10922–10926. https://doi.org/10.1016/j.matpr.2021.01.948
Patel A, Kethavath A, Kushwaha NL, Naorem A, Jagadale M, Sheetal KR, Renjith PS (2023) Review of artificial intelligence and internet of things technologies in land and water management research during 1991–2021: a bibliometric analysis. Eng Appl Artif Intell 123:106335. https://doi.org/10.1016/j.engappai.2023.106335
Patrício DI, Rieder R (2018) Computer vision and artificial intelligence in precision agriculture for grain crops: a systematic review. Comput Electron Agric 153:69–81. https://doi.org/10.1016/j.compag.2018.08.001
Puppe F (2012) Systematic introduction to expert systems: knowledge representations and problem-solving methods. Springer Science & Business Media
Rangarajan AK, Purushothaman R, Ramesh A (2018) Tomato crop disease classification using pre-trained deep learning algorithm. Proc Comput Sci 133:1040–1047. https://doi.org/10.1016/j.procs.2018.07.070
Riekert M, Zimpel T, Hoffmann C, Wild A, Gallmann E, Klein A (2020) Towards animal welfare monitoring in pig farming using sensors and machine learning. Gesellschaft für Informatik e.V.
Ryan M (2022) The social and ethical impacts of artificial intelligence in agriculture: mapping the agricultural AI literature. AI Soc. https://doi.org/10.1007/s00146-021-01377-9
Sharma A, Jain A, Gupta P, Chowdary V (2021) Machine learning applications for precision agriculture: a comprehensive review. IEEE Access 9:4843–4873. https://doi.org/10.1109/ACCESS.2020.3048415
Subeesh A, Mehta CR (2021) Automation and digitization of agriculture using artificial intelligence and internet of things. Artif Intell Agric 5:278–291. https://doi.org/10.1016/j.aiia.2021.11.004
Venkatesh V, Raj P, Anushia Devi R (2022) Chapter Ten—a smart framework through the Internet of Things and machine learning for precision agriculture. In: Raj P, Saini K, Surianarayanan C (eds) Advances in computers. Elsevier, pp 279–306
Wu Z, Chen Y, Zhao B, Kang X, Ding Y (2021) Review of weed detection methods based on computer vision. Sensors 21(11):3647. https://doi.org/10.3390/s21113647
Wu Y, Liu M, Peng Z, Liu M, Wang M, Peng Y (2022) Recognising cattle behaviour with deep residual bidirectional LSTM model using a wearable movement monitoring collar. Agriculture 12(8):1237. https://doi.org/10.3390/agriculture12081237

Wu D, Han M, Song H, Song L, Duan Y (2023) Monitoring the respiratory behavior of multiple cows based on computer vision and deep learning. J Dairy Sci 106(4):2963–2979. https://doi.org/10.3168/jds.2022-22501

Yi J, Krusenbaum L, Unger P, Hüging H, Seidel SJ, Schaaf G, Gall J (2020) Deep learning for non-invasive diagnosis of nutrient deficiencies in sugar beet using RGB images. Sensors 20(20):5893. https://doi.org/10.3390/s20205893

Yousaf A, Kayvanfar V, Mazzoni A, Elomri A (2023) Artificial intelligence-based decision support systems in smart agriculture: bibliometric analysis for operational insights and future directions. Front Sustain Food Syst 6

Yousefi DM, Rafie AM, Al-Haddad S, Azrad S (2022) A systematic literature review on the use of deep learning in precision livestock detection and localization using unmanned aerial vehicles. IEEE Access 10:80071–80091

Zhai Z, Martínez JF, Beltran V, Martínez NL (2020) Decision support systems for agriculture 4.0: survey and challenges. Comput Electr Agric 170:105256. https://doi.org/10.1016/j.compag.2020.105256

Zhang L, Jia J, Gui G, Hao X, Gao W, Wang M (2018) Deep learning based improved classification system for designing tomato harvesting robot. IEEE Access 6:67940–67950

Chapter 2
Remote Sensing in Precision Agriculture: Current Status and Applications

Dilip Kumar, Asit Kumar Pradhan, Rajni Jain, Vikas Kumar, Sneha Murmu, Ipsita Samal, and Himanshu Shekhar Chaurasia

Abstract Farmers all around the world are always looking for ways to increase their profits. Using cutting-edge technology helps farmers increase the size of their businesses. The science of combining cutting-edge sensors and analysis tools to increase agricultural yields and support management decisions is popular as precision agriculture (PA). One of the highly utilized precision agriculture tools today is remote sensing technology. The study of measuring radiation emitted from and reflected by an object or phenomenon is known as remote sensing (RS). The assessment of crop condition and yield forecasts, acreage estimates of particular crops, crop pests, and disease detection, wildlife management, water supply information and management, weather forecasting, and livestock surveys are all major applications of remote sensing technology. The science of combining cutting-edge sensors and analysis tools to increase agricultural yields and support management decisions is known as precision agriculture (PA). It makes extensive use of data and information to raise crop yields and quality while better utilizing agricultural resources. Many engineers, scientists, and large-scale crop growers use remote sensing technology, which is one of the essential elements of precision agriculture. Precision crop management systems can now incorporate remote sensing technologies' applications that have developed to the point that data from imaging is utilized to guide a number of policy-level choices about sustainable development, poverty reduction, and food security.

D. Kumar (✉) · R. Jain · V. Kumar
ICAR-National Institute of Agricultural Economics and Policy Research, New Delhi, India
e-mail: dilip.bhu.11051@gmail.com

A. K. Pradhan
ICAR-National Rice Research Institute, Cuttack, India

S. Murmu
ICAR-Indian Agricultural Statistics Research Institute, New Delhi, India

I. Samal
Sri Sri University, Cuttack, India

H. S. Chaurasia
ICAR-Central Institute for Research on Cotton Technology, Mumbai, India

K. Pandey et al. (eds.), *Artificial Intelligence and Smart Agriculture*,
Advances in Geographical and Environmental Sciences,
https://doi.org/10.1007/978-981-97-0341-8_2

Keywords Image processing · Interpretation and analysis · Precision agriculture · Remote sensing · Satellite · Sensor

2.1 Introduction

The most fundamental requirements of humanity are met by agriculture, which is a major driver of economic growth in many countries: food and fiber. In 2050, it is anticipated that there will be 9.8 billion people on the planet. To satisfy the demands of the growing population, food production is increasing exponentially. Yet, the goal of sustainable agriculture is likely to be seriously threatened by the scarcity of land and water, climate change, and a rise in extreme occurrences (Karthikeyan et al. 2020). The face of agriculture has changed as a result of technological advancements over the past century, like the Green Revolution. The Green Revolution or third agricultural revolution, which took place between 1960 and 1980, improved crop varieties, synthetic fertilizers, herbicides, and irrigation, which increased crop yield and food security, particularly in poor countries. The primary goals of site-specific management of agricultural inputs are to raise crop production's profitability, enhance product quality, and safeguard the environment (Adamchuk et al. 2004). In the twenty-first century, sustainable agricultural systems heavily rely on precision agriculture (PA). Precision agriculture was started in the middle of the 1980s by utilizing newly available technologies to advance the administration of fertilizers by altering rates and blends as needed within areas (Robert 2002). For the past ten years of the second millennium, precision agriculture has gained a lot of attention in the agricultural sector, yet within-field spatial variability has long been recognized as a reality (Stafford 2000). The use of precision agriculture has the potential to completely transform farming, crop quality, environmental protection, the standard of living on farms, food safety, and the rural economy of the nation (Srinivasan 2006; Pandey et al. 2020). For instance, improved crop production decisions can be made using new information technology. Using data gathered by sensors installed in agricultural equipment, precision farming aims to decrease production costs while improving control and resource usage efficiency. PA has been utilized not just in crop production but also in viticulture, horticulture, pastures, as well as the raising and handling of cattle. It strives to boost economic gains while lowering agriculture's energy use and environmental effect. A general phrase for gathering data on an object by "remote" inspection, i.e., without coming into physical contact with it, is "remote sensing." This chapter provides a succinct rundown of some fundamental concepts. Sensing is the process of gathering information about anything, such as temperature, pressure, an image, from a remote location or from something with which you are not directly in physical contact. In other words, one of the ways frequently employed for gathering physical data to be included into a Geographic Information System is remote sensing (GIS). Remote sensors gather information from earthly objects without coming into contact with them directly using the GIS data collecting and processing technology known as remote sensing.

In remote sensing, sensors on board aircraft or satellites are used to gather data, typically in the form of raster data, which is a grid-like pattern of pixels. Land use mapping, weather forecasting, environmental research, studying natural hazards, and resource exploration are some of the uses for remote sensing. Precision farming, a crucial component of modern agriculture, greatly benefits from remote sensing. Obtaining field data on crops and their growing environment could lead to precise management. The spatial and temporal variation in crop and soil status could be used to optimize field management practices like sowing, fertilizing, irrigating, and harvesting. In this chapter, we give an overview of remote sensing systems, data sources, and types, methods and techniques, types of remote sensing processes, their components, merits and demerits of RS, and applications, particularly in precision agriculture, such as yield estimation, irrigation management, nutrient management, disease and pest management, and a synthesis table of vegetation indices used for a variety of applications in PA.

2.2 Remote Sensing Data Availability

Source data is the first step for any research work. In a remote sensing system, there are a lot of data sources available online, some are free and some are paid. Different types/characteristics of data present in data sources are used in different aspects of research work. Massive amounts of data have been gathered by remote sensing devices for decades, but storing and analyzing them using standard software and desktop computing capabilities are impractical (GEE) (Amani et al. 2020). Data sources can be downloaded from an available online source of data. Some of the famous and useful data sources with data types and also an accessible link for downloading purposes are given in Table 2.1.

2.3 Software for RS Application

2.3.1 Software

Data from remote sensing is processed by remote sensing software. Similar to graphics software, remote sensing programs allow the creation of geographic information from satellite and airborne sensor data (Zhong et al. 2018). There are lots of software for remote sensing programs, and here is an important list of software as well as their symbols to the identification (Figs. 2.1 and 2.2).

Table 2.1 Top available public source of remote sensing data

	Data source	Data type	Link
1	GLOVIS	DOQs, EO-1 ALI, EO-1 Hyperion, GLS, Landsat 4–5 TM, Landsat 7 ETM+, Landsat 8 OLI/TIRS, Sentinel-2, ISRO ResourceSAT 1 and 2-AWIFS Sensor, ISRO ResourceSAT 1 and 2-LISS-3 Sensor, GeoEye's OrbView-3	https://glovis.usgs.gov/
2	NASA earth observation	Datasets are available in the form of JPEG, PNG, Google Earth, and GeoTIFF	https://neo.gsfc.nasa.gov/
3	USGS Earth Explorer	Full access to NASA's Land Data Products and Services such as Hyperion's hyperspectral data, disperse Radar data, and MODIS and AVHRR land surface reflectance	https://earthexplorer.usgs.gov/
4	ESA's sentinel data	Complete, free, and open access to Sentinel-1, Sentinel-2, and Sentinel-3 user products	https://sentinels.copernicus.eu/web/sentinel/home
5	NASA earth data	Communicates with a plethora of satellites such as NASA DC, GPS satellites, SMAP, JASON, METEOSAT, ALOS, TRMM, Aura, Aqua	https://www.earthdata.nasa.gov/
6	NOAA class	Class has a distinct online data library system, a pool of free top-quality and valuable geographic datasets	https://www.avl.class.noaa.gov/saa/products/welcome
7	NOAA digital coast	Get imagery, land cover, elevation, socio-economic and benthic data	https://coast.noaa.gov/digitalcoast/
8	IPPMUS Terra	Integrates population census data from around the world with global environmental data	https://terra.ipums.org/
9	LANCE	Near Real-Time (NRT) data and imagery from the AIRS, AMSR2, MISR, MLS, MODIS, OMI, and VIIRS instruments	https://www.earthdata.nasa.gov/esds

(continued)

Table 2.1 (continued)

	Data source	Data type	Link
10	VITO vision	Coarse vegetation data from PROBA-V, SPOT-Vegetation, and METOP	https://www.vito-eodata.be/PDF/portal/Application.html#Home
11	NOAA data access viewer		https://coast.noaa.gov/dataviewer/#/
12	Bhuvan		https://bhuvan-app3.nrsc.gov.in/data/download/index.php

Source geoawesomeness.com

SOFTWARE	
Geomatica PCI Geomatics	Opticks (Open Source)
SAGA GIS (Open Source)	Orfeo toolbox (Open Source)
TNTmips, (MicroImages)	RemoteView
ERDAS IMAGINE	SOCET SET
ENVI	IDRISI
Google Earth	ECognition
GRASS GIS	ArcGIS
OpenCV (Open Source)	SNAP

Fig. 2.1 Remote sensing popular software list. *Source* gisgeography.com

2.3.2 File Format

Remote sensing needs specialized file formats that contain sensor image data, geo-referencing information, and sensor metadata for their applications. In given table, there are some of the more popular remote sensing **file formats** applicable in remote sensing fields (Table 2.2).

2.3.3 Process

The interest of the target interacts with the incoming radiation during the remote sensing process. Imaging and a non-imaging system are used to accomplish this.

Fig. 2.2 Some popular software logo for remote sensing

Each stage of the remote sensing process, which consists of several steps, is crucial for its efficient execution (Aggarwal 2004).

(A) *Illumination* or *Energy Source in Phase 1*: The starting step is to acquire an energy source, such as our main energy source, the sun, that allows electromagnetic energy to reach the research area.
(B) *Radiation and the Atmosphere in Phase 2*: The second element is realizing how the atmospheric conditions in the research region will interact with the energy (or electromagnetic radiation) as it flows from its source.
(C) *Interaction with the Target in Phase 3*: Realizing that the interconnection depends on the characteristics of the energy and field of study is the third step.
(D) *The Sensor Records Energy in Phase 4*: The fourth stage involves having a remote sensor collect and update the electromagnetic-related radiation after the energy has been discharged.
(E) *Transmission, reception, and processing in Phase 5*: The fifth stage involves sending the remote sensor to an analysis point so that it can be sent with the electromagnetic radiation data and transformed into an image.
(F) *Analysis and Interpretation in Phase 6*: The photos are then converted into an electronic format in the sixth stage so that intelligence can be gathered through analysis.
(G) *Implementation in Phase 7*: When the analysis can use the intelligence to obtain understanding, uncover fresh information, or assist in resolving certain challenges, the seventh and final phase is complete (Fig. 2.3).

Table 2.2 Remote sensing application-supported file formats

File format for raster type		File format for vector type	
File formats	Full form	File formats	Full form
ADRG	Arc-digitized raster graphics	00	Arc export
BMP	Bit map image	COVERAGE	Arc/INFO coverage
ECW	Enhanced compression wavelet	DWG	AutoCAD drawing file
Exif	Exchangeable image file format	DXF	Data interchange file
GEOSPOT	Spot Image Corp. (georeferenced image)	HPGL	Hewlett Packard graphic language
GEOTIFF	Georeferenced tiff image	GDB	Geodatabase
GIF	Graphics interchange format	MDB	Personal Geodatabase
IGS	Image in grayscale	MIF/MID	MapInfo data transfer files
IMG	Erdas Imagine image	SVG	Scalable vector graphics
JPEG (JPG)	Joint photographic experts' group	ODG	OpenDocument graphics
MRSID	Multiresolution seamless image database	DGN	Microstation design file
PCX	Personal computer EXchange	SDTS	Spatial data transfer system
PICT	Image originator Apple computer	TIGER	Topological integrated geographic encoding and referencing
PNG	Portable network graphics	SHP	Shapefile
PSD	Adobe Photoshop document	VPF	Vector product format
RLC	Run length compressed	VML	Vector markup language
SDTS	Spatial data transfer standard	CDR	CorelDRAW
TGA (TARGA)	A simple run length encoded image format	HVIF	Haiku vector icon format
TIFF	Tagged image file format	AMF	Additive manufacturing file format

Source https://gisgeography.com/gis-formats/

2.4 Techniques and Tools for RS Applications

2.4.1 Remote Sensing Type

Satellites/sensors are used in remote sensing to gather data about a specific object or location. Data collecting techniques for remote sensing might be passive or active. Nowadays, practically every application that requires observed data from the top or oblique views uses unmanned aerial vehicle sensors and platforms. The necessary

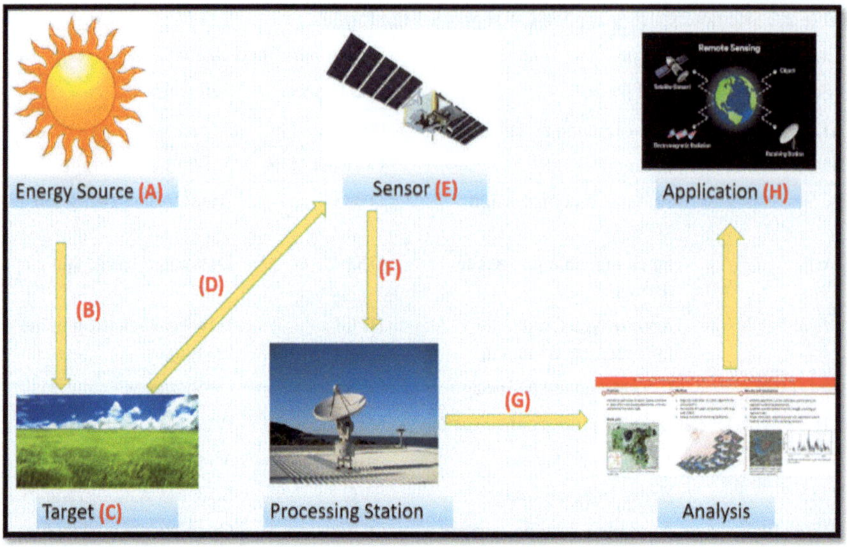

Fig. 2.3 Flow diagram for a process of remote sensing

RS data processing and analysis methods are still mostly ad hoc to applications, despite their intention to be a general remote sensing instrument (Yao et al. 2019).

2.4.1.1 Active Sensors

Active systems generate their own, controllable illumination energy. Active systems have some advantages over passive sensors, such as the fact that they may gather valuable data even in the absence of ideal weather conditions or solar lighting on surfaces.

2.4.1.2 Passive Sensors

The popular form of the sensor for vegetation sensing is the passive sensor. This is due in part to the fact that passive sensor systems are typically simpler to construct (designed merely to accept energy) and in part because certain wavelengths of the solar spectrum contain data that can be particularly useful for observing the characteristics of plants and their canopies. Passive methods have a significant drawback in that they frequently need sunshine to collect reliable data (Table 2.3; Figs. 2.4 and 2.5).

Table 2.3 Pros/cons of remote sensing sensor type

Sensor		Advantages	Disadvantages
Passive	Visible spectrum Camera	Very rich in content Easy to interpret Low price No inference problems with the environment	Not well suited to darkness condition
	Infrared Camera	Ability to measure the temperature Independent of the light source	Sensitive to weather condition
	Sonar	Possibility to measure the distance and speed of the target objects	Difficulties in interpreting the output signal returned by themselves Acquisition price Sensitive to weather condition
Active	Laser	Having a higher accuracy both in lateral and longitudinal direction	Acquisition price
	Radar	Image can be acquired day and night Can operate in different environmental condition without any storage limitations	High price

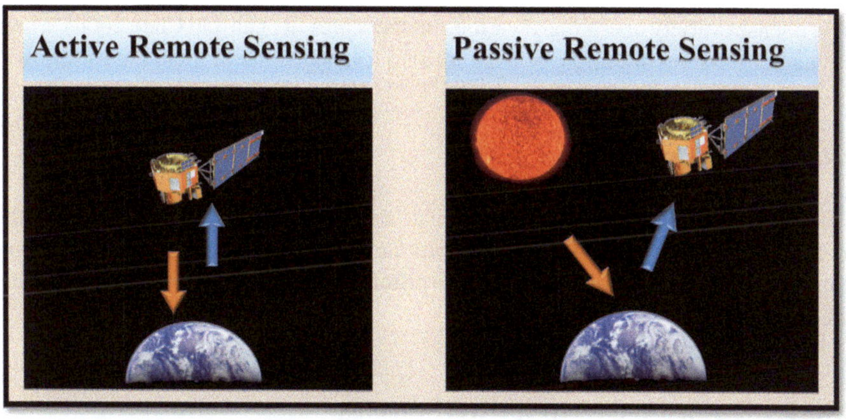

Fig. 2.4 Active and passive remote sensings

Active Sensors	Passive Sensors
❖ LiDAR	❖ Aerial Photography
❖ Radar	❖ FLIR
▪ InSAR	❖ Geodetic Survey
▪ PSInSAR	❖ Hyperspectral Imaging
▪ SAR	❖ Long-Wave Infrared
▪ SRT	❖ Multispectral Imaging
	❖ Near Infrared Surveys
	❖ Radiometrics
	❖ SWIR
	❖ Stereo Satellite Imagery

Fig. 2.5 Active and passive remote sensing sensors. *Source* www.analyticssteps.com

2.4.2 Technique for Remote Sensing

Remote sensing techniques and technology have substantially advanced to encompass a sensor variety operating at various image scales (Rogan and Chen 2004). There are numerous sorts of remote sensing techniques, and some of them are as follows:

2.4.2.1 Photogrammetry

This technique will make it possible to create accurate maps and measurements of the features of the earth. In addition to applications in forestry, geography, geology, planning, soil science, etc., this technology is primarily utilized for creating topographic sheets.

2.4.2.2 Thermal and Multi-spectral Scanning

The sensors used in this method measure radiant temperatures while sensing is carried out in the thermal infrared band, which has a wavelength range of 3–14 µm. A wider range of the spectrum, from the photographic to the infrared band, can be used during multi-spectral scanning.

2.4.2.3 Microwave Sensing

Using the microwave band of the electromagnetic spectrum, this kind of sensing enables securing extremely valuable environmental and mineral resource information. These sensors can function even in inclement weather like haze, light rain, snow, fog, smoke, etc. An active microwave sensor is radar. A microwave radiometer, however, is a passive sensor. Laser light is used in Light Detection and Ranging (LIDAR) (Tsang et al. 1985).

2.4.2.4 Satellites for Earth Resources

This is the most recent method for studying the world from space. In order to help with weather and crop forecasts, mineral prospecting, pollution detection, commercial fishing, etc., sensors are launched into space. Examples of this type of remote sensing technology include the American LANDSAT and the Indian INSAT series. The processes of invention and development are continuously influenced by science and technology (Fu et al. 2020).

2.4.2.5 The Photographic Camera

A straightforward passive sensor is the most widely used sensor technology. The development of photography technologies directly influenced many of the historical advancements in remote sensing. The design of camera systems resembles that of the human eye. Both have an enclosed chamber with a lens at one end and a light-sensitive material (film in the case of a camera and the retina in the case of an eye) at the other. An iris is employed in both systems to regulate the quantity of light that can reach the film or retina. To limit the amount of time that light may strike the film, a shutter is positioned in a camera between the lens and the film. To limit the wavelength of light that is allowed to strike the film, filters can be placed in front of a lens.

2.4.2.6 The Use of Electro-optic Radiometers

A radiometer is a device used to gauge the strength of electromagnetic radiation in a range of wavebands, from microwave wavelengths to ultraviolet. Because they measure electromagnetic radiation using optical methods and electronic detectors, the radiometers we have been talking about are known as electro-optic sensors. If they are installed in a scanner device, images can be formed even if they can only capture a single data value for their view region. Optical techniques cannot be utilized to detect this range of wavelengths due to the nature of microwave radiation.

2.4.2.7 Electromagnetic

Polarity, wavelength, and intensity are the characteristics of electromagnetic energy measured by microwave radiometers. These characteristics offer helpful details about an object's composition and structure. The domains of atmospheric and oceanographic research have seen the majority of applications for passive microwave radiometers. Moreover, it has shown to be a useful tool for measuring soil moisture, a crucial factor in studying vegetation.

2.4.2.8 Thermal, Infrared, and Visible Imaging Apparatus

One can build a sensor that can capture a 2D image of an area by integrating a number of detectors or radiometers into detector arrays. For imaging sensors, there are three fundamental designs: frame, push broom, and mechanical scanner.

2.4.3 *Platform*

A platform is referred to as a sensor carrier for remote sensing. Platforms are the buildings or automobiles that remote sensing equipment is mounted on. The number of attributes that may dictate how a specific sensor is used depends on the platform on which it is housed. Recent years have seen a rise in interest in the scientific field of ground-based/terrestrial radar interferometry (GBRI) (Pieraccini and Miccinesi 2019).

2.4.3.1 Ground-Based

Remote sensing utilizes a wide range of ground-based systems. Towers, cranes, tripods, and handheld equipment are a few of the more popular ones. Although they are also employed for long-term monitoring of terrestrial characteristics, permanent ground platforms are mainly used for monitoring atmospheric phenomena. When a sufficiently stable, long-term platform is required for research initiatives, towers and cranes are frequently used as support structures.

2.4.3.2 Airborne

The only non-ground-based platforms used for early remote sensing studies were airborne platforms. The most popular aerial platform at the moment is an airplane. The vast majority of commercial and military aircraft is employed for remote sensing tasks. Based on their limitations at higher altitudes, aircraft are categorized into three groups in this section: low, mid, and high. Helicopters are typically utilized for low-altitude operations where hovering is necessary. As they are so expensive to run, helicopters are normally only utilized when absolutely necessary. A rapidly growing category of aviation is ultralight aircraft. These affordable, portable, small aircraft can take off and land where larger ones cannot and are frequently moveable. They can only fly at slower speeds and at lower altitudes. Ultralight aircraft may be an acceptable substitute for bigger aircraft, provided that the requirements of the remote sensing requirement are not very stringent. Drones and remote-control aircraft are another kind of aircraft that have been in use for a long time. Remotely controlled airplanes are frequently utilized in situations when flying could be too risky.

2.4.3.3 Satellite

The most stable platform in the air is a satellite because it is sent into space. Currently, there are well over a hundred remote sensing satellites in orbit, and new ones are being launched annually. In addition to serving as a remote sensing satellite, the Space Shuttle is a special spacecraft that can be utilized again and again for various missions. The geometry and timing of a satellite's orbit can be used to classify it (Figs. 2.6 and 2.7).

2.4.4 Orbits

The satellite's orbits are coordinated to pass over the equator at the same local sun time each time. The satellites stay at the same relative position to the sun throughout all of its orbits by doing this. Altitude is another aspect of an orbit. For the past 25 years, lightning observations from satellites in low Earth orbit have produced estimates of the global flash frequency by season and latitude as well as details on daily changes (Christian et al. 1989).

2.4.4.1 Geostationary

A geostationary satellite orbit is one that observes the same area of the Earth's surface from a very high altitude (about 36,000 km). This enables the satellites to continuously monitor and gather data over particular regions. These orbital configurations are frequently used by communications and weather satellites.

2.4.4.2 Sun Synchronous

A geosynchronous satellite is positioned in an Earth-centered orbit which is the same as the Earth's rotation on its axis. As there are 360° in a circle and 365 days in a

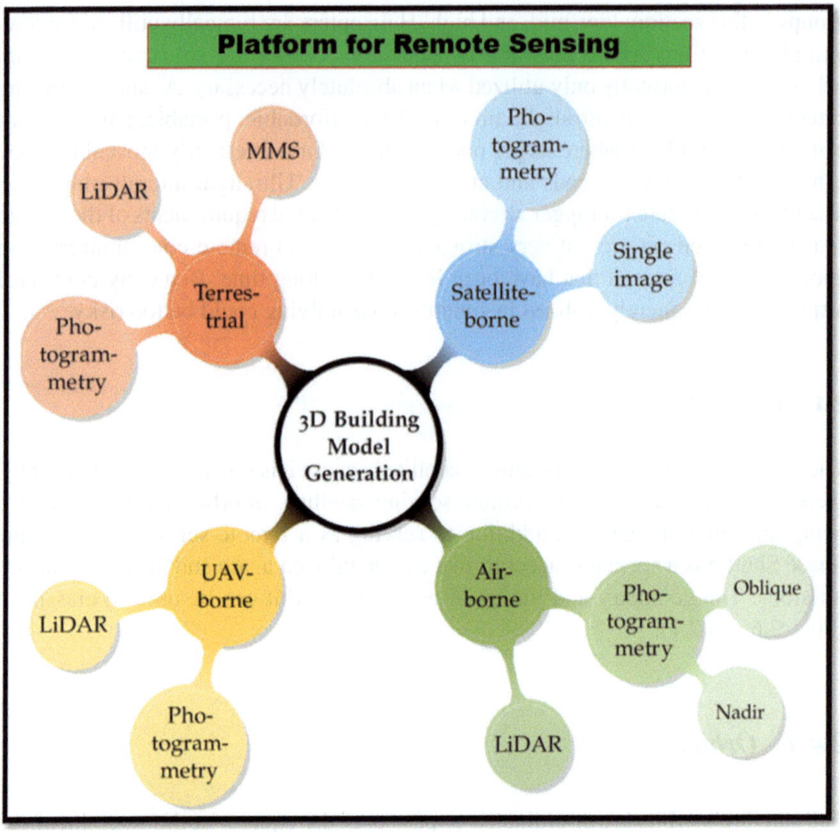

Fig. 2.6 Division of platform in model generation

year, the satellite's orbit must change by around 1 degree per day. Between 700 and 800 km is the altitude at which these satellites orbit.

2.4.4.3 Polar Orbiting

A full revolution of a satellite in a polar orbit, which travels north–south over the poles, takes around an hour and a half. The majority of satellites in polar orbits are located at lower altitudes. This satellite is primarily used for Earth mapping, observation, and time-lapse photography of the planet (Fig. 2.8).

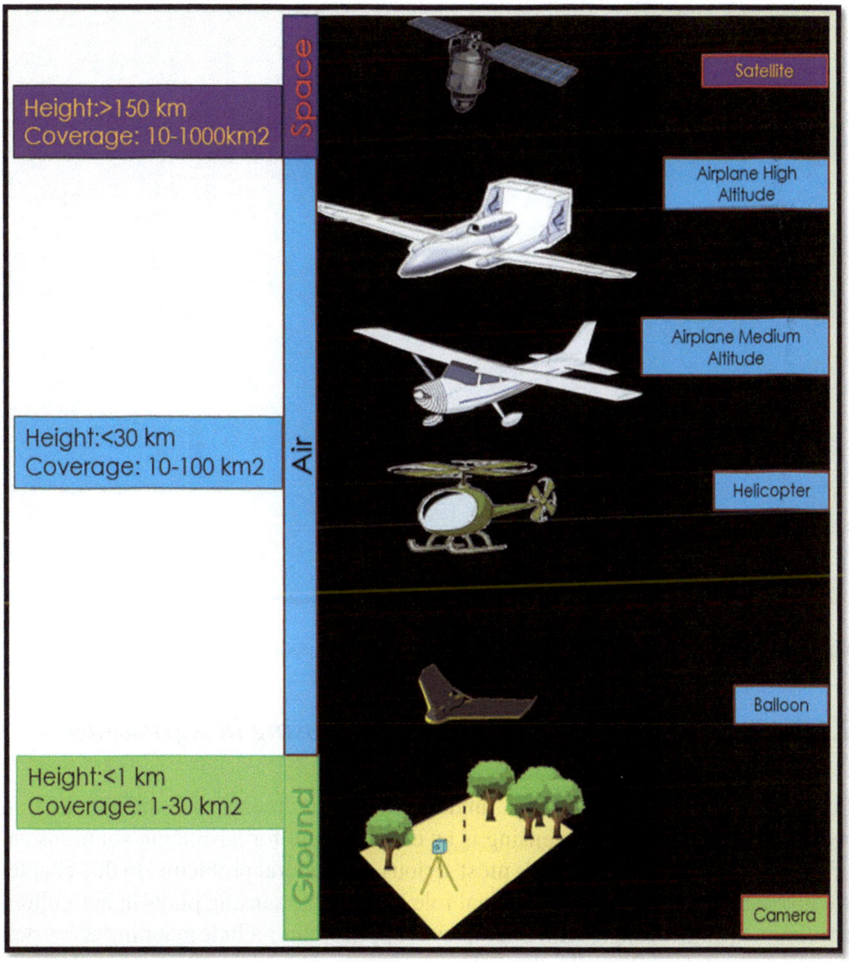

Fig. 2.7 Division of platform on height and coverage as well as appliance basis

2.5 Application of Remote Sensing

2.5.1 Application of Remote Sensing

Satellites play a massive role in the expansion of numerous technologies. Weather, forestry, agriculture, surface changes, and biodiversity are some major areas where remote sensing can be applied (Fig. 2.9).

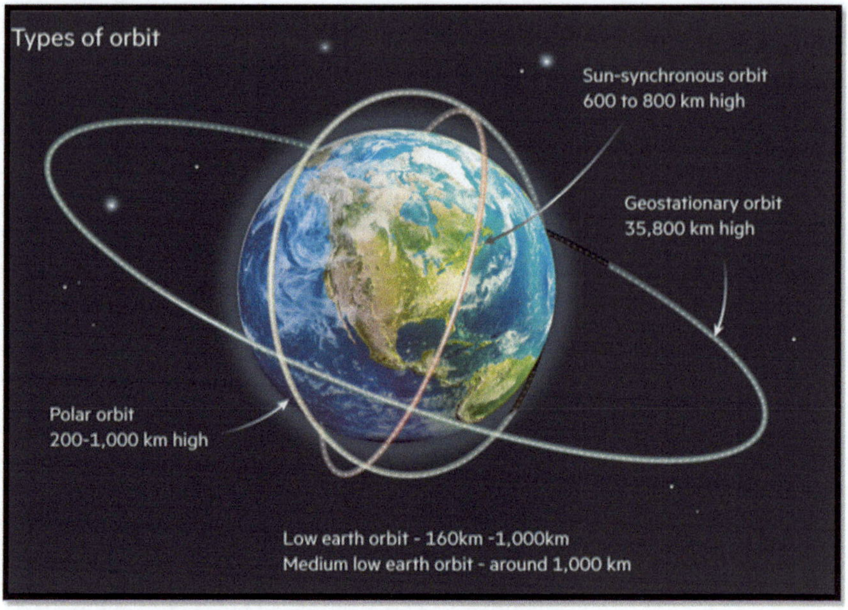

Fig. 2.8 Type of orbits used in remote sensing. *Source* Dristiias/analysis/chandrayaan-mission

2.5.2 Various Applications of Remote Sensing in Agriculture

Food and fiber are the two most fundamental requirements of humanity that are met by agriculture. Remote sensing is an essential tool for advancing solutions and offers useful insights into today's most serious agricultural problems. In this chapter, we give a brief overview of the crucial role that remote sensing plays in agricultural development and management. Examples of applications include mapping ecosystem dispersion and describing the three-dimensional organization of agricultural sectors (2020) (Fig. 2.10).

Fig. 2.9 Application area of remote sensing

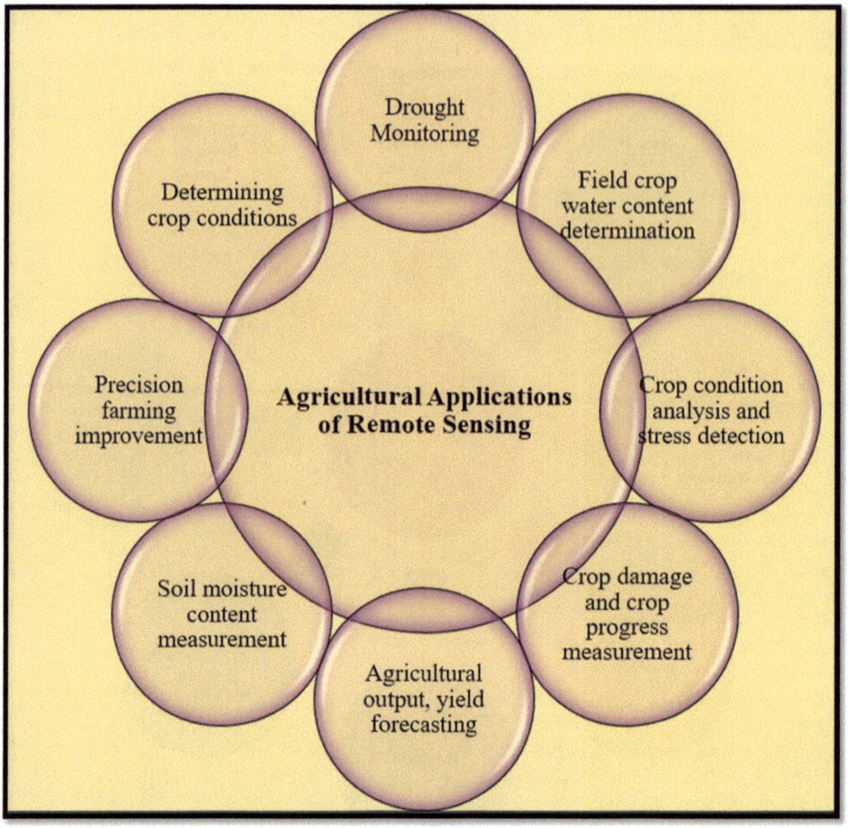

Fig. 2.10 Agricultural application of remote sensing

2.6 Conclusion

A gulf exists between technological capability and scientific understanding of the relationship between input supplies and output products as a result of the interest in and introduction of precision farming (PF). Precision agricultural innovation is fundamentally reliant on remote sensing technology. Although the technology's potential has previously been shown, substantial delivery in reality is challenging because it requires extensive commercial use to reap the rewards. To maximize production with the least amount of input and lessening environmental contamination is the primary goal of remote sensing in precision farming. Discussions have taken place regarding the scenario, current state, and adoption plans for RS at a small farm in India. There have also been some typical Indian case studies offered.

References

Adamchuk VI, Hummel JW, Morgan MT, Upadhyaya SK (2004) On-the-go soil sensors for precision agriculture. Comput Electron Agric 44(1):71–91

Aggarwal S (2004) Principles of remote sensing. Satell Remote Sens GIS Appl Agric Meteorol 23(2):23–28

Amani M, Ghorbanian A, Ahmadi SA, Kakooei M, Moghimi A, Mirmazloumi SM, Brisco B (2020) Google earth engine cloud computing platform for remote sensing big data applications: a comprehensive review. IEEE J Sel Top Appl Earth Obs Remote Sens 13:5326–5350

Bagyaraj M, Ramkumar T, Venkatramanan S, Gurugnanam B (2013) Application of remote sensing and GIS analysis for identifying groundwater potential zone in parts of Kodaikanal Taluk, South India. Front Earth Sci 7(1):65–75

Christian HJ, Blakeslee RJ, Goodman SJ (1989) The detection of lightning from geostationary orbit. J Geophys Res Atmos 94(D11):13329–13337

Fu W, Ma J, Chen P, Chen F (2020) Remote sensing satellites for digital earth. In: Manual of digital earth. Springer, Singapore, pp 55–123

Hrkal Z (2001) Vulnerability of groundwater to acid deposition, Jizerske Mountains, northern Czech Republic: construction and reliability of a GIS-based vulnerability map. Hydrogeol J 9(4):348–357

Karthikeyan L, Chawla I, Mishra AK (2020) A review of remote sensing applications in agriculture for food security: crop growth and yield, irrigation, and crop losses. J Hydrol 586:124905

Lechner AM, Foody GM, Boyd DS (2020) Applications in remote sensing to forest ecology and management. One Earth 2(5):405–412

Mary IA, Ramkumar T, Venkatramanan S (2012) Application of statistical analysis for the hydrogeochemistry of saline groundwater in Kodiakarai, Tamilnadu, India. J Coastal Res 28(1A):89–98

Meng J, Wu B, Du X, Zhang F, Zhang M, Dong T (2011) A review and outlook of applying remote sensing to precision agriculture. Remote Sens Land Resour 23(3):1–7

Pandey K, Kumar S, Malik A, Kuriqi A (2020) Artificial neural network optimized with a genetic algorithm for seasonal groundwater table depth prediction in Uttar Pradesh, India. Sustainability 12(21):8932

Pieraccini M, Miccinesi L (2019) Ground-based radar interferometry: a bibliographic review. Remote Sens 11(9):1029

Robert PC (2002) Precision agriculture: a challenge for crop nutrition management. In: Progress in plant nutrition: plenary lectures of the 14th international plant nutrition colloquium. Springer, Dordrecht, pp 143–149

Rogan J, Chen D (2004) Remote sensing technology for mapping and monitoring land-cover and land-use change. Prog Plan 61(4):301–325

Sishodia RP, Ray RL, Singh SK (2020) Applications of remote sensing in precision agriculture: a review. Remote Sens 12(19):3136

Srinivasan A (ed) (2006) Handbook of precision agriculture: principles and applications. CRC Press

Stafford JV (2000) Implementing precision agriculture in the 21st century. J Agric Eng Res 76(3):267–275

Tsang L, Kong JA, Shin RT (1985) Theory of microwave remote sensing

West H, Quinn N, Horswell M (2019) Remote sensing for drought monitoring & impact assessment: progress, past challenges, and future opportunities. Remote Sens Environ 232:111291

Yao H, Qin R, Chen X (2019) Unmanned aerial vehicle for remote sensing applications—a review. Remote Sens 11(12):1443

Zhong Y, Wang X, Xu Y, Wang S, Jia T, Hu X, Zhao J, Wei L, Zhang L (2018). Mini-UAV-borne hyperspectral remote sensing: from observation and processing to applications. IEEE Geosci Remote Sens Mag 6(4):46–62

References

[illegible]

Chapter 3
Revolutionizing Agriculture: Geospatial Technologies and Precision Farming in India

Wasim Ayub Bagwan

Abstract Geospatial technologies have emerged as a crucial component in the agriculture sector of India, driven by the advancements in electronic technology and the decreasing costs of computer and mobile devices. Indian agriculture confronts challenges such as severe climate events, pest invasions, nutrient management, and carbon emissions. To address these challenges, the adoption of new farming techniques is essential. The integration of new digital platforms into traditional agricultural practices has proven to be highly beneficial. These platforms assist farmers in various aspects of their agricultural practices, leading to reduced crop distress, enhanced yield, and improved overall efficiency. With the aid of digital agriculture, real-time information about farmland requirements can be gathered and analyzed. Geospatial technologies play a critical role in handling this data, providing valuable insights and decision support at various scales. By leveraging these technologies, farmers can make informed decisions and take appropriate actions to optimize productivity and market value. The Digital Agriculture Mission 2021–2025 has been designed with a specific focus on delivering agri-focused solutions that directly benefit farmers. The mission aims to increase farmers' income, enabling them to contribute significantly to the nation's economic growth and food security. The mission recognizes the potential of digital technologies in transforming the agriculture sector and seeks to harness their power through strategic initiatives and partnerships. Over the past two decades, the development and research in precision agriculture have made significant contributions to the Indian agriculture sector. Precision agriculture leverages spatial and temporal data related to crops and the environment, enabling farmers to achieve greater productivity and market value. By utilizing advanced technologies and analytical tools, precision agriculture facilitates efficient resource management, optimal decision-making, and targeted interventions. This chapter provides an in-depth technological review of the utilization of geospatial technologies in the agriculture sector of India. By understanding the potential and benefits of geospatial

W. A. Bagwan (✉)
Department of Environmental Science, Krishna Institute of Allied Sciences, Krishna Vishwa Vidyapeeth (Deemed to be University), Karad 415539, Maharashtra, India
e-mail: wasim.bagwan16@gmail.com

K. Pandey et al. (eds.), *Artificial Intelligence and Smart Agriculture*, Advances in Geographical and Environmental Sciences,
https://doi.org/10.1007/978-981-97-0341-8_3

technologies, stakeholders can effectively leverage them to drive the transformation and growth of the agriculture sector in India.

Keywords Digital agriculture · e-Governance · Geographical information system · Imaging in agriculture · Precision agriculture · Remote sensing

3.1 Introduction

Agriculture is an important sector that has a significant socio-economic impact in India. From sowing to harvest, each step contributes to and affects food production in the changing agriculture scenario. The use of digital technologies is becoming increasingly important in the agriculture sector to enhance productivity and keep up with population expansion. Geospatial technology has become widely adopted in India across various sectors, including agriculture, broadcastings, oil and gas, environmental management, forestry, public safety, infrastructure, and logistics. As stakeholders from these sectors increasingly recognize the value and cost-effectiveness of using geospatial tools and technologies, the geospatial industry is expected to experience significant growth in the coming years (Smriti Rao et al. 2022). GIS technology in agriculture aided in efficient management of land resources resulting in increased production and reduced costs. Smallholders and marginal farmers, who hold about 85% of farmers globally, benefit from reduced vulnerability and marginalization. Agricultural mapping, with geospatial technology, helps monitor and manage farmlands for irrigation and soil conditions, facilitating agricultural and rural development. Precise mapping of geographic and geologic features allows scientists and farmers to develop more efficient farming techniques, leading to higher-quality food production (Paul 2018). Through the use of various agricultural technologies, the yield can be increased. Internet of Things (IoT), Drones/Unmanned Aerial Vehicles (UAV), fluorescence imaging, Infrared Thermography (IRT), robotics, hyperspectral and wireless sensor networks, and many more technologies are used in the agricultural sector (Parihar et al. 2021). Recent developments in GIS and its supporting technologies, such as remote sensing (RS) and the global positioning system (GPS), enable quick and site-specific spatial data analyses for district-wide crop-soil suitability, transport network accessibility, planting and harvesting schedules, identification of areas to drought susceptibility, flooding, and erosion, digital soil mapping of soil fertility, estimation of cropped area, and crop yield forecasting (Paningbatan 2004). Satellite and drone technologies, coupled with digital tools like artificial intelligence (AI), machine learning (ML), and deep learning, facilitate remote sensing for monitoring crop growth and assessing soil moisture. These digital tools leverage computing techniques to rapidly process extensive data, resulting in expedited time to market (Sridhar et al. 2022; Pandey et al. 2023). Further, the new innovation taking place under the theme of precision agriculture (PA) utilizes the available resources more efficiently with the help of modern technologies and helps the farmers in the decision-making process and ensures food security. This system has no adverse impact on the environment and tracks the agriculture changes in a better way to understand real-time conditions at the farm and early preparation for forecasted meteorological impacts. In PA, multiple disciplines are involved such as computer sciences, electronics, RS, GIS, environmental modeling, entomology, agriculture meteorology, and others. Digital technologies like AI, ML, remote sensing, big data, blockchain, and IoT are transforming agriculture. While some countries have embraced these innovations, India is in the early stages. To accelerate adoption, India plans to promote digital agriculture through PPP. (PPP) (IBEF 2021). As a response to climate change, innovations in agriculture and food systems are being developed and introduced. This chapter provides an overview of the latest technologies in Indian agriculture and their benefits for farmers. It explores the primary categories of digital agriculture innovations and geospatial technologies that maximize agricultural benefits. The chapter assesses the significance of new technologies, services, and research in modern agricultural practices for farmers at different scales. By understanding these factors, we can address the challenges faced by Indian farmers through digital agriculture innovations.

3.2 Study Area

India possesses a distinctive culture that is considered one of the oldest and most magnificent cultures in the world. Since independence in 1947, India has made significant socio-economic progress across all fronts. The country spans an area of 3,287,263 km^2, encompassing the snow-laden mountain range of Himalaya to the lush tropical rainforests of the south. As the seventh largest nation globally, India stands out from other Asian countries due to its unique geographic features such as the mountains and sea, which contribute to its diverse topography. The Great Himalayas form the northern boundary, and as it stretches southwards, it gradually narrows down into the Indian Ocean, flanked by the Bay of Bengal to the east and the Arabian Sea to the west. The Indian climate regime can be generally categorized as tropical monsoon. According to the Indian Meteorological Department (IMD), the year can be divided into four certified seasons: winter (December–early April), summer or pre-monsoon (April–June, or April–July in northwestern India), monsoon or rainy (June–September), and post-monsoon (October–December) (knowindia.india.gov.in 2022). When considering to temperature regime, and climate of 1981–2010, India has four zones, viz. TRC1: Tropics, lowland, TRC2: Tropics, highland, TRC3: Subtropics, warm, and TRC10: Very cold. And, from the moisture regime scenario (1981–2010), in India, three zones were identified viz. M1: Desert/Arid, M2: Semi-arid/dry, and M3: Sub-humid/moist (Fischer et al. 2021). Figure 3.1 shows the locations of the research centers working in the field of precision agriculture.

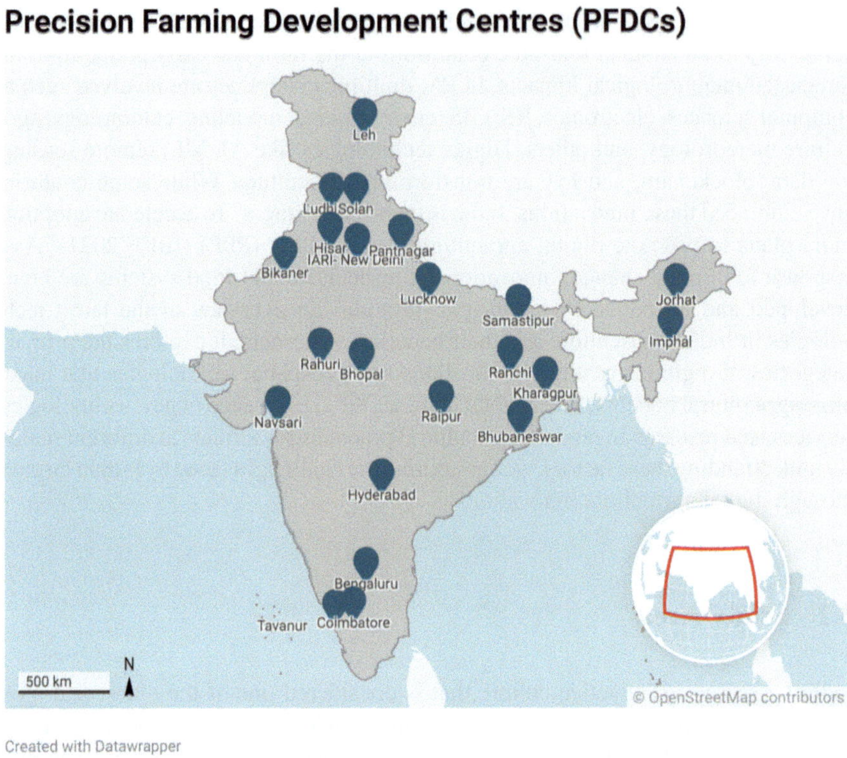

Fig. 3.1 Area map of the study along with the PFDC location; recreated using the source: https://www.ncpahindia.com/pfdc-mandate (To create this map web portal was used)

3.3 Climate Action Plans

The National Action Plan on Climate Change (NAPCC) has been implemented and is already operational as part of efforts to tackle climate change in the Indian agriculture system (Padhee and Whitbread 2022). As part of the XII Five Year Plan, the Government of India (GoI) has announced 8 National Missions under the National Action Plan on Climate Change (NAPCC). The National Mission on Strategic Knowledge for Climate Change (NMSKCC) is an initiative led by the Department of Science and Technology (DST) aimed at enabling the country to manage the effects of climate change. Specifically, the NMSKCC project focuses on enhancing the resilience of agriculture to climate change risks by developing adaptation technologies and updating simulation models to assess the impact of climate change and adaptation strategies. To support this project, current knowledge on climate change adaptation presented in numerous institutes and universities in the country was collected and analyzed. The implementation of climate change adaptation strategies by farmers can be facilitated through the acquisition of strategic knowledge. To prioritize the most effective approaches, a methodology was developed that considers factors such as importance, urgency, no-regret characteristics, co-benefits, and mitigation effects. This enables a systematic and informed decision-making process to address the challenges posed by climate change in agricultural systems (Gupta and Pathak 2016).

3.4 India's Population and Food Security

With a projected 1.67 billion people, India will surpass China as the most populous nation in the world by 2050 (sdg.iisd.org 2022). The agricultural sector and its associated industries play a crucial role in India's economy, providing a significant source of employment. Approximately, 70% of rural households rely on agriculture as their primary livelihood. Among the farming population, small and marginal farmers account for 82%. India is a major producer of food grains, with an estimated production of 275 million tons in 2017–18. The country leads in pulses production, contributing 25% to the global output, as well as consumption (27%) and importation (14%). With a cow population of 190 million in 2012 and an annual milk production of 165 million tons in 2017–18, India holds the title of the world's largest milk producer. It is also a significant producer of jute, pulses, rice, wheat, sugarcane, cotton, groundnuts, fruits, and vegetables, ranking among the top global contributors in these categories with 10.9% for fruits and 8.6% for vegetables (FAO 2022).

3.5 Digital Agriculture

India's Digital Agriculture Mission 2021–2025, initiated in September 2021, aims to expedite the integration of cutting-edge technologies in agriculture, encompassing AI, blockchain, remote sensing, robots, and drones. In collaboration with IBM, NITI Aayog has developed an AI-driven crop production forecast model as a part of this mission. The model offers real-time data and recommendations to farmers, empowering them to enhance crop yield, soil quality, and disease management. By aligning with the mission's objectives, this model holds immense potential to revolutionize India's agricultural sector (SIRU 2022). Digital India initiatives have significantly enhanced the country's digital profile, which encompasses a wide range of areas such as e-commerce, electronic and system design industries, mobile and Internet penetration, digital identity coverage, digital services and transactions, rural empowerment/capacity building, digital payments, and digital literacy (https://www.local2030.org 2022). The National e-Governance Plan in Agriculture (NeGPA) is focusing on new and developing technologies in the area of digital agriculture. Pilot projects were approved during the most recent fiscal year (2021–22 NeGPA) in ten states (agricoop.nic.in 2022). The Digital Agriculture Mission provides funding to states and union territories in India for projects that utilize advanced technologies such as AI, ML, blockchain, IoT, robotics, and more. In addition, funding is provided for adapting and migrating existing web and mobile applications from state-level platforms to a new platform that leverages digital technologies (pib.gov.in 2022).

3.6 Precision Agriculture

In precision agriculture, along with yield-related agronomic factors, the social condition has a driving role. It encircles yield-limiting factors concerning time, mapping of the interested elements in the area of interest, design of management strategies at a given spatial extent, and monitoring economic and environmental benefits in the entire process (Gassner et al. 2013). Farmers that base their judgments solely on common sense, experience, neighborhood rumors, and instinct run the risk of experiencing high risk and low returns. It is crucial to have timely access to pertinent information. For small and marginal farms, ICT offers the ability to fill in informational gaps (FAO and ICRISAT 2022). Precision agriculture (PA) is a data-centric approach that enables the management of field variability to optimize productivity, profitability, and sustainability. The integration of new algorithms enhances the decision-making process, improving various aspects of crop production. Advancements in space technology and IT have opened up new possibilities for the agricultural sector. Therefore, it is crucial to familiarize ourselves with emerging technologies and tailor them to suit the specific needs of the farming industry (Bhattacharyay et al. 2020). Precision agriculture (PA) utilizes modern technologies and novel algorithms to identify and prioritize field-based issues in agriculture. By directing farmers in a scientific manner, PA assists in addressing local challenges related to cropping and agronomic practices. It enables farmers to make informed decisions and implement targeted solutions, thereby improving productivity and efficiency in agriculture.

3.7 Artificial Intelligence in Agriculture

Globally, the use of agricultural automation is becoming increasingly popular, and artificial intelligence (AI) has emerged as a key technology in this area. AI-powered solutions not only enable farmers to produce more using fewer resources but also enhance crop quality and shorten the time required to bring products to market. Recent advancements in computer vision, mechatronics, AI, and ML have revolutionized the field of remote sensing technologies. These technologies now enable the identification and control of plants, weeds, pests, and diseases in agriculture. Through the integration of these cutting-edge technologies, farmers can efficiently monitor and manage their crops, leading to improved productivity and reduced losses. These technologies also offer an opportunity to develop innovative planting techniques for precise fertilization. By leveraging AI solutions, farmers can increase yields, improve product quality, and gain faster market access (NITI 2021). Scientists from International Crops Research Institute for the Semi-Arid Tropics (ICRISAT), Hyderabad, have developed an application named Plantix which uses machine learning algorithms and scientific image data to work on pests and diseases. This free application has great accuracy of 85% and in the beginning was useful for wheat, rice, groundnut, cotton, and tomato and available in eight Indian languages with users over 12 million (ICRISAT 2022). Currently, the Plantix application has broad crop coverage of 30 major crops, 500 plant damage detection, 12 Indian languages, and more than 15 million users (plantix.net 2022).

3.8 Digital Crop Yield Estimation

Currently, Microsoft Corporation is collaborating with farmers to provide farm advisory services in the Indian state of Andhra Pradesh utilizing the implementation of Cortana Intelligence Suite, which includes machine learning and Power BI, which has enabled the conversion of data into intelligent actions. Through the use of an AI-based sowing application that recommends sowing dates, cultivable land preparation, fertigation based on soil analysis, seed treatment, and selection, optimization of sowing depth, and FYM requirement and application, the pilot project has achieved a 30% increase in average crop yield per hectare (Trendov et al. 2019).

3.9 Crop Insurance and Detection of Crop Damage

Under the Pradhan Mantri Fasal Bima Yojna (PMFBY) crop insurance program of the Government of India, the primary parameter used for determining insurance claims is the estimation of crop production at the lowest administrative level specified. This means that the amount of compensation payable to a farmer in case of crop loss or damage would be determined based on the estimated yield of the insured crop at the relevant administrative level. Crop Cutting Experiments (CCE), a proven methodology, has been used up to now to estimate crop output. The current methodology used for yield estimation relies on statistical data and utilizes random numbers for the allocation and selection of plots to conduct the Crop Cutting Experiments (CCE). However, the demand for a large number of CCE with high precision to achieve accurate assessment at lower administrative levels, such as the village or village panchayat level, is a concern as it may not be practically feasible (FAO and ITU 2018). Such schemes have been playing a great role at the micro-level to help the marginal farmer pay the right price for the farmer's crop. A study was conducted by Singh et al. (2017) to investigate the use of RS data for analyzing hail damage in agricultural fields. The study found that the spectral radiance values in the red band of RS imagery were particularly effective in identifying moderate to severe levels of hail damage. These findings suggest that RS data can be utilized as a valuable tool in assessing and monitoring hail damage in agricultural areas, aiding in timely and accurate damage assessment for effective mitigation and management strategies.

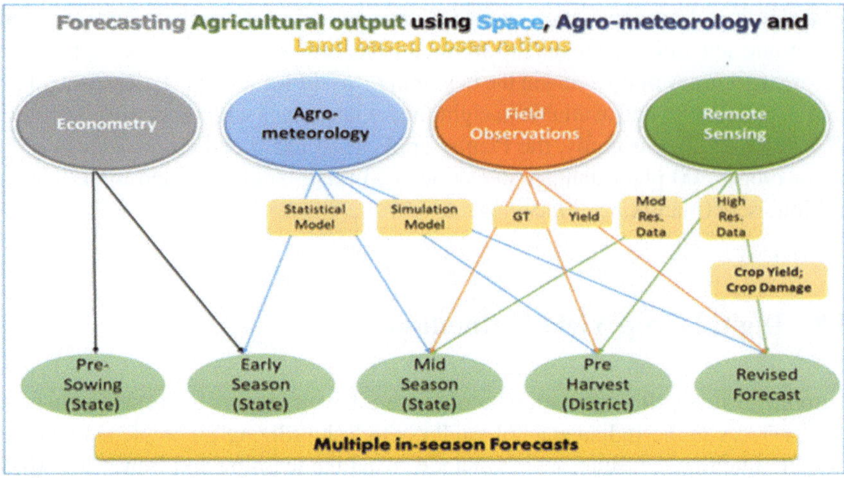

Fig. 3.2 Schematic illustration of the FASAL framework (*source* https://www.ncfc.gov.in/fasal.html)

3.10 FASAL

The Ministry of Agriculture and Farmers Welfare's new initiative with name 'Forecasting Agricultural Output using Space, Agro-meteorology and Land-based Observations (FASAL)' is a functioning unit that regularly predicts the crop output of nine major crops in the country at the national, state, and district levels. The Space Applications Centre, ISRO has developed the procedures used for this forecasting (NCFC 2022). Figure 3.2 shows the illustration of the FASAL framework.

3.11 Remote Sensing in Agriculture

Remote sensing (RS) is a science that uses measurements taken without direct physical touch with the things being analyzed to infer information about materials and objects. Two categories—sensor type and sensor platform—can be used to categorize RS systems for PA and agriculture in general. Satellites, aerial platforms, and ground-based stations are commonly utilized locations for installing sensors. (Awais et al. 2022). By using weather data, satellite-based crop health monitoring, and soil moisture analysis, the full crop life cycle may be mapped. This aids bankers in calculating cropped acreage, forecasting yields, determining the availability of ground and surface water, and analyzing price swings. It can speed up the current crop loan procedure and significantly increase its overall effectiveness (APRACA 2022).

3.12 Crop Monitoring and Pest Surveillance

National Innovations on Climate Resilient Agriculture (NICRA), a strategic-cum-applied research program, utilizes ICT for pest surveillance and real-time database development on pests and weather across different agro-climatic zones in India. It offers web-enabled pest forecasting for major pests and selects locations in crops such as rice, pigeon pea, groundnut, and tomato. The program aims to understand the impact of changing climate on pest dynamics in agriculture. An Integrated Pest Management (IPM), which is a holistic and knowledge-intensive approach, crop-pest surveillance data needs to be processed in a timely manner to facilitate need-based management actions that can be disseminated to farmers by extension functionaries. Information and Communication Technologies (ICT) not only allow for the assimilation of pest database over time and space but also facilitate quick processing of data to make decisions on pest management based on the available knowledge base and critical inputs. The e-pest surveillance approach to pest surveillance using ICT is systematic and rigorous, enabling pre-emptive actions to mitigate impending pest outbreaks and minimize yield losses. This can showcase the success of IPM and increase the pace of its implementation in a wider area. ICT can serve as a vehicle for launching area-wide IPM (Vennila 2016; Vennila et al. 2016). The Department of Revenue, Government of Maharashtra, has developed a digital platform called 'E-Peek Pahani' to monitor the crop and yield condition in Maharashtra state. This application utilizes a mobile app with GPS technology, enabling farmers to register their crops and upload crop photos. By doing so, accurate information about the types of crops being grown and expected yields is provided to the government in real-time. The implementation of this system across the state will not only support farmers but also enable the government to make informed pre-planning decisions based on reliable data. Figure 3.3 shows the interface of *E-Peek Pahani* Android application.

3.13 Digital Intervention in Agriculture

The Ministry of Agriculture (MoA) and Farmers Welfare seek to increase farmers' knowledge and awareness. Therefore, a thorough ICT strategy has been devised to not only reach out to farmers in an easy and better way but also for planning and monitoring of programs so that policy decisions can be made more rapidly and farmers can profit right away. Various ICT techniques have been developed to empower various groups in rural areas (MoA 2022).

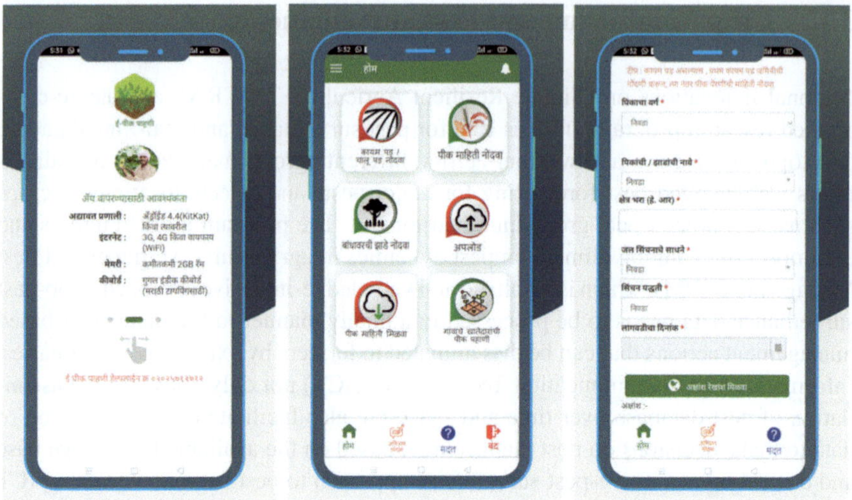

Fig. 3.3 *E-Peek Pahani* (Marathi word), an application developed by Department of Revenue, Government of Maharashtra

3.14 IBM's AgroPad as a Possibility

The implementation of AI technology has enabled farmers to monitor their soil and water health efficiently. IBM's AgroPad is a paper-based device that is the size of a business card and utilizes a microfluidics electronic chip to perform chemical analysis of the sample on the spot, approximately in 10 s. The color of each circle corresponds to the concentration of a particular chemical in the sample. Upon placing a drop of water or soil on the AgroPad, the results of a colorimetric test can be seen on the card's reverse side. Farmers can then use their smartphones to capture a single picture of the AgroPad and instantly receive the chemical test results for the water or soil sample (Beriya 2020).

3.15 Digital Soil Mapping

The recent advancement in the use of cartography techniques has helped the farmer to visualize soil nutrient properties to a larger extent. The spatial information about soil components at the landscape level is made possible by digital soil mapping (DSM). It is correct that creating a traditional soil map takes a lot of effort, time, skill, and synthesis from the soil surveyor. The majority of the data used in DSM is already used by skilled soil surveyors. They make use of their understanding of the connections between soil and topography, lithology, land use and vegetation, and past and present climate when locating their observations or combining them into a map (Arrouays et al. 2020). Airborne Visible/Infrared Imaging Spectrometer—Next Generation (AVIRIS-NG) is a next-generation imaging spectrometer that can measure reflected radiance at 5 nm intervals in the spectral range of visible/short-wave infrared (VSWIR) from 380 to 2500 nm (avirisng.jpl.nasa.gov 2022). A framework for digital soil mapping using AVIRIS-NG hyperspectral data, SRTM data (30 m), and Sentinel-2 data was developed to predict soil attributes at the taluka (block) level (ICAR 2022). The DSM technique was applied in India to predict soil organic carbon and inorganic carbon stocks at a 250 m resolution. To pinpoint the region's most susceptible to desertification and land degradation, quantitative soil knowledge is crucial. Information of this kind aids in simulating soil losses and identifies areas most susceptible to degradation (Moharana and Dharumarajan 2022). DSM is enabling scientists to enhance their understanding of soil nutrient profiles, while GIS technologies offer valuable geostatistical insights. By employing this technique, soil scientists can effectively distinguish between high-enriched and nutrient-deficient soils, thereby facilitating informed decision-making in agricultural practices.

3.16 Information, Education, and Communication

In India, the government platform known as mKisan offers various mobile applications for agriculture, gardening, animal husbandry, and related fields. Another organization called Digital Green is also operating in India, which concentrates on improving and cost-effectively enhancing agricultural extension services (Trendov 2019). Additionally, the International Crops Research Institute for the Semi-Arid Tropics (ICRISAT) is involved in several digital agriculture initiatives, such as Krishi Vani, an agro-advisory service based on voice messages, and Krishi Gyan Sagar (KGS), an Indian extension model that uses ICT (Stephenson et al. 2021). Along with the above-mentioned technologies, the Govt. of India launched Kisan Call Centres (KCCs) in January 2004 with the aim of the project is to answer farmers' queries on a telephone call in their dialect https://mkisan.gov.in/aboutkcc.aspx [Kisan Call Center: Dial 1800 180 1551].

3.17 Imaging and UAV in Agriculture

Remote sensing-based applications in the agriculture sector have a crucial role in implementing the non-destructive method in agriculture research. A thermal imager is a cost-effective and affordable method of tracking crop stress and is an important component of real-time irrigation. However, thermal imaging systems are limited in their data collection and data processing with spatiotemporal dimensions (ICAR 2022). Forty drone start-ups are working in India to improve technology and lower the cost of agriculture drones so that farmers may afford them and use them. The state of Maharashtra has been enticing drone businesses to collaborate with them.

A Memorandum of Understanding (MoU) was recently signed by the Maharashtra government and the World Economic Forum (WEF) Centre for the Fourth Industrial Revolution to examine the potential uses of drones for various government programs (Pathak et al. 2020). Drones may do crop scouting (counting the number of plants per unit area) without entering the paddock. By scanning the field with infrared cameras and optical sensors and providing real-time data, drones can aid in efficient crop surveillance. Farmers can take proactive steps to enhance the state of the plants in the field based on this knowledge. Agricultural drones can be used for a soil analysis to assess soil moisture content, topography, soil erosion, and soil fertility. Since drones have reservoirs that can be filled with seeds, they can be utilized for sowing and spreading seed balls. Drones with heat sensors can locate the dry areas of a field, allowing irrigation decisions to be made quickly for accurate and timely application. Additionally, it will assist in pollination in the production of hybrid rice and application of granular fertilizers to the crop field uniformly at lower cost and could help to give a synoptic view of natural disasters like drought, fires, flood, etc. (Vijaykumar et al. 2022). Drone technology is driving a transformative shift in the field of agriculture, delivering significant advancements in six distinct areas. Firstly, drones are capable of generating precise 3D maps that facilitate early soil analysis. Secondly, the utilization of drone-planting systems reduces planting costs by an impressive 85%. Thirdly, drones enable crop spraying at a remarkable five times faster rate, ensuring even coverage. Fourthly, they enhance crop monitoring through the provision of time-series animations. Fifthly, drones aid in the identification of dry or deficient areas in the field, facilitating improved irrigation practices. Lastly, by employing drone-carried devices, farmers can track plant health and swiftly detect diseases by scanning crops using visible and near-infrared light (TERI 2019). Notably, the UAV-based imageries play a major role in crop health monitoring, yield estimation and to facilitate the irrigation. The data captured by drones swiftly covers large area thus saving the time, labor requirement, and as compared to traditional crop health identification, it has potentiality to improve the efficiency and productivity of the farmland on real-time scale.

3.18 IoT and Agriculture

Agriculture has been undergoing modernization for a long time, and the introduction of the Internet of Things (IoT) is currently transforming the way India perceives agriculture. In the past few decades, agriculture has seen significant changes in its practices and utilization of modern techniques, along with the integration of cutting-edge technology. With the help of smart IoT and automation, farming has become innovative, strategic, technology-driven, and efficient (Indiaai.gov.in 2022). IoT facilitates the exchange of real-time data through embedded sensors, with multiple applications including weather conditions and crop monitoring. This reinvention enables near real-time and precise assessment of field conditions, thereby strengthening decision-making capacity.

3.19 Maharashtra Project on Climate Resilient Agriculture (PoCRA)

Maharashtra Project on Climate Resilient Agriculture is now reached in three years and a half years of its operation. The project is being carried out in 5142 communities, including 932 that are afflicted by salinity in the Purna river basin and spread across 15 districts of Maharashtra state. The deployment of digital initiatives in climate-resilient agriculture benefits both the people using it and the government's implementation machinery. Digital technologies make it possible for users to share and access information at the point of use and for government officials to efficiently monitor, track, and approve operations, which raises the quality of these services. As a result, by reducing and reengineering operating procedures, efficient IT efforts lower the cost, effort, complexity, and time required by organizational operations. Users can access services more easily, reliably, consistently, and transparently by using well-designed digital tools. Government employees can improve service delivery by saving time spent looking for, gathering, and collecting information. Cloud-based services now make cutting-edge technology like geospatial and satellite data, image processing, artificial intelligence and machine learning, sensors, and the Internet of Things available to everyone. With the right technology setup, real-time agricultural advisories based on weather forecasts, market prices, insect incidences, approaching calamities, etc. can be sent. Farmers that successfully implement digital projects can optimize their crop cycles, increasing their income and preparing them to deal with the effects of climate change while lowering their expenses and risks (mahapocra.gov.in 2021). Figure 3.4 shows the flowchart of digital architecture.

The GIS dashboard provides weather forecasts and geospatial information, enabling project participants to access location-specific weather updates, project activity status, maps, crop water balance estimates, micro-level planning, FFS information, and Soil Health Card (SHC) information, among other things. This dashboard is useful in enhancing implementation and providing analytical data on key indicators. Users of the GIS Dashboard have the flexibility to generate maps according to their specific requirements (mahapocra.gov.in 2022).

3.20 Conclusion and Future Direction

Digital technologies and their expansion in the agriculture sector improve the positive attention toward agriculture and help small landholders from risk and vulnerability of the agriculture productivity. Nowadays, the affordability of interned data and its reachability in rural areas further expanded the horizons of the information to the major agriculture producers and their allied sectors. The multiple source of data and information through various modes like mobile application, monitoring system, and information broadcasting has a great help in all-around preparedness from mental to climate resilience has been enhanced through this digitalization of agriculture to

Fig. 3.4 Schematic diagram of digital architecture implemented by PoCRA (mahapocra.gov.in 2021)

become its precision nature. Precision farming helps farmers to adapt to the local agro-climatic condition and to plan according to it, this would have a huge role to step at the right and accurate crop productivity. Digital technologies could make farmers smarter, and climate-proof and help them to find new solutions and innovative ideas in the field of precision agriculture. Geospatial technologies have immense potential in the future of Indian agriculture, particularly in conjunction with AI and IoT. By leveraging real-time geospatial data, farmers can make more informed decisions about crop management, irrigation, and soil health. With the help of AI and machine learning, geospatial data can be processed and analyzed more efficiently, providing more accurate insights and predictions. IoT sensors and devices can be integrated with geospatial technology to gather more precise data on weather, soil moisture, and other environmental factors. Overall, the integration of geospatial technology with AI and IoT has the potential to revolutionize the Indian agriculture system by enabling farmers to make data-driven decisions and optimize their resources. With the help of these technologies, India can improve agricultural productivity, reduce wastage, and ensure food security for its growing population. By harnessing the potential of UAVs in agriculture, these aspiring entrepreneurs can offer services such as aerial mapping, data analysis, and precision farming, contributing to the growth and development of the agricultural sector while also stimulating economic activities in their communities. UAVs in agriculture create job opportunities, enabling youth to start businesses in rural areas.

Conflict of Interest The author states that he has 'no conflicts of interest'.

References

APRACA (2022) Satellites and AI for agriculture lending: SatSure SAGE. https://www.apraca.org/wp-content/uploads/2020/09/Agriculture-LendingImplementation-Support_SatSure.pdf

Arrouays D, McBratney A, Bouma J, Libohova Z, Richer-de-Forges AC, Morgan CLS et al (2020) Impressions of digital soil maps: the good, the not so good, and making them ever better. Geoderma Reg 20:e00255. https://doi.org/10.1016/j.geodrs.2020.e00255

Awais M, Li W, Cheema MJM, Zaman QU, Shaheen A, Aslam B et al (2022) UAV-based remote sensing in plant stress imagine using high-resolution thermal sensor for digital agriculture practices: a meta-review. Int J Environ Sci Technol. https://doi.org/10.1007/s13762-021-03801-5

Beriya A (2020) Digital agriculture: challenges and possibilities in India, ICT India working paper, no. 35, Columbia University, Earth Institute, Center for Sustainable Development (CSD), New York, NY

Bhattacharyay D (2020) Future of precision agriculture in India, 1st edn. In: Protected cultivation and smart agriculture. New Delhi Publishers. https://doi.org/10.30954/NDP-PCSA.2020.32

FAO (2022) FAO in India. https://www.fao.org/india/fao-in-india/india-at-a-glance/en/

FAO & ICRISAT (2022) Digital agriculture in action. FAO. https://doi.org/10.4060/cc0017en

FAO & ITU (2018) E-agriculture in action: drones for agriculture

Fischer G, Nachtergaele F, van Velthuizen H, Chiozza F (2021) Global agro-ecological zones (GAEZ v4) model documentation. FAO & IIASA, 303 p. https://www.fao.org/3/cb4744en/cb4744en.pdf

Gassner A, Coe R, Sinclair F (2013) Improving food security through increasing the precision of agricultural development. In: Precision agriculture for sustainability and environmental protection

Gupta A, Pathak H (2016) Climate change and agriculture in India: a thematic report of national mission on strategic knowledge for climate change (NMSKCC) under national action plan on climate change (NAPCC)

https://www.local2030.org (2022) Digital India for sustainable development goals pathways for progress people, planet, prosperity, peace & partnerships. https://www.local2030.org/library/674/SDG-Vision-for-Digital-India.pdf

IBEF (2021) Digital agriculture—the future of Indian agriculture. https://www.ibef.org/blogs/digital-agriculture-the-future-of-indian-agriculture

ICAR (2022) CRP on farm mechanisation and precision farming. https://ciae.icar.gov.in/aicrps/crp-on-farm-mechanisation-and-precision-farming
ICRISAT (2022) https://www.icrisat.org/artificial-intelligence-to-track-pests-and-diseases-in-india/
Indiaai.gov.in (2022) http://indiaai.gov.in/article/how-iot-can-transform-agriculture-in-india
knowindia.india.gov.in (2022) India at a Glance. https://knowindia.india.gov.in/profile/india-at-a-glance.php
mahapocra.gov.in (2021) Digitizing agriculture for climate resilience—PoCRA. https://mahapocra.gov.in/docs/PoCRA-DigitizingAgricultureForClimateResilience.pdf
mahapocra.gov.in (2022) Project on climate resilient agriculte, annual progres report 2021–22. https://mahapocra.gov.in/assets/docs/mne/PoCRA_Annual%20report__March%202022.pdf
MoA (2022) Ministry of agriculture: digital agriculture division. https://agricoop.nic.in/en/divisiontype/digital-agriculture-division
Moharana P, Dharumarajan S (2022) Digital soil mapping for precise land management. Indian Farming 72(5):29–32
NCFC (2022) Mahalanobis National Crop Forecast Centre (MNCFC). https://www.ncfc.gov.in/fasal.html
NITI (2021) Artificial intelligence intelligent inputs revolutionising agriculture. https://www.niti.gov.in/sites/default/files/2021-09/IntelligentInputsRevolutionisingAgriculture.pdf
Padhee A, Whitbread A (2022) Indian agriculture: the route post-CoP 26. https://www.downtoearth.org.in/blog/climate-change/indian-agriculture-the-route-post-cop-26-81154
Pandey K, Singh KG, Singh A (2023) Multi-sensors based smart nutrient reuse management system for closed soilless culture under protected cultivation. Comput Electron Agric: 107495
Paningbatan EPJ (2004) Geographic information system (GIS) application in soil science. Philippines Univ. Los Banos, College, Laguna (Philippines). Dept. of Soil Science
Parihar G, Saha S, Giri LI (2021) Application of infrared thermography for irrigation scheduling of horticulture plants. Smart Agric Technol 1:100021. https://doi.org/10.1016/j.atech.2021.100021
Pathak H, Kumar GAK, Mohapatra SD, Gaikwad BB, Rane J (2020) Use of drones in agriculture: potentials, problems and policy needs. Publication no. 300 ICAR-NIASM, pp 13 + iv
Paul S (2018) How GIS is enabling the agricultural sector. Geospatial world. https://www.geospatialworld.net/blogs/gis-in-agriculture/
plantix.net (2022) Home page https://plantix.net/en/
Singh SK, Saxena R, Porwal A, Neetu SSR (2017) Assessment of hailstorm damage in wheat crop using remote sensing. Curr Sci 112(10):2095. https://doi.org/10.18520/cs/v112/i10/2095-2100
SIRU (2022) Strategic investment research unit. Digitalisation of Agriculture in India. https://www.investindia.gov.in/team-india-blogs/digitalisation-agriculture-india
Smriti Rao P, Barthwal A, Khatana R (2022) Role of geospatial technology in agriculture. Krishijagran. https://krishijagran.com/featured/role-of-geospatial-technology-in-agriculture/
Sridhar A, Balakrishnan A, Jacob MM, Sillanpää M, Dayanandan N (2022) Global impact of COVID-19 on agriculture: role of sustainable agriculture and digital farming. Environ Sci Pollut Res 30(15):42509–42525. https://doi.org/10.1007/s11356-022-19358-w
Stephenson J, Chellew T, von Köckritz L, Rose A, Dinesh D (2021) Digital agriculture to enable adaptation: a supplement to the UNFCCC NAP technical guidelines. CCAFS working paper no. 372. Wageningen, the Netherlands: CGIAR Research Program on Climate Change, Agriculture and Food Security (CCAFS)
TERI (2019) ICDL 2019 digital transformation for SDGs
Trendov N, Varas S, Zeng M (2019) Digital technologies in agriculture and rural areas—status report. Rome. Licence: cc by-nc-sa 3.0 igo
Vennila S (2016) Information and communication technology for effective integrated pest management. Curr Sci 110(3):287–288

Vennila S, Lokare R, Singh N, Ghadge S, Chattopadhyay C (2016) Crop pest surveillance and advisory project of Maharashtra—a role model for an e-pest surveillance and area wide implementation of integrated pest management in India. ICAR- National Research Centre for Integrated Pest Management, New Delhi 110 012 India, 56 p

Vijaykumar S, Madireddy H, Bhusarapu SC, Kumar R, Sundaram R (2022) Drone application in rice cultivation: experiences from ICAR-IIRR trails. 72(12):03–06

Chapter 4
Assessment of Water Use Efficiency for Sustainable Water Management by Integration of Remote Sensing and Machine Learning

K. Krupavathi, M. Raghu Babu, A. Mani, P. R. K. Parasad, and L. Edukondalu

Abstract The present study focused on estimation of the irrigation water application and water use efficiency in Krishna Central Delta (KCD) by integrating the remote sensing image processing and machine learning along with ancillary data. The results obtained from remote sensing methods were evaluated by field method of estimation of water use efficiency of a crop. The estimated water use efficiency was varied between 0.202 kg/m^3 at Mopidevi and 0.350 kg/m^3 at Penamaluru. The standard deviation from mean is 0.066. The coefficient of variation among all mandals is 0.227. The spatial maps prepared indicated the variation in rainfall, water application efficiency in the study area and were used to suggest the sustainable water management strategies.

Keywords Water use efficiency · Remote sensing · Machine learning · Sustainable agriculture

K. Krupavathi (✉)
NTR College of Agricultural Engineering, ANGRAU, Bapatla 522101, India
e-mail: k.krupavathi@angrau.ac.in

M. Raghu Babu
College of Agricultural Engineering, ANGRAU, Madakasira, India

A. Mani
ANGRAU, Lam, Guntur, India

P. R. K. Parasad
Agricultural College, ANGRAU, Bapatla 522101, India

L. Edukondalu
College of Food Science and Technology, ANGRAU, Pulivendula, India

K. Pandey et al. (eds.), *Artificial Intelligence and Smart Agriculture*, Advances in Geographical and Environmental Sciences,
https://doi.org/10.1007/978-981-97-0341-8_4

4.1 Introduction

India is an irrigation infrastructure (irrigation projects)-rich country and attempts are further increasing to convert rainfed area into irrigated agriculture for sustainable food production. In Andhra Pradesh, out of the 41.54 lakh ha of irrigated area, about 19.99 lakh ha of land is irrigated by canals (Anonymus 2020) and the cost for one hectare of land in irrigation projects is about Rs. 97,033 per acre (Ninth Plan, 1997–2002). Further, the present focus of the Government is to increase funds to irrigation projects at a large scale in the state to meet the demands for the sustainable irrigation facilities. Under such huge investments on creation irrigation source, the need to estimate and monitor irrigation potential utilization on a regular basis in irrigation projects is essential (Levidow et al. 2014). The performance of many irrigation systems is below their potential due to poor design, poor construction, lack of maintenance (Elshaikh et al. 2018). Hence, there is a need for study of irrigation systems.

Under climate change and water scarcity conditions, sustainable agriculture needs critical management of the water and may be achieved by adopting improved irrigation application, water use efficiency. Quantifying water use efficiency (WUE) will help stakeholders to understand issues of productivity and to make decisions effectively while utilizing water resources (Kumar et al. 2014). Through the technological intervention, WUE at command area level can be estimated accurately (Krupavathi et al. 2022a). Hence, it is attempted to quantify WUE for the entire command area by using the most advanced remote sensing techniques in Krishna Central Delta (KCD) and is planned to explain the detailed investigations carried out with the objectives like estimation of irrigation area, detailed crop classification, extraction and precise mapping of paddy crop in the study area from decision tree approach, estimation and mapping of crop yield, and mapping of water use efficiency (kg m^{-3}) maps. Most advanced neural network is adopted for the estimation of crop yields.

4.2 Materials and Methods

4.2.1 Study Area

The present study area selected is KCD which comprises two major canals of Prakasam barrage, namely Bandar and Krishna Eastern Bank canal, with an ayacut of 274,834.09 acres in 18 mandals of Krishna district. The major crops growing were Rice, Turmeric, Sugarcane, Maize, Vegetables, and Pulses. The land utilization pattern shows that 1539.342 km^2 is under agriculture followed by 402.971 km^2 under aquaculture and 57.648 km^2 is under horticulture (Krupavathi et al. 2022b). This location of the study area is shown in Fig. 4.1.

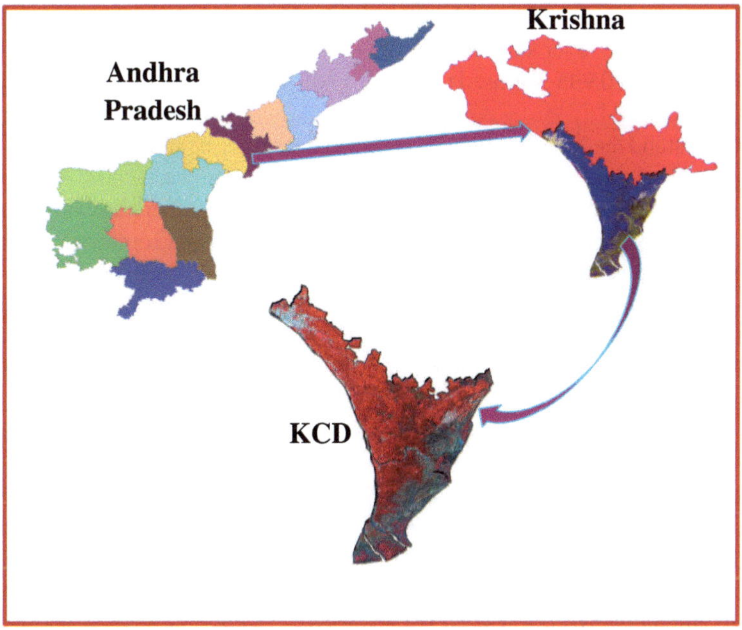

Fig. 4.1 Location details of Krishna Central Delta

4.2.2 Estimation of Total Irrigated Area and Detailed Crop Classification Using Decision Tree Approach Using Satellite Data

In the present study, high-resolution freely available Landsat-8 images were downloaded for the years 2016 and 2017. The single date image in peak vegetation stage in each season was used to generate irrigated crop mask of the area in that respective season. The satellite imageries of Digital Numbers (DN) in the level-1 products were converted to Top of Atmosphere (TOA) reflectance using the radiance rescaling factors as per the respective metadata file.

The normalized difference between the NIR and RED reflectance bands is known as NDVI which indicates the responsive changes of vegetation cover and greenness. In equation NDVI is expressed as

$$\mathrm{NDVI} = \frac{\rho_{\mathrm{NIR}} - \rho_{\mathrm{RED}}}{\rho_{\mathrm{NIR}} + \rho_{\mathrm{RED}}}, \tag{4.1}$$

where

ρ_{NIR} NIR band reflectance (0.870 μm).
ρ_{RED} RED band reflectance (0.66 μm).

NDVI images were developed for all days selected in a season. All the derived NDVI images were layer stacked. The ISODATA clustering method with a convergence threshold of 0.995 was used to extract 15 distinct classes from the layer-stacked Landsat-8 images.

To understand the NDVI profiles of each separate class, a random 30 sample points were picked and generated seasonal NDVI class trends from each class. Although the shape of the crop profile is similar for all 30 sample points, there are variations due perhaps to atmospheric changes between scenes or errors in geometric registration. Zonal mean values of each class were calculated from zonal attributes. Based on the generated zonal means, NDVI profiles of each class were identified for its crop. The zonal means were extracted for 15 zones with zonal attributes, and with reference to these NDVI trends and visual observation, 15 classes were finally grouped into six most suitable crop classes.

Unsupervised classification was also carried out on each image and generated probable crop maps for them. Differencing and thresholding techniques were applied on three NDVI images to separate the irrigated area. Permanent vegetation, settlements, water bodies, and bare lands were removed first from the image based on class in unsupervised classification and NDVI values. The NDVI of cropped areas in *kharif* and *rabi* images was extracted based on threshold values (for example, irrigated land in *kharif*, $0.325 < \text{NDVI} < 0.735$).

An additional condition was applied to separate paddy from the other crop cover type, and LSWI was compared with NDVI (Xiao et al. 2006). To validate the classified map, after completion of classification, ground truth data was collected for *Kharif* cropping season and also compared with the total cropped area as per district statistics.

4.2.3 Estimation of Crop Yield Maps Using Artificial Neural Network

The development of a crop yield model involved the selection of remote sensing indices such as Surface Temperature (Ts), NDVI, Absorbed Photo Synthetically Active Radiation (APAR), Water Stress Index (WSI), and the average crop yield data from the past five years. These indices served as inputs for an Artificial Neural Network (ANN) model. APAR, which represents the portion of radiation absorbed by crop leaves for photosynthesis, can be calculated using the formula: APAR = FAPAR * PAR. FAPAR is crucial for the estimation of pixel level crop yields based on light use efficiency. Canopy Surface Temperature (Ts), indicating the intensity of evaporation and the radiated sunlight onto leaves, is computed as:

$$\text{Ts} = (\text{BT}/1) + W * (\text{BT}/14{,}380) * \ln(\varepsilon), \tag{4.2}$$

where Ts depends on the ToA brightness temperature (BT) in degrees Celsius, calculated using thermal constants (K_1 and K_2) from the metadata file. The NDVI/Ts, also known as the Crop Water Stress Index (CWSI), serves as another indicator of crop yield. CWSI is a normalized index used to quantify stress, mitigating the influence of other factors.

Ts, NDVI, APAR, and CSWI maps were extracted from remote sensing images captured at various crop growth stages. Attribute values for these parameters were linked to ground truth points and exported as .csv files in Excel format, which were subsequently used as input data for the neural network algorithm.

4.2.3.1 Design of ANN Model

The data extracted from the thematic maps was divided into two distinct sets: one for training and the other for testing. In Fig. 4.2, you can see the current three-layered architecture of the Multi-Layer Feedforward Perceptron (MLFF) in action. In the input layer, previously computed factors, including NDVI, Ts, APAR, CWSI, and mean yield, were designated as neurons. The output layer contains a single neuron responsible for predicting crop yield.

To optimize the model's performance, the hidden layers were analyzed to determine the ideal number of neurons, ranging from 1 to 30. The network's weights, labeled as Wij (representing connections from input layer neurons to hidden layer

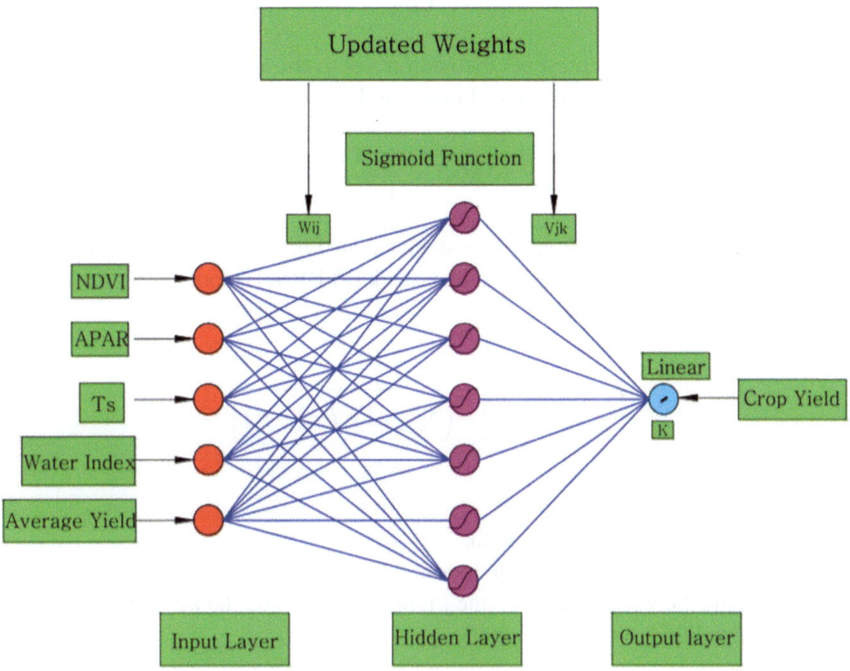

Fig. 4.2 Architecture of the proposed FFBPNN model

neurons) and Vjk (representing connections from hidden layer neurons to the single output neuron, where $k = 1$), were fine-tuned during the training process. These weight factors were retained for making predictions on new input data and for assessing the relationship between crop growth factors and their corresponding yields.

Sigmoidal transfer functions were used in the hidden layers, while linear activation functions were employed in the output layer. This neural network was implemented using the MATLAB programming language package.

Data from the input layer is transmitted to the hidden layer. Within the hidden layer, the model multiplies these inputs by appropriate weights and calculates the sum. Before sending the result to the output layer, a sigmoid activation function is applied.

The mathematical expression for the linear activation function is given by:

$$x = \sum_{i=1}^{n} w_i x_i + b = w_1 x_1 + w_2 x_2 + w_3 x_3 + \cdots + w_n n + b. \tag{4.3}$$

The output, denoted as y, is determined by the activation function:

$$y = f(x) = f\left(\sum_{i=1}^{n} (w_i x_i + b)\right). \tag{4.4}$$

In this equation, 'f' represents the neuron's activation or transfer function, with the transfer function being a sigmoid function defined as:

$$y = f(x) = \frac{1}{1 + \mathrm{e}^{-x}}. \tag{4.5}$$

The neuron activation function is depicted in Fig. 4.2. The final form of the Feedforward Backpropagation Neural Network (FFBPNN) model, after substituting the weights, is presented as:

$$Y = \frac{1}{1 + \exp\left[-\left(\sum_{j=1}^{q} V_{jk} H_{ij} - \gamma\right)\right]}, \tag{4.6}$$

where

Y represents the yield per unit area.
q signifies the number of hidden nodes.
V_j represents the weight between the jth hidden node and the kth output node.
γ is the threshold of the output node.

As previously explained, the neural network, following the training process, discerns the functional connection between the given input parameters and the desired target output. The training commences with the initialization of random weights using the 'LM' algorithm, which fine-tunes the error function, as described by Krupavathi et al. (2022a). The network not only generates and retains these weights but also retains the functional relationship for subsequent steps.

Crop yield estimation models have been crafted for paddy crops at the regional level, specifically for the kharif season. To assess the model's performance, 25% of the data is allocated for testing. Prior to the training phase, all input data parameters are normalized, and the output yields are subsequently denormalized to the required units.

4.2.3.2 ANN Model Validation

To evaluate the effectiveness of the generated Feedforward Backpropagation Neural Network (FFBPNN) models for crop yield, a range of statistical metrics was employed. These metrics encompass the root mean squared error (RMSE), the coefficient of determination (R-squared), the mean absolute error (MAE), as well as the ratio between the average predicted output and the target yield values (R-ratio). In addition to these, a relative error was calculated for each individual data point.

4.2.4 Quantification and Mapping of Water Use Efficiency

For quantification of water use efficiency, the study area is divided into three parts (Fig. 4.3) mainly based on water utilization and cropping pattern as upper reach, middle reach, and tail reach. The upper reach consisting of five mandals is mainly depending on ground water with a little use of canal irrigation. The middle reach consists of seven mandals with the main irrigation source of canal irrigation with less utilization of ground water. The tail end area consists of six mandals. The water source in this tail end area is purely canal water except at Avanigadda. The area is mainly irrigated by KEB canal in Koduru and Avanigadda, Nagayalanka.

Field data was collected during ground truth survey on the questionnaire prepared, to acquire complete information regarding their current situation of water application, cultivation practices on the farm (Irrigation method, crop, depth of application, soil type, time of application etc.). Data in about plots of 50 farmers' fields with different irrigation systems and crops was considered and studied.

The components of quantified water use and yield of each crop were used for estimating water use efficiency of representative fields. Then aggregated at the mandal level to estimate water use efficiency of each mandal. Spatial distribution maps of water use efficiency for each crop were generated with the calculated crop-wise indicator values of water use efficiencies in ArcGIS 10.3 environment.

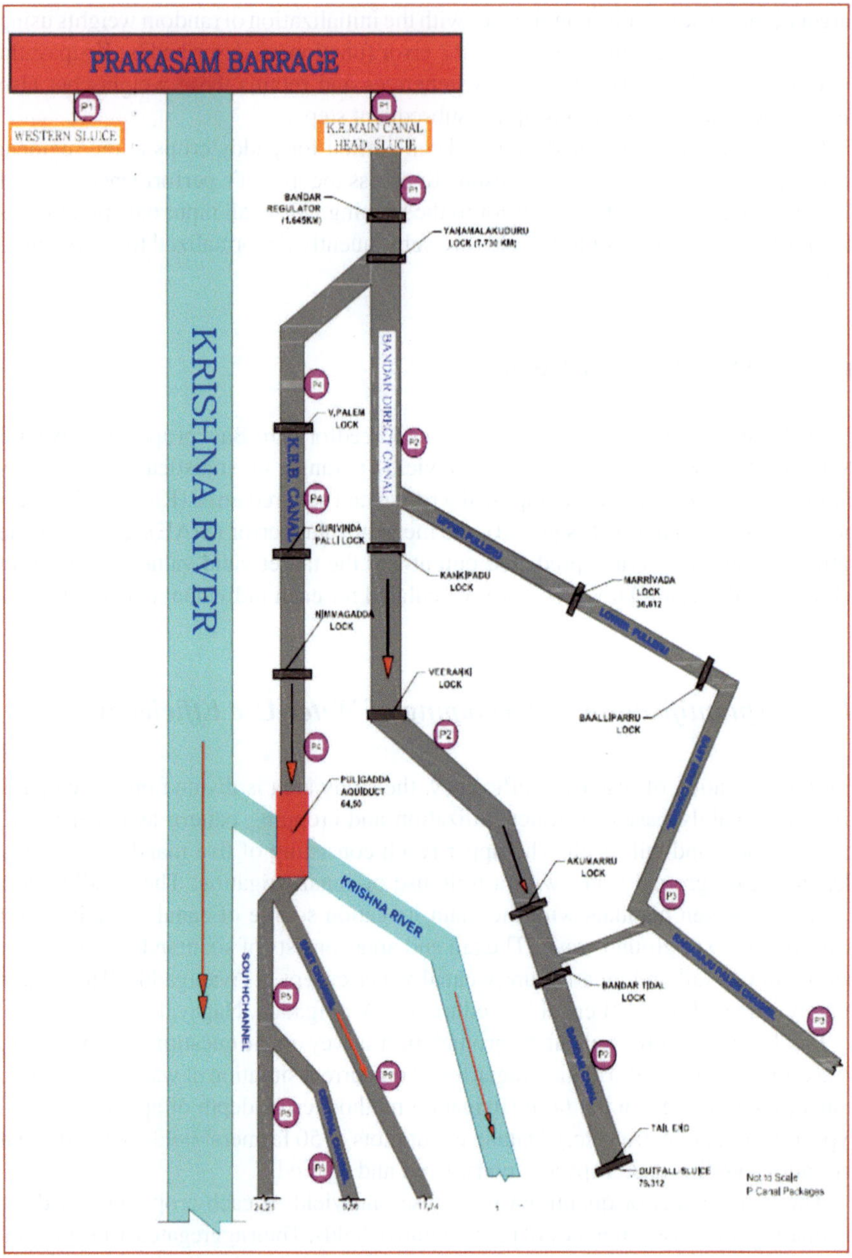

Fig. 4.3 Line diagram of main canals in KCD. *Source* Water Resources Department, KCD, Vijayawada

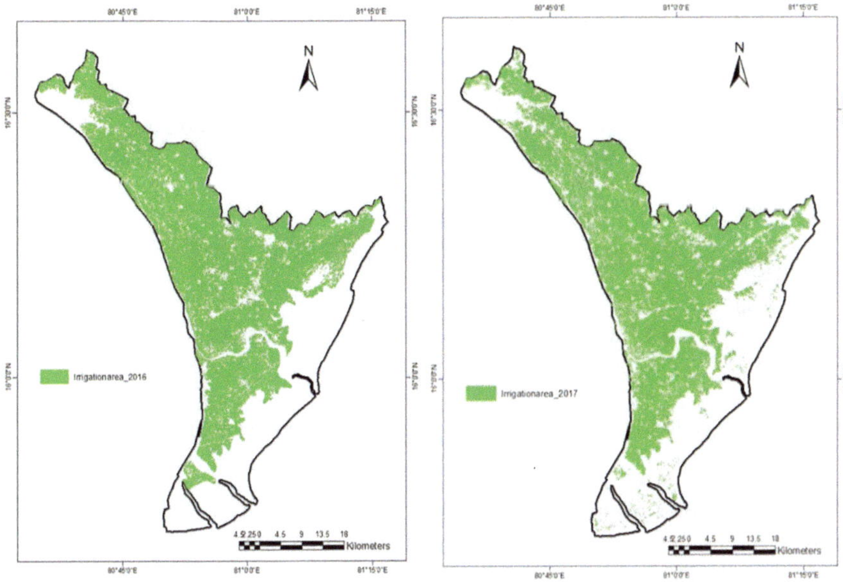

Fig. 4.4 Satellite-derived irrigated area maps of KCD in *kharif* season during 2016 and 2017

4.3 Results

4.3.1 Total Irrigated Area in KCD

The derived total irrigated area maps for different years under study are presented in Fig. 4.4. The results indicated that there was a good extent of irrigated area every year. The highest irrigated area was 130005 ha in 2016. The irrigated area was 122,192 ha only in the year 2017 under KCD. The spatial maps of irrigated area (Green tone) in 2017 showed less intense and also patches of fallow were also observed. The same was confirmed with irrigation department and also with ground truth data during October 2017.

4.3.2 Crop Classification of Different Crops in KCD

Each paddy class has specified phenology depicted by NDVI magnitude. The NDVI profile was observed in reaching to a peak in the month of November then falling (Fig. 4.5). The NDVI time-series also facilitated in the separation of early rice and late rice based on factors such as when a crop calendar begins and the magnitude of NDVI. The similar results expressed by Muller et al. (2015) and Zhang et al. (2015) used temporal Landsat-8 data for crop differentiation. Teluguntla et al. (2015) used

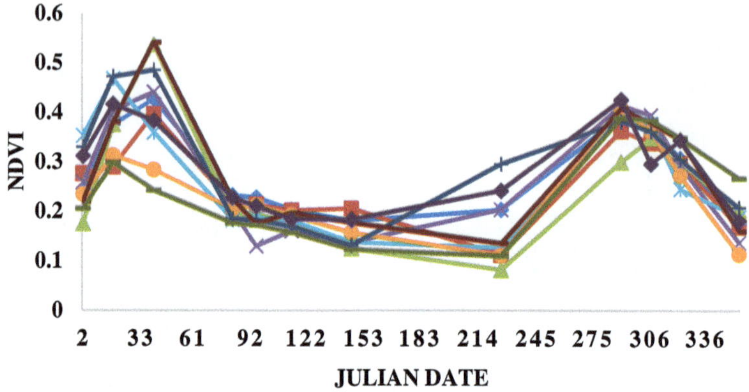

Fig. 4.5 NDVI profiles of areas with paddy in *kharif* followed by pulses in *rabi*

MODIS data. Most of the area was paddy followed by pulses. The transplantation of rice was taking place in July–August and harvesting of the crop is taking place in November and December months.

It was found that there was a misclassification in the paddy group with swampy lands. To extract paddy fields exactly, LSWI + Δ > NDVI was applied. Xiao et al. (2006) reported that the value of Δ = 0.05 and 0.03 gave best results. In the present study area, the best threshold value of Δ = 0 was selected based on trial-and-error approach to eliminate swamp lands, and the above rule with Δ = – 0.4 was used after the harvest period.

It was noticed that in the southern parts of the area, i.e., in Koduru, Nagayalanka, etc., paddy transplantation starts in late August or September. This is due to delayed water availability and also chronic salinity problems in the tail ends. Due to seawater intrusion, the ground water was contaminated and the farmers were completely depending on canal water only. Time-series NDVI as shown in Fig. 4.6 is matching with the visual observation and also with ground truth. As per the field survey, the farmers were not taking second crop as the ground water irrigated pulses got complete failure due to salinity. All these areas were surrounded by Aquaculture which is also other reason for the second crop failure. The complete reverse trend is observed in Fig. 4.7. The time-series spectral profile corresponding to a double paddy crop is grown on bore wells. It is clear from the figure that detection of two season's. In the first one, the start of season was in the middle of June and the end in November and the second season began between December and January and the end of season occurred in late April. The main water source in this area is ground water near to barrage at Vijayawada and banks of river Krishna.

The final crop classified map of KCD is shown in Fig. 4.8. The map shows main six crop classes. Green represents only *Kharif*-rice and *Rabi* pulses, and dark green represents double crop rice (i.e., both *Kharif* and *Rabi* rice). Magenta color represents late *Kharif* paddy and *Rabi* fallow, blue is the short duration vegetable crops, and sugarcane, perineal crops, and orchards are represented by purple color.

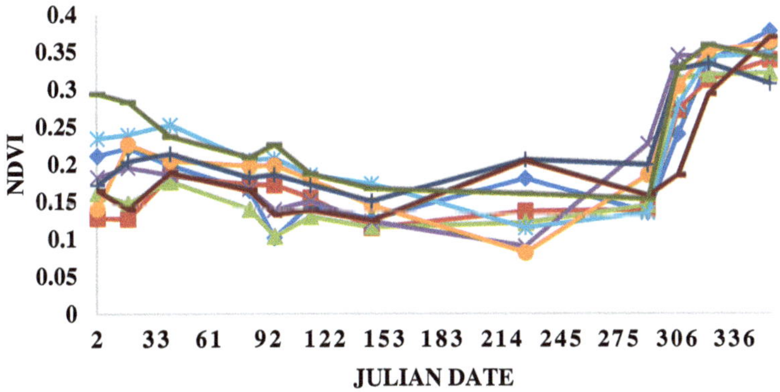

Fig. 4.6 NDVI profiles of areas with late paddy in *kharif* and no crop in *rabi*

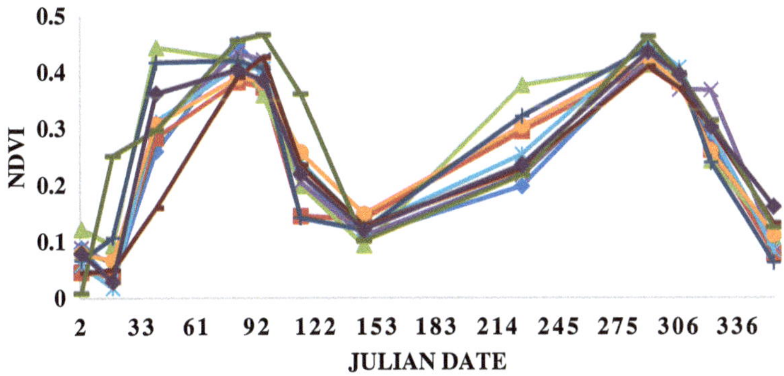

Fig. 4.7 NDVI profiles of areas with double paddy in *kharif* and *rabi*

4.3.3 Estimated Crop Yields of Paddy in KCD by Using ANN

The mean relative error between observed and predicted crop yields exhibited significant variation across different regions. The lowest mean relative error, measuring just 0.133%, was observed in Challapalle, while the highest, at 6.166%, occurred in Pedana (as depicted in Fig. 4.9). This study underscores the notable enhancement in prediction accuracy, surpassing 95%, thanks to the application of neural network techniques.

In Table 4.1, find the statistical performance indicators of the model during both the training and testing phases in the kharif season. The *R*-squared (R^2) values within the range of 0.948–0.992 for training in 2016 and 2017, and the same range for testing, which is 0.947–0.989, is presented.

Following the calibration and validation of the model, the parameter weights are utilized to create maps depicting crop yield per unit area for the kharif season (Jiang et al., 2014), as illustrated in Fig. 4.10. The outcomes reveal a consistently high yield

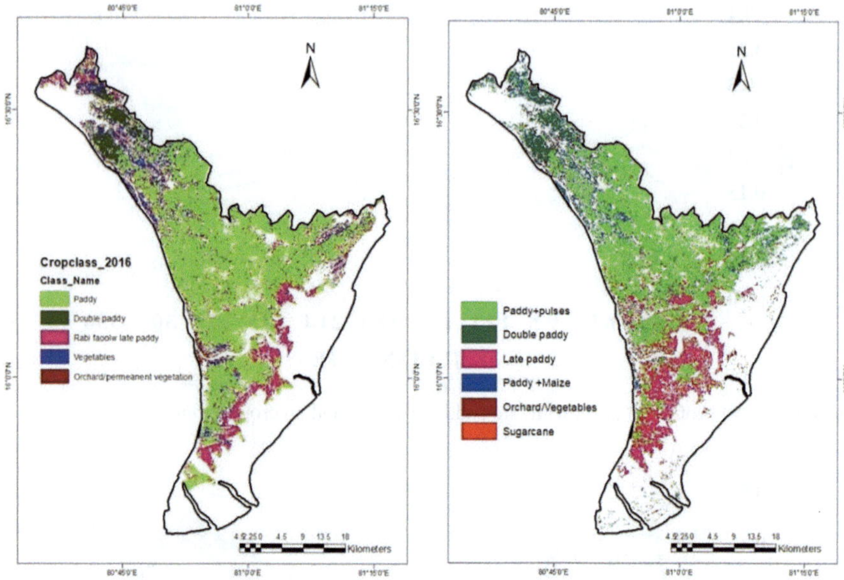

Fig. 4.8 Classified crop map of KCD in 2016 and in 2017

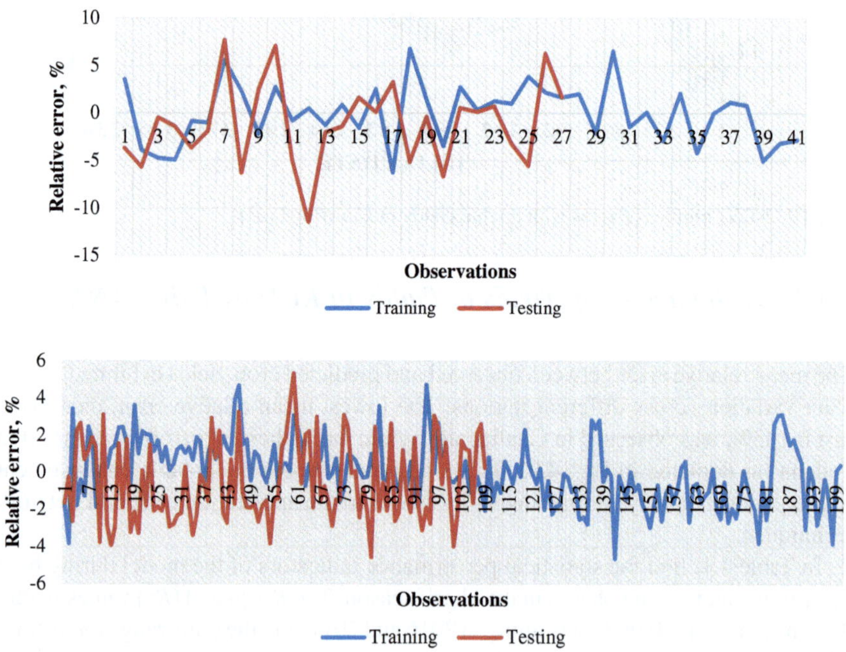

Fig. 4.9 Relative errors of developed FFBPNN during training and testing during 2016 and 2017

Table 4.1 Statistical analysis of neural network training and testing of paddy crop in *kharif* season in different years

	Training				Testing			
Year	RMSE	R_{ratio}	MAE	R^2	RMSE	R_{ratio}	MAE	R^2
2016	0.068	1.002	0.054	0.948	0.084	1.020	0.063	0.947
2017	0.017	0.998	0.013	0.992	0.027	1.000	0.024	0.989

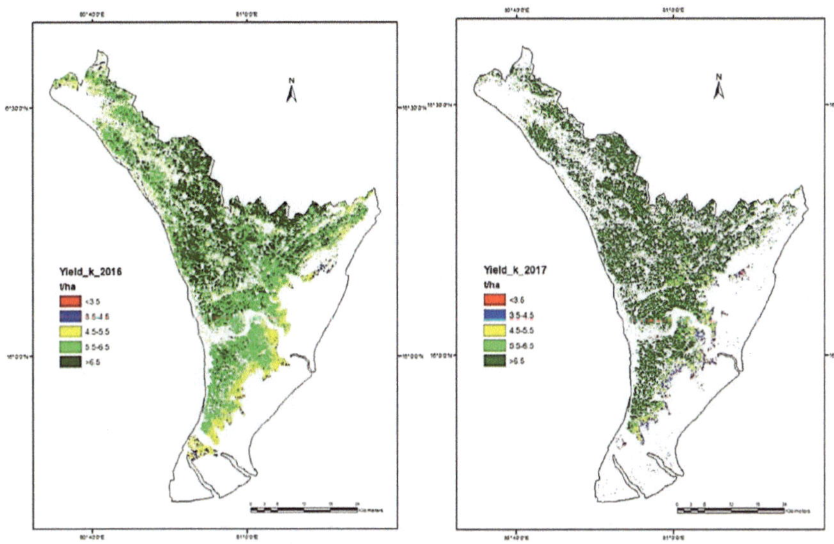

Fig. 4.10 Yield per unit area of paddy crop in *kharif* season during 2016 and 2017

in the upper region of the area, with lower yields observed in the fields at the tail end of the KCD, relative to other locations within the study area.

4.3.4 Mapping of Water Use Efficiency of Major Crops

The irrigation water applied to crops mandal wise, average yield, water use efficiency for each mandal for two years are estimated and presented in Tables 4.2 and 4.3. The total water applied varied from 1082.50 ha in Kankipadu to 1974.45 ha-mm in Mopidevi in *kharif* 2017. Less water application was observed at Koduru, Machilipatnam, Nagayalanka. More water application was observed at Challapalle, Pamarru, Avanigadda, Mopidevi, Ghantasala.

The WUE was higher at upper reach and lower reach as well. Middle reach showed lower WUE. The WUE varied from 0.202 kg/m^3 at Mopidevi to 0.358 kg/m^3 at Penamaluru. The highest is followed by 0.372 kg/m^3 at Pammarru and 0.350 kg/

Table 4.2 Effective rainfall, total water applied, and water use efficiency of paddy crop in *kharif* during 2017

S.No.	Name_of_mandal	Crop name	Rainfall, mm	Effective rainfall, mm	Water applied, ha-mm	Total water applied, ha-mm	Yield, kg	WUE, kg/ha-mm	WUE, kg/m^3
1	Avanigadda	Paddy	366.55	320.485	1386.000	1706.485	3985.904	2.336	0.234
2	Challapalle	Paddy	495.00	476.728	1506.954	1983.681	4140.788	2.087	0.209
3	Ghantasala	Paddy	467.70	447.474	1255.795	1703.268	3889.429	2.284	0.228
4	Guduru	Paddy	402.13	300.400	864.000	1164.400	3850.153	3.307	0.331
5	Kankipadu	Paddy	476.58	458.807	623.700	1082.507	4457.002	4.117	0.412
6	Koduru	Paddy	308.23	263.689	965.250	1228.939	3944.615	3.210	0.321
7	Machilipatnam	Paddy	256.50	224.730	1039.500	1264.230	3196.175	2.528	0.253
8	Mopidevi	Paddy	486.00	467.504	1506.954	1974.458	3992.752	2.022	0.202
9	Movva	Paddy	432.65	417.688	1356.300	1773.900	3767.854	2.124	0.212
10	Nagayalanka	Paddy	381.25	350.858	950.400	1301.258	4156.498	3.194	0.319
11	Pamarru	Paddy	324.90	319.965	1386.000	1705.965	3864.454	2.265	0.227
12	Pamidimukkala	Paddy	256.45	232.954	879.056	1112.011	4135.954	3.719	0.372
13	Pedana	Paddy	274.80	249.589	950.400	1199.989	3422.762	2.852	0.285
14	Penamaluru	Paddy	509.57	490.605	741.654	1232.259	4405.844	3.575	0.358
15	Vuyyuru	Paddy	533.45	513.374	648.000	1161.374	4062.239	3.498	0.350
16	Vijayawada	Paddy	621.45	596.031	668.250	1264.281	4101.168	3.244	0.324
17	Thotlavalluru	Paddy	296.74	272.660	1130.215	1402.876	4225.884	3.012	0.301
Standard deviation				115.261	320.878	332.460	310.983	0.657	0.066
Mean				379.014	1063.790	1442.80	3996.102	2.898	0.290
Coefficient of variation				0.304	0.302	0.2300	0.078	0.227	0.227

Table 4.3 Effective rainfall, total water applied, and water use efficiency of paddy crop in *kharif* during 2016

S. No.	Mandal name	Rainfall, mm	Effective rainfall, mm	Water applied, ha-mm	Total water applied, ha-mm	Yield, kg	WUE, kg/ha-mm	WUE, kg/m^3
1	Avanigadda	468.420	402.979	1386.00	1788.97	3373.114	1.880	0.188
2	Challapalle	367.300	330.005	1506.95	1836.95	3894.386	2.120	0.212
3	Ghantasala	342.300	327.813	1255.79	1583.60	3991.735	2.520	0.252
4	Guduru	341.900	327.057	792.000	1119.05	2768.944	2.474	0.247
5	Kankipadu	502.230	447.684	693.000	1140.68	4200.886	3.682	0.368
6	Koduru	395.100	319.858	707.850	1027.70	2489.741	2.422	0.242
7	Machilipatnam	373.400	305.216	891.000	1196.21	3099.531	2.591	0.259
8	Mopidevi	413.120	351.867	1506.95	1858.82	3173.396	1.707	0.170
9	Movva	432.650	417.688	1582.30	1999.99	3467.854	1.733	0.173
10	Nagayalanka	428.350	372.680	712.800	1085.48	2956.298	2.723	0.272
11	Pamarru	336.000	268.769	1386.00	1654.76	3462.033	2.092	0.209
12	Pamidimukkala	405.320	354.125	879.056	1233.18	4288.601	3.477	0.347
13	Pedana	313.400	298.441	950.400	1248.84	2327.559	1.863	0.186
14	Penamaluru	601.560	550.701	518.400	1069.10	4109.358	3.843	0.384
15	Vuyyuru	549.560	491.453	648.000	1139.45	4280.810	3.756	0.375
16	Vijayawada	415.320	366.072	712.800	1078.87	4000.596	3.708	0.370
17	Thotlavalluru	562.130	503.525	1130.21	1633.74	4681.813	2.865	0.286
Standard deviation			79.5620	353.742	337.431	688.419	0.758	0.076
Mean			378.584	1015.26	1393.85	3562.744	2.675	0.267
Coefficient of variation			0.210	0.348	0.242	0.193	0.283	0.283

m^3 at Vuyyuru. The range of water use efficiencies for flooded rice was on par with Bouman and Toung (2001). In the year 2016, the highest water use efficiency was observed at Penamaluru as 0.384 kg/m^3 followed by 0.375 kg/m^3 at Vuyyuru. The lowest water use efficiency was observed at 0.1733 kg/m^3 at Movva due to the application of more water and 0.1863 kg/m^3 due to less yields. In farmer's interaction, it was found that less yields due to late and insufficient water supply at Pedana. For the middle portion where ensured water supply and ground water dependent areas have similar range of water applied and water use efficiencies.

The spatial WUE maps of paddy in *kharif* season during 2016, 2017 are shown in Fig. 4.11. It shows high WUE at north mandals, where cultivation is done by ground water, followed by tail ends where less water is available for cultivation. The middle portions where plenty of water is available showed less WUE with the high application of water with normal yields. The trend of results for the year 2017 is similar with 2016 except at tail ends. The WUE was high at tail end mandals, especially Nagayalanka and Koduru and Guduru. Overall, the yields and water use

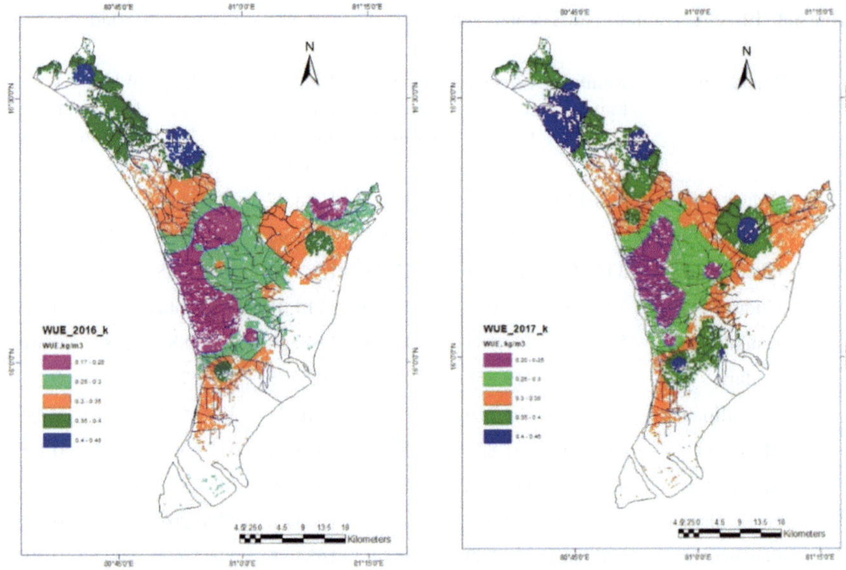

Fig. 4.11 Spatial maps of water use efficiency of paddy crop in *kharif* during 2016, 2017

efficiency were high in 2017 compared to 2016. The spatial variation of cv is 0.283 and 0.227 in 2016 and 2017, respectively. The spatial variation results are similar to results obtained by Bastiaanssen et al. (1998).

4.4 Discussion

The irrigated area of the paddy in 2017 was less than the 2016 of the *kharif* season. The methodology followed in deriving spatial maps of irrigated area and crop classification was highly accurate based on the validation with the ground truth data. The input parameters selected for derivation of yield maps are crucial for the high accuracy. The results showed that the constant higher yields in the upper part of the study area, lower yields at the tail-end fields of the KCD. Multiple factors like salinity, salt water up coning, low water, and late release of canals are contributing the low yields at the tail ends of KCD. From this study, it was clear that the neural network application has increased the prediction accuracy of the crop yield to more than 95%.

It was clear from the study that there is a high spatial variation of water application and water use efficiency in case of paddy in *kharif* season. By conjunctive use of ground and canal water in the middle reaches enhances the possible availability of canal water to saline patches of the study area at lower reaches. It is also important to reduce the canal seepage in these areas. In the upper reaches, as the plenty of ground water is available due to seepage from reservoir, it is recommended to divert canal water fully tail end fields. Metering of water releases is also recommended in this command area to arrest excess usage of water. The study is highly recommended in all canal command areas for improving water use efficiency. It will also help in water saving in the canal command. The application of machine learning algorithms in the canal commands for different studies is highly recommended.

4.5 Conclusions

It was clear from the study that there is a high spatial variation of water application and water use efficiency. Relatively low WUE was observed at middle parts of the study area with high application of water for normal yields. The farmers in the delta are applying water without considering rainfall and available soil moisture leading to inefficient water use. The decision tree model based on NDVI profiles developed from zonal means gave high accurate crop classification with high level of confidence compared to previous methods. The crop yields were accurately estimated with proper selection of input parameters and training of neural network models. The spatial yield maps developed by using FFBPNN models are useful for yield prediction accurately at regional scale. The output maps of this study are very useful for planning water releases and implementing water saving measures under a set of given location specific conditions. There is a need to conduct similar studies on water use efficiency by introducing machine learning algorithms for increasing water use efficiency in canal commands.

References

Anonymous (2020) Agricultural statistics at a glance 2019–20, Andhra Pradesh. http://www.apdes.ap.gov.in

Bastiaanssen WGM, Thiruvengadachari S, Sakthivadivel R, Molden DJ (1998) Satellite remote sensing for estimating productivities of land and water. Water Resour Dev 15(1/2):181–196. https://doi.org/10.1080/07900629949005

Bouman BA, Tuong T (2001) Field water management to save water and increase its productivity in irrigated lowland rice. Agric Water Manag 49(1):11–30. https://doi.org/10.1016/j.agwat.2003.09.002

Elshaikh AE, Jiao X, Yang SH (2018) Performance evaluation of irrigation projects: theories, methods, and techniques. Agric Water Manag 203:87–96. https://doi.org/10.1016/j.agwat.2014.07.012

Jiang D, Yang X, Clinton N, Wang N (2014) An artificial neural network model for estimating crop yields using remotely sensed information. Int J Remote Sens 25(9):1723–1732. https://doi.org/10.1080/0143116031000150068

Krupavathi K, Raghubabu M, Mani A, Parasad PRK, Edukondalu L (2022a) Field-scale estimation and comparison of the sugarcane yield from remote sensing data: a machine learning approach. J Indian Soc Remote Sens 50(2):299–312. https://doi.org/10.1007/s12524-021-01448-w

Krupavath K, Raghu Babu M, Mani A (2022b) Comparative evaluation of neural networks in crop yield prediction of paddy and sugarcane crop. In: The digital agricultural revolution: innovations and challenges in agriculture through technology disruptions, pp 25–55. https://doi.org/10.1002/9781119823469.ch2

Kumar A, Dubey OP, Ghosh SK (2014) GIS based irrigation water management. Int J Res Eng Technol 3(14):62–65

Levidow L, Zaccaria D, Maia R, Vivas E, Todorovic M, Scardigno A (2014) Improving water-efficient irrigation: prospects and difficulties of innovative practices. Agric Water Manag 146:84–94. https://doi.org/10.1016/j.agwat.2014.07.012

Muller H, Rufin P, Griffiths P, Siqueira AJB, Hostert P (2015) Mining dense Landsat time series for separating cropland and pasture in a heterogeneous Brazilian savanna landscape. Remote Sens Environ 156:490–499. https://doi.org/10.1016/j.rse.2014.10.014

Teluguntla P, Ryu D, George B, Walker J, Malano H (2015) Mapping flooded rice paddies using time series of MODIS imagery in the Krishna River Basin, India. Remote Sens 7(7):8858–8882. https://doi.org/10.1016/j.isprsjprs.2018.07.017

Xiao X, Boles S, Frolking S, Li C, Babu JY, Salas W, Moore B III (2006) Mapping paddy rice agriculture in South and Southeast Asia using multi-temporal MODIS images. Remote Sens Environ 100:95–113. https://doi.org/10.1016/j.rse.2005.10.004

Zhang L, Tian J, He H, Ren X, Sun X, Yu G, Lu Q, Lv L (2015) Evaluation of water use efficiency derived from MODIS products against eddy variance measurements in China. Remote Sens 7:11183–11201. https://doi.org/10.3390/rs70911183

Chapter 5
Pathways for Intensified Agriculture Through Sustainable Water Management: Need of an Hour

Pooja Purushotham, G. Sahana, Gaurav Rakhonde, Namburi Karunakar Reddy, Anusha M. Nayak, and K. M. Pravalika

Abstract The United Nations' 2030 agenda has its primary goals as the abolition of poverty, hunger, and paucity access to clean water, all of which are essential for a substantial chunk of the continuation of the life of the population. Growing populations, global economic development, and resource sustainability are all related to these goals. By the end of the year 2050, it has been predicted that the economic efficiency of the farming sector will need to boost by about 25–110%. The production system must adjust itself regarding the ramifications of global warming which includes, alteration in the pattern of rainfall, extended dry spells, extreme weather conditions, changes in soil humidity, in addition to satisfying the sustainable intensification of agriculture. Agriculture can use a multitude of cutting-edge technological and non-technological options to increase water productivity. Technologies can be used to improve rainwater collection and use, ways to supervise crops and the soil, as well as on-farm warehousing. Alternative technologies may bring about the intended shift through social and institutional innovations. To put into practise equitable use of water, it is necessary to incorporate risk-aware planning and management techniques into climate change adaptation and mitigation plans. Integrated rainfed production methods like irrigated crop cultivation, crop-livestock systems, and ecological agriculture can be used to adapt to the weather in the local area. Future water management strategies will need to go beyond current technological advancements. Particularly in view of the predicted effects of climate change, adequate institutional reforms are

P. Purushotham (✉) · G. Rakhonde · N. Karunakar Reddy · A. M. Nayak
Department of Plant Pathology, University of Agricultural Sciences, GKVK, Bangalore 560065, India
e-mail: poojapurushotham17@gmail.com

G. Sahana
Department of Agronomy, University of Agricultural Sciences, GKVK, Bangalore 560065, India

K. M. Pravalika
Department of Agronomy, College of Agriculture, University of Agricultural Sciences, GKVK, V C Farm, Mandya, Bangalore 560065, India

K. Pandey et al. (eds.), *Artificial Intelligence and Smart Agriculture*,
Advances in Geographical and Environmental Sciences,
https://doi.org/10.1007/978-981-97-0341-8_5

also necessary to increase the resilience of rural communities and maintain family food security.

Keywords Sustainable · Water management · Population · Climate change · Technology · Irrigation

5.1 Introduction

Water is widely recognized as potentially vital resource for all living things. Although water resources are indispensable for producing services as well as goods related to agriculture and industry, it is important ecological element (Chartzoulakis and Bertaki 2015). It goes further to include an enormous contribution public well-being and environmental safeguarding. Interestingly, just 2.7% of the global water supply is fit for drinking, and a mere 30% of that might be exploited to fulfil both human and cattle needs (Ertek and Yilmaz 2014). Since water is the most important resource, elevated levels of water stress, along with droughts happening more often as well as being worse, have decreased the availability of freshwater resources in dry and semi-arid areas, primarily owing to global warming dynamics (Nazari et al. 2018; Pandey et al. 2020).

Despite decrease in the abundance of resources for freshwater, the global prerequisite for freshwater has steadily risen over the course of the past 60 years for a multitude of plausible explanations, including quick growing population and growing economy (Kaur et al. 2010), urbanization and industrial development (Biswas 2010), shifts in land use (Sophocleous 2004), intensive cultivation practises (Tilman et al. 2002), and deteriorating the environment (Vorosmarty et al. 2000) (Fig. 5.1). At the same time, by 2050, it is anticipated that there will be 9.8 billion people on Earth, with more than half of them living in urban areas (UN 2017). As a result, more food will need to be produced in the future as both populations and economies continue to expand, meaning that by 2050, almost 40% more water is required.

As a consequence, water scarcity will likely become a key issue in the future (Fig. 5.2) (Doungmanee 2016). Freshwater resources are expected to be drastically reduced by virtue of global warming, and dry spell will probably happen more frequently, more intensely, and for longer periods of time (IPCC 2014). By 2025, it is anticipated that two-thirds of the global inhabitants will be experiencing scarcity of water, despite the fact that global warming is ahead having an influence on the time-based and geographic arbitrariness of surface and groundwater abundance (IPCC 2014). In areas where agricultural systems rely extensively on irrigation, the prerequisite for irrigation water will increase as freshwater resources become more constrained.

Currently, the world's largest user of six water resources is irrigated agriculture, which uses over 70% of all water withdrawn (FAO 2013). Intensive irrigation-based agriculture devours over 90 per cent of freshwater in developing nations wherein the rural economies rely solely on agricultural products (Samian et al. 2015), and the

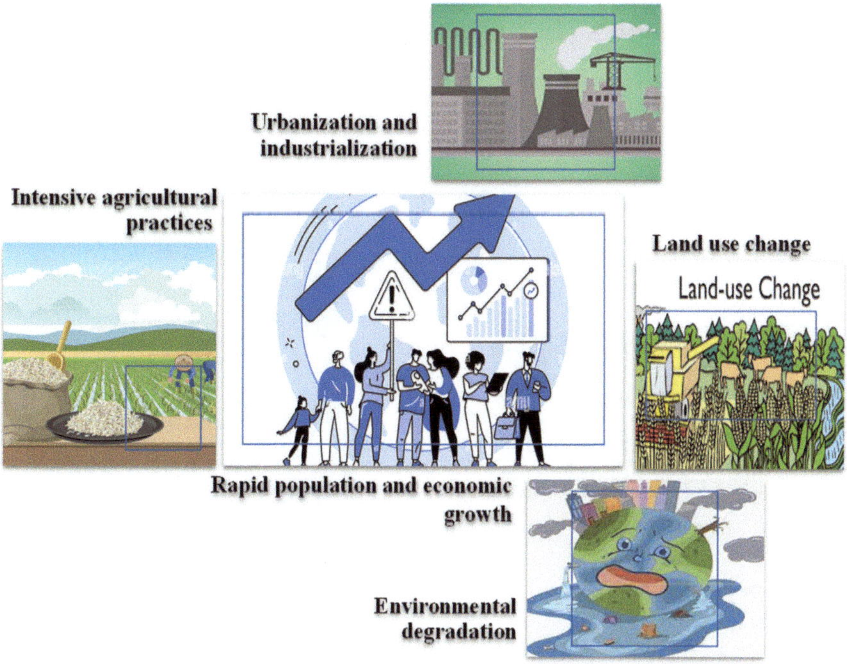

Fig. 5.1 Reasons for increased global demand for freshwater during the last 60 years

aforementioned pattern is anticipated to persist as long as there are renewable water resources available (Rosegrant et al. 2009). In dry and partially arid locations, when there is a dearth of water which is an unavoidable reality for the agricultural field, it is essential to have oversight on agricultural water usage to guarantee the long-term viability of the water supply in the planet.

To alleviate the problem faced in water scarcity, water resource conservation and utilizing irrigation water sustainably are necessary. Without thorough preparation, a steadfast strategy of logical irrigation development will significantly jeopardize the nation's food security (Kang et al. 2016), exacerbate rural poverty, and cause more forced migration (Keshavarz et al. 2013; Madani 2014; Moazediet al. 2011).

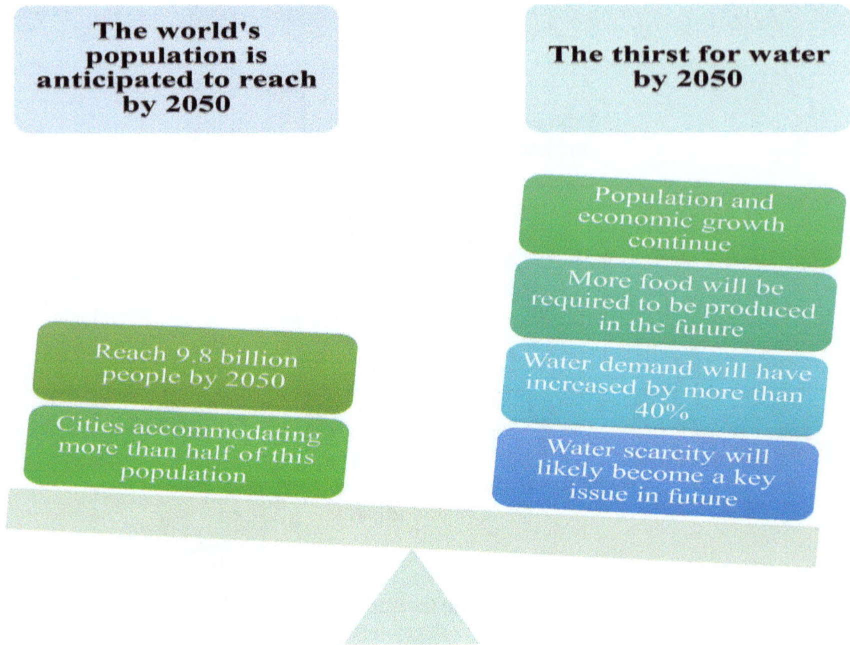

Fig. 5.2 Relation representing the world's population and demand for water by 2050 (UN 2015, 2017)

5.2 Needs of Sustainable Intensification of Agricultural Water

- To encounter the rising requirement for food, agriculture must be sustainably intensified. However, merely increasing production would not guarantee food security. Additionally, crucial are the allocation of equitable food and the maintenance of ecosystem amenities. In regions with high-water demand and challenging transboundary water agreement negotiations, the accessibility of water to link up rising food prerequisites dominates discussions concerning agricultural water management (FAO 2011).
- According to predictions, the need for agricultural production would increase globally by 70% by 2050 as a result of rising incomes and population (FAO 2011). Despite the fact that crop production would need more water, the positive aspects of other water uses and the positive aspects of agriculture are mutually exclusive.
- Just as intersectoral competition for water intensifies, agricultural intensification will necessitate more irrigation, better water use efficiency, and higher yields. Extraneous factors' components like changes in climate, sector competition, and socioeconomic changes exacerbate physical restrictions on water resources.

- Up until 2050, the agricultural sector is expected to expand by more than 80% thanks to higher productivity on already-cultivated lands (FAO 2011). Legitimate management of water and land is therefore required to close the productivity gaps that now exist. Policymakers have a difficult time determining how much water can actually be dedicated to beneficial uses without having a negative effect on crucial environment (Vaux 2012).
- Falkenmark and Rockstrom (2010) noted four ways to elevate supplies of water to the minimum required when both nourishment and water are on hand needs outstrip the potentiality of currently cultivated areas and perennial pastures to produce. One of these is increasing the area held for production aimed at maximizing the use of rainfall, another is collecting and storing rainwater from nearby lands, a third is raising crop and water productivity, and a fourth is maximizing international trade in agricultural goods.

5.3 Sustainable Water Management Practices

The mission of legitimate agricultural water management is to balance water obtainability and needs with regard to quality, quantity, and location over time, furthermore with regard to cost and environmental impact (Chartzoulakis and Bertaki 2015). Below is a brief description of them.

5.3.1 Land Levelling

Successful crop growth depends upon the levelness of the field. Largely, management options that are frequently overlooked are proper land levelling. The main advantages of land levelling are better crop establishment, uniform water coverage of the field, uniform crop stand and maturation, a 40% reduction in weed growth, an increase in a farming area, a yield increase of 10–20% on average, and a 10–15% reduction in farm operation times. It results in uniform plant growth, which raises yields and improves the prudent consumption of water while increasing the potency of water application. Additionally, it leads to more effective use of pesticides and fertilizers (Singh 2014).

5.3.2 Soil Moisture Retention

Drought effects are lessened when the soil preserves water. Mulching, a build-up of organic content in the soil, and conservation tillage are all methods for keeping moisture in the soil.

Fig. 5.3 Material types used in mulching

5.3.2.1 Mulching

Farmers occasionally use the traditional practice of mulching to preserve soil moisture. Mulching is possible using a variety of materials, including soil (dust mulch), pebbles (stone mulch), or substances that are organic or inorganic and come from either natural (such as leftover crop residues from fields) or artificial (such as plastic sheets and soil conditioners) sources (Fig. 5.3). However, choosing the mulch material is dependent on factors such as price, availability, handling ease, and effectiveness (Sharma and Kumar 2014).

5.3.2.2 Accumulation of Biomass in Soils

Accelerated or retained organic content in the topsoil layers enhances aggregation of soil, restricts crust development or securing soil surface, and intensifies the soil's ability to retain water (Chartzoulakis and Bertaki 2015).

5.3.2.3 Conservation Tillage

Traditional tillage, which frequently involves a combination of primary and secondary tillage activities, always has the effect of accelerating the disintegration of soil organic matter (SOM) by breaking fragments of soil and disclosing them to weathering agents. This makes the soil looser and more susceptible to being wiped away by downpour of water. Conservation tillage was found to be an efficient replacement for conventional tillage in order to prevent soil loss driven by water or wind, enhance and retain soil health, and reduce farming expenses. Conservation tillage is also less intensive than conventional tillage and preserves a minimum of 30% of the top layer of soil wrapped by remnants after cultivation (Sharma and Kumar 2014).

5.3.3 Agricultural Practices

Agricultural practises like soil management, fertilizer use, and control of pest and diseases are linked to environmentally friendly supervision of water in agriculture. Wide range of conventional and contemporary coping with soil and crops that can help conserve water. These methods include reducing runoff, improving soil infiltration rates, increasing water holding strength, reducing evaporation and soil loss (Pereira et al. 2009). To prevent further environmental pollution, effort must be done to use chemicals for reduction of pest and weeds in a sensible manner (Fig. 5.4).

5.3.3.1 Tillage

With respect to soil management, there are three main techniques: soil surface tillage, contour tillage, and conservation tillage, which includes no-tillage and reduced tillage and retains crop residues on the soil during planting. Because it helps to maintain high levels of organic material in the soil, conservation tillage is very efficient in enhancing infiltration of soil and lessening erosiveness. The operation of mulching the soil's surface with crop residues reduces evaporation losses, slows water evaporation losses, enhances infiltration conditions, and shades the soil. It also aids with weed control, reducing wasteful water use.

5.3.3.2 Mulching

It restricts raindrops from striking the soil directly, thereby preventing crusting and sealing processes. Intensifying or sustaining the carbon and biological content of the top layer of soil promotes enhanced structure, less crust development or securing soil surface, and advancement in the soil's magnitude to store water. Liming, which is

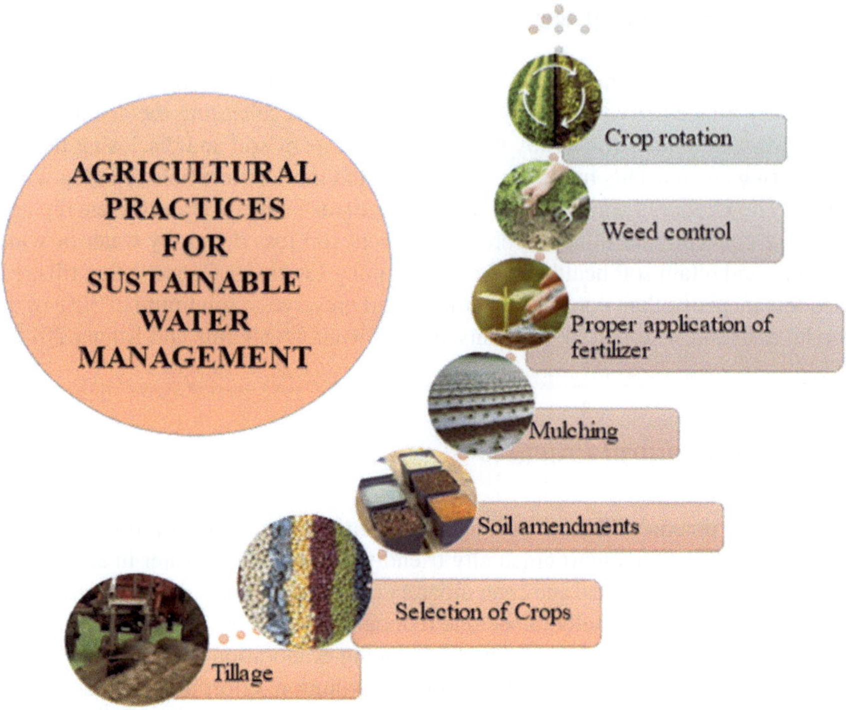

Fig. 5.4 Agricultural practices that enhance sustainable water management

applied to high pH soils similarly to applying gypsum, can control acidity. This procedure encourages a slight increase in soil water availability which can be attributed to greater amount of rooting, greater crop development, and enhanced soil aggregation.

5.3.3.3 Selection of Crop

Management of crops is choosing the right crops and crop patterns while taking seasonality of rainfall and water use efficiency for various types of crops into consideration. Utilization of varieties of crop that are resistant to water stress rather than highly profitable but more fragile ones. Utilizing crops or varieties with a short cycle lowers the total amount of water utilized.

5.3.3.4 Other Practices

Implementing crop rotation techniques reap the benefit of the soil's remaining moisture. Logical and reasonable fertilizer use by means of soil and leaf analysis, the kind and amount of fertiliser should be applied in accordance with the crop's needs.

Employing effective weed control methods to reduce weed competition for water and to check transpiration losses.

5.3.4 Methods of Irrigation

Irrigation methods refer to the ways in which irrigation water is applied to the ground. The type of soil, the topography of the land, the crops grown, the water readily accessible for irrigation, besides other location-specific changes, all have an impinge upon the choice of the best irrigation technique. To increase productivity per applied unit of water, it is critical to adhere to the recommended guidelines. Despite extensive research on these aspects, the suggestions for selecting the optimal irrigation water supply technique are frequently disregarded.. By using the right technique, an efficiency of 10–15% in the usage of water feasible is achieved. Different methods of irrigation used in sustainable water management are described below (Fig. 5.5). Advantage and limitations of various irrigation methods are given in Table 5.1.

Fig. 5.5 Different types of irrigation methods used in sustainable water management

Table 5.1 Advantages and limitations of different irrigation methods

Sl. No.	Types of irrigation	Advantages	Limitations
1	Furrow irrigation	Water flows in a small channel, with low capital and maintenance costs	Inadequate water control, soil erosion, runoff, and high labour costs
2	Check basin irrigation	With potential design, it is more efficient and requires less labour than a furrow	Require levelling of ponded water and sloping fields
3	Border irrigation	There is less labour and runoff than with a furrow, and smoother depth administration	Water flows throughout the full surface of soil
4	Sprinkler irrigation	Efficient water management, can be automated and irrigated often, fits irregular topography of fields, increased uniformity, and low labour	High initial and ongoing capital and maintenance costs; and may interfere with intercultivation
5	Drip irrigation	Water control is excellent, and regular applications are made sure	Higher establishment costs, to avoid clogging of pipes clean water or treatment and filtration are required

5.3.4.1 Check Basin and Irrigation Along Border Strips

Numerous techniques for surface irrigation have been developed over time. The most well-liked irrigation technique among them is check basin irrigation. Applying water in the right way is crucial if you want to reduce the negative effects of irrigation. The type of soil, the topography of the land, the types and quantities of the crops that will be grown, and other unique to the site deviations all influence the best irrigation technique. To increase productivity per applied unit of water, it is critical to adhere to the recommended guidelines. Although significant research is conducted on these factors, the recommendations to choose the best irrigation water application method are frequently disregarded. Growers engage in surface irrigation, wasteful use of this valuable natural resource. By using the right technique, it is simple to intensify the fruitful use of water by 10%–15%.

5.3.4.2 Surface Irrigation

Oldest and widely used method of irrigation, as it merely relies on employing gravity and the land's contours, water is distributed over a field. The slope and run length amend the water distribution. It can only be used if the region or area is instinctively inclined and possesses sufficient water; instead, it requires abundant labour. The field area nearest to the inlet receives greater opportunity time and gives rise to the increased depth of infiltration, whereas the downfield farthest from the inlet receives the least. It includes uncontrolled flooding, level basins, furrow methods, and border strip methods. Low application efficiencies, water waste, a rise in the water table, and waterlogging are drawbacks of surface irrigation techniques.

5.3.4.3 Furrow Irrigation

Row crops and vegetables are typically watered using the furrow irrigation technique, which is pertinent for the rate of soil infiltration ranging 0.5 and 2.5 cm per hour. It works best with streams that are between 1 and 2 l s^1 in size and slopes between 0.2 and 0.5%. Many field crops that currently receive water supply by flooding, check reservoir irrigation, or border strip irrigation easily changed to a raised bed irrigation system or a modified form of furrow irrigation.

5.3.4.4 Surge Flow Irrigation

Irrigating via border strips and furrows is severely constrained by exorbitant water intake and substantial percolation losses. Surge flow irrigation reduces uptake and percolation losses, increases efficiency, and conserves water by applying intermittent water multiple times over time spans. It accelerates water front advance and reduces water needed for advance phase. Researches have shown that it improves the effectiveness of water distribution, lowers infiltration rates, and minimizes percolation wastage by increasing cumulative infiltration levels.

5.3.4.5 Localized Irrigation

Competent ways to water crops are widely acknowledged to be localized irrigation (Keller and Bliesner 1990). Localized irrigation systems apply water directly to the growing plant's region, reducing water loss due to soil evaporation (micro-sprayers, trickle, or drip irrigation). To concentrate the water application near zone of root, it is applied close to the base of crop. Water is typically applied frequently, in small amounts, and at low flow rates. The high-efficiency rates that can be attained—up to 90% in some cases—are the main benefits of localized irrigation systems. High productivity could lead to significant water savings. Higher harvests and improved standards are frequently obtained by carefully controlling the application of nutrients and water. Given that neither the foliage nor the entire soil surface is wet, control of weeds and pests may be improved.

5.3.4.6 Drip or Trickle Irrigation

A localized irrigation technique called drip or trickle irrigation uses water at very low flow rates. It is also called pressurized irrigation. The pressures needed for drip irrigation typically range from 0.7 to 1.4 kg/cm^2 (10 and 20 psi). With regards to water savings and crop yields, the pressurized system of irrigation has multiple benefits over-irrigation techniques of surface gravity. Most soil types and topographic types can benefit from drip irrigation. Situations with scarce water supplies or high-water costs are very well suited to it. Only tree crops and crops with elevated value are capable of covering the substantial expenses associated with them. These pressurized irrigation systems are particularly helpful for soils with a very coarse texture or undulating, uneven terrain. These methods encourage the use in irrigation commands for particular conditions.

5.3.4.7 Subsurface Drip Irrigation (SDI)

SDI uses underground tubes for water application, with soil matrix vacuum for extraction. As the adjacent soil becomes saturated, water seeps everywhere. SDI may provide water conservation, improved fertilizer effectiveness, homogeneous, and exceptionally effective delivery of water, eradication of surface infiltration difficulties and evaporation losses, versatility for offering prevalent and light irrigations, diminished disease and weed issues, and lesser operating pressure. Subsoil irrigation can benefit nearly all crops, particularly highly valuable fruits and vegetables, turfs, and panoramas. Crop installation depth varies, with vegetables 15–20 cm, tree 30–50 cm with high initial costs, and increased clogging risk (Chartzoulakis and Bertaki 2015).

5.3.4.8 Fertigation

The practice of applying fertilizers via irrigation systems, also known as fertigation, has become widespread in contemporary irrigated agriculture. Fertigation can be done with localized irrigation systems, which have a high potential for water application efficiency. Thus, the soluble fertilizers are given to the saturated portion of the soil via the system of irrigation at concentrations required by crops. When irrigation design or operation is inadequate, potential drawbacks include uneven chemical distribution, over-fertilization when irrigation is not contingent on actual plant needs, and soluble fertilizers are used excessively.

5.3.5 Scheduling Irrigation

Scheduling irrigation involves selecting water application schedules for crops. The best irrigation schedule should outline the optimum quantity of water to utilize at the correct time interval for every irrigation. The main practical aspects of irrigation, such as calculating how much water should be applied and when it should be applied to a particular crop, are greatly influenced by the type of soil and climatic conditions. It is essential to maximizing irrigation system performance and sustainability, as well as for water conservation and agricultural production optimization. Certain criteria need to be taken care of while scheduling the irrigation.

5.3.5.1 Estimates and Measurements of Soil Water

By regulating the plant's water status, soil water directly influences plant growth. In this method, the soil moisture content is estimated to determine the shortage in available soil moisture at which irrigation is suggested considering the predetermined soil moisture content to fill the field with soil moisture.

To assess soil water availability for plant growth, estimate the quantity and evaluate the effectiveness of soil water potential. Accuracy depends on sampling techniques and chosen locations for point observations, as soil water variability affects the information (Peyremorte and Chol 1992). Soil texture, appearance, water potential measurement, and water content measurements (tensiometers, soil spectrometers, and pressure transducers) (TDR) (Fig. 5.6) by remote sense are all used to estimate and measure soil water for irrigation scheduling.

5.3.5.2 Variables Associated with Climate

Local or regional irrigation programmes frequently make use of climatological parameters. The reference evapotranspiration (ETo) for a particular region is calculated using weather information along with empirical equations, after being locally calibrated, to give precise estimates. Estimate crop evapotranspiration using appropriate crop coefficients. Real-time data processing is a possibility, as well as more typically, information processed using past data. These methods include measuring evaporation for the purpose of calculating ETo, estimating crop evapotranspiration using climatic data and remote sensing ET (Allen et al. 1998).

5.3.5.3 Crop Stress Parameters

Crop stress parameters can detect an irrigation time signal from the plant without specifying irrigation depths, thereby avoiding the need to guesstimate soil water parameters. Plant personality, which responds quickly to the soil, water, plant, and evaporative demand of the atmosphere, could be linked to a lack of plant water and may serve as a criterion for timing irrigation to crops. As a result of the proper sampling, this message may come from specific plant tissues, or it may come from the entire canopy. Plant stress parameters are crucial for consistent irrigation depths. Crop water stress indicators include leaf potential, stem/fruit size changes, xylem flow estimation, vegetation cover heat, and satellite images (Deumier et al. 1995; Jackson et al. 1981; Itieret al. 1992).

Fig. 5.6 Tensiometers, soil spectrometers, and transducers to determine soil texture and appearance, soil water content measurement, and soil water potential measurement

5.3.6 Inadequate Irrigation Techniques

Prior to now, water supply restrictions were not taken into consideration while identifying the requirement for irrigation of crops. The irrigation timeline was subsequently established to satisfy the water requirements of the entire crop. However, as the need for water for industry and municipalities increased, the portion of water accessible for agriculture is steadily reduced in drought-prone and semi-drought-prone locations. Correspondingly, existence of particularly non-availability of water to produce the highest yields. The adoption of irrigation strategies that are independent of a crop's overall water needs is thus recommended for exclusively effective and intellectual use of water. These monitoring techniques included subsurface irrigation, inefficient irrigation, partial root drying, and Regular Deficit Irrigation (RDI). Crops are permitted to endure a certain level of water shortage and decreased yields under regulated deficit irrigation (RDI). During RDI, a particular amount of water stress is imposed on the plant, either temporarily or over the entire growing season. The primary goal of RDI is to strengthen vegetative growth control and crop water use efficiency (WUE) by removing irrigation practices that do not significantly affect yield. With the benefits of using the water saved to irrigate additional plants for which water might otherwise be inadequate using traditional irrigation techniques, the yield loss may be insignificant.

5.3.7 *Partial Root Drying (PRD)*

Partial root drying is an innovative irrigation method that exposes a portion of a plant's root system to a dry environment while irrigating the other. It was first used on grapevines. On a 7–14 day cycle, the root system's wet and dry sides alternate. PRD balances vegetative and reproductive growth by using plants' biochemical responses to water stress. Vines close stomata to store moisture, using abscisic acid (ABA) as their first defense. As soil water supply declines, ABA is produced in roots and transferred to leaves through transpiration, causing stomata to close more tightly, reducing water loss. Half closure of stomata results in an improvement of WUE. Grapevine drip irrigation, grapevine subsurface irrigation, and furrow irrigation have been used successfully with PRD in pear, citrus, and grapevines. (Dry et al. 1999; Clancy 1999).

5.3.8 *Water Harvesting*

Rainwater harvesting includes gathering, centering, and enhancing the resourceful utilization rainwater lowering wasteful depletion like runoff, evaporation, and seepage. Simply stated, it is a method for saving rainwater and runoff for later use. Runoff is collected and used, as opposed to being left to erode the land. Runoff collection lessens soil erosion, stormwater runoff, flooding, and sediment, metal, and pesticide contamination of surface water. By gathering and storing rainwater, a large volume of water for irrigation can be made available. Rainwater collection can be used for domestic uses, supplemental irrigation, groundwater recharge, and pre-sowing irrigation. With this approach, yields and production reliability can both be significantly increased. Rainwater collection and storage are crucial in the study area due to evaporation and runoff losses. Rainwater can be a low-cost resource that can be channelled and stored with the installation of a minimal infrastructure (Velasco-Munoz et al. 2019). The monitoring of rainfall is another important factor that is supposed to be implemented attention to while estimating the water requirements for crops based on annual rainfall. The duration, intensity, and frequency of the rain should all be tracked during this rainfall monitoring process. Data collection enables climate prediction model for rainwater infrastructure development for agricultural use.

Utilizing rainwater where and when it falls on the land, where it is both available and usable, is known as in situ water harvesting. Land levelling, land shaping, micro-catchment water harvesting, raised sunken bed technology, and terracing are options for on-site rainwater harvesting.

Ex situ water harvesting requires a suitable storage facility and conveyance system to catch and hold excess runoff, stream flow, spring water, and roof-top water (Velasco-Munoz et al. 2019).

5.3.9 *Irrigation with Water of Poor and Marginal Quality*

Agriculture faces competition from domestic, industrial, and power sectors, requiring marginal water usage, with water allocation predicted to decline 10–15% over 20–30 years (Singh 2014). Three deviations from standard irrigation techniques are necessary when using waters of poor quality:

(a) Choosing crops that can withstand salt conditions.
(b) Boosting the administration of water and, in certain instances, implementing cutting-edge irrigation technology.
(c) Maintaining soil-physical characteristics for crop water and leaching requirements (LR) (Oster 1994).

5.3.10 *Desalination of Water*

The process of desalinating seawater involves removing the salts from the water. Although there are various desalination techniques, reverse osmosis is the most frequently employed. In this procedure, pressure is used to push water through semi-permeable membranes. The water molecules move while the salts are kept in the membranes. Water that has been desalinated is perhaps utilized in agriculture.

5.3.11 *Multiple Uses of Water*

Traditional land productivity approach must shift to improving water productivity systemically, as water shortages are expected to escalate in the future. Using water in multiple ways offers opportunities to improve productivity of water, allow for diversification of products into high-value markets, reduce threat, optimize the utilization of resources, and ensure a consistent stream of revenue and use water of poor or marginal quality wisely. IFS agriculture improves water productivity, reduces external inputs, and increases farmers' earnings.

5.3.12 Water Pricing Policy

If not properly managed, the rising demand for water is anticipated to outpace the supply. Integrative efforts in each area relating to water, such as development, distribution, consumption, and management, are essential because of the disparity in readily accessible sources of water over the course of time and space. Water pricing ensures efficient use and allocation using volumetric metering and accounting techniques. Promoting progressive in nature, yearly, excessive water use and short-term drought surcharge rates will help people conserve water. Additionally, a scheme of raising block tariff charges that deters water use above a crop's absolute minimum needs must be put in place. It will serve as the cornerstone for encouraging conservation, minimizing losses, and allocating resources. Additionally, it might have an impact on how crops are grown, how money is distributed, how effectively water is managed, and how much additional income there is that may be utilized to fund and maintain water projects.

5.3.13 Smart Water Management Systems

Smart water management involves integrating systems and measures to monitor, control, and regulate water resource use and quality. IoT water management equipment and software tools optimize production, distribution, and consumption. Advancements in information technology and sensors, precession farming, GIS, simulation tools, and remote sensing create new opportunities for water resource management. These tools aid in informed decision-making and can be effectively used in monitoring and evaluating watershed development studies.

5.3.14 Precision Irrigation

Precision irrigation is a sustainable, efficient method for watering plants and in small measured doses to give it the best growing conditions. Being precise about the uniform application of water has long been a goal of irrigation science, so the term "precision irrigation" is not new. Taking into account the spatial aspect of water management, or applying water based on the site-specific requirements of locate a specific field location, is the new dimension of precision irrigation. Precision irrigation system components are data acquisition, interpretation, and control.

Data acquisition and Interpretation: precision irrigation systems must recognize and measure soil and crop conditions, ensuring accurate interpretation. However, inadequate decision-support systems hinder the adoption of precision agriculture and real-time data interpretation (McBratney et al. 2005).

Control: a precision irrigation system optimizes inputs and manages irrigation at appropriate temporal and spatial scales, determining water application in different depths. Automatic controllers with real-time data are crucial for reliable and accurate control of irrigation applications.

5.3.15 Soil Moisture Sensors for Irrigation Scheduling

Soil moisture sensors measure liquid per volume using volumetric water content (VWC) sensors, expressed as percentages. Soil water tension, or matric potential sensors, measures the force required by plant roots to draw water from soil particles. It rises as water is removed, measured in centibars (cb) or atmospheric pressure bars. Soil moisture sensors are classified into two types: volumetric water content sensors and soil tension sensors embedded in the soil profile.

5.3.16 Remote Sensing

Compared to an in situ measurement, remote sensing can easily and effectively provide information on crop water content. Due to the size of most irrigation fields which can range from ten to one hundred hectares—even considering ground-level measurements may be impractical. Remote sensing aids irrigation water management by detecting crop needs and stress. Crop water stress index (CWSI) determines crop water status by comparing canopy temperature to surrounding air and ambient pressure. Additional indices include canopy temperature variability, temperature stress day, canopy-air temperature difference, and stress degree day (Mendez-Barroso et al. 2008).

5.4 Technical, Social, and Institutional Challenges in Implementing Sustainable Water Management Strategies

Land ownership structure, water use rights, regulation, technical expertise, low rainfall, and lack of environmental knowledge contribute to noncompliance and inadequacy in innovations (Ochoa-Noriega et al. 2022).

- Agricultural water management interventions offer potential benefits for rural lives and the economy, but institutional capacity is limited in African nations. Shifting focuses from relief to development requires long-term, comprehensive, environmentally sound, and carefully monitored interventions (Amede et al. 2014).

- Low rates of adoption of effective interventions have been caused by constraints like the paucity of materials (Douxchamps et al. 2012, 2014; Botoni and Reij 2009; Ouedraogo et al. 2010).
- Major technical, social, and environmental issues were cited as the cause of this (Wondimkun and Tefera 2006). They determined that the major reasons for the subpar performance of confined reservoirs were issues with design and infrastructure. Although individual farmers implement and manage in-field methods for managing rainwater, small reservoir construction and management and active and stakeholder participation are crucial for effect (Faulkner et al. 2008). Additionally, it is frequently necessary to support construction with outside funding (Venot et al. 2012).
- Limited market incentive for farm investments and inadequate market access hinders effective small-scale irrigation schemes (Poussin et al. 2015). There are biophysical, economic, and social aspects to the factors driving changes in the management practices of rainwater (Douxchamps et al. 2015; Poussin et al. 2015). Targeting and expanding effective interventions continue as a difficult task (Barron 2013). About which interventions and in which settings they are most likely to be successful, little is known.
- Targeting Agricultural Water Management1 Interventions (TAGMI), a decision-support tool for locating potential sites for expanding effective interventions. The authors estimate success probabilities for small-scale irrigation, small reservoirs, and soil and water conservation using TAGMI's map-based visualization, highlighting differences in adoption across districts (Barron et al. 2015)
- Water resource resilience management is hindered by weak institutional ties, policies, and investments. National and regional institutions are working on land and water management, but opportunities for sharing insights and policymaking are limited. Innovation platforms provide a platform for stakeholders to collaborate (Amede and Sanginga 2014).
- Stakeholder involvement is crucial for innovation success and consistently interested in collaborating to learn new things and solve problems. It is impossible for the research community to create innovation on its own (Greenough 2013). Successful integrated water resource management requires interactions between multiple actors, often beyond daily concerns (Amede 2015; CPWF 2013). Strategies like companion modelling in the Volta basin improve interactions and group decision-making by encouraging discussion among actors about natural resource control (CPWF 2013).

5.5 Future Prospects

The challenge of providing sustainable livelihoods for humanity requires changes in water, food, energy, and other goods and services. This requires changes in manufacturing and recycling practices. To ensure sufficient water for an expanding population, life support systems, and other goods and services, goals such as "more crop per drop," "more jobs per drop," "better environment per drop," and improved nutrition per drop are crucial. Fulfilling unresolved human requirements and desires is also a growing challenge, while maintaining the functioning of water and terrestrial ecosystems. Identifying and evaluating solutions to these challenges are a topic worth a collaborative multidisciplinary research effort (Falkenmark and Rockstrom 2004).

The technological shifts and developments that are probably to benefit the water sector's quick and efficient adaptation are cybernetics, artificial intelligence, and instantaneous information technology which are transforming various aspects of life. Nanotechnology, desalination and low-cost fusion energy, has become cost-effective for coastal urban areas. Improved sanitation, wastewater treatment, and recycling technologies have also been developed. Biotech pest control approaches have increased agricultural yields. Ecological engineering is being used for habitat preservation, reversing species extinctions, and combating invasive species.

Groundwater resource mapping and sustainable extraction levels are also being improved. Energy technology, such as solar, space-based energy, and algae, is being used to conserve species. Biotechnology is being used for population feeding and species conservation. Space-based environmental monitoring systems and real-time feedback to predictive models are also being developed. Geoengineering is being used to reverse global warming, and most weather and climate events can be predicted accurately and reliably. Renewable energy is being replaced by fossil fuels, leading to low-carbon societies.

5.6 Conclusion

Since population growth is concentrated in urban areas, meeting future urban water and food demands may put a strain on public water systems and agricultural producers. Urban water management must be sustainable in an effort to avoid loss and guarantee a sufficient supply of finite resources. Furthermore, to be environmentally sound, water management techniques like controlled drip irrigation systems, sheet mulching, swales and basins, and rainwater harvesting have been shown to save large amounts of water. By adjusting water quantity and quality to the needs of plants, preserving soil moisture, and reducing reliance on high-quality potable water, successfully generates water and energy savings along with environmental and nutrition education.

Urban agriculture can be used as an alternative to the significant reliance on current large-scale food production in addition to increasing water productivity. It is possible that conventional agriculture uses a lot of water-intensive and environmentally damaging resources. Additionally, food processing and distribution use a lot of energy and water. Research is essential for assessing the implications of potential future paths and avoiding compromises due to our diverse dreams and goals. However, translating research findings into policy-relevant language is crucial. WRR papers should be translated into language that motivates, entertains, and informs policymakers and the general public, making them easily understandable.

Personnel required to build sustainable world are master developers, designers, builders, and operators in shaping society's built environment, as well as guardians of natural resources. There are innovators and integrators of ideas and technology across public, private, and academic sectors. They also serve as risk and uncertainty supervisors and leaders in discussions and decisions about public environmental and infrastructure policy. As a community, it is crucial to determine, establish, and execute technological, economic, political, and social measures for a sustainable, secure future through studies, engineering, science, and governance.

References

Allen RG, Pereira LS, Raes D, Smith M (1998) Crop evapotranspiration—Guidelines for computing crop water requirements—FAO Irrigation and drainage paper 56. FAO, Rome, p 300(9), D05109

Amede T, Awlachew S, Mati B, Yitayew M (2014) Managing rainwater for resilient dryland systems. In: Melesse A, Abtew W, Setegn S (eds) Nile River basin: ecohydrological challenges, climate change and hydropolitics. Springer, Switzerland

Amede T, Sanginga P (2014) Innovation platforms for sustainable land management in East African landscapes: stewardship, incentives and challenges. J Soil Water Conserv 69(4):127–132

Amede T (2015) Technical and institutional attributes constraining the performance of small scale irrigation in Ethiopia. Water Res Rural Dev

Barron J(2013) A targeting and outscaling decision support tool. CPWF Volta brief. Stockholm Environment Institute, Stockholm

Barron J, Kemp-Benedict E, Morris J, de Bruin A, Wang G, Fencl A (2015) Mapping the potential success of agricultural water management interventions for smallholders: where are the best opportunities? Water Resour Rural Dev (this issue)

Biswas A (2010) Water for a thirsty urban world. Brown J World Aff 147–162. Retrieved from http://www.thirdworldcentre.org/thirstyurban.pdf

Botoni E, Reij C (2009) La transformation silencieuse de l'environnement et des systèmes de production au Sahel: impacts des investissements publics et privés dans la gestion des ressourcesnaturelles. Amsterdam and Ouagadougou

Chartzoulakis K, Bertaki M (2015) Sustainable water management in agriculture under climate change. Agric Agric Sci Procedia 4:88–98

Clancy T (1999) Riverina has the capacity to deliver diverse requirements. Austr Vitic 3(6):38–49

CPWF (Challenge Program on Water and Food) (2013) Summary of CPWF research in the Volta river basin

Deumier JM, Leroy P, Peyremorte P (1995) Tools for improving management of irrigated agricultural crop system

Doungmanee P (2016) The nexus of agricultural water use and economic development level. Kasetsart J Soc Sci 37:38–45

Douxchamps S, Ayantunde A, Barron J (2012) Evolution of agricultural water management in rainfed crop-livestock systems of the Volta basin. International Water Management Institute, Colombo

Douxchamps S, Ayantunde A, Barron J (2014) Taking stock of forty years of agricultural water management interventions in smallholder systems of Burkina Faso. Water Resour Rural Dev 3:1–13

Douxchamps S, Ayantunde A, Panyan EK, Ouattara A, Kabore K, Karbo N (2015) Agricultural water management and livelihoods in the crop-livestock systems of the Volta basin. Water Resour Rural Dev

Dry PR, Stoll M, Mc Carthy MG, Loveys BR (1999) Using plant physiology to improve the water use efficiency of horticultural crops. In: III international symposium on irrigation of horticultural crops, vol 537, pp 187–197

Ertek A, Yilmaz H (2014) The agricultural perspective on water conservation in Turkey. Agric Water Manag 143:151–158

FAO (2011) The state of the world's land and water resources for food and agriculture (SOLAW)—managing systems at risk. Food and Agriculture Organization of the United Nations, and London. Earthscan, Rome

FAO (2013) FAO statistical yearbook 2013. Food and Agriculture Organization of the United Nations, Rome, p 289. Retrieved from http://www.fao.org/docrep/018/i3107e/i3107e00.htm

Falkenmark M, Rockstrom J (2004) Balancing water for humans and nature: the new approach in ecohydrology. Earthscan, London

Falkenmark M, Rockstrom J (2010) Back to basics on water as constraint for global food production: opportunities and limitations. In: Garrido A, Ingram H (eds) Water for food in a changing world, vol 2. Contributions from the Rosenberg international forum on water policy. Routledge, London, pp 103–116

Faulkner JW, Steenhuis T, De Giesen NV, Andreini M, Liebe JR (2008) Water use and productivity of two small reservoir irrigation schemes in Ghana's Upper East Region. Irrig Drain 57:151–163

Greenough KM (2013) Stories of change and innovation in the Volta Basin Development Challenge Programme. Challenge Program on Water and Food-Volta Basin

Intergovernmental Panel on Climate Change (IPCC) (2014) Climate change and health: on the latest IPCC report. Lancet 383(9924):1185–1189

Idso SB, Reginato RJ, Reicosky DC, Hatfield J (1981) Determining soil-induced plant water potential depressions in alfalfa by means of infrared thermometry. Agron J 73(5):826–830

Itier B, Flura D, Belabbes K (1992), An alternative way for CWSI calculation to improve relative evapotranspiration estimates-results of an experiment over soybean. In: International symposium on irrigation of horticultural crops, vol 335, pp 333–340

Jackson RD, Idso SB, Reginato RJ, Pinter PJ Jr (1981) Canopy temperature as a crop water stress indicator. Water Resour Res 17(4):1133–1138

Kang S, Hao X, Du T, Tong L, Su X, Lu H, Li X, Huo Z, Li S, Ding R (2016) Improving agricultural water productivity to ensure food security in China under changing environment: from research to practice. Agric Water Manag 179:5–17

Kaur B, Sidhu RS, Vatta K (2010) Optimal crop plans for sustainable water use in Punjab. Agric Econ Res Rev 23:273–284

Keller J, Bliesner RD (1990) Sprinkle and trickle irrigation. 3(5):86–96

Keshavarz M, Karami E, Vanclay F (2013) Social experience of drought in rural Iran. Land Use Policy 30(1):120–129

Madani K (2014) Water management in Iran: what is causing the looming crisis? J Environ Stud Sci 4:315–328

McBratney A, Whelan B, Ancev T, Bouma J (2005) Future directions of precision agriculture. Precision Agric 6(1):7–23

Mendez-Barroso LA, Garatuza-Payan J, Vivoni ER (2008) Quantifying water stress on wheat using remote sensing in the Yaqui Valley, Sonora, Mexico. Agric Water Manag 95(6):725–736

Moazedi A, Taravat M, Jahromi HN, Madani K, Rashedi A, Rahimian S (2011) Energy-water meter: a novel solution for groundwater monitoring and management. In: Proceedings of the world environmental and water resources congress 2011. Palm Springs, California. ASCE, pp 962–969

Nazari B, Liaghat A, Akbari MR, Keshavarz M (2018) Irrigation water management in Iran: implications for water use efficiency improvement. Agric Water Manag 208:7–18

Ochoa-Noriega C, Velasco-Munoz JF, Aznar-Sanchez JA, Lopez-Felices B (2022) Analysis of the acceptance of sustainable practices in water management for the intensive agriculture of the Costa de Hermosillo (Mexico). Agronomy 12(1):154

Oster JD (1994) Irrigation with poor quality water. Agric Water Manag 25(3):271–297

Ouedraogo M, Dembele Y, Some L (2010) Farmer perceptions and adaptation options to rainfall change: evidence from Burkina Faso. Sécheresse 21:87–96

Pandey K, Kumar S, Malik A, Kuriqi A (2020) Artificial neural network optimized with a genetic algorithm for seasonal groundwater table depth prediction in Uttar Pradesh, India. Sustainability 12(21):8932

Pereira LS, Cordery I, Iacovides I (2009) Coping with water scarcity: addressing the challenges. Springer

Peyremorte P, Chol P (1992) Pilotage automatique optimise des irrigations localisees. In: Proceedings of the workshop on real time sensing and control of automated irrigation systems, ICID, 16th European regional conference, Budapest (Hungary), p 26

Poussin JC, Renaudin L, Adogoba D, Sanon A, Tazen F, Dogbe W (2015) Performance of small reservoir irrigated schemes in the Upper Volta basin: case studies in Burkina Faso and Ghana. Water Resour Rural Dev. https://doi.org/10.1016/j.wrr.2015.05.001

Rosegrant MW, Ringler C, Zhu T (2009) Water for agriculture: maintaining food security under growing scarcity. Annu Rev Environ Resour 34:205–222

Samian M, Naderi-Mahdei K, Saadi H, Movahedi R (2015) Identifying factors affecting optimal management of agricultural water. J Saudi Soc Agric Sci 14:11–18

Sharma PK, Kumar M (2014) Status and management of water in rainfed agriculture. Effic Water Manag Sustain Agric 41

Singh AK (2014) Sustainable management of water resources: issues and strategies. Effic Water Manag Sustain Agric 29:1

Sophocleous M (2004) Global and regional water availability and demand: prospects for the future. Nat Resour Res 13(2):61–75

Tilman D, Cassman KG, Matson PA, Naylor R, Polasky S (2002) Agricultural sustainability and intensive production practices. Nature 418:671–677

UN Water (2010) Water stress versus water scarcity. Retrieved from http://www.un.org/waterforlifedecade/scarcity.html

United Nations (2015) The Millennium Development Goals Report 2015. Retrieved from https://www.un.org/development/desa/en/news/population/world-population-prospects-2017.html

United Nations (2017) World population projected to reach 9.8 billion in 2050, and 11.2 billion in 2010. Retrieved from http://www.undp.org/content/dam/undp/library/MDG/english/UNDPMDGReport2015.pdf&

Vaux HJ (2012) Water for agriculture and the environment: the ultimate trade-off. Water Policy 14(Suppl. 1):136–146

Venot JP, de Fraiture C, Acheampong EN (2012) Revisiting dominant notions: a review of costs, performance and institutions of small reservoirs in Sub-Saharan Africa. IWMI Research Report 144. International Water Management Institute, Colombo, p 31

Velasco-Munoz JF, Aznar-Sanchez JA, Batlles-delaFuente A, Fidelibus MD (2019) Rainwater harvesting for agricultural irrigation: an analysis of global research. Water 11(7):1320

Vorosmarty CJ, Green P, Salisbury J, Lammers RB (2000) Global water resources: vulnerability from climate change and population growth. Science 289(5477):284

Wondimkun Y, Tefera M (2006) Household water harvesting and small scale irrigation schemes in Amhara Region. In: Awulachew SB, Menker B, Abesha D, Atnafe T, Wondimkum Y (eds) Best practices and technologies for small scale agricultural water management in Ethiopia. Proceedings of a MoARD-MoWR-USAID-IWMI symposium, Addis Ababa, Ethiopia, 7–9, pp 11–17

Chapter 6
Crop Yield Prediction Using Artificial Intelligence and Remote Sensing Methods

Rahul Banerjee, Bharti, Pankaj Das, and Sadaf Khan

Abstract Agriculture is a major part of the economy in the majority of developing and third world nations. An accurate and well-timed crop yield prediction will not only help in crop management, crop insurance but will also facilitate policy and decision-makers to frame apt strategies and policies regarding food security to combat hunger and eventually achieve zero hunger, one of the most important Sustainable Development Goals. Over the past decades, crop yield has been predicted through mathematical, statistical, and survey-based models. Artificial intelligence (AI)-based methods such as Artificial Neural Network (ANN), Support Vector Machine (SVM), Random Forest, and Deep learning methods could be a potential replacement to statistical modelling, because it produces precise results and can handle complexity and nonlinearity in data much more effectively. There are several fields in agriculture in which remote sensing can be advantageous, viz. crop yield forecasting, soil property detection, crop type classification, and meteorological data assessment. This chapter provides a framework of the existing methodologies of crop yield prediction and aims to describe the recent crop yield prediction techniques based on artificial intelligence and remote sensing approaches. It also focuses on the potential advantages of artificial intelligence methods on crop yield prediction at field and regional levels.

Keywords Crop yield prediction · Prediction using mathematical models · Yield optimization · Mathematical models · Remote sensing · Artificial intelligence techniques

R. Banerjee (✉) · Bharti · P. Das · S. Khan
Division of Sample Surveys, ICAR-IASRI, Library Avenue, New Delhi 110012, India
e-mail: rahuliasri@gmail.com

S. Khan
e-mail: skhan18@myamu.ac.in

K. Pandey et al. (eds.), *Artificial Intelligence and Smart Agriculture*,
Advances in Geographical and Environmental Sciences,
https://doi.org/10.1007/978-981-97-0341-8_6

6.1 Introduction

In India, accurate and timely predictions of crop yield are crucial for the government, agricultural industries, agro-based traders, and farmers. These predictions aid the government in making policy decisions regarding acquisition, dispensation, buffer-stocking, import/export, value fixation, and marketing of agricultural products. The industries and traders rely on these predictions for effective planning of their operations, while farmers use them to maximize their yields and increase their profits.

Currently, the crop yield predictions provided by the Directorate of Economics and Statistics, Ministry of Agriculture (DESMOA), are subjective and prone to personal bias, and objective crop cutting experiments are not useful as they are conducted after the harvest is over. Therefore, there is a need for the development of an objective crop yield prediction technique. AI and remote sensing have emerged as promising tools for yield prediction and optimization. They can be used to analyze data and identify areas of potential improvement in soil fertility, water availability, and pest infestation. Additionally, AI and remote sensing can help farmers to identify the best planting and harvesting times, manage water and fertilizer use efficiently, monitor crop health, and optimize resources, all of which can lead to reduced environmental impact. Overall, the use of AI and remote sensing in yield prediction and optimization has the potential to greatly benefit modern agriculture by providing farmers with the necessary tools to make informed decisions about crop selection and management, thereby maximizing yields and reducing environmental impact.

6.2 Materials and Methods

6.2.1 Methodologies for Crop Yield Prediction Using Mathematical Models

a. **Composite weather index-based models**: One of the most key determinants influencing crop yield is between and within the season variations in weather parameters. The effect of the weather variables varies with different stages of crop growth. Fisher (1924) and Hendricks and Scholl (1943) have described some of the models that require much lesser number of parameters to be estimated and also take into account the distribution of weather over the entire crop season. The entire crop season is generally bifurcated into 5/7 days interval with the assumption that the effect of change in weather variables in successive periods is not random, but it follows certain specific mathematical distribution and can generally be expressed as polynomial functions of time.

$$Y = B_0 + B_1X_1 + B_2X_2 + \cdots + B_nX_n + \varepsilon, \tag{6.1}$$

where the dependent variable Y denotes the yield and the explanatory variable denotes the Xw weather variable in wth period $w = 1,2,..,n$ and ε denotes the error term. Fisher has suggested the use of $k = 5$ in most of the practical situations. The model mentioned in Eq. (6.1) was used by Sir. R. A. Fisher to study the effect of rainfall on the yield of the wheat crop. The above-mentioned technique was further modified by Hendricks and Scholl. They considered a second degree polynomial model to express the effects in successive periods. With this assumption, the model suggested by Hendricks and Scholl is obtained as follows:

$$Y = B_0 + b_0 \sum_{w=1}^{n} X_w + b_1 \sum_{w=1}^{n} wX_w + b_2 \sum_{w=1}^{n} w^2 X_w + \varepsilon, \tag{6.2}$$

where B_0, b_0, b_1, and b_2 are the respective model parameters. Hendricks and Scholl's model was further refined by representing the impact of changes in weather variables on yield in the wth period as a second-degree polynomial in terms of the corresponding correlation coefficient between yield and the weather variables (Agrawal et al. 1980; Agrawal and Jain 1982; Agrawal et al. 1983; Jain et al. 1980). These models were subsequently further refined (Agrawal et al. 1986).

b. **Discriminant function-based models**: Discriminant function analysis is also an alternative technique to create district level yield prediction models using weather variables. The long 25–30 year data are usually trifurcated into three groups namely, congenial, normal, and adverse, depending upon crop yield. These data from the groups are used to fit linear and quadratic discriminant analysis models Rai and Chandrahas (2000). This approach was found to obtain accurate yield forecasts, with only about 1% deviation from the observed yield. In the earlier study, the forecast of wheat yield in Kanpur district of Uttar Pradesh has been done using discriminant function analysis.
c. **Water balance technique methods**: This method has been found to be useful for rainfed crops such as Rice, Sorghum, and Maize (Saksena et al. 2001). In this method, water deficit/surplus is calculated for weekly intervals. Weighted stress indices are then prepared in phases with the incorporation of apt weights to water surplus and water deficit depending upon crop growth stage. These stress indices act as explanatory variables in the prediction model along with the time trend. Equation 6.3 is used to determine the predicted soil moisture in the root zone at the end of ith week.

$$\theta_i = \theta_{i-1} + \xi_i - \varphi_i, \tag{6.3}$$

where ξ_i is the rainfall obtained in ith week, φ_i is the water requirement of the crop in ith week; $\xi_i = ki \times evapi$, where k_i is the crop coefficient and evap$_i$ is the pan evaporation in ith week. Stress due to water surplus was taken as '1' if θ_i > water holding capacity. The stress due to water deficit $(\theta_i) = \frac{1-AE_i}{\varphi_i}$. A range between 1 and 11% has been recorded for deviations of forecasts from observed yields using this method for future years.

d. **Model based on growth indices**: The explanatory variables in these models are growth indices based on plant characteristics observed at two or more points in time throughout the crop growth stages. These explanatory variables can be thought of as weighted accumulations of observations on plant characteristics made at various points in time during the various periods. The weights are the crop yield and plant character correlation coefficients. The model can be pictured as follows:

$$Y = \beta_0 + \sum_i \beta_i G_i + \varepsilon, \tag{6.4}$$

where $G_i = \sum_{w=1}^{n_2} r_{iw} X_{iw}$, represents the index of the ith character, w period identification, n_1 and n_2 the initial and final periods considered for developing the index. r_{iw} is the value of the correlation coefficient between yield and ith character in the wth period, and X_{iw} is the value of the ith character in the wth period.

e. **Probability-based Markov chain models**: In this method, the entire crop growth process is demarcated into phenological stages. The predicted yield distributions are obtained using transition probability matrices used to predict crop yield. Higher Markov chain models are also used for the above-mentioned purpose (Jain and Agrawal 1992a; Agrawal and Jain 1996, Jain and Ramasubramanian 1998; Ramasubramanian and Jain 1999; Ramasubramanian et al. 2004).

f. **Logistic models**: The pattern of dry matter buildup and the panicle/ear weight of the crop are frequently modelled using logistic models. The curve and the value of the study variables at harvest are fitted using partial crop season data, and this curve is then utilized to predict yield. The model's form is expressed as follows:

$$Y_t = \frac{\alpha^{\left(\frac{t_m}{t_f}\right)^{\frac{1}{2}}}}{(1 + \beta\rho^t)} + e_t, \tag{6.5}$$

where *Yt* represents the dependent growth variable, t is the independent time variable, *tm* represents the time of maturity, and *tf* represents the time of forecast, $a > 0$, $\beta > 0$, $0 < r < 1$. This method's predictions have been shown to be rather close to those of the observations. (Jain et al. 1992b).

g. **Detrended yield approach:** By using weekly data on weather variables (weather indices) as explanatory variables and time series data on crop yield as the dependent variable, crop yield forecast models are developed in this method. The yield is then detrended using the model parameter estimates, and the ARIMA model is used to predict the crop's yield using the detrended yield. This method has been used in five districts of Uttar Pradesh, viz. Lucknow, Kanpur, Banda, Jhansi, and Faizabad, to predict the yield of the wheat crop (Banerjee 2015; Gurung et al. 2017). The model can be conceptualized as follows:

Consider ξ_{iq} to be the value of the ith; ($i = 1(1)p$) weather variable at the qth week ($q = 1(1)n$), and let r_{iq} be the value of the simple correlation coefficient between weather variable X_i at the qth week and yield over a period of g years.

The generated variables are given by: $Z_{ij} = \frac{\sum_{q=1}^{n} r_{iq}^{j} \xi_{iq}}{\sum_{q=1}^{n} r_{iq}^{j}}$; $j = 0, 1$. The following model is then fitted to study the effect of individual weather variables: $Y = a_0 + a_1 Z_{i0} + a_2 Z_{i1} + cT + \varepsilon$, where Y is yield, T is variable expressing time effect, a_0, a_1, a_2, and c are the unknown parameters to be evaluated from the model, and ε is the error term which is supposed to be distributed with a zero expectation and a constant variance σ^2.

6.2.2 *Methodologies for Crop Yield Prediction Using Remote Sensing*

The use of satellite-based remote sensing techniques to estimate crop yield is growing in popularity. The production of important agricultural crops like rice, wheat, sugarcane, and maize is tracked via space-borne satellite remote sensing. The vegetation indices, which are morphological, biochemical, and biophysical variables, are investigated as crop growth metrics. The spectrum data from satellite remote sensing have been used in agricultural programmes for a variety of crop research, issues with crop inventories, precise yield estimates, and the construction of models. The creation of algorithms was the main emphasis of the remote sensing techniques, which depended on weather predictions, clouds, and radiation of different wavelengths. Recently, the emphasis has switched to data use.

The methods that are based on remote sensing can be further divided into those that are detailed in the following subsections:

a. **Regression-based methods:** Regression and machine learning approaches are used to accurately and quantitatively process data such as crop type, edaphic parameters, and meteorological conditions in order to estimate using this technology. In order to predict future yields using fresh data from the vegetative features, a regression relationship is built between the crop yield and vegetable indices.
b. **Physics-based methods:** For the purpose of modelling, physics-based approaches use biophysical processes like leaf optics. A radiative transfer PROSPECT model developed by Jacquemound and Baret (1990) makes use of data on leaf morphology, pigment content, water content, and dry matter. In the wavelength range of 400–2500 nm, this model simulates leaf optics. These days, it also has updated variants called PROSAIL-4 and PROSAIL-5.
c. **Assimilation of RS data in crop simulation models:** A blend of a physics-based method and a crop simulation model can solve difficulties unique to each model. In the first, certain crop-specific soil parameters are used. The latter, however, also makes use of meteorological information and farming techniques in addition

to the agronomic and edaphic conditions. When crop models are applied to wide fields, they present unique issues due to the enormous amount of data needed. The use of remote sensing techniques can solve these issues. These two can be combined in one of two ways to provide an effective output: Either the crop growth variables may be replaced by the simulated variables, or the former may be used to adjust the latter, so changing the output.

d. **RS of plant photosynthetic activity:** Crop monitoring over vast areas has been made possible by satellite remote sensing. It employs a variety of spectrum wavelengths. Based on relationship between reflectance and plant structure, the GPP inference is made. Since biomass is a crucial component of yield, any parameter pointing in its direction can be used to forecast production. The optically based vegetative indices could result in a calibration inaccuracy, which would result in inaccurate yield forecasts. It is difficult to forecast production in ecosystems where light utilization efficiency varies seasonally and where carbon assimilation varies as a result of climate change.
e. **Microwave data-based methods:** The microwave data use microwave with wavelengths between 1 and 1 m for environment forecasting. These wavelengths can pass through clouds, haze, dust, and fog cover, making it possible to obtain data in all weather conditions other than heavy downpours. The active microwave and the passive microwave are the two different types. On an object's surface during the day, the active radiation is captured, whereas the passive radiation is transmitted and not dependent on light source. Given that one is based on backscattering and the other is based on the optical depth of vegetation, the two may be complimentary to one another.

Estimating crop yield is a crucial component of farming. Various methods for estimating crop yields are utilized globally. However, the researchers cannot agree on a single approach that satisfies all criteria and may be used as a general model. Each technique has unique operating limitations. Specific goals, the area under examination, and the level of precision needed are taken into account while choosing a method. The most promising of them is models based on remote sensing that vary the supplemental techniques. It remains the most popular technique in spite of its own shortcomings. Algorithms based on artificial intelligence are supporting and enhancing current techniques and giving farmers and stakeholders a greater understanding of the agricultural systems. Figure 6.1 represents a thematic map showing a classification of vegetation index using remote sensing data, the green colour represents good health plants, and as the green colour fades to yellow and red, the health status of the plant reduces. The grey colour in the map represents barren land.

Fig. 6.1 Thematic remote sensing map showing variations of crop health within fields. Available at: https://seos-project.eu/agriculture/agriculture-c01-p05.html

6.2.3 Sub-section Methodologies for Crop Yield Prediction Using Artificial Intelligence

Crop yield prediction using artificial intelligence (AI) can help farmers to increase their crop yields and maximize their profits. AI can help farmers to predict yields accurately, which can lead to better planning, improved decision-making, and more efficient use of resources. AI can also help farmers to assess the risk of a crop failure and develop strategies to mitigate the risk. AI can also be used to identify optimal planting times, assess soil conditions, and collect real- time data on crop health and pest control. By using AI, farmers can also identify and control weeds and pests and can reduce the amount of water and fertilizer needed for their crops. AI can also help farmers to reduce their environmental impact by reducing the amount of chemical pesticides used in their crops. By using AI, farmers can also save time and effort by automating certain tasks.

Artificial intelligence (AI) is an interdisciplinary field of computer science which studies the ability of machines to perform tasks traditionally requiring human intelligence. AI encompasses a wide range of techniques and algorithms, including machine learning, deep learning, natural language processing, and computer vision. AI systems are designed to learn from data, identify patterns, and make decisions in complex environments. AI systems are now being used in a variety of industries, from health care to robotics, and are beginning to be applied to the agricultural sector for the purpose of crop yield prediction.

The use of artificial intelligence (AI) in crop yield prediction is a rapidly growing field, with the potential to revolutionize the way farmers manage their fields. AI-based approaches have the potential to provide more accurate predictions of crop yields than existing methods, as well as to help farmers to better understand the factors influencing their yields. AI-based approaches to crop yield prediction involve a combination of data collection, data analysis, and machine learning. Data collection typically includes collecting information about soil type, weather, irrigation, and other environmental conditions. Data analysis involves analyzing the collected data to identify patterns and trends in the data, which can then be used to make better predictions. Finally, different learning algorithms are used to develop predictive models that are able to accurately predict crop yields.

AI techniques have the potential to revolutionize the way farmers predict and manage crop yields, leading to higher yields and better profits. One of the most popular AI-based methods for crop yield prediction is the use of machine learning algorithms. Machine learning algorithms use historical data to learn patterns in the data and make predictions about future crop yields. These models can be used to predict yields for a range of different crops, based on the data provided. Another AI-based methodology for crop yield prediction is the use of deep learning algorithms. Deep learning algorithms use Artificial Neural Networks to learn complex patterns in the data and make predictions. These models are often used in conjunction with other AI-based methods, such as machine learning algorithms, to provide more accurate predictions. Finally, AI-based methods are also being used to develop predictive models based on satellite images. These models use satellite data to detect changes in the crop environment and predict future yields. This type of model is particularly useful for predicting yields in areas with limited or no access to traditional farming data. Following are some of the commonly used AI-based methods related to prediction of crop production:

a. **Machine learning**: Machine learning algorithms such as Decision Trees, Support Vector Machines, Neural Networks, Random Forests can be used to predict crop yield using data such as crop type, soil fertility, water availability, soil type, climate.
b. **Remote sensing**: Remote Sensing technology can be used to predict crop yield by gathering satellite imagery and data on crop growth, soil moisture, temperature, etc., from different sources.
c. **Predictive analytics**: Predictive analytics can be used to analyze historical data in order to identify patterns and trends that can be used to predict crop yield.
d. **Image analysis**: Image analysis techniques can be used to analyze aerial and satellite imagery to determine crop health and yield.
e. **Natural language processing**: Natural language processing techniques can be used to analyze agricultural reports, journals, and news articles to identify data points that can be used to predict crop yield.
f. **IoT sensors**: IoT sensors can be deployed in agricultural fields to collect data such as soil moisture, temperature, humidity that can be used to predict crop yield.

In short, AI-based methods have emerged as powerful tools for predicting crop yields. ML, DL, evolutionary algorithms, and fuzzy logic are all AI-based methods that have been used for crop yield prediction. These methods have a range of applications, including supporting decision-making and optimizing production. AI-based methods have the potential to revolutionize the way crop yields which are predicted and managed and could have a significant impact on agricultural production and food

Table 6.1 Advantages and disadvantages of the models

Approach	Advantages	Disadvantages
Machine learning	• Assumption free • Self-adaptive • Parallel processing • Robust to outlier • Capable to handle missing data	• Data-driven • Requires huge amount data for training • Model complexity • High error susceptibility
Remote sensing	• Large area coverage • Easy and quick collection of data over a variety of scales and resolutions and expensive method for data analysis	• High maintenance • Cost, need technical expertise to operate and maintain • Data acquisition is complex
Predictive analytics	• Improves decision-making • Risk management • Competitive intelligence • Cost efficient	• Need consistent data update • Complete data • Objective orient decision-making
Image analysis	• Increased efficiency • Enhanced security • Improved image quality • Enhanced security	• Need better resolution • Quality of data • Model complexity • Time-consuming
Natural language processing	• Easy to implement • Capable to perform large-scale analysis • More accurate analysis • Cost reduction	• Complexity of model • Error prone • Need technical knowledge
IoT sensors	• Real-time monitoring • Increased efficiency • Cost-effective • Improved data accuracy • Easy customization	• Vulnerable to security breaches • Data theft and other security risks • Compatibility issues with existing system • High operational and maintenance cost • IoT sensors are limited in their range (may not work effectively in remote areas or locations with poor connectivity)

security. Figure 6.2 represents a thematic representation of how remote sensing data can be combined with AI techniques for the development of crop yield maps.

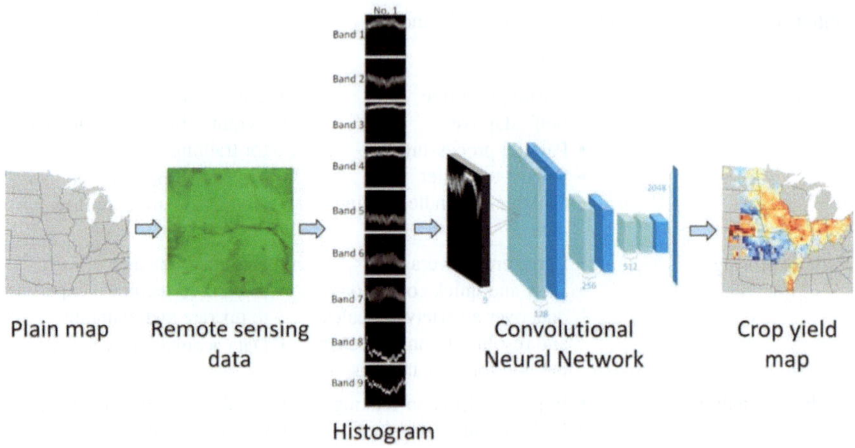

Fig. 6.2 Thematic representation of crop yield maps using remote sensing and AI. Available at: http://sustain.stanford.edu/crop-yield-analysis

6.3 Recent Advances

Crop yield prediction techniques have been developed over the decades to provide guidelines for the decision-makers at regional and national levels. The initiative has started in western nations but later spread throughout the world. In India, National Sample Survey Office (NSSO) under the Ministry of Statistics and Programme Implementation is mainly in authority for compiling data with the motive of predicting the yield of crops. The yield forecast for the Kharif crop is still being prepared annually in July–August, with the results being reviewed in December and January. The advance estimate for the Rabi crop is estimated in December and January, and the results are checked in April after information regarding the production of the respective crop is available. The traditional techniques used in crop yield prediction suffer from shortcoming; therefore, remote sensing and artificial intelligence-based prediction can be very effective tool. In past decades, several researchers have tried to predict the crop yield using remote sensing and artificial intelligence techniques. A systematic literature review was done by Klompenburga et al. (2020) to summarize and synthesize the techniques and features applied in crop yield prediction studies. According to them, the most used features were temperature, rainfall, and type of soil, and the most applied techniques were Artificial Neural Networks (ANNs). This section provides a review of some recent work done on the use of remote sensing and artificial intelligence in crop yield prediction. However, it is challenging to create a chronology of the prediction techniques due to their interdependency and interconnectivity.

In recent years, remote sensing and machine learning techniques are being extensively used globally to predict the yield of food grains. In the USA, MODIS data were used to predict the yield of maize and soybeans at the county level. It was tested whether MODIS data could detect inter-annual variations in crop yields. The findings showed that including MODIS-derived phenology metrics improved model performance and two-band enhanced vegetation index offered the most reliable basis for yield predictions (Douglas and Mark 2013). Advanced sensing techniques along with machine learning were used to predict the yield of wheat in Bedfordshire, U.K (Pantazi et al. 2016). In India, various machine learning techniques are being used in prediction of the yield of major food grains such as composite model for forecasting the yield of rice (Agrawal and Jain 1982); Support Vector Machine technique in prediction of the yield of rice crop (Gandhi et al. 2016); machine learning and remotely sensed data prediction of corn yield (Khanal et al. 2018); satellite remote sensing-based machine learning technique to predict wheat yield (Meraj et al. 2022).

Recently, machine learning-based techniques such as deep learning, image processing are also being rapidly gaining importance in prediction of the yield of fruit crops. Several researches attempted to predict the yield of fruit based on morphological and other characters. Bharti et al. (2022) applied principal component technique for the selection of significant input morphological variables in the Artificial Neural Network model with various hidden layers for predicting yield of apple fruit crop. Chen et al. (2019) in a study demonstrated that yield of individual apple trees can be predicted by using multi-source remote sensing data and ensemble learning. Das et al. (2023) demonstrated how hybrid machine learning approaches can be applied for lentil yield prediction. Machine learning and remote sensing techniques were also used to study crop phenotyping and management in the strawberry fruit crop (Zheng et al. 2021). Remote sensing techniques are very helpful in providing a large set of data that describe the condition as well as health of crop. This information in turn will be helpful to generate advanced management practices. An investigation into the research works related to prediction of orchard yield (He et al. 2022) from the year 2010 to 2021 indicated that vegetation indices were mostly used as input variables in yield prediction, while machine learning techniques were among the most used algorithms. Vegetation indices for the yield prediction of agricultural crops thorough machine learning techniques specially neural network were explored (Panda et al. 2010; Douglas and Mark 2013; Johnson et al. 2016). Many researcher also made an attempt for early forecasting of fruit yields such as in apple (Črtomir et al. 2012), in strawberry (Cheng et al. 2017). Researcher has also conducting the research to predict the yield of vegetables and other crops using various machine learning techniques like processing tomato yield prediction using remotely sensed aerial images and crop growth models (Koller and Upadhyaya 2005); yield prediction of bitter melon (Villanueva and Salenga 2018); yield prediction of robusta coffee using Random Forest, extreme learning machine and multiple regression techniques (Kouadio et al. 2018); prediction of soybean yield by integrating machine learning with satellite and weather data (Schwalbert et al. 2020).

6.4 Discussion and Future Prospects

Artificial intelligence (AI) and remote sensing-based yield prediction techniques have a promising future in agriculture. These technologies have the potential to revolutionize how we monitor and manage crops, ultimately leading to increased yields and more sustainable farming practices. One major advantage of using AI and remote sensing for yield prediction is the ability to collect and analyze vast amounts of data quickly and accurately. With the help of machine learning algorithms, AI can learn from past data to make predictions about future crop yields, which can be used to optimize planting strategies, fertilizer application, and irrigation practices. Remote sensing technology, which involves using satellites and other sensors to gather information about the Earth's surface, can provide valuable data on crop health and growth patterns. These data can be combined with AI algorithms to predict crop yields, detect nutrient deficiencies or disease outbreaks, and make more informed decisions about crop management. Another advantage of using AI and remote sensing-based techniques for yield prediction is the ability to monitor crops in real-time, which can help farmers respond quickly to changes in weather patterns, pest infestations, or other factors that could affect crop yields. Overall, the future prospects for AI and remote sensing-based yield prediction techniques are very promising. As technology continues to advance, we can expect to see even more sophisticated algorithms and sensors that will allow farmers to monitor and manage their crops more effectively, leading to higher yields and more sustainable agricultural practices.

6.5 Conclusion

Crop yield is a complex trait that is influenced by a large number of parameters. The most important parameters influencing crop yield can broadly classified into weather variables and crop inputs. The dependence of yield on these parameters is also complex and often nonlinear in nature. This makes crop yield prediction a very daunting task. There are several organizations concerned with the task of crop yield prediction in India and abroad. Traditionally yield prediction has been achieved using several mathematical and analytic techniques. But, the lacunae of this technique is that majority of the models are based upon the assumption of a linear relationship between the yield and other factors, and often in the absence of a linear relationship, the estimates obtained are imprecise and erroneous. Some of the estimates are also obtained based upon observation and personal judgement which are prone to subjective and personal biases. With the advancement in remote sensing and Artificial Technique methodologies, a vast avenue remains in front of us to use these methodologies for better and efficient predictions. The chapter discusses some of the available methodologies of yield prediction using remote sensing and AI techniques; also a tabular review of some of the recent works done in this domain has been mentioned.

Conflict of Interest The authors declare no conflict of interest.

References

Agrawal R, Jain RC (1982) Composite model for forecasting rice yield. Indian J Agric Sci 52(3):189–194

Agrawal R, Jain RC (1996) Forecast of sugarcane yield using eye estimate along with plant characters. Biom J 38(5):731–739

Agrawal R, Jain RC, Jha MP (1983) Joint effects of weather variables on rice yields. Mausam 34(2):177–181

Agrawal R, Jain RC, Jha MP (1986) Models for studying rice crop weather relationship. Mausam 37(1):67–70

Agrawal R, Jain RC, Singh D (1980) Forecasting of rice yield using climatic variables. Indian J Agricult Sci 50(9):680–684

Banerjee R (2015) Pre-harvest forecasting of crop yield using detrended yield: a new approach. M.Sc. thesis, PG School, IARI, New Delhi

Bharti, Das P, Banerjee R, Ahmad T, Devi S, Verma G (2022) Artificial neural network based apple yield prediction using morphological characters. Horticulturae 9:436. https://doi.org/10.3390/horticulturae9040436

Cakir Y, Kirci M, Gunes E (2014) Yield prediction of wheat in south-east region of Turkey by using artificial neural networks. In: Proceedings of the 3rd international conference on agro-geoinformatics, agro-geoinformatics. https://doi.org/10.1109/Agro-Geoinformatics.2014.6910609

Chen Y, Lee WS, Gan H, Peres N, Fraisse C, Zhang Y, He Y (2019) Strawberry yield prediction based on a deep neural network using high-resolution aerial orthoimages. Remote Sens 11:1584. https://doi.org/10.3390/rs11131584

Cheng H, Damerow L, Sun Y, Blanke M (2017) Early yield prediction using image analysis of apple fruit and tree canopy features with neural networks. J Imag 3–6. https://doi.org/10.3390/jimaging3010006

Črtomir R, Urška C, Stanislav T, Denis S, Karmen P, Martin P, Marjan V (2012) Application of neural networks and image visualization for early forecast of apple yield. Erwerbs-Obstbau 54:69–76. https://doi.org/10.1007/s10341-012-0162-y

Das P, Jha GK, Lama A, Parsad R (2023) Crop yield prediction using hybrid machine learning approach: a case study of lentil (*lens culinaris medik.*). Agriculture 13(3):596. https://doi.org/10.3390/agriculture13030596

Douglas KB, Mark AF (2013) Forecasting crop yield using remotely sensed vegetation indices and crop phenology metrics. Agric For Meteorol 173:74–84. https://doi.org/10.1016/j.agrformet.2013.01.007

Fisher RA (1924) The influence of rainfall on the yield of wheat at Rothamsted. Philos Trans Roy Soc B 213:89–142

Gandhi N, Petkar O, Armstrong L, Tripathy A (2016) Rice crop yield prediction in India using support vector machines. In: Proceedings of the 13th international joint conference on computer science and software engineering (JCSSE), pp 1–5. https://doi.org/10.1109/JCSSE.2016.7748856

Gurung B, Panwar S, Singh KN, Banerjee R, Gurung SR, Rathore A (2017) Wheat yield forecast using detrended yield over a sub-humid climatic environment in five districts of Uttar Pradesh India. Indian J Agric Sci 87(1):87–91

Hendrick WA, Scholl JC (1943) Technique in measuring joint relationship—the joint effects of temperature and precipitation on crop yield. N Carolina Agric Exp Sta Tech Bull 74

Jacquemound S, Baret, F (1990) PROSPECT: a model of leaf optical properties spectra. Remote Sens Environ 34(2):75–91. https://doi.org/10.1016/0034-4257(90)90100-Z
Jain RC, Agrawal R (1992a) Probability model for crop yield forecasting. Biometrical J 34(4):501–511
Jain RC, Ramasubramanian V (1998) Forecasting of crop yields using second order Markov Chains. J Indian Soc Agric Stat 51(1):61–72
Jain RC, Agrawal R, Jha MP (1980) Effect of climatic variables on rice yield and its forecast. Mausam 31(4):591–596
Jain RC, Agrawal R, Singh KN (1992b) A within year growth model for crop yield forecasting. Biometrical J 34(7):789–799
Johnson M, Hsieh W, Cannon A, Davidson A, Bédard F (2016) Crop yield forecasting on the Canadian Prairies by remotely sensed vegetation indices and machine learning methods. Agric For Meteorol 218–219:74–84. https://doi.org/10.1016/j.agrformet.2015.11.003
Khanal S, Fulton J, Klopfenstein A, Douridas N, Shearer S (2018) Integration of high resolution remotely sensed data and machine learning techniques for spatial prediction of soil properties and corn yield. Comput Electron Agric 153:213–225. https://doi.org/10.1016/j.compag.2018.07.016
Klompenburga TV, Kassahuna A, Catalb C (2020) Crop yield prediction using machine learning: a systematic literature review. Comput Electron Agric 177:105709. https://doi.org/10.1016/j.compag.2020.105709
Koller M, Upadhyaya SK (2005) Prediction of processing tomato yield using a crop growth model and remotely sensed aerial images. Trans ASABE 48(6):2335–2341
Kouadio L, Deo RC, Byrareddy V, Adamowski JF, Mushtaq S, Nguyen VP (2018) Artificial intelligence approach for the prediction of robusta coffee yield using soil fertility properties. Comput Electron Agric 155:324–338. https://doi.org/10.1016/j.compag.2018.10.014
He L, Fang W, Zhao G, Wu Z, Fu L, Li R, Majeed Y, Dhupia J (2022) Fruit yield prediction and estimation in orchards: a state-of-the-art comprehensive review for both direct and indirect methods. Comput Electron Agric 195:106812. https://doi.org/10.1016/j.compag.2022.106812
Meraj G, Kanga S, Ambadkar A, Kumar P, Singh SK, Farooq M, Johnson BA, Rai A, Sahu N (2022) Assessing the yield of wheat using satellite remote sensing-based machine learning algorithms and simulation modeling. Remote Sens 14(13):3005. https://doi.org/10.3390/rs14133005
Panda SS, Ames DP, Panigrahi S (2010) Application of vegetation indices for agricultural crop yield prediction using neural network techniques. Remote Sensing 2:673–696. https://doi.org/10.3390/rs2030673
Pantazi XE, Moshou D, Alexandridis T, Whetton RL, Mouazen AM (2016) Wheat yield prediction using machine learning and advanced sensing techniques. Comput Electron Agric 121:57–65. https://doi.org/10.1016/j.compag.2015.11.018
Rai T, Chandrahas (2000) Use of discriminant function of weather parameters for developing forecast model of rice crop, project report, IASRI, New Delhi
Ramasubramanian V, Jain RC (1999) Use of growth indices in Markov Chains models for crop yield forecasting. Biom J 41(1):99–109
Ramasubramanian V, Agrawal R, Bhar LM (2004) Forecasting sugarcane yield using multiple markov chains, project report, IASRI, New Delhi
Saksena A, Jain RC, Yadav RL (2001) Development of early warning and yield assessment models for rainfed crops based on agro-meteorological indices, project report, IASRI, New Delhi
Schwalbert RA, Amado T, Corassa G, Pott LP, Prasad PV, Ciampitti IA (2020) Satellite-based soybean yield forecast: integrating machine learning and weather data for improving crop yield prediction in southern Brazil. Agric For Meteorol 284. https://doi.org/10.1016/j.agrformet.2019.107886

Villanueva MB, Salenga MLM (2018) Bitter melon crop yield prediction using machine learning algorithm. Int J Adv Comput Sci Appl 9(3). https://doi.org/10.14569/IJACSA.2018.090301
Zheng C, Abd-Elrahman A, Whitaker V (2021) Remote sensing and machine learning in crop phenotyping and management, with an emphasis on applications in strawberry farming. Remote Sens 13:531. https://doi.org/10.3390/rs13030531

Part II
Technology, Tools and Techniques

Chapter 7
Role of Artificial Intelligence in Revolutionizing Agricultural Technology

Ali Raza, Yongguang Hu, Siham Acharki, Noman Ali Buttar, Muhammad Zubair, Neyha Rubab Syed, Dinesh Kumar Vishwakarma, and Ahmed Elbeltagi

Abstract The agricultural sector faces several challenges that hinder productivity, including poor soil management, insect and parasite invasion, output impedance, and a lack of understanding among agricultural producers. A new generation of technologies, such as artificial intelligence (AI) and machine learning (ML), has recently been

A. Raza (✉) · Y. Hu (✉) · N. A. Buttar
School of Agricultural Engineering, Jiangsu University, Zhenjiang 212013, China
e-mail: alir3201@gmail.com

Y. Hu
e-mail: deerhu@ujs.edu.cn

S. Acharki
Department of Earth Sciences, Faculty of Sciences and Technologies of Tangier (FSTT), Abdelmalek Essaadi University, 93000 Tetouan, Morocco

N. A. Buttar
Department of Agricultural Engineering, Khwaja Fareed University of Engineering and Information Technology, Rahim Yar Khan 64200, Pakistan

M. Zubair
School of Transportation, Southeast University, Nanjing 21009, China
e-mail: 233217035@seu.edu.cn

N. R. Syed
School of Energy and Environment, Power Engineering and Engineering Thermophysics, Southeast University, Nanjing 21009, China

D. K. Vishwakarma
Department of Irrigation and Drainage Engineering, Govind Ballabh Pant University of Agriculture and Technology, Pantnagar, Uttarakhand 263145, India
e-mail: dinesh.vishwakarma4820@gmail.com

A. Elbeltagi
Agricultural Engineering Department, Faculty of Agriculture, Mansoura University, Mansoura 35516, Egypt
e-mail: ahmedelbeltagy81@mans.edu.eg

K. Pandey et al. (eds.), *Artificial Intelligence and Smart Agriculture*, Advances in Geographical and Environmental Sciences,
https://doi.org/10.1007/978-981-97-0341-8_7

introduced into the agricultural sector in an attempt to overcome these challenges. This study highlights the adaptability, excellence, efficiency, and frugality of AI in agribusiness, which can be seen in its potential benefits, such as early detection of crop diseases and improper weeding, and precision farming to regulate individual plants and crops. Furthermore, this study aims to provide an overview of AI's applications in managing weeds, diseases, crops, and land, as well as the benefits and drawbacks of these applications, and methods for increasing productivity through the use of intelligent systems. In summary, the integration of AI in agriculture can help address the challenges facing the sector, leading to increased output and profitability for farmers. Nonetheless, its successful integration requires careful planning, education, and training.

Keywords Artificial intelligence · Management and practices · Digital agriculture · Applicability and pragmatism · Limitation and challenges

7.1 Introduction

Throughout the history of mankind, agriculture has been one of the oldest and most dynamic occupations. The increased number of people throughout the world and the diminishing availability of land have made it more critical than ever for everyone to think creatively and to come up with new ways of farming in order to use the land to grow more crops and to increase productivity. This will lead to an increase in yield, which in turn will result in an increase in productivity. The agricultural industry is being aided by artificial intelligence in a number of different ways to enhance quantity and quality in the production of crops, viz., weather forecasting, soil moisture monitoring, real-time irrigation, manage pests, agriculture robotics, grow conditions, health monitoring of crops, disease diagnosis, coordinate farmers' data, intelligent spraying, help with the workload, and many advance agricultural tasks across the whole food supply chain by helping to produce more healthy crops, track soil, manage pests, and accomplish growing conditions.

As a result of the weather, climate, and soil moisture, etc., there are dozens of dynamic and intricate environmental variables that play a significant role in affecting the production of agricultural products and have a significant impact on the agricultural industry. These conditions have minimal influence on agricultural operations and cannot be entirely predicted (van Mourik et al. 2021). The agriculture sector has welcomed the use of artificial intelligence into their practice in order to improve and enhance their efficiency and reduce the adverse effects on the environment. The mainstream of agriculturally based startups is focusing on technology that enhances the efficiency of the production process by using technology based on machine learning (ML) and artificial intelligence (AI) (Dhanya et al. 2022). In order to ensure that farmers can make the right decisions for their crops, they are now using artificial intelligence (AI) to recognize insights in data like solar radiation, temperature, precipitation, and wind speed (Markuna et al. 2023; Vishwakarma et al. 2023). One of the main advantages of using artificial intelligence in agriculture and allied is that it will support farmers to improve their progress while at the same time eliminating jobs. Artificial intelligence can be used to produce, harvest, and sell essential crops in the most efficient manner with the highest degree of accuracy.

Compared to rain-fed agriculture, proper irrigation can increase agricultural output and income (Jaramillo et al., 2020). A further benefit of irrigated agriculture is that it is dependable and offers a wide range of crops with a higher value in a given area (Asmamaw et al. 2021). Irrigated agriculture also consumes 20% of all farmland, yields 40% of global crop production, and makes a significant global impact on food security (Meybeck and Redfern 2016; Zongo et al. 2022). Agriculture accounted for around 92% of the freshwater withdrawn worldwide in 2016, and the Food and Agriculture Organization (FAO) predicts that by 2050, the water demand for agriculture will increase by 11% as a result of population growth. (World Bank 2022). Earlier in the study, Liu et al. (2022) claimed that since water is an ongoing resource, it could last till 2050 if water resources are managed properly and mitigation strategies are applied to reduce risks in the agriculture sector by controlling green land to be developed for urban purposes. The major cause of water depletion in most regions has been identified as irrigation, which has been cited as the main contributor to the global water crisis because of the large amount of water extracted worldwide—roughly 70% (Aryalekshmi et al. 2021). Throughout several areas in the country, water shortages have been attributed to an inadequate and inappropriate water management system (Entezari et al. 2019). In the future, irrigated agriculture is expected to become more productive while consuming a lesser amount of water in order to increase the yield (de Jong 2021). Agricultural water is lost to evapotranspiration to a significant extent, which leads to a decrease in crop yield. It is important to make use of in-situ measures of plants and soils in order to determine crop water demands based on evapotranspiration, including soil moisture measurement, plant moisture measurement, and soil and plant physical parameters through artificial intelligence (Liou and Kar 2014; Eliades et al. 2022).

Inadequate guidance on soil fertility, climate variations that happen throughout the process, and the ultimate harvesting time all contribute to a substantial proportion of agricultural output being lost. The country's diversified topography results in a multiplicity of harvest periods and crop variations. Such intricacies and localization increase the adaptability and dynamism of Asian agriculture. Surveying, mapping, and monitoring technology supported by mechanics can assist farmers increase their yield. Currently, data analysis, sensors, and robots are utilized in digital agriculture for automating traditionally laborious processes. Farmers should put their energy into providing services such as independent harvesting, field reconnaissance, and weed management. Sensor technique was utilized in Agriculture 3.0, Information and Communication Technology (ICT) was incorporated in Agriculture 4.0, and Agriculture 5.0 implemented artificial intelligence and robotics technology.

A major concern for all regions is agriculture automation as it is a new and evolving challenge that needs to be addressed. Among the most significant developments are object recognition, algorithms for task planning, digitalization, and sensor optimization. A worldwide food supply crisis may occur due to the gap between population expansion and agricultural output increase, says a United Nations assessment. It is predicted that production will need to increase by a factor of 2.5 to keep up with demand. A process that requires minimal human input is referred to as "automated" (Kamilaris and Prenafeta-Boldú 2018). Several automated agricultural operations, such as crop monitoring and yield prediction, benefit from the use of machine vision. Automation of machine vision has, by all accounts, solved many issues for humans (Möller 2010). Machine vision systems have made use of image sensors for material recognition and classification. Several researchers have looked at how machine vision might be used to streamline the processes involved in managing grain and seed production in agriculture (Abdullahi et al. 2015). The long-term survival of every economy depends on agriculture. Its role varies from country to country, but it is crucial for future structural and economic transformation (Mogili and Deepak 2018; Shah et al. 2019). Earlier, crop production was agriculture's sole objective. Production, marketing, and distribution of cover crops and livestock have all expanded during the last 20 years. As a result of recent developments in agriculture, the economy as a whole has benefited from increased GDP, the creation of state commerce, decreased unemployment, and an increased supply of raw resources for other industries (FAO 2017). To create application systems that can mimic and even surpass the capabilities of human intellect, artificial intelligence draws on a wide range of scientific fields, including computer science, psychology, neuroscience, computer engineering, behavioral science, and bioscience. By modelling the way data interacts in human intelligent cognition, artificial intelligence aims to create a computer system capable of resolving problems and generating responses in a manner similar to that of humans (Jani et al. 2020).

Agriculture 5.0 technology, which includes data analysis, integrated robotics, and sensors, is being increasingly used to convert labor-intensive tasks into ongoing mechanized ones, resulting in the growing popularity of digital farming. This chapter focuses on current advancements in agricultural robotics, such as real-time automated irrigation systems, weeding and hoeing operations, crop monitoring approaches, and fertigation and chemical spraying strategies (Kim et al. 2008). For the development of virtual farms, many robotics, human–robot interaction, and environmental restoration using ground-based sensors have been suggested as methods of improvement (Ahir et al. 2020; Gandhi et al. 2020; Kundalia et al. 2020; Pandey et al. 2023). The development of mobile robots networks that coordinate their efforts to optimize farming resources and reveal previously hidden information is a hot issue in agricultural field robotics research (Murugesan and Sivarajan 2018; Shamshiri et al. 2018). Combining real-time monitoring by machine learning technologies created by AI with the exploration capabilities of autonomous robots could potentially facilitate the provision of more information on crop health and its surrounding environment. Before the emergence of these new technologies, robots' screens could only show the recorded images from their cameras. Owing to AI, they are now able to chart their surroundings, keep track of objects, and offer continuous feedback on their analyses (Wang et al. 2007).

Due to the continuous increase in requirement for food, farmers are finding that traditional methods of producing food are not enough and they must resort to using increasingly harmful pesticides to serve the growing demand. In addition, it has significant effects on agriculture, as well as on the soil, as the earth as a whole still remains unfertile as a result of it. It is evident that agriculture is affected by a number of factors, such as crop pests, the lack of adequate storage space, pesticide control, plant management, irrigation, and water management, and a number of factors that affect agriculture and are all important issues that need to be addressed. I believe that artificial intelligence will be able to overcome all of these problems in the future. This chapter summarizes the methods in which AI has been used to control weeds, diseases, crops, and agricultural land management. The application's advantages and disadvantages are addressed, along with suggestions for enhancing efficiency using intelligent systems. As a result of this research, the current and future prospects of agricultural sustainability have been highlighted, as well as the benefits of technological advancements.

7.2 Use of Automation in Agriculture

7.2.1 Artificial Intelligence (AI) Replaced Traditional Methods

Improvements in irrigation, crop output, crop monitoring, crop establishment soil content sensing, and weeding are just a few examples of how artificial intelligence (AI)-based solutions are helping the agricultural industry become more efficient overall (Kim et al. 2008). Valuable applications are being carried out by embedded robots that are being developed in the agriculture sector using artificial intelligence. Due to the growing global population, the agriculture sector has a difficulty, but artificial intelligence can give a superior solution. Artificial intelligence-based technological solutions have assisted farmers in increasing the quality of their agricultural yields while decreasing their time to market. The average farm of the future is projected to produce 41,000,000 data points daily by the year 2050. The production and administration of agriculture have benefitted immensely from automation. According to Dursun and Ozden (2011), autonomous agricultural equipment is being developed as a solution to the inefficiency of traditional farming tools. This technology's primary goal is to substitute human labor while providing substantial benefits to manufacturers of all sizes (Manivannan and Priyadharshini 2016). Pedersen et al. (2008) reported that manufacturing in this industry has benefited greatly from the application of robotics. Linked robot can independently tend to gardens, doing things like watering and weeding. They can also protect crops from harmful environmental conditions, enhance accuracy, and monitor each plant's productivity (Ahir et al. 2020; Gandhi et al. 2020; Kundalia et al. 2020) In 1794, Eli Whitney (1765–1825) developed a machine that revolutionized the cotton industry by dramatically accelerating the separation of cotton seeds from cotton fiber. In a single day, it generated fifty pounds of cotton. As a result, self-sufficient farm robots have been created. Nowadays, self-sufficient farm robots are capable of accurately planting seeds by pinpointing their exact location and ensuring they remain in place once in contact with the ground. The entire growth history of the plant was automatically recorded (Griepentrog et al. 2005). Diverse biosensors for monitoring plant growth and detecting illness have been created (Tothill 2001). Laser welding technology has replaced manual weeding by severing weed cells with computer-controlled, mobile-focused infrared light. Automatic irrigation systems, which utilize water more efficiently, were also installed (Griepentrog et al. 2006).

7.2.2 Effective Watering System in Agriculture

More than 85% of all freshwater is put to use in agriculture. And that proportion is growing as the world's population and appetite rise as well. To ensure effective use of irrigation water supplies, additional measures are required. One such measure involves replacing manual watering with automatic irrigation scheduling based on soil moisture data. This approach considers several environmental factors such as humidity, wind speed, solar radiation, as well as crop characteristics, including development stage, plant density, soil quality, and pests. The autonomous irrigation technology employed also takes into account plant evapotranspiration.

The basic goal of machine-to-machine (M2M) technology is to make it easier for devices in an agricultural setting to talk to one another and share data with a central server or cloud. They also programmed a robot to automatically detect changes in humidity and temperature using boards from the Arduino and Raspberry Pi3 development platforms. The Arduino microcontroller constantly polls for new information and converts analogue signals to digital form. The irrigation water pump is controlled by a signal from the sign's Raspberry Pi3 to an Arduino. Shekhar et al. (2017) recorded the sensor's reading as water was distributed to the field. Jha et al. (2019) developed a real-time automated irrigation system using Arduino technology for more efficient watering and less labor investment. Savitha (2018) devised a novel automatic watering system using remote sensors and Arduino technology that could potentially boost yields by 40%. Another automated watering system is equipped with moisture, temperature, pressure, and molecular sensors, as well as digital cameras, to assess crop growth and field conditions. The data from each of these gadgets is transmitted wirelessly to a network access point, where it is digitized and sent to a multiplexer. Subsurface drip irrigation emitters are placed directly beneath the crop to minimize water loss due to evaporation and runoff. Soil moisture sensors and raindrop sensors are employed to evaluate the fields' moisture levels. This system is fueled by solar energy and a wireless network. The farmer receives SMS readings from the soil moisture and raindrop sensor through GSM module and his mobile phone. Therefore, farmers can accept the request to turn the water supply ON and OFF. To enable farmers to easily control the water supply in their field, regardless of weather conditions, a soil moisture sensor is utilized (Varatharajalu and Ramprabu 2018). This sensor is placed in the soil near the plant's roots, and it provides an accurate reading of the soil moisture level to the irrigation controller. Soil moisture sensors, meanwhile, can significantly reduce water consumption (Qualls et al. 2001; Dukes et al. 2009). Farmers may adjust the moisture threshold in accordance with the soil's capacity using moisture sensors, which then trigger the controller to water only when the zone's moisture content drops below the specified point. The system evaluates the moisture threshold value of each zone and authorizes irrigation as required when the scheduled watering time approaches. For each zone, there is a designated start and stop time and duration (Yong et al. 2018). According to Gebregiorgis and Savage (2006), the mass permittivity (dielectric constant) of the soil can be used to calculate soil moisture levels. An innovative approach proposes a mechanized system that utilizes dielectric soil moisture sensors for real-time automatic irrigation management (Kuyper and Balendonck 1997). Dielectric properties are a promising way to measure various factors (Zhen et al. 2010), although the accuracy of electricity transport through soil is affected by factors including tensiometers and electrical resistance (Hanson et al. 2000). Soil's dielectric strength for conducting electricity is mostly constant. Since soil contains several elements including air, water, and minerals, the dielectric constant must be calculated by adding up the total commitment of each of these groups. Automated watering can be accomplished using moisture detectors and microcontrollers (Rajpal et al. 2011).

7.3 Application of ML/AI in Agricultural Management

In agriculture, the use of artificial intelligence is one of the most powerful tools for helping to facilitate a wide range of technological advancements. The evolution and advancement of this can be categorized into several aspects, including the provision of consulting services, data analytics, the Internet of Things, use of cameras and other sensors, as well as software and hardware developments. In the future, artificial intelligence (AI) and machine learning (ML) will be able to run better understandings into agricultural production by examining different types of data such as weather, soil moisture, crop growth, and temperature as well as analyzing various data sources. As shown in Table 7.1, a number of significant applications of AI are discussed in the agriculture sector.

7.3.1 Soil Management

Maintaining healthy soil is a cornerstone of sustainable farming. In order to maximize agricultural output and protect the soil resources, it is essential to have a good understanding of the various types and conditions of soil. To improve soil performance by the application of actions, methods, and treatments, traditional soil survey techniques can be used to investigate the possibility of contaminates in urban soils (De Kimpe and Morel 2000). Soil porosity and aggregate structure can be enhanced

Table 7.1 Summary of ML/AI studies in various domain of agriculture (Eli-Chukwu 2019)

References	Domain	Method	Implications	Considerations
Li and Yost (2000)	Soil management	MOM	Reduces nitrate leaching while increasing production	Takes time and is confined to nitrogen
López et al. (2008)		Fuzzy Logic: SRC-DSS	Can identify soil based on related dangers	Takes large amounts of data and just a few examples have been analyzed
Montas and Madramootoo (1992)		DSS	Reduces erosion and sedimentary yield	Requires big data for training
Tajik et al. (2012)		ANN	It is capable of predicting soil enzyme activity. It is able to accurately predict and classify soil structure on the basis of its properties	It only measures a small number of soil enzymes in the soil. The focus of this approach is more on classifying the soil rather than improving its performance
Levine et al. (1996)			It is capable of predicting the mean temperature of the soil on a monthly basis	The soil performance is considered to be influenced only by temperature
Bilgili (2011)			It predicts soil texture	In order to train the system, big data is necessary. The scope of implementation is restricted in certain areas
Zhao et al. (2009)			Able to predict soil moisture	In the long run, this prediction will fail, because weather conditions are difficult to predict due to their unpredictable nature
Elshorbagy and Parasuraman (2008)			In this application, soil texture is successfully reported	There is no improvement in soil texture, nor does it offer a solution to bad soil texture
Chang and Islam (2000)			Besides being cost-effective, it is also time-saving and has a 92% accuracy rate	This requires a lot of data

(continued)

Table 7.1 (continued)

References	Domain	Method	Implications	Considerations
Behrens et al. (2005)			The ability to estimate the soil nutrients following erosion can be achieved with this tool	The estimate is only restricted to only HM
Plant (1992)	Crop management	CALEX	The ability to formulate guidelines for scheduling crop management activity	It takes more time
Lal et al. (1992)		PROLOG	Organizes and removes farm tools that are less frequently used	Location-specific
Dahikar and Rode (2014)		ANN	This tool predicts the yield of crops	Crop yields are only taken into consideration as a result of weather conditions
Pilarski et al. (2002)		ROBOTICS-Demeter	Up to 40 hectares of crop can be harvested with this machine	It is expensive: It uses a lot of fuel
Van Henten et al. (2002)		ROBOTICS	A harvesting success rate of 80% can be achieved by this machine	It is difficult to pick accurately and at a slow pace
Song and He (2005)		ANN	Detection of crop nutrition disorders has a success rate of over 90%	The symptoms that were considered were a limited number
Papageorgiou (2011)		FUZZY Cognitive Map	Improve the decision management of cotton crops by predicting cotton yields	The process is relatively slow
Dai et al. (2011)		ANN	The soil moisture and salinity of the soil can be used to predict the response of crops	The only factors taken into account in this study are the soil temperature and texture
Yang et al. (2003)		ANN and Fuzzy Logic	Reduces the number of insects attacking crops and the damage they cause	The ability to distinguish between crop and weed is lacking
Ji et al. (2007)		ANN	Achieves high levels of accuracy in predicting rice yields	It is time-consuming, and it can only be done in specific climates

(continued)

Table 7.1 (continued)

References	Domain	Method	Implications	Considerations
Balleda et al. (2014)	Disease management	Computer vision system (CVS), genetic algorithm (GA), ANN	Fast and efficient way of working. It is capable of multitasking	Good species may be adversely affected as a result of dimension-based detection
Balleda et al. (2014)		Rule-Based Expert, Data Base (DB)	The results produced in the test environment are accurate and reliable	Data bases are inefficient when implemented in a large-scale environment
Jesus et al. (2008)		Fuzzy Logic (FL), Web GIS	It is cost-effective, eco-friendly, and a good investment	Inefficiency due to scattered distribution. Takes time to locate and disperse data. The location of the data is determined by a mobile browser
Kolhe et al. (2011)		FL Web-Based, Web-Based Intelligent Disease Diagnosis System (WIDDS)	Good accuracy. Responds swiftly to the nature of crop diseases	Limited usage as it requires Internet service. Its potency cannot be ascertained as only four seed crops were considered
Kolhe et al. (2011)		FL and TTS converter	Resolves plant pathological problems quickly	Requires high-speed Internet. Uses a voice service as its multimedia interface
Munirah and Rozlini (2013)		Expert system using rule-base in disease detection	Having diseases diagnosed earlier allows for faster treatment as a result of quicker diagnosis. Due to its preemptive approach, this program is extremely cost-effective	This is a time-consuming process. Monitoring is essential to ensure that the preventive measures have not been able to build immunity to the pests
Liu et al. (2006)		ANN, GIS	An accuracy rate of 95% is achieved	Requires an Internet connection. The access to the Internet will not be available to some farmers in rural areas

(continued)

Table 7.1 (continued)

References	Domain	Method	Implications	Considerations
Siraj and Arbaiy (2006)		FuzzyXpest provides pest information for farmers. It is also supported by Internet services	Predictions with a high degree of precision	Requires an Internet connection
Patel et al. (2012)		Web-Based Expert System	An excellent performance	It is a web-based and Internet-based software
Wang et al. (2008)		ANN	Predicts with a success rate of above 90%	ANNs do not kill or reduce the effects of infections in any way
Tobal and Mokhtar (2014)	Weed management	ANN, GA	The performance is high. Eliminates the need for trial and error	This requires a lot of data
Saxena and Gautam (2012)		Optimization using invasive weed Optimization (IVO), ANN	Enhance performance and cost-effectiveness at the same time	Adapting to new data is a challenging task
Brazeau (2018)		Mechanical Control of Weeds. ROBOTICS. Sensor machine learning	It eliminates resistant weeds and saves time as well	Quite expensive. It is likely that soil productivity will be reduced as a result of the constant use of heavy machines
Pérez-Ortiz et al. (2016)		UAV, GA	Detects and monitors weeds in an effective and timely manner	It is not very effective at controlling weeds. An expensive investment
Stigliani and Resina (1993)		Saloma expert system for evaluation, prediction and weed management	Adaptability and prediction levels are both very high	In order to achieve this, big data expertise as well as usage expertise is required
Karimi et al. (2006)		Support Vector Machine (SVM), ANN	A quick detection of crop stress will prompt the appropriate remedies to be applied at the correct time on a particular site	The device only detects nitrogen levels below a certain threshold

(continued)

Table 7.1 (continued)

References	Domain	Method	Implications	Considerations
Gerhards and Christensen (2003)		Digital image analysis (DIA), GPS	The accuracy rate and success rate of this product are above 60%	Despite the fact that it is successful after 4 years, it took a lot of time to achieve, so it could be considered time-consuming
López-Granados (2011)		UAV	Weeds are detected at a high rate within a short period of time, which is very impressive	The cost of this process is really high, and it requires a huge amount of human expertise and experience
Yang et al. (2002)		Learning Vector Quantization (LVQ), ANN	Detection of weeds in a short period of time with high weed recognition rate	Data input method plays an important role in determining the performance of artificial intelligence

through the application of compost and manure. As a result of this increase in soil aggregation, organic matter is being incorporated, which also serves as a preventative measure against soil crust formation. To reduce soil physical deterioration, it is possible to employ alternate tillage strategies. Improving soil quality requires the incorporation of organic materials (Pagliai et al. 2004). Vegetable and other food crop yields are severely impacted by several soil-borne diseases that must be managed by soil preparation and application (Abawi and Widmer 2000). Understanding that different soils have varying resistance to change and capacities for recovery makes sensitivity to soil degradation implicit when evaluating the long-term viability of land management practices (Lasanta and Vicente-Serrano 2012).

To lessen nitrate leaching, management-oriented modelling (MOM) creates plausible management options, runs simulations on each one, and then selects the one that most fulfils the user's weighted numerous conditions. Strategically, MOM employs hill climbing and best-first techniques goal to find the nearest path from beginning nodes to targets (Li and Yost 2000). There are three steps in the process of building the Soil Risk Characterization Decision Support System (SRC-DSS), the acquisition of knowledge, the conceptual design, and the implementation of the system, which require engineering skills, as indicated by López et al. (2008). An artificial neural network (ANN) model is used to forecast the soil texture (percentage of sand, clay, and silt concentrations) using current coarse resolution soil maps and hydrographic data which derived from a digital elevation model (DEM) as inputs (Zhao et al. 2009). Meanwhile, remote sensing devices equipped with a higher-order neural network (HONN) are utilized for assessing and approximating soil moisture dynamics (Elshorbagy and Parasuraman 2008).

7.3.2 Crop Management

The process of managing crop involves various stages, starting from sowing seeds to harvesting, securing the crop for storage, and distributing the produce to customers. It might be defined as anything that helps crops flourish and provide more food. An in-depth familiarity with the timing and ideal soil type for various crop varieties is unquestionably the key to maximizing crop yields. Precision crop management (PCM) is increasingly being adopted by farmers to minimize environmental damage and maximize profits. However, the lack of timely and distributed data on crop and soil conditions has hindered PCM's progress (Moran et al. 1997). In order to make up for water scarcity brought on by factors such as inadequate irrigation, climate, or soil, farmers must use diverse crop management strategies. Because of their adaptability, decision rule-based crop management systems are preferred. Selecting a cropping strategy is highly dependent on the timing, intensity, and prediction of drought (Debaeke and Aboudrare 2004).

Anticipating weather patterns accurately plays a vital role in making informed decisions that lead to high-quality agricultural yields (Aubry et al. 1998). The operational behavior of a farm system may be evaluated with the help of PROLOG by inputting information such as weather, equipment capabilities, manpower availability, and personnel, tractors, and tools allowed. It also estimates farm-wide and individual crop output, gross revenue, and net profit (Lal et al. 1992). The best crop may be predicted with the help of the crop prediction method by analyzing a wide range of soil parameters and weather forecasts. A variety of factors were considered by Dahikar and Rode (2014) study including the physical and chemical characteristics of soil (such as type of soil type, density, porosity, hydraulic conductivity, soil depth, Ph, etc.), macro and micronutrients (such as nitrogen, phosphate, potassium, organic carbon, calcium, magnesium, sulfur, manganese, copper, iron), and meteorological parameters (such as temperature, precipitation, and humidity). The computer-controlled rowing machine, Demeter, uses two video cameras and a global positioning system sensor to cut crop rows, reposition itself in the field, and identify unexpected obstacles, allowing it to plan and execute harvesting operations for a whole field (Pilarski et al. 2002). The cucumber harvesting robot uses artificial intelligence to control various hardware and software components of the robot, including the autonomous vehicle, the manipulator, the end effector, two computer vision systems that detect fruit and the environment and generate 3D images, and a collision-free motion control system that can be used during harvesting (Van Henten et al. 2002). For each site, field-specific precipitation and meteorological data can be utilized. Adjusting neural network parameters influences the precision of rice yield forecasts. To reduce model optimization time, the many hidden nodes and/or learning rate can be reduced for smaller datasets (Ji et al. 2007).

7.3.3 Disease Management

Maximum crop yields depend on effective disease control during agricultural harvests. Plant and animal illnesses pose significant barrier to increasing yields, and factors such as rainfall, drought, wind, and temperature contribute to the development of diseases. In large-scale agriculture, regulation of the repercussions is a serious task because of these factors and the inconsistency of the causative effect of certain diseases. A farmer's best bet for preventing disease and minimizing losses is a method known as "integrated disease control and management," which employs various techniques (e.g., physical, chemical, and biological) to combat the problem. The time and money that are invested in this process are both inefficient and wasteful, highlighting the need for an AI-based strategy to manage and control infectious diseases. An expert system explanation block (EB) (Balleda et al. 2014) that uses an innovative method of rule promotion based on fuzzy logic offers insightful results for managing crop diseases, and its explanation block provides a clear understanding of its reasoning. Additionally, a text-to-speech converter is necessary to implement a TTS user interface for real-time, web-based interaction (Kolhe et al. 2011). In a study published in 2013, Munirah and Rozlini (2013) created a rule-based inference engine utilizing forward-chaining inference techniques to assist in the diagnosis and treatment of diseases.

7.3.4 Weed Management

Weeds are invasive plant species that lower crop yields and quality in agricultural settings (Tu et al. 2001). Weeds drastically lower agricultural output by contending with crops for most essential resources such as spacing between crops, water, nutrients, and sunlight. Both inter-row and intra-row weeds can cause problems for farmers cultivating row crops, with intra-row weeds having been found to decrease crop output by as much as 33% (Peruzzi et al. 2007). Management strategies that are efficient, punctual, and well-planned are, thus, crucial. There are a number of strategies for managing weeds in the inter-row space, but controlling them between the rows is a far more difficult operation. Manual weeding is the most efficient method, but it also costs the most. In addition, manual techniques require stooping often and might expose workers to toxic plants on occasion. As a result of the dangers they provide to workers' health, manual labor has been all but eliminated in developed countries (Tu et al. 2001; Chandel et al. 2018; Chethan et al. 2018a, b).

Before an automated weed management system can be developed, it is necessary to differentiate between weed seedlings and crop seedlings (Bhagyalaxmi et al. 2016; Chang and Lin 2018). To tell the difference between the carrot and ryegrass seedlings, a method was developed. The unique approach developed to evaluate leaf shape based on this basic morphological attribute was used to put this idea into practice. This approach may identify plants from weeds with an accuracy between 52 and

75% by evaluating the variance in leaf size. The use of digital imagery allowed for the creation of a new weeding technique. A neural network that can organize its own activity was used to develop this idea, but its accuracy and cost-effectiveness fell short of expectations. As a result, a ANN-based system was developed that can distinguish between species with a 75% accuracy rate that was capable of identifying them from each other (Aitkenhead et al. 2003). Even while manual techniques requiring direct contact with weeds were formerly used, several automated technologies have emerged in recent decades. The efficiency of robotic weeding is influenced by the density and weeds' placement. Intra-row weeding used spring mechanism or duck foot tines in the past, this loosened the soil and made the roots more accessible to weed wiping, which made weed wiping easier; however, tillage is not advised since it may decrease crop-soil interaction (Griepentrog et al. 2006). Thus, according to (Heisel et al. 2001), methods like laser therapy and microspraying were developed to avoid damaging the roots while tending to the soil. Nakai and Yamada (2014) discussed how agricultural robotic machines can be employed to rid weeds' fields and how new methods can be created to control the robots' positions in the field. In order to eliminate weeds and control the robot's position, a laser range finder (LRF) method was used to determine the location of the robot. Multiple cameras and sensors gave the robot excellent eyesight. A gray-level vision was used to create a row structure to assist the robot in navigating the rows, while a color-based vision was used to identify individual weeds. This method was first trialed in a greenhouse to manage weeds along a crop row. The robots were guided along the row structure using vision-based technologies to eliminate weeds and identify crop plants from non-crop plants.

Weed regularly affects the anticipated profit and productivity of farmers (Neil Harker 2001). A survey reveals that weed infestations might reduce the production of dry beans and corn by 50% (Neil Harker 2001). Approximately, 48% of wheat output is lost owing to weed competition (Khan and Haq 2002; Fahad et al. 2015). These losses can sometimes reach 60% (Rao et al. 2014). A research on the effects of weeds on soybean output revealed an 8–55% decline (Datta et al. 2017). According to a research, yield losses in sesame harvests range between 50 and 75% (Mruthul et al. 2015). The extent of these losses depends on various factors such as the duration of crop exposure to weeds (Swanton et al. 2015; Jha et al. 2017) and the weeds' geographical heterogeneity (Milberg and Hallgren 2004). Moreover, cannabis has both good and harmful consequences on the ecology, with some weeds causing flooding during hurricanes, paving the way for fires, producing allergic reactions, and posing health risks.

One such approach involves using unmanned aerial vehicles (UAVs) to capture imagery, partition the image, and develop a classification model for weed and crop pixels. Crop row recognition methods can aid in distinguishing between weed and crop pixels, which can be challenging due to their spectral similarity. Online weed detection utilizing digital image analysis captured by a UAV, computer-based decision making, and GPS-controlled patch spraying can be used to manage weeds in various crops such as sugar-beet, maize, winter wheat, and winter barley. Studies show that drones can effectively identify tomato and weed sites, with the drone in one study traveling at a speed of 1.2 km/h and requiring 58.10 and 37.44 ms to identify the tomato and weed sites, respectively, for the spray controller.

Although herbicides have been for decades to control weeds, they have not been effective in preventing yearly crop losses exceeding $500 million in western Canadian field crops (Swanton et al. 1993). To overcome this issue, a more professional weed control approach is required (Neil Harker 2001). One such approach involves using unmanned aerial vehicle (UAV) imagery to partition the image and develop a classification model for weed and crop pixels. Crop rows recognize can aid in distinguishing between weed and crop pixels (Pérez-Ortiz et al. 2016), which can be challenging due to their spectral similarity. To manage weeds in crops such as sugar-beet, maize, winter wheat, and winter barley, it is possible to use online weed detection through digital image analysis captured by a UAV (drone), computer-based decision making, and GPS-controlled patch spraying (Gerhards and Christensen 2003). In a study by López-Granados (2011), a drone traveling at a speed of 1.2 km/h required 58.10 and 37.44 ms to identify tomato and weed sites, respectively, for the spray controller.

7.4 Discussion on Practical Application of AI in Agriculture

As a result of the advancement of artificial intelligence, a vast array of agriculture is using it to improve the productivity and efficiency of their operations. A number of agriculture-related issues, including forecasted weather data, crop selection, soil preparation, seed selection, seed sowing, irrigation (when and how much), crop growth, fertilizing, harvesting, disease detection, monitoring crop and soil health, fruit and plant categorization, as well as crop management, are being addressed by artificial intelligence approaches. Artificial intelligence refers to the processing of data in a way similar to what the human brain does, creating patterns that can be used for making decisions.

In agriculture, drone technology has already become an integral part of precision farming operations that are carried out by farmers. In order to achieve the highest possible yield for their crops, farmers can utilize drone data to their advantage. Despite the fact that AI has enormous potential for application in agricultural production, the vast majority of farms throughout the world are still unaware of high-tech artificial intelligence solutions. The development of AI-driven robots requires significant research and development investments along with ongoing maintenance to ensure their smooth operation for the foreseeable future. The agriculture sector may greatly benefit from the use of data analytics in terms of improved production and lower costs. Farming can be improved by combining artificial intelligence with big data by synthesizing accurate crop demand categorization and incorporating it into real-time crop planning. Therefore, agriculture techniques such as irrigation, fertilizer, crop protection, and harvesting will be more precise and will require less guessing in the future, thus reducing the risk of errors. Farmers' expert systems could be made more valuable with the use of location-specific agrometeorological data, as well as sensor-based soil, plant, water, and pest conditions to make the extension system and methodology more effective and efficient. The adoption of AI approaches may also facilitate a paradigm change for millions of farmers in our country, ensuring that location-based advice services are replaced by personalized and context-specific advice services, allowing them to achieve better results.

In order for such solutions to be of value to farmers, it is essential that they understand their applicability, benefits, and future before they are put into use. In spite of the fact that farmers feel pride in their traditional occupation, they are reluctant to switch from their traditional farming practice to a new one unless the technology promises higher returns on investment and is easy to use.

7.5 Conclusions

As a conclusion, agricultural artificial intelligence is still at the beginning stages of its development and there is still a long way to go. Although there are many serious issues associated to crop productivity that can be addressed using this technology, there are also serious environmental concerns that must be addressed. Advancements in agricultural technology have brought significant changes to the farming industry. United Nations forecasts indicate that, with the global population set to reach 9.7 billion by 2050, only 4% of the land will be available for agriculture (FAO 2017). To meet the needs of an additional two billion people, food production will have to increase by 60%, according to FAO (2017). Traditional processes, in comparison, are incapable of servicing this tremendous demand. Farmers and agribusinesses are required to develop solutions that simultaneously boost production and decrease waste. As a result, artificial intelligence is increasing relevance in the technological growth of the agriculture business. To feed two billion more people by 2050, the global food supply must rise by 50%. To achieve this, the use of artificial intelligence in agriculture is becoming increasingly important. AI-based solutions can increase productivity improve crop yield, quality, and time to market.

Lack of proper irrigation systems, weeds, and extreme weather conditions, among others, impede the agricultural business. However, with the assistance of technology, performance may be improved, enabling the resolution of these issues. For instance, remote sensors for monitoring soil moisture content and GPS-assisted automatic watering can enhance crop growth. Precision weeding techniques have been limited by the substantial loss of crops during the process. However, the use of autonomous robots is becoming increasingly common and can offer a solution to this issue. Artificial intelligence can be applied in agriculture to identify limitations in resources and labor availability. Traditional methods needed for obtaining agricultural data such as soil quality and plant height involve time-consuming manual testing, which requires a significant large amount of labor. The explored technologies would enable rapid and non-destructive high-throughput phenotyping, as well as flexible and beneficial activities, on-demand information access, and geographical objectives.

In combination with machine learning algorithms, satellite imagery, and images taken by drones and satellites, artificial intelligence-enabled technologies are able to predict weather conditions, analyze crop sustainability, and evaluate farms for the presence of diseases and pests and inadequate plant nutrition utilizing data such as temperature, precipitation, wind speed, and solar radiation. AI applications are available to farmers with Wi-Fi connectivity, which can provide them with a farm plan tailored to their specific needs. By utilizing AI-driven solutions that unlock the potential for further growth in production and income without depleting the world's priceless natural resources, farmers can meet the global demand for a higher food supply and greater profitability in the near future.

7.6 Challenges and Limitations

Agricultural management can benefit from expert systems because they can provide advice that is site-specific, integrated, and easily comprehended. Nevertheless, the use of expert systems in commercial agriculture is still uncommon despite significant progress in artificial intelligence within the industry. Although artificial intelligence has made considerable strides in the agriculture business, its influence on agricultural operations remains below average when compared to its potential and impact in other industries. There are still several obstacles that need to be overcome to fully leverage the benefits of artificial intelligence in agriculture.

Intelligent or expert systems have the ability to perform tasks quickly and accurately, making them valuable tool. The vast majority of systems lack response speed, accuracy, or both. The choice of task strategy used by a user is impacted by system delay. A cost function that considers the effort required to synchronize input system availability and the level of accuracy provided can govern strategy choice. Users can choose between autonomous performance, pacing, and monitoring to reduce effort and improve precision. Additionally, input data volume is a statistic for measuring an intelligent agent's effectiveness, especially for real-time AI system that must constantly monitor a vast amount of data while discarding irrelevant information. Subject matter experts must have an in-depth understanding of the system's task, and only highly relevant data should be used to enhance the system's speed and precision. The development of an agricultural expert system requires the participation of professionals from multiple agricultural disciplines as well as producers who will use the system.

The execution technique of any expert system is its greatest asset. Due to the utilization of large amounts of data, the process of searching up and training must be adequately described for speed and precision. However, most artificial intelligence (AI) systems are Internet-based, which limits or restricts their deployment, particularly in remote or rural areas. To aid farmers, the government could develop a web service-enabled device that is compatible solely with their artificial intelligence systems and has a lower price. Additionally, providing a "how to use" orientation (training and re-training) can also considerably aid farmers in adjusting to AI's employment on farms. Flexibility is also a significant feature of any dependable AI system. While AI techniques have made significant progress in applying to some discrete tasks, the main focus of AI-based robotics technology is the integration of subsystems within an integrated environment. This necessitates the adaptability of the subsystems themselves and comprehensive options for handling extra user data from the field expert.

7.6.1 Future Research Direction

Younger farms are more likely to be engaged in and committed to artificial intelligence. The introduction of the cutting-edge technologies has been slow and methodical. The agricultural sector is increasingly turning toward precision farming, which will regulate individual plants and crops. Setting the stage for long-term agricultural growth and prosperity requires the application of deep learning and other forms of automation. Finally, a wider variety of items and production methods is possible as a result of increased agricultural output. Artificial intelligence methods utilizing a computer network can assist in the early identification of crop disease and improper weeding. While we cannot provide crops with a controlled environment like a greenhouse, wireless technologies and sensors can enable weather monitoring without human intervention. Robots can be used by farmers for crop weeding, seeding, planting, harvesting, fertilizing, irrigation, and shepherding. Drones fitted with thermal cameras give continuous real-time data of the farm, allowing irrigation to begin and preventing water floods or crop failure.

A new generation of artificial intelligence (AI) technologies is bringing innovative and precise solutions to the major agricultural issues that farmers around the world face in the future. Artificial intelligence will be able to solve almost anything, from pest control to weather predictions to the assistance with farm work, in the near future. The next couple of years will see a number of exciting developments in artificial intelligence related directly to the farming industry.

By 2050, the world's population is expected to reach nine billion, calling for a 70% increase in food production. Underutilized regions may only provide around 10% to this increased output, with the rest being fulfilled by the intensification of current production. The use of state-of-the-art technological solutions to boost agricultural productivity is still essential in this setting. Current agricultural intensification strategies involve considerable energy inputs, yet the market wants high-quality food. The advent of robotics and RAS promises to revolutionize the business world. These technologies will have a large effect on enormous parts of the economy with weak productivity, such as agriculture (food production from the farm to the retail shelf).

AI is assisting farmers with automating their farming operations, as well as moving them toward precision agriculture for improved crop output and quality while using fewer resources to do so. With the development of future technological advancements, businesses will be able to enhance goods or services based on AI in the future, such as training data for agriculture, autonomous drones, and automated manufacturing machines, enabling world governments to address challenges related to food production for a growing population in the future. In the agriculture industry, AI will be beneficial and effective because it can maximize resources utilization and efficiency, and to a significant extent, solve the problem of oversaturation of both resources and labor. In addition to these applications, this technology will also play an important role in the research and development of agriculture and horticultural practices in the future.

Acknowledgements The authors are grateful to the reviewers for their comments and valuable suggestions.

Author Contributions A.R and Y.H conceptualized the book chapter. A.R. completed the original draft preparation. All the coauthors edited and reviewed the chapter. All authors have read and agreed to the published version of the book chapter.

Funding The research was supported by Key R&D program of Jiangsu Provincial Government (BE2021340) and the Priority Academic Program Development of Jiangsu Higher Education Institutions (PAPD-2018–87).

Not applicable.

Ethics approval and consent to participate

Institutional Review Board Statement Not applicable.

Informed Consent Statement Not applicable.

Data Availability Statement Data available on reasonable request.

Conflicts of Interest The authors declare no conflict of interest.

References

Abawi GS, Widmer TL (2000) Impact of soil health management practices on soilborne pathogens, nematodes and root diseases of vegetable crops. Appl Soil Ecol 15:37–47

Abdullahi HS, Mahieddine F, Sheriff RE (2015) Technology impact on agricultural productivity: A review of precision agriculture using unmanned aerial vehicles. In: Wireless and Satellite Systems: 7th International Conference, WiSATS 2015, Bradford, UK, July 6–7, 2015. Revised Selected Papers 7. Springer, pp 388–400

Ahir K, Govani K, Gajera R, Shah M (2020) Application on virtual reality for enhanced education learning, military training and sports. Augment Hum Res 5:1–9. https://doi.org/10.1007/s41133-019-0025-2

Aitkenhead MJ, McDonald AJS, Dawson JJ et al (2003) A novel method for training neural networks for time-series prediction in environmental systems. Ecol Model 162:87–95. https://doi.org/10.1016/S0304-3800(02)00401-5

Aryalekshmi BN, Biradar RC, Chandrasekar K, Ahamed JM (2021) Analysis of various surface energy balance models for evapotranspiration estimation using satellite data. Egypt J Remote Sens Sp Sci 24:1119–1126

Asmamaw DK, Janssens P, Desse M et al (2021) Deficit irrigation as a sustainable option for improving water productivity in Sub-Saharan Africa: the case of Ethiopia. A critical review. Environ Res Commun 3:102001

Aubry C, Papy F, Capillon A (1998) Modelling decision-making processes for annual crop management. Agric Syst 56:45–65

Balleda K, Satyanvesh D, Sampath N, et al (2014) Agpest: an efficient rule-based expert system to prevent pest diseases of rice and wheat crops. In: 2014 IEEE 8th International Conference on Intelligent Systems and Control (ISCO). IEEE, pp 262–268

Behrens T, Förster H, Scholten T, Steinrücken U, Spies ED, Goldschmitt M (2005) Digital soil mapping using artificial neural networks. J Plant Nut Soil Sci 168(1):21–33. https://doi.org/10.1002/jpln.200421414

Bhagyalaxmi K, Jagtap KK, Nikam NS et al (2016) Agricultural robot: irrigation system, weeding, monitoring of field, disease detection. Int J Innov Res Comput Commun Eng 4:4403–4409

Bilgili M (2011) The use of artificial neural networks for forecasting the monthly mean soil temperatures in Adana, Turkey. Turk J Agric For 35(1):83–93. https://doi.org/10.3906/tar-1001-593

Brazeau M (2018) Fighting weeds: can we reduce, or even eliminate, herbicides by utilizing robotics and AI. Genetic Literacy Project, North Wales

Chandel AK, Tewari VK, Kumar SP, et al (2018) On-the-go position sensing and controller predicated contact-type weed eradicator. Curr Sci 114:1485–1494. https://doi.org/10.18520/cs/v114/i07/1485-1494

Chang C-L, Lin K-M (2018) Smart agricultural machine with a computer vision-based weeding and variable-rate irrigation scheme. Robotics 7:38. https://doi.org/10.3390/robotics7030038

Chang DH, Islam S (2000) Estimation of soil physical properties using remote sensing and artificial neural network. Remote Sens Environ 74(3):534–544. https://doi.org/10.1016/S0034-4257(00)00144-9

Chethan CR, Tewari VK, Nare B, Kumar SP (2018a) Transducers for measurement of draft and torque of tractor-implement system—a review. Agric Mech Asia, Africa Lat Am 49:82

Chethan CR, Chander S, Kumar SP (2018b) Dynamic strength based dryland weeders-ergonomic and performance evaluation. Ind J Weed Sci. https://doi.org/10.5958/0974-8164.2018.00081.3

Dahikar SS, Rode SV (2014) Agricultural crop yield prediction using artificial neural network approach. Int J Innov Res Electr Electron Instrum Control Eng 2:683–686

Dai X, Huo Z, Wang H (2011) Simulation for response of crop yield to soil moisture and salinity with artificial neural network. Field Crops Res 121(3):441–449. https://doi.org/10.1016/j.fcr.2011.01.016

Datta A, Ullah H, Tursun N et al (2017) Managing weeds using crop competition in soybean [*Glycine max* (L.) Merr.]. Crop Prot 95:60–68

de Jong IH, Arif SS, Gollapalli PKR et al (2021) Improving agricultural water productivity with a focus on rural transformation. Irrig Drain 70:458–469

De Kimpe CR, Morel J-L (2000) Urban soil management: a growing concern. Soil Sci 165:31–40

Debaeke P, Aboudrare A (2004) Adaptation of crop management to water-limited environments. Eur J Agron 21:433–446

Dhanya VG, Subeesh A, Kushwaha NL, Vishwakarma DK, Kumar TN, Ritika G, Singh AN (2022) Deep learning based computer vision approaches for smart agricultural applications. Artif Intel Agric 6211–6229. https://doi.org/10.1016/j.aiia.2022.09.007

Dukes MD, Shedd M, Cardenas-Lailhacar B (2009) Smart irrigation controllers: how do soil moisture sensor (sms) irrigation controllers work? Univ Florida, Inst Food Agric Sci FL, USA, p 1

Dursun M, Ozden S (2011) A wireless application of drip irrigation automation supported by soil moisture sensors. Sci Res Essays 6:1573–1582

Eliades M, Bruggeman A, Djuma H et al (2022) Quantifying evapotranspiration and drainage losses in a semi-arid nectarine (*Prunus persica* var. nucipersica) field with a dynamic crop coefficient (Kc) derived from leaf area index measurements. Water 14:734

Eli-Chukwu NC (2019) Applications of artificial intelligence in agriculture: a review. Eng Technol Appl Sci Res 9:4377–4383

Elshorbagy A, Parasuraman K (2008) On the relevance of using artificial neural networks for estimating soil moisture content. J Hydrol 362:1–18

Entezari A, Wang RZ, Zhao S et al (2019) Sustainable agriculture for water-stressed regions by air-water-energy management. Energy 181:1121–1128
Fahad S, Hussain S, Chauhan BS et al (2015) Weed growth and crop yield loss in wheat as influenced by row spacing and weed emergence times. Crop Prot 71:101–108
FAO (2017) Food and Agriculture Organization of the United Nations. The state of food and agriculture leveraging food systems for inclusive rural transformation. FAO, pp 1–181
Gandhi M, Kamdar J, Shah M (2020) Preprocessing of non-symmetrical images for edge detection. Augment Hum Res 5:1–10. https://doi.org/10.1007/s41133-0190030-5
Gebregiorgis MF, Savage MJ (2006) Determination of the timing and amount of irrigation of winter cover crops with the use of dielectric constant and capacitance soil water content profile methods. South African J Plant Soil 23:145–151. https://doi.org/10.1080/02571862.2006.10634746
Gerhards R, Christensen S (2003) Real-time weed detection, decision making and patch spraying in maize, sugarbeet, winter wheat and winter barley. Weed Res 43:385–392
Griepentrog HW, Nørremark M, Nielsen H, Blackmore BS (2005) Seed mapping of sugar beet. Precis Agric 6:157–165. https://doi.org/10.1007/s11119-0051032-5
Griepentrog HW, Noerremark M, Soriano JF (2006) Close-to-crop thermal weed control using a CO_2 laser. In: Proceedings: CIGR World Congress, Agricultural Engineering for a Better World, Bonn, Germany, 3rd–7th September
Hanson B, Peters D, Orloff S (2000) Effectiveness of tensiometers and electrical resistance sensors varies with soil conditions. Calif Agric 54:47–50. https://doi.org/10.3733/ca.v054n03p47
Heisel T, Schou J, Christensen S, Andreasen C (2001) Cutting weeds with a CO_2 laser. Weed Res 41:19–29. https://doi.org/10.1046/j.1365-3180.2001.00212.x
Jani K, Chaudhuri M, Patel H, Shah M (2020) Machine learning in films: an approach towards automation in film censoring. J Data Inf Manag 2:55–64. https://doi.org/10.1007/s42488-019-00016-9
Jaramillo S, Graterol E, Pulver E (2020) Sustainable transformation of rainfed to irrigated agriculture through water harvesting and smart crop management practices. Front Sustain Food Syst 4:437086
Jesus J, Panagopoulos T, Neves A (2008) Fuzzy logic and geographic information systems for pest control in olive culture. In: 4th IASME/WSEAS International Conference on Energy, Environment, Ecosystems & Sustainable Development, Algarve, Portugal, June, pp 11–13
Jha P, Kumar V, Godara RK, Chauhan BS (2017) Weed management using crop competition in the United States: a review. Crop Prot 95:31–37
Jha K, Doshi A, Patel P, Shah M (2019) A comprehensive review on automation in agriculture using artificial intelligence. Artif Intell Agric 2:1–12. https://doi.org/10.1016/j.aiia.2019.05.004
Ji B, Sun Y, Yang S, Wan J (2007) Artificial neural networks for rice yield prediction in mountainous regions. J Agric Sci 145:249–261
Kamilaris A, Prenafeta-Boldú FX (2018) Deep learning in agriculture: a survey. Comput Electron Agric 147:70–90. https://doi.org/10.1016/j.compag.2018.02.016
Karimi Y, Prasher SO, Patel RM, Kim SH (2006) Application of support vector machine technology for weed and nitrogen stress detection in corn. Comput Electr Agric 51(1–2):99–109. https://doi.org/10.1016/j.compag.2005.12.001
Khan M, Haq N (2002) Wheat crop yield loss assessment due to weeds. Sarhad J Agric
Kim Y, Evans RG, Iversen WM (2008) Remote sensing and control of an irrigation system using a distributed wireless sensor network. IEEE Trans Instrum Meas 57:1379–1387. https://doi.org/10.1109/TIM.2008.917198
Kolhe S, Kamal R, Saini HS, Gupta GK (2011) An intelligent multimedia interface for fuzzy-logic based inference in crops. Expert Syst Appl 38:14592–14601
Kundalia K, Patel Y, Shah M (2020) Multi-label movie genre detection from a movie poster using knowledge transfer learning. Augment Hum Res 5:1–9. https://doi.org/10.1007/s41133-019-0029-y
Kuyper MC, Balendonck J (1997) Application of dielectric soil moisture sensors for real-time automated irrigation control. III International Symposium on Sensors in Horticulture 562:71–79

Lal H, Jones JW, Peart RM, Shoup WD (1992) FARMSYS—a whole-farm machinery management decision support system. Agric Syst 38:257–273

Lasanta T, Vicente-Serrano SM (2012) Complex land cover change processes in semiarid Mediterranean regions: an approach using Landsat images in northeast Spain. Remote Sens Environ 124:1–14. https://doi.org/10.1016/j.rse.2012.04.023

Levine ER, Kimes DS, Sigillito VG (1996) Classifying soil structure using neural networks. Ecol Model 92(1):101–108. https://doi.org/10.1016/0304-3800(95)00199-9

Li M, Yost RS (2000) Management-oriented modeling: optimizing nitrogen management with artificial intelligence. Agric Syst 65:1–27

Liou Y-A, Kar SK (2014) Evapotranspiration estimation with remote sensing and various surface energy balance algorithms—a review. Energies 7:2821–2849

Liu X, Liu Y, Wang Y, Liu Z (2022) Evaluating potential impacts of land use changes on water supply–demand under multiple development scenarios in dryland region. J Hydrol 610:127811. https://doi.org/10.1016/j.jhydrol.2022.127811

López EM, García M, Schuhmacher M, Domingo JL (2008) A fuzzy expert system for soil characterization. Environ Int 34:950–958

López-Granados F (2011) Weed detection for site-specific weed management: mapping and real-time approaches. Weed Res 51:1–11. https://doi.org/10.1111/j.1365-3180.2010.00829.x

Manivannan L, Priyadharshini MS (2016) Agricultural robot. Int J Adv Res Electr Electron Instrum Eng 5:153–156

Markuna S, Kumar P, Ali R, Vishwakarma DK, Kushwaha KS, Kumar R, Singh VK, Chaudhary S, Kuriqi A (2023) Application of innovative machine learning techniques for long-term rainfall prediction. Pure Appl Geophys 180(1):335–363. https://doi.org/10.1007/s00024-022-03189-4

Meybeck A, Redfern S (2016) Knowledge and information for sustainable food systems. In: Joint FAO/UNEP Workshop on Knowledge and Information for Sustainable Food Systems Rome (Italy) 10–11 Sep 2014. FAO/UNEP

Milberg P, Hallgren E (2004) Yield loss due to weeds in cereals and its large-scale variability in Sweden. F Crop Res 86:199–209

Mogili UMR, Deepak B (2018) Review on application of drone systems in precision agriculture. Proc Comput Sci 133:502–509. https://doi.org/10.1016/j.procs.2018.07.063

Möller J (2010) Computer vision—a versatile technology in automation of agricultural machinery. J Agric Eng 47:28–36

Montas H, Madramootoo CA (1992) A decision support system for soil conservation planning. Comput Electr Agric 7(3):187–202. https://doi.org/10.1016/S0168-1699(05)80019-5

Moran MS, Inoue Y, Barnes EM (1997) Opportunities and limitations for image-based remote sensing in precision crop management. Remote Sens Environ 61:319–346

Mruthul T, Halepyati AS, Chittapur BM (2015) Chemical weed management in sesame (*Sesamum indicum* L.). Karnataka J Agric Sci 28:151–154

Munirah MY, Rozlini M (2013) An expert system development: its application on diagnosing oyster mushroom diseases. In: 2013 13th International Conference on Control, Automation and Systems (ICCAS 2013). IEEE, pp 329–332

Murugesan R, Sivarajan S (2018) Industry 4.0 for sustainable development. In: Annual technical volume of mechanical engineering division board, vol 3

Nakai S, Yamada Y (2014) Development of a weed suppression robot for rice cultivation: weed suppression and posture control. Int J Electr Comput Electron Commun Eng 8:1736–1740

Neil Harker K (2001) Survey of yield losses due to weeds in central Alberta. Can J Plant Sci 81:339–342

Pagliai M, Vignozzi N, Pellegrini S (2004) Soil structure and the effect of management practices. Soil Tillage Res 79:131–143

Pandey K, Singh KG, Singh A (2023) Multi-sensors based smart nutrient reuse management system for closed soilless culture under protected cultivation. Comput Electron Agric 204:107495

Papageorgiou EI (2011) Learning algorithms for fuzzy cognitive maps—a review study. IEEE Trans Syst Man Cybern Part C (Appl Rev) 42(2):150–163. https://doi.org/10.1109/TSMCC.2011.2138694

Patel M, Virparia P, Patel D (2012) Web based fuzzy expert system and its applications–a survey. Int J Appl Inf Syst 1(7):11–15

Pedersen SM, Fountas S, Blackmore S (2008) Agricultural robots—Applications and economic perspectives. In: Service robot applications. IntechOpen

Pérez-Ortiz M, Gutiérrez PA, Peña JM, et al (2016) Machine learning paradigms for weed mapping via unmanned aerial vehicles. In: 2016 IEEE symposium series on computational intelligence (SSCI). IEEE, pp 1–8

Peruzzi A, Ginanni M, Fontanelli M et al (2007) Innovative strategies for on-farm weed management in organic carrot. Renew Agric Food Syst 22:246–259

Pilarski T, Happold M, Pangels H et al (2002) The demeter system for automated harvesting. Auton Robots 13:9–20

Plant RE, Horrocks RD, Grimes DW, Zelinski LJ (1992) CALEX/Cotton: an integrated expert system application for irrigation scheduling. Trans ASAE 35(6):1833–1838. https://doi.org/10.13031/2013.28803

Qualls RJ, Scott JM, DeOreo WB (2001) Soil moisture sensors for urban landscape irrigation: effectiveness and reliability 1. JAWRA J Am Water Resour Assoc 37:547–559. https://doi.org/10.1111/j.1752-1688.2001.tb05492.x

Rajpal A, Jain S, Khare N, Shukla AK (2011) Microcontroller-based automatic irrigation system with moisture sensors. In: Proceedings of the International Conference on Science and Engineering. pp 94–96

Rao AN, Wani SP, Ladha JK (2014) Weed management research in India-an analysis of past and outlook for future

Savitha M, UmaMaheswari OP (2018) Smart crop field irrigation in IOT architecture using sensors. Int J Adv Res Comput Sci 9:302–306. https://doi.org/10.26483/ijarcs.v9i1.5348

Saxena V, Gautam A (2012) Machine learning and artificial intelligence in food industry. Int Res J Modernization Eng Technol Sci 3(9):585–603

Shah G, Shah A, Shah M (2019) Panacea of challenges in real-world application of big data analytics in healthcare sector. J Data Inf Manag 1:107–116. https://doi.org/10.1007/s42488-019-00010-1

Shamshiri RR, Weltzien C, Hameed IA, J Yule I, E Grift T, Balasundram SK, Pitonakova L, Ahmad D, Chowdhary G (2018) Research and development in agricultural robotics: a perspective of digital farming. Int J Agric Biol Eng 11(4):1–11. https://doi.org/10.25165/j.ijabe.20181104.4278

Shekhar Y, Dagur E, Mishra S, Sankaranarayanan S (2017) Intelligent IoT based automated irrigation system. Int J Appl Eng Res 12:7306–7320

Siraj F, Arbaiy N (2006) Integrated pest management system using fuzzy expert system. In: Proceedings of knowledge management international conference & exhibition (KMICE), 6–8 June 2006 Legend Hotel Kuala Lumpur, Malaysia. Universiti Utara Malaysia, Sintok, pp 169–176. ISBN 9833282903

Song H, He Y (2005) Crop nutrition diagnosis expert system based on artificial neural networks. In: Third international conference on information technology and applications (ICITA'05), 2005 Jul 4, vol 1, pp 357–362. IEEE. https://doi.org/10.1109/ICITA.2005.108

Stigliani L, Resina C (1993) SELOMA: expert system for weed management in herbicide-intensive crops. Weed Technol 7(3):550–559. https://doi.org/10.1017/S0890037X00037337

Swanton CJ, Harker KN, Anderson RL (1993) Crop losses due to weeds in Canada. Weed Technol 7:537–542

Swanton CJ, Nkoa R, Blackshaw RE (2015) Experimental methods for crop–weed competition studies. Weed Sci 63:2–11

Tajik S, Ayoubi S, Nourbakhsh F (2012) Prediction of soil enzymes activity by digital terrain analysis: comparing artificial neural network and multiple linear regression models. Environ Eng Sci 29(8):798–806. https://doi.org/10.1089/ees.2011.0313

Tobal A, Mokhtar SA (2014) Weeds identification using evolutionary artificial intelligence algorithm. J Comput Sci 10(8):1355–1361

Tothill IE (2001) Biosensors developments and potential applications in the agricultural diagnosis sector. Comput Electron Agric 30:205–218

Tu M, Hurd C, Randall JM (2001) Weed control methods handbook: tools & techniques for use in natural areas

Van Henten EJ, Hemming J, Van Tuijl BAJ et al (2002) An autonomous robot for harvesting cucumbers in greenhouses. Auton Robots 13:241–258

van Mourik S, van der Tol R, Linker R et al (2021) Introductory overview: systems and control methods for operational management support in agricultural production systems. Environ Model Softw 139:105031

Vishwakarma DK, Kuriqi A, Abed SA, Kishore G, Ansari NA, Pandey K, Kumar P, Kushwaha NL, Jewel A (2023) Forecasting of stage-discharge in a non-perennial river using machine learning with gamma test. Heliyon 9(5):e16290. https://doi.org/10.1016/j.heliyon.2023.e16290

Varatharajalu K, Ramprabu J (2018) Wireless irrigation system via phone call & SMS. Int J Eng Adv Technol 8:397–401

Wang Y, Huang L, Wu J, Xu H (2007) Wireless sensor networks for intensive irrigated agriculture. In: 2007 4th IEEE Consumer Communications and Networking Conference. IEEE, pp 197–201

Wang X, Zhang M, Zhu J, Geng S (2008) Spectral prediction of Phytophthora infestans infection on tomatoes using artificial neural network (ANN). Int J Remote Sens 29(6):1693–1706. https://doi.org/10.1080/01431160701281007

World Bank (2022) World Bank Data: Annual Freshwater Withdrawals, Agriculture (% of Total Freshwater Withdrawal). https://data.worldbank.org/indicator/ER.H2O.FWAG.ZS?end=2016&start=2011&view=chart. Accessed 01 Oct 2022

Yang CC, Prasher SO, Landry JA, Ramaswamy HS (2003) Development of a herbicide application map using artificial neural networks and fuzzy logic. Agric Syst 76(2):561–574. https://doi.org/10.1016/S0308-521X(01)00106-8

Yang CC, Prasher SO, Landry J, Ramaswamy HS (2012) Development of neural networks for weed recognition in corn fields. Trans ASAE 45(3). https://doi.org/10.13031/2013.8854

Yong W, Shuaishuai L, Li L et al (2018) Smart sensors from ground to cloud and web intelligence. IFAC-Papers Online 51:31–38. https://doi.org/10.1016/j.ifacol.2018.08.057

Zhao Z, Chow TL, Rees HW et al (2009) Predict soil texture distributions using an artificial neural network model. Comput Electron Agric 65:36–48

Zhen X, Wengang Z, Changjun S, et al (2010) The measurement of soil water content using the dielectric method. In: 2010 World Automation Congress. IEEE, pp 241–245

Zongo B, Barbier B, Diarra A et al (2022) Economic analysis and food security contribution of supplemental irrigation and farm ponds: evidence from northern Burkina Faso. Agric Food Secur 11:1–18

Chapter 8
Weather Intelligence for Climate-Resilient Agriculture

V. A. Blessy, K. K. Shaheemath Suhara, Anu Varughese, M. R. Namitha, P. R. Anjitha Krishna, Ardra Wilson, and K. Sreeja

Abstract Climate-smart agriculture (CSA) is considered the best solution to cope with the adverse impacts of climate change by adopting resilient practices. By adopting CSA practices, farmers can adapt to climate changes through proper monitoring and forecasting of weather conditions. Therefore, this study aims to discuss different approaches and advancements in weather monitoring and forecasting systems relevant to CSA. Various weather monitoring systems based on remote sensing, IoT, and other techniques are discussed. Automatic weather stations are also explored as they help record, store, and retrieve real-time weather data. In addition, different regional climate models are presented that provide predictions of climate and regional-level climatic data for ungauged watersheds. This chapter also highlights different weather forecasting methods at various temporal and spatial scales, including Artificial Neural Networks (ANN) and Machine Learning (ML), as well as various soft computing tools. Overall, this study provides insights into the importance of weather monitoring and forecasting systems in agriculture and the potential for using modern technologies to address climate change impacts.

Keywords IoT · Climate models · ANN · Climate-smart agriculture · Weather monitoring

---

V. A. Blessy (✉)
ICAR—Indian Institute of Soil and Water Conservation, Regional Centre, Uthagamandalam, Tamilnadu, India
e-mail: vablessy@gmail.com

K. K. S. Suhara
Tamil Nadu Agricultural University, Coimbatore, Tamilnadu, India

A. Varughese · M. R. Namitha · A. Wilson · K. Sreeja
KCAET Tavanur, Kerala Agricultural University, Thrissur, Kerala, India
e-mail: anu.varughese@kau.in

P. R. A. Krishna
Dayalbagh Educational Institute (Deemed University), Agra, Uttar Pradesh, India

K. Pandey et al. (eds.), *Artificial Intelligence and Smart Agriculture*, Advances in Geographical and Environmental Sciences,
https://doi.org/10.1007/978-981-97-0341-8_8

8.1 Introduction

The world population has increased four times within a short span of around 100 years, from less than 2 billion in 1900 to approximately 8 billion today. It is predicted that it may reach around 11 billion by the end of this century (Adam 2021). Agricultural production needs to be increased accordingly to meet the requirements of the growing population. This becomes more difficult under the changing climate scenario. The Inter-governmental Panel on Climate Change (IPCC) reported that the average air temperature in the world increased by 0.74 ± 0.18 °C in the twentieth century (Sadiq and Qureshi 2010). In most cases, an increase in temperature, erratic rainfall, heat waves, floods, etc., affect crop yield and nutritional quality. However, in certain areas and crops, it may increase productivity due to factors such as increased temperature, higher carbon dioxide levels, and favorable rainfall patterns.

Due to the negative effects on ecosystems, agricultural productivity, water resources, and socio-economy on a global as well as regional basis, climate change poses a threat to modern society (Nazripouya et al. 2022). Therefore, agricultural practices need to be reoriented or modified to cope up with the effects of climate change. This adaptation is often referred to as climate-smart agriculture (CSA) (Lipper et al. 2018). The Food and Agricultural Organization (FAO) proposed the idea of CSA to meet the food demand or manage the food security of the growing population in the long run (FAO 2016; Lipper et al. 2018).

CSA helps to increase production based on two main objectives, viz., to attain sufficient production under the new scenario without causing deterioration of natural resources and to attain sustainable production with the intention of decreasing greenhouse gas emissions (Mizik 2021). Weather monitoring and forecasting are thus the major pillars on which the planning of climate-smart agriculture is to be constructed. Systematic and timely investigation of changes in atmospheric conditions is termed weather monitoring. Weather forecasting, on the other hand, involves using meteorological parameters to predict what is likely to happen within a few days. Proper weather monitoring helps in the suitable planning of farming operations and reduces input costs. Furthermore, real-time weather monitoring using sensors and automatic weather monitoring systems help in precise predictions and optimal application of inputs. (Marcu et al. 2019; Dayananda et al. 2022; Pandey et al. 2023).

Climate models are used for long-term predictions of various climate parameters based on earth-atmosphere interactions (Vautard et al. 2020). These models help in projecting climate scenarios for a longer period of time. Weather forecasting, on the other hand, provides short-term predictions based on meteorological parameters. It utilizes recent advances in science and technology to help farmers and agriculturists plan cropping patterns, schedule irrigation, control pest and disease incidence, and make decisions on field preparation, timely application of fertilizers, and chemical spraying (Elbeltagi et al. 2022; Kushwaha et al. 2022).

Advancements in weather monitoring and forecasting have been significant in recent years, with the development of technology, remote sensing, and satellite-based systems providing real-time weather information (Ullo and Sinha 2021). Additionally, machine learning (ML) and artificial intelligence (AI) techniques are being employed to enhance the accuracy and reliability of weather forecasting (Kumari et al. 2021). These advancements have improved the precision and reliability of weather information, which is critical for successful climate-smart agriculture.

In light of the above discussions, this study aims to investigate in detail the various advancements developed for weather forecasting and monitoring as part of climate-smart agriculture.

8.2 Weather Monitoring

The term weather stands for atmospheric conditions prevailing at a place during a short span of time. The purpose of weather monitoring is to continuously monitor environmental parameters such as temperature, dew point, humidity, wind speed and direction, and air pressure over a period of a few hours to several days. Apart from these, monitoring soil moisture, crop health, and livestock health are essential components in the real-time management of modern agricultural farms (Tabassum and Hossain 2018; Talaviya et al. 2020; Singh et al. 2022a, b) for irrigation scheduling, water management during critical crop growth stages, and improving farm productivity.

Although weather data and weather forecasts are available globally through satellite schemes and climate models, more reliable and accurate short-term localized data are still lacking for agriculture applications (Adoghe et al. 2017). Abrupt changes in weather parameters are expected from climate change impacts, which lead to frequent crop failures from floods, drought stress, and disease outbreaks, making farming more resource and energy-intensive. The efficiency of any irrigation practice fundamentally depends on knowledge of the time and quantity of irrigation (Hachimi et al. 2022). Timely prediction of diseases reduces yield loss and declines the chemical load on the environment from the farm sector (Akhter and Sofi 2022). Accurate and stable data acquisition is vital for deciding the harvesting time, farm product transmission to the market, and controlled environment agriculture (Kim et al. 2022).

The conventional weather monitoring system comprises a set of instruments, including a rain gauge (recording or non-recording type), wind vane, anemometer, thermometer, barometer, hygrometer, soil thermometer, and lux meter. These instruments are expensive and require skilled labor for their operation. Regular maintenance and data collection are also necessary, adding an extra daily task for the users. These stations may be located in fixed locations, such as airports or other locations with a high demand for weather information, or mobile, such as on ships or aircraft. Government agencies or private organizations often operate traditional weather stations, and the data collected by these stations are frequently used to produce weather forecasts and warnings.

Autonomous weather stations (AWS) are weather observation systems that use automated sensors to measure and report weather conditions. These systems are designed to operate continuously without the need for human intervention. AWS typically include sensors for measuring temperature, humidity, atmospheric pressure, wind speed direction, and precipitation. Some systems may also include sensors for measuring solar radiation, soil moisture, and other weather-related variables. AWS have various applications, including agriculture, aviation, construction, energy, and transportation. Moreover, AWS can generate essential data that is useful for weather forecasting, monitoring climate change, and various other applications.

The factors contributing to the increasing popularity of autonomous weather stations (AWS) compared to traditional weather monitoring systems can be summarized as follows:

1. Accuracy and reliability: AWS use automated sensors to measure and report weather conditions, which can be more accurate and consistent than traditional weather stations that rely on human observers.
2. Continuous operation: AWS can operate continuously without the need for human intervention, which can be especially useful in remote or hazardous locations.
3. Cost-effectiveness: It is more cost effective than traditional weather stations, as they do not require the same level of human resources and can operate continuously without the need for breaks or shift changes.
4. Ease of installation and maintenance: AWS can be easier to install and maintain than traditional weather stations, as they do not require a power source or regular maintenance from human technicians.
5. Wide range of applications: AWS can be used in a variety of applications, including agriculture, aviation, construction, energy, and transportation.

8.2.1 AWS and Its Recent Advancements

An AWS is an upgraded self-powered version of traditional weather station comprising meteorological sensors, data logger, controller system (to establish connectivity), communication system, software (data analysis by human–computer interaction) and battery-powered by solar/wind energy. The AWS integrates the measurement and transmission of weather data of land, atmosphere, or ocean for research and other industrial applications. It allows accessing information about real-time weather changes in remote locations more frequently with more reliability and reduced energy consumption by making use of any wireless communication technologies (Wi-Fi, cellular networks, Bluetooth, satellites, wide area networks, etc.). Ioannou et al. (2021) described different categories of AWS, as shown in Table 8.1.

The most common infrastructure of an AWS in an agricultural system stores data in internal memory and uses General Packet Radio Services (GPRS) for data transfer, which relies on network availability provided by an operator. The system will also have an external memory, which allows for downloading of sensor data at prescribed

Table 8.1 Categorization of automatic weather stations by World Meteorological Organization

S. No	Category	Concerned parameters
1	Light AWS	Precipitation, air temperature
2	Basic AWS	Air temperature, relative humidity, wind speed, wind direction, precipitation, atmospheric pressure
3	Extended AWS	Additional to basic AWS, solar radiation, sunshine duration, soil temperature, evaporation
4	AWS with automation of visual observations	Additionally, cloud base height, present weather

time intervals to a computer during the maintenance of the station (Munizaga et al. 2016).

Cloud computing provides possibilities for securely storing and processing whole data from AWS in a single platform. Cloud computing platforms are applications using powerful servers (physical or virtual machines) which provide worldwide accessibility to authenticated users through the Internet, overcoming the limitations of managing multiple systems (server, database, etc.). A user can use a computer or a selected processor (hardware-based) for logging into the cloud platform. IoT enables interaction between the cloud and other devices in weather stations for data collection and processing or allows the user to control the devices remotely (Mullapudi et al. 2020). In order to monitor the microclimate at an agricultural farm, Kohila and Raja (2019) designed and tested an effective, affordable, and power-efficient cloud-based weather station. With this technology, air quality monitoring facilities can also be expanded. AWS powered by renewable sources is also presented in different studies. The climate station made by Valencia et al. (2022) observed weather data as well as the charging status of the power source and electricity generated by the photovoltaic system using open-source hardware. Leelavinodhan et al. (2021) developed and validated energy neutral weather monitoring station for wind data collection by employing an asynchronous optimization algorithm, which was reported to reduce energy use by 60%.

The data from a weather station could be combined with image processing techniques and machine learning without the requirement of an Internet facility (through edge analytics). It helps predict rainfall to decide irrigation schedules for rural farmers in developing countries without network connections (Ambildhuke and Banik 2022). The machine learning (ML) technique is based on the information gained from the training of dataset (past experience) and used to teach a machine which is then used for data prediction. Internet of Things (IoT) technology collects and transmits data through device networks and Internet connectivity, whereas ML aims at making wise decisions for better management of agriculture systems. The capability of ML mainly depends on the right choice of algorithm and classifier.

8.2.2 IoT-Based Weather Monitoring System

A typical weather monitoring system in smart farms utilizes a network of sensors (WSN-wireless sensor network) for getting the update on soil moisture, susceptibility to rain, humidity, and temperature. The system then analyses and transfers the database to a user at a remote location. Many research papers are available on sensors-aided automatic irrigation systems for smart water management at agriculture farms.

The IoT-based weather monitoring technique collects, monitors, and controls environmental parameters, including rainfall, temperature changes, soil moisture, relative humidity and CO_2 level, etc., and sends the information to a web page via the Internet utilizing cutting-edge electronic sensors. The data is then stored and plotted as graphical statistics for real-time weather monitoring. Inaccurate definitions of changes in the climate system are common, but using an IoT module, we can characterize more approximate changes in an environment and update them in the cloud (Islam 2019). IoT-based automatic weather stations are characterized by reliability, quality control, convenient transmissions, interoperability through standard protocols, and flexible scalability.

8.2.2.1 System Architecture

The discussion in this section focuses on the hardware and software development components of the IoT approach. The hardware development process involves building circuits and creating a prototype, while the software portion includes IoT coding, circuit schematics, circuit simulation, and data collection (Pauzi and Hasan 2020).

The general hardware components used in the existing weather monitoring systems are a power supply module, microcontroller (Arduino Uno, Node MCU, Raspberry Pi, etc.), Wi-Fi module, LCD with adapter module, sensors, IoT application (Thing Speak, Google Sheet with IFTTT platform, etc.). The system architecture of an IoT-based weather monitoring system used by Prasanna and Dubey (2020) is shown in Fig. 8.1.

Microcontroller: Microcontrollers like Arduino UNO, Node MCU, Raspberry Pi, etc., are used to create the alarm system that provides warning and alerting. The microcontroller, the main processing unit, controls the system implementation. For example, the Arduino board is an open-source physical computing platform that transforms analog data produced by the sensor into digital data. It then updates the data online through a Wi-Fi module attached to it (Islam 2019). Patil and Rane (2016) developed Raspberry Pie (ARM11)-based microcontroller program using Python coding for monitoring weather parameters and transmitting instant information to users through google cloud.

Power supply module: The system can be powered by USB or an adapter.

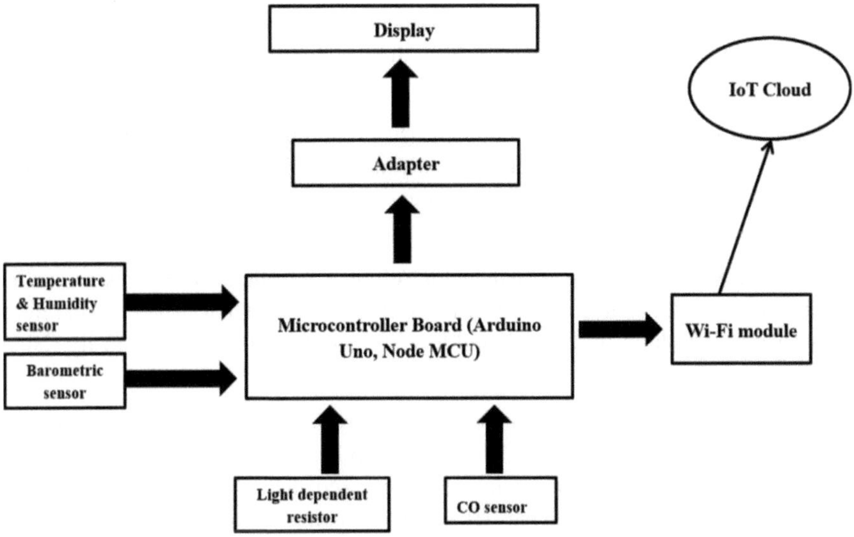

Fig. 8.1 System architecture of an IoT-based system (Prasanna and Dubey 2020)

Wi-Fi module: The weather data is sent to the IoT cloud and can be viewed from anywhere by connecting to the Internet. A Wi-Fi module enables any microcontroller to connect to a Wi-Fi network.

Sensors: Sensor devices are strategically placed to measure and collect data in order to predict the behavior of a specific area of interest. In weather monitoring systems, sensors are used to measure the weather parameters like temperature, humidity, carbon monoxide (CO) levels, rain, etc., as analog data. This analog data is transformed into digital data by the controller module. The common sensors used in weather monitoring are temperature-humidity sensors, CO sensors, barometric pressure sensors, raindrop modules, air quality sensors, Light Dependent Resistor (LDR), etc.

Temperature-humidity sensor: This sensor measures temperature and humidity and tracks weather information. The DHT11 is a highly efficient and low-cost sensor (Prasanna and Dubey 2020). DHT11 is suitable for readings of 0–50 °C temperatures with an accuracy of 2 °C and readings of 20–80% humidity with an accuracy of 5% (Girija et al. 2018).

Carbon Monoxide (CO) sensors: The CO sensors are suitable for detecting CO concentrations in the atmosphere. Girija et al. (2018) used the MQ-6 sensor to acquire CO data. The MQ-6 is a CO sensor with a quick response time and can detect CO concentrations in air between 20 and 2000 ppm.

Air quality sensor: It is a sensor that detects the concentration of various gases, such as smoke, oxygen, ammonia, in the atmospheric air.

Barometric pressure sensor: This sensor collects the barometric pressure of the area. The BMP180 pressure sensor was used in the weather monitoring system setup by Prasanna and Dubey (2020).

Raindrop Module: This sensor module measures rainfall intensity and detects rain. It has both analog and digital output. When the moisture limit is exceeded, this module analyses the moisture using an analog output pin and produces a digital output (Islam 2019).

LDR: It is used as a light-sensitive detector. A light-controlled variable resistor called an LDR alters its resistance in response to the amount of light that strikes it. Therefore, resistance will be high in the dark and low in the case of low light intensity (Islam 2019).

Data display: The LCD, OLED display, or the Blynk app can be used by the user to access or display the sensor data. The Blynk app is an IoT utility that uses wireless connectivity to display sensor data. The processed data stored in the cloud (IoT platforms like 'If This Then That' (IFTTT) and Thing Speak, for example) can be available to users using mobile phone applications like the Blynk app or LCD or any Buzzer/alarm/ LED display or can be saved in google sheets format.

8.2.2.2 Applications of IoT in Weather Monitoring

The microcontroller collects the sensor data and uploads it to the IoT applications like the Thing Speak website, IFTTT platform, etc., for weather data analysis. IoT applications enable real-time data visualization, aggregation, storage, and analysis in the cloud.

Thing Speak is an open-source application and programming interface for the IoT that allows users to save and retrieve data from things through a local area network or the Internet. The development of sensor recording software, position tracking software, and a social network of things with status updates is made possible by Thing Speak (Girija et al. 2018). Thing Speak analyses and presents the uploaded data using MATLAB numerical computing software. 'If This Then That' (IFTTT) is one of the IoT platforms that can store data in the Google sheet format from any IoT device for easy data analysis.

IoT makes sensor devices much smarter by enabling communication and sharing monitoring information. Through Wi-Fi facility and WSN, IoT provides a simple user interface for displaying, analyzing, and decision-making for distant farms; anything available anywhere without space or time limitations (Akhter and Sofi 2022). IoT-based weather monitoring systems are intelligent options that uneducated farmers could even operate through Android mobile connectivity (Jawad et al. 2017; Khanna and Kaur 2019). Mobile applications providing accurate weather data for up to 14 days with expert advice for farming are available now (Devapal 2019). Low Power Wide Area Networks (LPWAN) and long-range WAN (LoRaWAN) are techniques widely used in IoT applications in agriculture for weather monitoring (Soy and Dilay 2021). Zigbee-based smart sending platforms are also common in weather station systems for smart farms (Haefke et al. 2011). Real-time monitoring of soil moisture, humidity, temperature, and/or quality of irrigation water is used for turning on/off a motor using Global System for Mobile Communication (GSM) or GPRS modules and microcontrollers (Dursun and Koc 2013; Pavithra and Srinath 2014; Udit and Vidhi 2020).

The availability of affordable and simple monitoring systems, which enable computer programming using a desktop or television, could improve remote accessibility to farms. Tabassum and Hossain (2018) designed a Raspberry Pi-based monitoring system for a greenhouse. Jagatheesan and Janaki (2022) proposed an IoT-based weather monitoring system for farms encompassing the application of a proteus simulation tool, a sensor network for temperature, pressure, rain and soil moisture, a Raspberry Pi computer, and Think Speak platform for the prediction of appropriate conditions for crop growth. An agile-based, portable, robust, low cost, and easy-to-use system was developed by Faid et al. (2022) by applying AI and IoT for monitoring multiple agriculture parameters (temperature, humidity, and pressure) and temperature forecasting with cloud server support for data backup. This system supported farmers and researchers (users) with a web-based graphical user interface (GUI) for crop management.

8.2.3 Remote Sensing Applications in Weather Monitoring

Remote sensing refers to the process of acquiring information about objects or areas using devices located at a distance. With advancements in sensor technology and digital processing, remote sensing has evolved from interpreting aerial images to analyzing satellite imagery, and from local area studies to global assessments.

Meteorology was the first scientific field to utilize space technology, with satellites providing details about the condition of the atmosphere to support weather forecasts. Today, remote sensing is a crucial input for data-driven smart agriculture. Remote sensing devices can monitor fields over time and collect data that can be used to make farming decisions. For example, they can help identify ideal weather conditions for growing crops. Weather data and other field data collected via various sensors provide stakeholders with much of the information they need during cultivation. Additionally, remote sensing data plays a significant role in precision agriculture, including irrigation scheduling, understanding plant stress, pest and disease infestation, measuring crop yield, and more. Aerial sensors, such as those on drones, can collect data about a specific field area.

8.2.3.1 Weather Data Collection Systems

1. *Weather balloons and weather stations*: Weather balloons and weather stations have historically been used to collect weather data. However, weather stations

have very few measuring options due to their fixed location and closeness to the earth's surface. On the other hand, weather balloons may soar far higher but have the drawback of being uncontrollable.

2. *Satellites*: Satellites are a potential additional source for weather data collection. They can gather information on water vapor and the development of clouds. Even though the primary objective of meteorological satellites is to detect the development and movement of clouds and storm patterns, they are also used for understanding snow cover, drought, atmospheric composition, etc. They are usually ineffective in monitoring temperature, humidity, or wind. Furthermore, clouds are an obvious barrier that prevents satellites from taking data below them. A gap in observational data often occurs since satellites can observe only a specific area of the earth's surface at a time.
3. *Radiosondes*: These are battery-powered systems used for weather monitoring and carried by weather balloons. The use of radiosondes is one way to close the data gap. Radiosondes are frequently used by National Meteorological and Hydrological Services (NMHS), but they are not always the ideal choice for many reasons. Helium, a naturally occurring gas that is becoming increasingly limited, is used to fill the equipment, which can be expensive to produce. Additionally, the wind can carry the devices up to 250 km from their launch location.
4. *Aircraft*: Weather data is also gathered by aircraft. Even though they are relatively expensive, they are frequently employed during storm events. However, a pilot sent into a storm also carries a significant danger. A meteorological drone is a better option for this.
5. *Weather Drones*: These drones are designed to gather weather data. The boundary layer, the lowest part of the atmosphere, is where they fly. They have specialized sensors that let them gather data on the atmosphere's temperature, humidity, and wind. The information gathered greatly enhances weather forecasting models. In addition, since they can maneuver better and endure rapid wind changes, weather drones are better suited to gather vertical data. Meteorologists will be able to provide far more precise hyper-local forecasts and now casts as a result of the weather forecast models' substantially enhanced accuracy.
6. *Meteodrones*: These drones, also known as the future mobile weather stations, were developed in Switzerland. Local factors influence meteorological occurrences, yet there is essentially little data accessible, especially for the lower atmosphere (approx. 2 km from the earth's surface). Future access to this data will be made possible by the meteodrones. In contrast to probes, which take readings in a very unpredictable manner due to their flying direction, weather drones can be manually managed close to where they are being used or from a distance. The drones can then be directed to fly locally or to locations that are far away.

8.2.3.2 Satellites—the Most Common Form of Weather Data Collection

Satellites offer meteorologists a data source that no other existing source can match for equivalent spatial and temporal coverage. One of the first civilian applications of satellite remote sensing was weather monitoring and forecasting. TIROS-1, launched by the National Aeronautics and Space Administration (NASA) in April 1960, was the first weather satellite to be regarded as successful. In order to continuously track global weather changes, other weather satellites were launched into near-polar orbits. Now it is possible to track the development and movement of weather systems continually (OECD 2014).

The satellite-based data regarding the earth-atmosphere system and its constituent parts have improved our knowledge and comprehension of the processes and dynamics within the earth-atmosphere system. In order to determine atmospheric temperature, winds, moisture, and cloud cover, two generations of Geostationary Operational Environment Satellite (GOES) satellites have been launched. On a worldwide scale, operational satellite systems regularly deliver useful data on atmospheric characteristics (Theis and Bendix 2011).

Disaster monitoring using remote sensing technology is another area relevant for weather monitoring in agriculture. Establishing the timeline, carrying out sensor planning, and building a representation model to monitor the information is essential in remote sensing-based weather monitoring, by which remote sensing data-processing techniques can extract information about disasters. However, meteorological disasters are highly unpredictable and difficult to monitor and predict. Despite the effectiveness of various remote sensing techniques, the process of monitoring meteorological disasters still has several shortcomings. Therefore, future studies in sensor development, information on representation models, multi-source data fusion, etc., are essential to providing a critical framework and direction for meteorological disaster process monitoring.

8.2.3.3 Impact of Remote Sensing on Smart Farming

1. *Planning the crop, dates of planting, and appropriate time of harvest*: Farmers may better plan their crop cycle with the use of remote sensing, which provides comprehensive information on optimum environmental and climatic conditions.
2. *Soil mapping and forecasting irrigation need*: Assessing soil health and moisture information on a particular plot of land relies heavily on soil mapping. It can be used to determine the best crop to be cultivated in the area, its type, and the amount of moisture it needs at various phases of development.
3. *Crop health assessment and yield prediction*: Crop health in a particular field may be continuously monitored to spot possible risks and respond to them before they can harm. It also acts as a trustworthy way to calculate the cultivation cycle's yield output and choose the best time to harvest based on the weather.

4. *Keeping a virtual eye on the farm*: With the help of remote sensing, all stakeholders can keep track of the crop as it moves from one stage to another, maintaining the standard and consistent quality needed by the operation.
5. *Monitoring drought and weather-related dependencies*: Satellite data can be used to monitor the weather trends in a certain area continuously.
6. *Evaluating historical land quality and degradation mapping*: Knowing the past performance of a piece of land is useful not just for farmers but also for banks and other lenders who need to allocate loans or determine crop insurance coverage.
7. *Protecting crops against pests and diseases*: Real-time alerts help rapidly identify the major threats due to pests or diseases and handle them immediately.

8.3 Weather Forecasting

Weather forecasting refers to the prediction of the atmospheric conditions of a place during a certain period of time, by applying the principles of science and technology augmented by various statistical and empirical techniques. The quantitative climatic data of the past and the current state are collected and analyzed using scientific principles to project the atmospheric processes that may evolve in the future.

Weather forecasting is crucial and inevitable since sudden or gradual changes in climate may adversely affect day-to-day activities and even lead to natural disasters. Therefore, humankind has been curious to foresee the weather for centuries, and they have attempted to predict the weather by all possible means. A person who has observed the signs of nature can interpret the wind direction, aspects of the sky, cloud cover, temperature, etc., and foretell the weather. The recognition of weather patterns helped in the formation of different weather prediction methods.

The scientific study of climatic variations started with the invention of various instruments for measuring meteorological parameters. The invention of the mercury barometer in 1643 by the Italian physicist and mathematician Evangelista Torricelli paved the way for the beginning of such meteorological research. Since then, the field of weather prediction has seen many breakthroughs through the ideas put forth by a number of climatologists and meteorologists. Nowadays, weather forecasts are calculated by dynamic computer models that simulate the real atmosphere using mathematical equations.

Weather scientists utilize two basic types of forecasting: deterministic and probabilistic. The prediction of specific events that will occur at an exact location is deterministic forecasting. For instance, forecasting when a hurricane or tornado will strike. On the other hand, predicting an anticipated weather event that may arise in a particular region is called probabilistic forecasting, e.g., predicting a storm that may last a few days.

Weather forecasting entails three main steps: (i) observation and analysis of data, (ii) extrapolation of the data to make projections for getting the state of the atmosphere in the future, and (iii) estimation of specific climatic variables (e.g., temperature, rainfall, etc.).

8.3.1 Uses of Weather Forecasting

Weather forecasting has immense importance in our day-to-day life as it is influential enough to protect human life and property and improve health, safety, and economic prosperity. Sudden or gradual changes in weather greatly influence agriculture and the farming sector. Proper knowledge of the weather patterns enables the farmers to fix the best schedule for cropping and farming operations. Moreover, forecasting rainfall and drought events benefit human life as it provides a precautionary time to safeguard life and properties. Today, technology has been developed to the extent that farmers get their forecasts on their smartphones using advanced weather forecasting mechanisms.

The anticipation of severe or hazardous weather events such as tsunamis, tornados, severe thunderstorms, cyclones, aids in safeguarding life and helps to path out from becoming victims of natural calamities. Severe weather alerts can be broadcasted through various media, viz. radio, television, and Internet, using an emergency alert system. Moreover, the fluctuations in climate and weather highly influence the tourism industry. Therefore, the increase in profitability and safety of tourism can be ensured through weather predictions. By checking the weather forecasts, tourists are able to pick out appropriate destinations for the best travel experiences and to avoid the possible risks due to bad weather conditions.

8.3.2 Types of Weather Forecasting

8.3.2.1 Short-Range Forecasting

The prophesying of weather variables up to a period of 12 h is referred to as very short-range forecasting. Since the forecasted event may occur within a short time span, the availability of an efficient information broadcasting system is a prerequisite. The accuracy of such forecasts is found to be high. The forecasts done for a period of 12–72 h are known as short-range forecasting. The anticipated weather conditions of an area over the forthcoming two or three days will be informed to the users to provide them time to take proper precautionary and protective measures against the upcoming weather fluctuation. Short-range forecasting is broadcast by a wide set of media (newspapers, radio, TV, Internet, etc.) and can convey fundamental information to farmers. Therefore, the potential usefulness of short-range forecasts is high. However, the accuracy of short-range weather forecasts generally decreases as the forecaster attempts to foretell the weather farther ahead in time.

(a) **Medium-range forecasting**

The forecasting that lasts 3 days to 2 weeks is termed medium-range forecasting. Medium-range forecasts have the potential to influence the areas of business budgeting and development. However, insufficient attention to medium-scale forecasting can lead to major consequences for the economic management and sustainability of the company.

(b) **Long-range forecasting**

Forecasting weather for a time longer than four weeks is termed long-range forecasting. Long-term weather forecasts are essential to make strategic decisions within an organization to make sustainable use of resources in an optimal manner to cope with fluctuating weather conditions. By checking long-term forecasts, the organizations will be able to make proper predictions and plan to tackle the expected changes.

8.3.2.2 Methods of Weather Forecasting

1. Synoptic method: In the synoptic method, the current weather conditions are analyzed and linked to past scenarios with the assumption that the current scenario would act similarly to an analogous condition that prevailed in the past.
2. Statistical method: In the statistical method, statistical relations are made through regression equations to interrelate different weather variables, and the subsequent climate is predicted using this relation.
3. Numerical Weather Prediction (NWP) techniques: NWP model aims to report the temporal development of weather by means of solving prognostic equations governing various processes in the atmosphere and oceans (Kalnay 2003). In this system, the fluctuations in the atmosphere can be expressed by a series of mathematical relations based on physical laws ruling airflow, air pressure, and other parameters. The method is best suited for medium-range weather forecasts. In addition, these numerical equations can be used for the generation of global climate models.

8.3.3 Global Climate Models (GCMs)

GCMs are documentation of physical processes that imitate the exchange of energy and matter through ocean–atmosphere-land interactions. The GCMs, popularly known as climate models or general circulation models, are mathematical representations of the earth's atmospheric interactions and serve as suitable tools for predicting the climate behavior of a place. GCMs work based on the general theories of fluid dynamics and thermodynamics. The GCMs can be used as effective tools to estimate future weather drifts by identifying and analyzing the thresholds and fluctuations in the available models.

Climate models divide the earth's surface into 3-D grids or cells, and the transfers of matter and energy between grid cells are modeled. The size of the grid cell defines the model resolution; the smaller the grid cells, the greater the number of cells and the higher the model's resolution. In addition, climate models can incorporate a time element in minutes, hours, days, or years. The smaller the time step provided, the higher the temporal resolution. Therefore, more computational expertise is needed for models with smaller grid cells and shorter time steps.

The generated climate models are tested for accuracy by running the model against the past and comparing the results with observed weather conditions. This process is known as hind-casting. A model that will run during hind-casting can be used to project the future climate.

8.3.3.1 Regional Climate Models (RCMs)

RCMs are higher-resolution numerical climate prediction models that simulate the various physical processes and atmosphere–ocean interactions on a regional scale. It can be defined as a limited area model (LAM) used for research on the climatic fluctuations of a particular region on a finer scale.

8.3.3.2 Downscaling Models

Downscaling of climate change models is the method of making use of large-scale climate models to predict the climate at a smaller temporal and spatial scale for befitting the objective of local-level planning and analysis. The study of the impacts of climate change leads to the necessity of generating localized climate information on a finer scale. One common issue with using GCMs is that they typically have a spatial resolution of 100–200 km. However, for impact studies that require localized climate information, a higher resolution of around 1–10 km or less may be necessary (Lanzanteet al. 2018). This can create a challenge when accurately predicting and assessing the impacts of climate change on specific regions or localities. The outputs from the large-scale GCMsare to be refined to a suitable form for individual sites or local scale through downscaling models. The downscaling procedure is used for the creation of different scenarios, modification of input data for impact modeling, the evaluation of effects based on present climate, and analysis of proper adaptation responses (Wilby and Dawson 2013). Downscaling methods are of two distinguishable types: dynamical downscaling (DD) and statistical downscaling (SD).

a. **Dynamic downscaling**

In dynamic downscaling models, the local-scale climate is estimated by pairing a meso scale higher-resolution model or regional climate model (RCM) with a global climate model (GCM) (Gaur and Simonovic 2019). RCMsare climate models that cover a limited area having a finer spatial resolution (1–50 km grid spacing). Thus, RCMs can be used to simulate the interaction between large-scale weather patterns with the local terrains. The bio-physical interactions in the land-atmospheric interfaces within the grids are considered for generating a higher-resolution local climate. The boundary conditions obtained from GCMs are fed as input to RCM for estimating the climate at a finer scale. Dynamic downscaling methods are used to address the issue of low spatial resolution in global climate models (GCMs) by modeling climate variability within a GCM grid. This is achieved by taking into account the geophysical variability present within a GCM grid cell (Gaur and Simonovic 2019). By doing so, dynamic downscaling can provide more localized climate information that is necessary for impact studies.

b. **Statistical downscaling**

In the statistical downscaling method, statistical relationships are developed between the climate data of low resolution obtained from large-scale global climate models and locally observed climate data. Then, higher-resolution climate data to fine scale for the proposed region are estimated and projected from these statistical relationships. The following assumptions are to be made before going for statistical downscaling:

i. The predictor and predict variables chosen for the downscaling must have a strong relationship.
ii. The variable chosen as the predictor must be simulated accurately by the global climate model. Otherwise, the downscaling model will carry over the error.
iii. The predictor variable must manage to represent the fluctuations in climate change in the predict and variable adequately. For instance, pressure above sea level is found to be a very good predictor of surface temperature. Yet, the use of pressure above sea level as a lone predictor of temperature can lead to unrealistic temperature values (Benestad et al. 2008).

The statistical relationships obtained as a result of statistical downscaling of global climate models can vary for different locations within a GCM grid due to differences in background climate, effects of the local landscape, and effects of local urbanization (Lowry 1977).

Statistical downscaling methods are classified into two based on the procedure of downscaling as either the PerfectProg or the Model Output Statistics (MOS) model. The relationship between predictor and predict and variables is obtained using observed predictors in the PerfectProg model. In contrast, the data obtained from GCM is directly used for downscaling in the MOS model. Sometimes, the raw data from GCM is bias corrected to remove the systematic biases in the GCM-modeled output before it is used for downscaling in the MOS models.

The PerfectProg models can be further classified into three categories: (1) regression-based, (2) weather pattern-based, and (3) weather generator-based methods (Wilby and Wigley 1997). The most popular PerfectProg model is the regression-based statistical downscaling model. In this model, a regression relationship is developed between predictor and predict and variables which in turn helps in predicting the climate to a finer scale. The method establishes a linear relationship (Eq. 8.1) between one large-scale climate indicator variable (x) and a local scale observed value of the same climate indicator variable (x) by assessing locally observed data and correlating it with GCM or RCM output.

$$y = \alpha + \beta x \tag{8.1}$$

The α and β values are estimated based on past time period and are used to derive downscaled climate projections for the future. A novel statistical downscaling model named physical scaling has been proposed recently, which attempts to build on the soundness of both statistical and dynamic downscaling.

8.3.4 Artificial Intelligence (AI)-Based Weather Forecasting

AI is a rapidly expanding area of computer science that has numerous uses in various aspects of weather forecasting. Artificial intelligence is the intelligence exhibited by robots, as opposed to the inherent intelligence of humans. Utilizing AI approaches helps speed up operations, make better use of data resources, and improve forecasting accuracy. AI models will first gather a big amount of data from numerous sources, and after that, the data will be analyzed using the models, which will require a lot of computing power. The difficulties that the weather algorithms demonstrate are related to prediction speed and accuracy.

A foundational element for the developing science of conceptual intelligence is thought to be soft computing. The supplements to soft computing models are fuzzy logic (FL), ML, ANN, probabilistic reasoning (PR), and evolutionary computation (EC). Different algorithms used in soft computing tools applicable to weather forecasting (Fig. 8.2) are explained in this section.

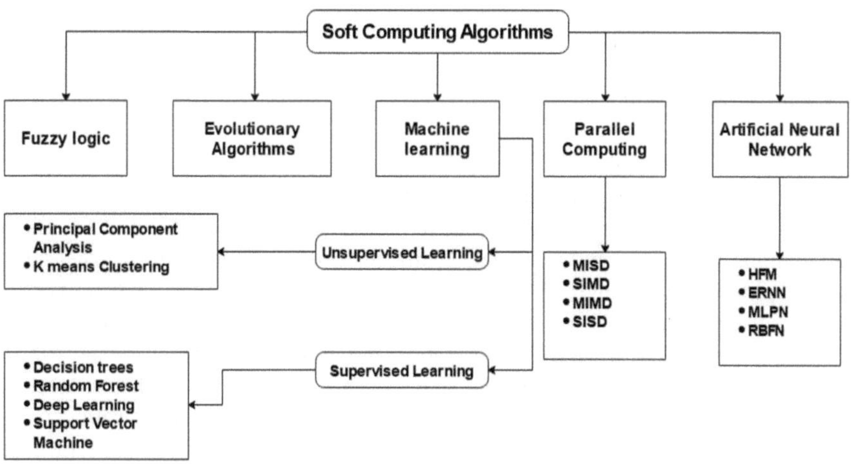

Fig. 8.2 Various algorithms used in soft computing

8.3.4.1 Machine Learning (ML)

ML is a field within AI that involves the study of the algorithms and mathematical models that computer systems assign to carry out particular tasks. For example, for time series-based weather forecasting, several researchers use various algorithms listed in Fig. 8.2 (Dritsas et al. 2022; Chavan and Momin 2017).

Machine learning can be divided into two categories based on the data's characteristics: supervised and unsupervised learning. In *supervised learning*, labeled data can be utilized as a training dataset to create a function that transforms inputs into outputs. In *unsupervised learning*, algorithms must choose other ways to divide or reduce a given dataset for future analysis because they lack labeled data to train on. The popular weather forecasting techniques in this group are principal component analysis and K-means clustering (PCA). The different algorithms that fall into these groups are listed in Fig. 8.2.

It should be noted that while the methods discussed in this chapter can be effective in various applications, certain techniques, such as RF, maybe more accessible for those with limited experience in machine learning. On the other hand, methods like DL may require a higher level of expertise to ensure proper implementation. Machine learning algorithms are a new approach mainly emphasized given in the twenty-first century. Different researchers found several applications of machine learning, such as the prediction of drought in an area and applications for assessing forest fires and assessment of deforestation, extreme events, and cyclones (Anandharajan et al. 2016). For time series-based weather forecasting by different algorithms, listed in Fig. 8.2. The predictive capacity of the ARIMA model is more while compared to other models for weather forecasting (Chavan and Momin 2017). In a similar study for prediction of weather classification by different machine learning techniques such as Support Vector Machine (SVM) and LR, for comparison, Random Forest (RF) and k-Nearest Neighbors (k-NN). This linear SVM model was observed with the highest accuracy (98.5%) for the prediction of weather (Dritsas et al. 2022).

8.3.4.2 Fuzzy Logic

Fuzzy logic is mathematical reasoning that attempts to address issues with an undefined and imprecise range of facts. It makes it simple to find a variety of accurate conclusions, essentially created to find the best answer to complicated situations using all the knowledge and input data available. The most effective solution finders are thought to be fuzzy logic. ANN simulates the human brain's neuronal network (a machine that can think like a human mind). So that, like the human brain, a computer, or other machines can learn new information and make judgments.

Based on different weather parameters, extrapolating future weather conditions is done using soft computing techniques; mainly, the neural network is trained using the different combinations of weather parameters. After training is done through the neural network, all predicted values of weather parameters apply to fuzzy logic algorithms, and future weather is predicted by different rules (Khajure and Mohod 2016).

8.3.4.3 Artificial Neural Network (ANN)

An artificial neural network is one of the best expert systems to develop a forecast model based on the time series of weather parameters like rainfall, temperature, wind speed, etc. The significant advantage of the ANN in weather forecasting is its capability to handle non-linear climate datasets compared to traditional techniques. The idea of ANN emerged from the human neural network system, consisting of several nodes and the linkages that connect the pair of nodes. Figure 8.3 shows a simple representation of ANN, where a_i represents the inputs into the neural networks, and w_i represents the weights. In the ANN, each input is multiplied by the corresponding weightage; the obtained products are added together and fed to the transfer function to get the output (plus bias) (Kakar et al. 2018).

Generally, two types of propagation methods are employed in the ANN: forward and backward propagation. In forward propagation, the data moves from the input layer to the output layer (one direction). In contrast, in backpropagation, initially, it runs forward propagation and compares the predicted output with the targeted output (bias), accordingly reassigning the weights for the forward run till it gives reliable results. The training algorithms are usually used to iterate for adjusting weights. In weather forecasting, backpropagation is generally adopted as it reduces biases.

The applications of ANN in forecasting different weather parameters were reported in the literature. For example, Baboo and Shereef (2010) developed a temperature-forecasting model based on backpropagation and recommended it as an alternative to traditional technologies. Furthermore, Lestari et al. (2021) developed a rainfall forecasting model using similar algorithms.

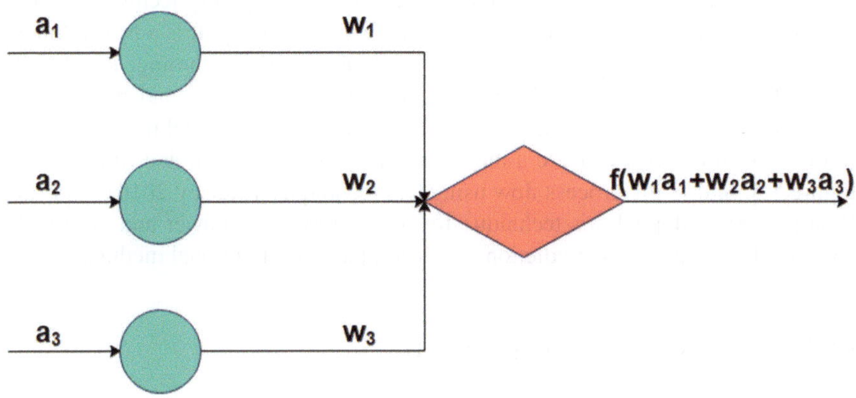

Fig. 8.3 Simple representation of ANN

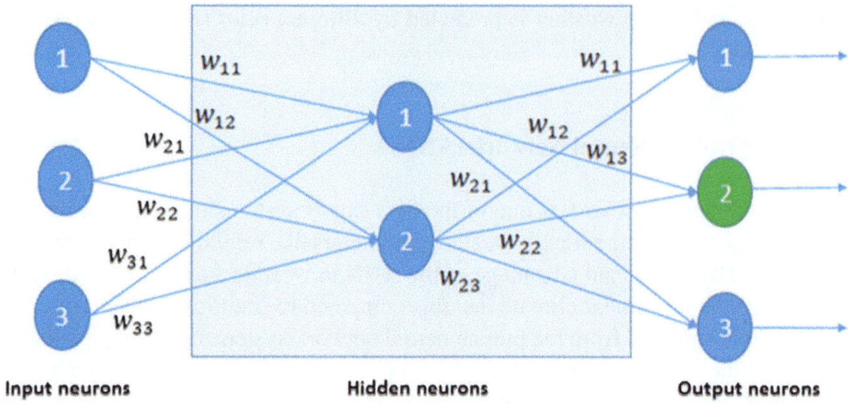

Fig. 8.4 Representation of MLPN (*Source* Aldana-Bobadila et al. 2018)

Different types of ANN techniques adopted for weather forecasting and reported in the literature are Multilayer Perceptron Networks (MLPN), Radial Basis Function Networks (RBFN), Elman Recurrent Neural Networks (ERNN), and the Hopfield model (HFM). Among these, MLPN has tremendous application in climate studies (Gardner and Dorling 1998; Hunasigi et al. 2022). It is a simple model that trains the data without prior assumptions about its distribution. Hence, it is the most commonly used neural network model. MLPN handles highly non-linear, even complex datasets. MLPN is a completely linked, feed-forward type ANN consisting of one input and output layer and, between them, one or more hidden layers (Fig. 8.4).

8.3.4.4 Evolutionary Algorithm (EA)

An evolutionary algorithm is a computer-based soft computing technology (optimization techniques) that mimics the natural evolution theory. It is a broad domain consisting of genetic algorithms, evolution strategies, programming, etc. It optimizes the solution through selection, recombination, mutation, and reproduction processes over a series of iterations. Examining the result and making the changes in the algorithm helps provide a more optimized output (Eiben and Smith 2015). Figure 8.5. Shows the process flow using GA techniques (McCoy 2010). Jursa and Rohrig (2008) adopted this technique for forecasting wind power and observed a considerable reduction in prediction errors compared to traditional methods.

8.3.4.5 Support Vector Machine (SVM)

It is a powerful tool for data mining (classifying the data). This algorithm can also be used for regression and identifying the pattern of the climate time series dataset. SVM stands out in situations when identifying rough and hidden patterns from large

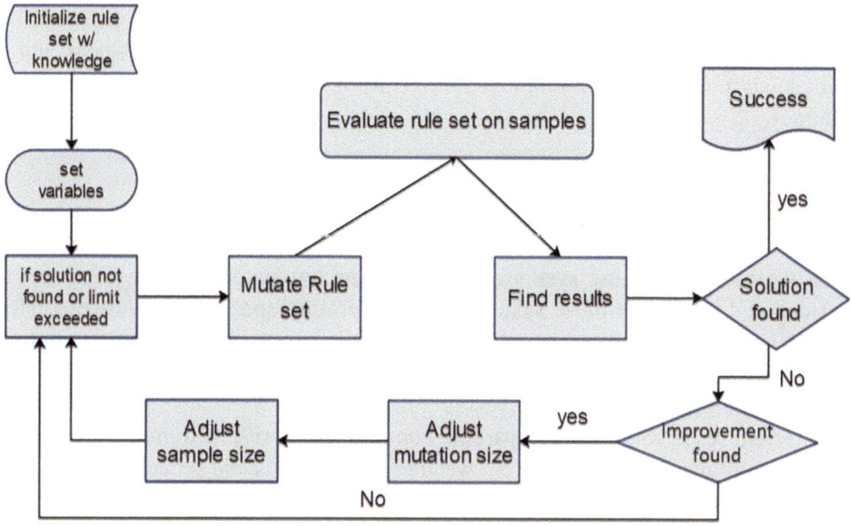

Fig. 8.5 Process flow using an evolutionary algorithm (McCoy 2010)

datasets. For example, Praba et al. (2018) used the SVM algorithm to know the probability of the occurrence of rainfall by training the weather datasets into heavy rain and not heavy rain conditions. The block diagram of the SVM algorithm used by Praba et al. (2018) is shown in Fig. 8.6.

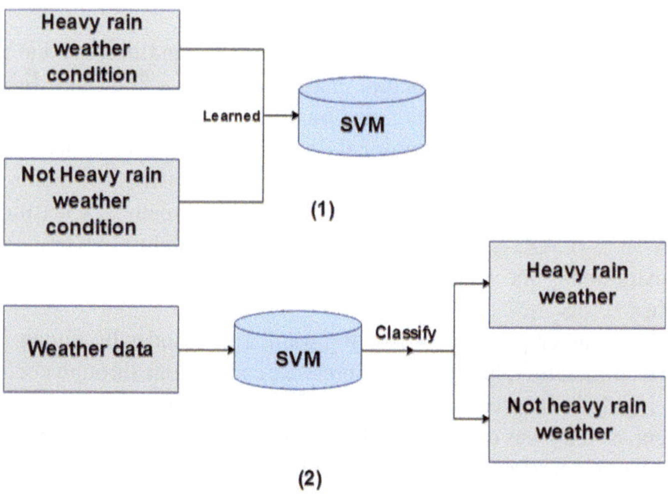

Fig. 8.6 Rainfall forecasting using SVM (Praba et al. 2018)

The SVM classifies the data into different classes based on the hyperplane by projecting the data into different dimensions. Kernel functions like linear, polynomial, sigmoid, etc., are employed in SVM to separate data with significant accuracy.

8.3.4.6 Parallel Computing

Most weather forecasting systems work by using computer systems. The important demand related to these systems is the operating speed of the processor for solving complex problems like numerical weather forecasting. Parallel computing is a process to improve the performance of computer systems by performing two or three instructions simultaneously.

Different researchers applied parallel computing algorithms to numeric weather prediction; it generally takes current weather observations and processes these data by different computer models to forecast future weather. For example, for flood modeling, one of the essential parameters is the spatial distribution of rainfall. Parallel algorithms (SIMD) are used along with two interpolation methods, Inverse Distance Weighing (IDW) and Ordinary Kriging (OK). The computational efficiency of parallel algorithms was compared with the serial algorithms, and it was found that the computational time of OK was 23% higher than the IDW method (Latifah and Nurhadiyatna. 2014).

8.3.4.7 Decision Trees (DT)

Decision trees are popular and effective technique for classification and prediction of climate. It is a type of tree structure that is similar to a flowchart. In a decision tree, each internal node represents a test on an attribute, each branch represents a test result, and each leaf node (or terminal node) represents a class label. This makes it easy to understand and interpret the decision-making process of the model. Each node may have two or more branches, depending on the method. Classification and Regression Trees (CART), for instance, produce trees with just two branches at each node. A binary tree is one such tree. A multiway tree is one that allows for more than two branches.

Predicting weather parameters like fog, rain, and thunder by giving input data like average temperature, humidity, and pressure can be done through decision trees. For observing different weather conditions, flood management, and natural disaster management, a study was done by combining machine learning algorithms and decision trees, where decision tree CART had a better classification accuracy (Gad and Hosahalli 2022). Similarly, for the correct prediction of solar irradiance, the accuracy of different models (machine learning along with three models, namely Logistic Regression (LR), Random Forest (RF), and DT for predicting solar irradiance was compared, in which decision tree has shown maximum accuracy (Singh et al. 2022b). DT can also be used to identify the hidden patterns in the metrological variables by which they can forecast the weather variables (Biruntha et al. 2022). For example, Saravanan (2022) conducted a study to predict the temperature using decision tree algorithms and found 92% prediction accuracy. In addition to this, time series studies related to weather parameters were also done by different researchers (Azari et al. 2022).

8.4 Conclusions

Climate change poses a significant challenge to agriculture and food security. CSA has emerged as a promising approach to mitigate the negative impacts of climate change and ensure sustainable agriculture. This chapter has provided an overview of different weather monitoring and forecasting systems relevant to climate-smart agriculture, including remote sensing, IoT, and automatic weather stations. Various weather forecasting methods, including ANN, ML, and soft computing tools, have also been discussed. While these advancements have increased the reliability and accuracy of weather monitoring and forecasting, there is still a need for further research and development in this area. Future studies could explore integrating different monitoring and forecasting techniques to improve the overall performance of climate-smart agriculture.

It is recommended that policymakers and researchers collaborate to develop effective strategies and policies to support adopting CSA practices at a larger scale, which will help enhance agricultural productivity and achieve food security. Climate change predictions at the regional level must be taken into account for the holistic development plan of any region. The engineers, administrators, and scientists working in this line must be updated with the latest climate-smart technologies and must work hand in hand for the benefit of the farmers. Impact-based forecasting and monitoring will help the farmers to plan well ahead and to act quickly.

References

Adam D (2021) How far will global population rise? Researchers can't agree. Nature 597(7877):462–465. https://doi.org/10.1038/d41586-021-02522-6

Adoghe AU, Popoola SI, Chukwuedo OM, Airoboman AE, Atayero AA (2017) Smart weather station for rural agriculture using meteorological sensors and solar energy. Proc World Congress Eng 1–4

Akhter R, Sofi SA (2022) Precision ariculture using IoT data analytics and machine learning. J King Saud Univ Comput Inf Sci 34(8):5602–5618. https://doi.org/10.1016/j.jksuci.2021.05.013

Aldana-Bobadila E, Kuri-Morales A, Lopez-Arevalo I, Rio-Alvarado AB (2018) An unsupervised learning approach for multilayer perceptron networks. Soft Comput. https://doi.org/10.1007/s00500-018-3655-2

Ambildhuke G, Banik BG (2022) IoT based portable weather station for irrigation management using real time paremeters. Int J Adv Comput Sci Appl 13(5):267–278

Anandharajan TRV, Hariharan GA, Vignajeth KK, Jijendiran R (2016) Weather monitoring using artificial intelligence. In: 2nd international conference on computational intelligence and networks (CINE), pp 106–111

Azari B, Hassan K, Pierce J, Ebrahimi S (2022) Evaluation of machine learning methods application in temperature prediction. Environ Eng 8:1–12

Baboo SS, Shereef IK (2010) An efficient weather forecasting system using artificial neural network. Int J Environ Sci Dev. https://doi.org/10.7763/ijesd.2010.v1.63

Benestad R, Hanssen-Bauer I, Chen D (2008) Empirical-statistical downscaling. World Scientific Publishing Company, Singapore, p 228

Biruntha S, Sowmiya BS, Subashri, R.,& Vasanth, M. (2022). Rainfall prediction using kNN and decision tree. In: International conference on electronics and renewable systems (ICEARS)

Chavan, G.,& Momin, B. (2017). An integrated approach for weather forecasting over Internet of Things: a brief review. In: International conference on I-SMAC (IoT in social, mobile, analytics and cloud) (I-SMAC), pp 83–88

Dayananda LPSSK, Narmilan A, Pirapuraj P (2022) An IoT based low-cost weather monitoring system for smart farming. Agric Sci Digest. https://doi.org/10.18805/ag.D-370

Devapal D (2019) Smart agro farm solar powered soil and weather monitoring systems for farmers. Mater Today Proc. https://doi.org/10.1016/j.matpr.2020.03.609

Dritsas E, Trigka M, Mylonas P (2022) A multi-class classification approach for weather forecasting with machine learning techniques. In: 17th international workshop on semantic and social media adaptation & personalization (SMAP). https://doi.org/10.1109/SMAP56125.2022.9942121

Dursun M, Koc F (2013) Linear switched reluctance motor control with PIC18F452 microcontroller. Turk J Electr Eng Comput Sci. https://doi.org/10.3906/elk-1110-67

Eiben AE, Smith JE (2015) Introduction to evolutionary computing. Natural Comput Ser. https://doi.org/10.1007/978-3-662-44874-8_3

Elbeltagi A, Raza A, Hu Y, Al-Ansari N, Kushwaha NL, Srivastava A, Vishwakarma DK, Zubair M (2022) Data intelligence and hybrid metaheuristic algorithms-based estimation of reference evapotranspiration. Appl Water Sci. https://doi.org/10.1007/s13201-022-01667-7

Faid A, Sadik M, Sabir E (2022) An agile AI and IoT-augmented smart farming: a cost effective cognitive weather station. Agriculture 12(1):35–65. https://doi.org/10.3390/agriculture12010035

FAO (2016) Climate change agriculture and food security. State of Food and Agriculture Report, FAO Rome. http://www.fao.org/3/a-i6030e.pdf

Gard I, Hosahalli D (2022) A comparative study of prediction and classification models on NCDC weather data. Int J Comput Appl 44(5):414–425

Gardner MW, Dorling SR (1998) Artificial neural networks (the multilayer perceptron)—a review of applications in the atmospheric sciences. Atmos Environ 32(14–15):2627–2636. https://doi.org/10.1016/S1352-2310(97)00447-0

Gaur A, Simonovic SP (2019) Introduction to physical scaling: a model aimed to bridge the gap between statistical and dynamic downscaling approaches. Trends Changes Hydroclim Var 199–273

Girija C, Harshalatha H, Shires AG, Pushpalatha HP (2018) Internet of things (IoT) based weather monitoring system. Int J Eng Res Technol NCESC

Hachimi CEL, Belaqziz S, Khabba S, Chehbouni A (2022) Towards smart big weather data management. Chem Proc 10(54):1–7. https://doi.org/10.3390/IOCAG2022-12240

Haefke M, Mukhopadhyay SC, Hartmut E (2011) A zigbee based smart sensing platform for monitoring environmental parameters. IEEE Instrum Measur Technol Conf. https://doi.org/10.1109/IMTC.2011.5944154

Hunasigi P, Jedhe S, Mane M, Shinde VP (2022) Multilayer perceptron neuralnetworkbased models for prediction of the rainfall and reference cropevapotranspiration for sub-humid climate of Dapoli, Ratnagiri District, India. ActaEcologica Sinica. https://doi.org/10.1016/j.chnaes.2022.09.004

Ioannou K, Karampatzakis D, Amanatidis P, Aggelopoulos V, Karmiris I. (2021) Low-cost automatic weather stations in the internet of things. Information

Islam MM (2019) Weather monitoring system using internet of things. Trends Tech Sci Res 3(3):65–69. https://doi.org/10.19080/TTSR.2019.03.555615

Jagatheesan M, Janaki G (2022) Weather monitoring system using IoT for smart farming. ECS Trans 107(1):17439. https://doi.org/10.1149/10701.17439ecst

Jawad H, Noordin R, Gharghan S, Jawad A, Ismail M (2017) Energy efficient wireless sensor networks for precision agriculture: a review. Sensors. https://doi.org/10.3390/s17081781

Jursa R, Rohrig K (2008) Short-term wind power forecasting using evolutionary algorithms for the automated specification of artificial intelligence models. Int J Forecast. https://doi.org/10.1016/j.ijforecast.2008.08.007

Kakar SA, Sheikh N, Naseem A, Iqbal S, Rehman A, Kakar AU, Kakar BA, Kakar HA, Khan B (2018) Artificial neural network based weather prediction using back propagation technique. Int J Adv Comput Sci Appl 9(8):462–470. https://doi.org/10.14569/ijacsa.2018.090859

Kalnay E (2003) Atmospheric modeling, data assimilation, and predictability. Cambridge University Press

Khajure S, Mohod SW (2016) Future weather forecasting using soft computing techniques. Proc Comput Sci 78:402–407

Khanna A, Kaur S (2019) Evolution of internet of things (IoT) and its significant impact in the field of precision agriculture. Comput Electr Agric 157(1):218–231. https://doi.org/10.1016/j.compag.2018.12.039

Kim J, Minagawa D, Saito D, Hoshina S, Kanda K (2022) Development of KOSEN weather station and provision of weather information to farmers. Sensors 22(6):1–27. https://doi.org/10.3390/s22062108

Kohila R, Raja K (2019) Arduino based agriculture weather monitoring and security system using GPRSL. Int J Adv Res Innov 7:82–85

Kumari M, Kumar A, Singh P, Singh S (2021) Multidisciplinary real-time model for smart agriculture based on weather forecasting using IoT, machine learning, big data and cloud. In: 5th International conference on information systems and computer networks (ISCON), Mathura. https://doi.org/10.1109/ISCON52037.2021.9702455

Kushwaha NL, Rajput J, Sena DR, Elbeltagi A, Singh DK, Mani I (2022) Evaluation of data-driven hybrid machine learning algorithms for modelling daily reference evapotranspiration. Atmos Ocean. https://doi.org/10.1080/07055900.2022.2087589

Lanzante JR, Dixon KW, Nath MJ, Whitlock CE, Adams-Smith D (2018) Some pitfalls in statistical downscaling of future climate. Bull Am Meteor Soc. https://doi.org/10.1175/BAMS-D-17-0046.1

Latifah AL, Nurhadiyatna A (2014) Parallel algorithms for spatial rainfall distribution. INKOM J Inf Control Syst Comput 8(1):29–35

Leelavinodhan PB, Vecchio M, Antonelli F, Maestrini A, Brunelli D (2021) Design and implementation of an energy efficient weather station for wind data collection. Sensors 21(1):3831–3849. https://doi.org/10.3390/s21113831

Lestari AA, Munir A, Suhardi (2021) Weather forecasting uses backpropagation algorithm artificial neural network model for agricultural planning in three villages at three sub-districts of Gowa Regency. IOP Conf Ser Earth Environ Sci 921(1). https://doi.org/10.1088/1755-1315/921/1/012013

Lipper L, McCarthy N, Ziberman D, Asfaw D, Branca G (2018) Climate smart agriculture building resilience to climate change. http://www.springer.com/series/6360

Lowry WP (1977) Empirical estimation of the urban effects on climate: a problem analysis. J Appl Meteorol 16:129–135

Marcu I, Voicu C, Drăgulinescu AMC, Fratu O, Suciu G, Balaceanu C, Andronache MM (2019) Overview of IoT basic platforms for precision agriculture. In: Poulkov V (eds) Future access enablers for ubiquitous and intelligent infrastructures. FABULOUS 2019. Lecture notes of the

institute for computer sciences, social informatics and telecommunications engineering, vol 283. Springer, Cham. https://doi.org/10.1007/978-3-030-23976-3_13
McCoy LE (2010) The development of an evolutionary algorithm to predict outcomes in meteorological trends, Thesis (M.Sc, Computer Science). San Diego State University
Mizik T (2021) Climate-smart agriculture on small-scale farms: a systematic literature review. Agronomy 11(6). https://doi.org/10.3390/agronomy11061096
Mullapudi RK, Nikitha K, Bhanu Sree M (2020) Weather monitoring using AWS cloud computing. Int J Adv Res Innov 8(2):205–208
Munizaga MA, Gomez R, Aviles M, Vasquez M, Coronel GCR (2016) A cloud computing based framework for storage and processing of meteorological data. In: International conference on technologies and innovation. Springer International Publishing, pp 90–101. https://doi.org/10.1007/978-3-319-48024-4_8
Nazripouya H, Meshram SG, Bahramloo Sepehri M, Ali SA, Kushwaha NL, Kiani-Harchegani M, Rodrigo-Comino J (2022) Assessment of spatiotemporal changes of the length and starting date seasons in the west of Iran. Acta Geophys **70**:2813–2825. https://doi.org/10.1007/s11600-022-00929-x
OECD (Organisation for Economic Cooperation and Development) (2014) The space economy at a glance. OECD Publishing. https://doi.org/10.1787/9789264217924-en
Pandey K, Singh KG, Singh A (2023) Multi-sensors based smart nutrient reuse management system for closed soilless culture under protected cultivation. Comput Electr Agric 204:107495
Patil MP, Rane KP (2016) Cloud based weather monitoring system. Int J Recent Innov Trends Comput Commun 4(5):446–450
Pauzi AF, Hasan MZ (2020) Development of IoT based weather reporting system. IOP Conf Ser Mater Sci Eng. https://doi.org/10.1088/1757-899X/917/1/012032
Pavithra DS, Srinath MS (2014) GSM based automatic irrigation control system for efficient use of resources and crop planning by using an Android mobile. IOSR J Mech Civil Eng 11(4):49–55. https://doi.org/10.9790/1684-11414955
Praba MSB, Martin AJ, Srivastava S, Rana A (2018) Weather monitoring system and rainfall prediction using SVM algorithm. Int J Res Eng Sci Manage 10:745–750
Prasanna RL, Dubey S (2020) IOT based crop monitoring system. Int J Res Appl Sci Eng Technol 8(6):2321–9653
Sadiq N, Qureshi MS (2010) Climatic variability and linear trend models for the five major cities of Pakistan. J Geogr Geol 2(1). https://doi.org/10.5539/jgg.v2n1p83
Saravanan MS (2022) Prediction of temperature for next three days using decision tree algorithm by comparing sliding window algorithm for better accuracy. ECS Trans 107(1):14097
Singh N, Jena S, Panigrahi CK (2022b) A novel application of decision tree classifier in solar irradiance prediction. Mater Today Proc 58:316–323
Singh DK, Sobti R, Jain A, Malik PK, Le DN (2022a) LoRa based intelligent soil and weather condition monitoring with internet of things for precision agriculture in smart cities. IET Commun 16(1):604–618. https://doi.org/10.1049/cmu2.12352
Soy H, Dilay Y (2021) A conceptual design of LoRa based weather monitoring system for smart farming. Eur J Sci Technol 28(1):906–910. https://doi.org/10.31590/ejosat.1011947
Tabassum S, Hossain A (2018) Design and development of weather monitoring and controlling system for a smart agro (farm). Intell Control Autom 9(1):65–73. https://doi.org/10.4236/ica.2018.93005
Talaviya T, Shah D, Patel N, Yagnik H, Shah M (2020) Implementation of artificial intelligence in agriculture for optimization of irrigation and application of pesticides and herbicides. Artif Intell Agric 4(1):58–73. https://doi.org/10.1016/j.aiia.2020.04.002
Thies B, Bendix J (2011) Satellite based remote sensing of weather and climate: recent achievements and future perspectives. Meteorol Appl RMetS. https://doi.org/10.1002/met.288
Udit S, Vidhi K (2020) Intelligent farming system with weather forecast support and crop prediction. In: IEEE 2020 5th international conference on computing, communication and security (ICCCS), Patna, India 1–6. https://doi.org/10.1109/ICCCS49678.2020.9277437

Ullo, S. L., & Sinha, G. R. (2021). Advances in IoT and Smart Sensors for Remote Sensing and Agriculture Applications. Remote Sensing. 13(13). https://doi.org/10.3390/rs13132585

Valencia JSB, Herrera MM, Perce JM (2022) Low cost climate station for smart agriculture applications with photovoltaic energy and wireless communication. HardwareX 11(1):1–16. https://doi.org/10.1016/j.ohx.2022.e00296

Vautard R, Kadygrov N, Iles C, Boberg F, Buonomo E, Bülow K, Coppola E, Corre L, van Meijgaard E, Nogherotto R, Sandstad M, Schwingshackl C, Somot S, Aalbers E, Christensen OB, Ciarlo JM, Demory ME, Giorgi F, Jacob D, Jones RG, Keuler K, Kjellstrom E, Lenderink G, Levavasseur G, Nikulin G, Sillmann J, Solidoro C, Sørland SL, Steger C, Teichmann C, Warrach-Sagi K, Wulfmeyer V (2020) Evaluation of the large EURO-CORDEX regional climate model ensemble. J Geophys Res Atmos 126(17). https://doi.org/10.1029/2019JD032344

Wilby RL, Dawson CW (2013) The statistical down scaling model (SDSM): insights from one decade of application. Int J Climatol 33:1707–1719

Wilby RL, Wigley TML (1997) Downscaling general circulation model output: a review of methods and limitations. Prog Phys Geogr 21(4):530–548

Chapter 9
Big Data Analytics for Improved Weather Forecasting and Disaster Management

Gaurav Y. Rakhonde, Shalaka Ahale, Namburi Karunakar Reddy, Pooja Purushotham, and Ananya Deshkar

Abstract Weather forecasting and agriculture is an interdisciplinary research programme based on advancements in several domains. Earth temperature is anticipated to rise by 2 °C by 2050, putting up to 50 million people at danger of starvation owing to agricultural consequences. Lot of models have been made to detect changes and warn of potentially dangerous circumstances in future. Data mining methods and methodologies are more effective than traditional ones. Convolutional Neural Networks and Long Short-Term Memory Networks envisage weather radar echo images with various radar echo shapes depicting various terrible weather situations moulding forecast to estimate future changes to avert agricultural and natural disaster circumstances and minimize economic loss and impact too. WSN-based systems also assist in avoiding such things. Internet of Thing-based Agribot (IOT-Agribot) is playing a crucial role in precision agriculture, natural disaster as well as in monitoring. Big data not only critical in agriculture but also important in tourism, airport operations, mining sector, and power generation in forecasting changes. Because of advancements in weather monitoring systems and the rapid increase in meteorological data, weather forecasting entered into new era. Using contemporary big data analytic (BDA) techniques and new tools and technology, now it is feasible to forecast in the beginning or change in severity of diseases and pests of crops. Deep learning is excellent application in numerous domains has spurred its usage in forecasting makes a significant advance. In future, transfer learning and standard data augmentation are insufficient, and generative adversarial networks (GANs) are employed to overcome data scarcity problem in agricultural computer vision. GAN may generate images with synthetic "real" cropping rather than just rotating or adding noise. Big

G. Y. Rakhonde (✉) · N. K. Reddy · P. Purushotham
Department of Plant Pathology, University of Agricultural Sciences, GKVK, Bengaluru, Karnataka 560065, India
e-mail: gauravrakhonde09@gmail.com

S. Ahale
Department of Plant Pathology, Punjab Agricultural University, Ludhiana, Punjab 141004, India

A. Deshkar
Department of Horticulture, University of Agricultural Sciences, GKVK, Bengaluru, Karnataka 560065, India

K. Pandey et al. (eds.), *Artificial Intelligence and Smart Agriculture*,
Advances in Geographical and Environmental Sciences,
https://doi.org/10.1007/978-981-97-0341-8_9

data is the future to secure the future in various aspects. Thus, this chapter includes the application and its future employment with its limitations to overcome.

Keywords BDA · Weather · Forecasting · Monitoring · Techniques

9.1 Introduction

Weather forecasting first appeared in nineteenth century, and its use is utmost important in various aspects till now. Evaluation of changes in the present condition of the atmosphere is weather forecasting which is dynamic and playing necessary role in people's daily lives which relies on temperature, rainfall, air pressure, wind speed and its direction, radiation, and humidity. The fluctuations in length of the yearly seasons and daily temperatures have a substantial influence on human and natural ecosystems and might aid in the prediction of negative effects with irreversible implications. Thus, a huge data is collected or created to predict weather and is jumbled, too. Prediction of weather using meteorological data is a complicated endeavour with too many variable elements and are responsible to alter weather conditions rapidly. To suggest weather forecasting algorithm, evaluation of its specific properties as intensity, data continuity, and multidimensional and chaotic behaviour are needed (Li and Liu 2005).

Weather forecasting evolved from human exhaustive efforts to a computer process, which necessitates high-tech equipment but influenced by a variety of circumstances. Geographical location, season, input data accuracy, weather classifications, time lead, and rationality time are some effective elements in it. Big data analytic (BDA) is a procedure of searching for hidden patterns, undiscovered relationships, and other relevant information to make improved judgments. It has capacity to improve decision-making processes in various fields unlike the conventional ways of processing enormous data is challenging and ineffective in organizing massive data with little human influence, which has a higher possibility of inaccuracy (Poornima and Pushpalatha 2018). To efficiently analyse data, there is a need to look for a best performing stand and utilization of straight forward data mining practice (Tsai et al. 2015). Analysis of large data to reveal ignored patterns and relevant information leads to improved results. Several sectors of society, including the meteorological institution, are attentive in big data these days. BDA will produce better weather forecasting findings and assist forecasters in forecasting weather more precisely. Numerous big data approaches and strategies have been planned to handle and evaluate massive meteorological data from various resources to attain aim and identify appropriate solutions. The limitations associated with old data organizing approaches and technologies can be overcome by exploiting BDA in weather forecasting (Fathi et al. 2021).

The necessity of exact prediction cannot be overstated consequently. Forecasting was once reliant on forecasters in the information era, but now accomplished via usage of technology and data. Atmospheric data sets include sunlight intensity, rainfall, humidity, radiation, air pressure data acquired from diverse sources. Processing this massive data necessitates usage of cutting-edge technology and software (Reddy and Babu 2017; Bendre et al. 2015). BDA has the capability of it and will give significant verdicts to employ and minimize the impacts with precision with minimum manual influence.

9.2 Weather Forecasting and Monitoring

Weather is the daily fluctuations in the atmosphere and data gathered in various meteorological stations from sea, land, radar, and other types of observation. These data are fed into various programmes and algorithms to identify distinct trends known as weather forecasting. The use of Numerical Weather Prediction (NWP) in 1970s and 1980s made great advances in forecasting manoeuvres (Xiao et al. 2012). These models have numerical equations and with the employment of computer programme to unravel changes to monitor in meteorological attributes. It could be done day to day or several days ahead of time. Weather forecasting is a vital activity in everyday activity since it may affect various areas as well as save many lives from unexpected fortunes (Leu et al. 2014; Mittal and Sangwan 2018).

9.3 Big Data

Volume of data expands very fast in comparison with computer technology. Big data entails new tools, analytics, and technology architectures to build up high value sources for organizations by collecting enormous volumes of hidden information through analytics. Diversity, volume, velocity, variability, and value are the primary characteristics. Data exist in a variety of sorts and forms. All data from various resources such as sensor device data, web log files, web pages, e-mail, documents, and social networking sites are completely diversified. The volume of data will display its immensity and expected to grow to as the time passes. The rate of data change, generation rate with its transport from one point to another is the velocity. This attribute is relevant to the data entry rate as well as the data stream rate. Variability takes into account the irregularity of the data stream. Maintenance of the data placed on user devices is extremely difficult, exclusively with the exploitation of social media (Fathi et al. 2021) (Fig. 9.1).

9.3.1 Big Data Categories

Structured, unstructured, and semi-structured data are the categories of big data. Unstructured data do not conform to any predetermined data model, structure, or

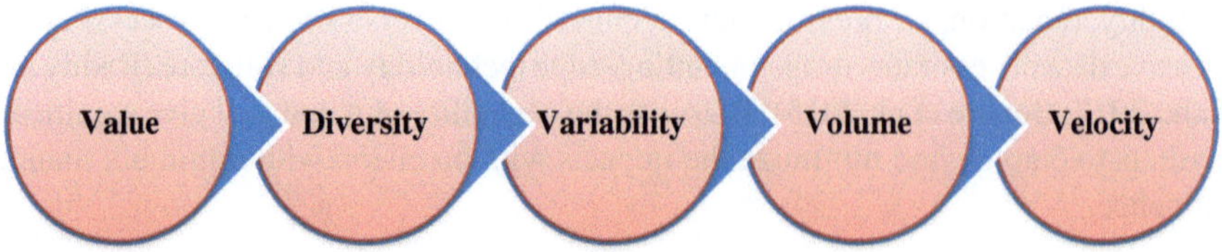

Fig. 9.1 Big data primary characteristics

format. As an outcome, it contains data that is difficult to find. Log data, X-ray photos, CCTV recordings, and movies are examples of unstructured data. Semi-structured one data have markers to recognize logical components, but unable to conform to standard data structures. They are distinguished by their uneven, incomplete, and implicit assembly. The combination of several types of structured data might result in semi-structured data (Pathak et al. 2018; Abkenar et al. 2021).

9.4 Applications of BDA

9.4.1 Agriculture

Undoubtedly, population expands, so does the demand for food production (Pandey et al. 2023). Weather forecasting and BDA should be utilized to advance the fineness of agricultural outputs. Weather forecasting assists farmers in recognizing erosion of soil, abiotic stress like drought and overwatering. Predicting rainfall allows farmers to make more exact agricultural decisions also anticipate food prices. Weather forecasting also assists grocery businesses in properly organizing stock control (Bendre et al. 2015).

Agricultural tragedies are destructive, unanticipated, and disastrous ones influenced by development of land, crops, and animals. Enormous data, its processing, and its management standards for application are major challenges in such area, while natural disasters remain a significant issue but the enormous data availability has unlocked novel strategies for dealing. As a result, BDA has radically altered how individuals respond to agricultural calamities. The data handling model employing BDA (DHM-BDA) investigates impact on disaster management impact on agriculture and offering efficient practical solutions. DHM-BDA is utilized to handle critical huge data, like climatic causes and associated successes, along with evolving technology difficulties in various disaster management stages. Furthermore, it contributes in monitoring, mitigation, allcviation, and acceptance of agricultural disasters with recovery and rebuilding progression. The simulation results were implemented and projected model outperforms other current techniques with respect to prediction ratio, decision making, data management, production ratio, and risk reduction ratio (Jaber et al. 2022).

9.4.1.1 Drought Monitoring

Droughts have been more frequent and intense in recent years with a significant impact on several regions in world. So, in drought monitoring also, utilizing big data analytics takes popularity. It is among the most destructive natural calamities, jeopardizes agricultural productivity, ecology, and socio-economic progress. Early warning, precise appraisal, and efficient forecast are critical, especially for the nations frequently threatened by this threat. Numerous new research is being conducted on big data with its uses to monitor drought ailment. In reality, big data handles data heterogeneity and helps in drought prediction, provides a perspective to many characteristics such as its geographical distribution, temporal distribution, and severity detection (Balti et al. 2020).

Zhang et al. (2017a, b) offered a large databased strategy to drought research and prediction in California. They employed lot of data sources, including satellite data, climate sensor data, meteorological data, drought, and water use efficiency. This approach is categorized as Web UI, Web Engine, Middle Tier. Web Engine has three layers. First level houses data storage portion followed by data mining, processing. REST API safeguards communication among Web client and server, involved in final level. The second Middle Tier component is storing user data to decrease employer bandwidth and server stress while attaining several commands simultaneously. Last level contains application interaction with user requests with processing.

BDA framework for remote sensing (RS), categorized into three levels. The RS big data acquisition unit (RSDU) initially assembles satellite data from many satellites orbiting around globe. Owing to air particles of gases and dust, the data is somewhat erroneous. They employ SPECAN or Doppler techniques to remove it and enhance the quality. Through downlink channel, assembled data is transferred towards base station for initial processing. The second level data processing unit (DPU) is a crucial architectural component for improved big data processing via filtering following load balancing with parallel processing. In last level, upper layer unit is essential to assemble, store results, and provide verdict based on DPU outcome (Rathore et al. 2015).

9.4.2 *Plant Disease*

Plant disease epidemics are on the rise, threatening food security. A global epidemic is currently endangering health worldwide. Thus, stable, healthy food supply will be essential to pull people out of poverty and improve health. Climate change, global food trade networks, pathogen spillover, and pathogen lineage development all are contributing to spread and boom of plant diseases, both endemic and newly emergent races. To solve these difficulties, a new set of tools is necessary, as well as disease monitoring and enhanced detection technologies such as pathogen sensors, predictive modelling, and data analytics (Ristaino et al. 2021). BDA and machine learning (ML) approaches have grown more significant in the agro-based industry in recent decades to meet the issues created by climate change affecting productivity. The examination of environmental, climatic, and cultural aspects enables to predict crop irrigation and nutritional needs, crop yield with improved quality to be increased while using less input.

Potato late blight is regarded as the most devastating diseases in the world responsible for Irish famine. Unanticipated epidemics could indeed result in major financial and yield losses. In Sardinia, the DSS LANDS anticipates late blight disease. Based on meteorological parameters provided by ARPAS weather stations, disease severity is predicted using a Feed-Forward Neural Network and Support Vector Machine Classification. The estimate precision for ANN was 96%, while for SVM classification was 98% precise (Fenu and Malloci 2019).

9.4.2.1 WSN in Disease Forecasting

All across the world, wireless sensor networks (WSN) have been employed for precision farming. Many researchers employed sensor networks to investigate their function in monitoring as well as automating farming chores such as disease predictions, among others, by utilizing real-time environmental data. When apple scab became an epidemic in late 1970s in Himachal Pradesh, India 5 Apple scab monitoring and research laboratories (ASMRL) were constructed in Mashobra, Kotkhai, Thanedhar, and Sarahan in Shimla province while Bajaura in Kullu province. At several sites, Apple scab infection cycle time was predicted with de Wit 7 day leaf wetness recorder, Reuter Stokes apple scab predictor and SWG Biomat. As part of Orchard 2000, a decision support system, apple black spot (ABS) employed in 1992 for farmers. Orchard 2000 was made up of application software (ABS monitor) with a user interface and an orchard meteorological database. VENTEM, PODEM, and NECTEM are PC-based apple disease alarm systems created by Horticulture Research International, New Zealand. Apple scab warning system and VENTEM was launched in English and German as version 4.0 in 1993 for commercial purpose. PODEM is used to treat powdery mildew, while NECTEM to treat European canker and fruit rot (Beresford and Mackay 2012; Laurenson and Manktelow 1992; Xu and Butt 1993).

9.4.3 Industry

Industries are using recent technical advancement in production, with the objective of increasing productivity while remaining adaptable and efficient. This revolutionary transition is changing the way manufacturer's function, which is propelled by different technological developments such as BDA, Internet of Things (IoT), artificial intelligence (AI), and cloud computing. BDA has been identified as an important industrial component that provides significant information for smart factory management. Big data and industries have the ability to reduce resource utilization and boost operations, therefore considerably contributing to achieve sustainable development. Big data in industry investigates the most recent discoveries in the field of big data and its applications (Bhuvaneswari 2022).

9.4.4 Tourism

Tourism is travelling and residing someplace outside of one's typical area for happiness, conferences, business, education, friends get together, etc. Choosing a tourist destination is heavily influenced by weather conditions. Climate change may have a direct impact on the destination. So, the comfort and security of visitors with their appealing goals are inextricably linked to weather predictions. Weather forecasting may also be used to assess assistances of this sector (Li et al. 2018). Hardly a tourism expert can afford to be without a copy of "Forecasting Tourism Demand". The tourist business has seen an unprecedented expansion in recent years, and being able to forecast future trends as precisely as possible is critical in the battle to stay one step ahead of the competition. Building on the popularity of "Practical Tourist Forecasting", this volume examines 13 forecasting methodologies and, in a user-friendly language, walks the reader through each method, highlighting its strengths and shortcomings and demonstrating how it might be applied to the tourism sector.

"Forecasting Tourism Demand" uses charts and tables to get prior knowledge so that they can organize a forecasting project, analyse time series and other information. Thus, choose the suitable forecasting model to predict and evaluate its outcomes. Ideal for corporate marketing managers and strategic planners, transportation planners, and economic policymakers who must forecast demand for their products among visitors. Executives who rely on projections made by others will find it quite useful in assessing the validity and reliability of predictions and forecasts. Those who study business trends will find it valuable in forecasting the future of what has been dubbed the world's largest industry (Frechtling 2012).

9.4.5 Construction

Temperature, wind, humidity, and damp weather all have an impact on building. Thus, weather forecasting is extremely useful in safeguarding the workforce, activities, and resources from the hazards of changing weather patterns. To put it another way, the sooner weather-related problems are detected, the sooner they can be addressed. As a result, weather forecasting aids in cost savings and worker safety (Hazyuk et al. 2012).

Since, industrial sector in addition to construction sectors also consume the majority of energy. So, implementing energy efficiency tactics, optimization techniques, and managing is a key step towards worldwide energy saving. The practice of ML devices in energy prediction is becoming more prevalent because to its ability to tackle complex nonlinear problems typically seen in the commercial and residential sectors. This approach proposes and investigates the use of two deep neural networks, feed forward, and recurrent in calculating manufacture capability, energy usage, and workshop situations based on production agendas, climatic conditions, building thermal attributes, and building behaviour, as well as their efficient use. When provided with new data, feed forward approach predicted building energy, workshop air temperatures, and humidity. These neural networks were proficient in using data from a simulation of medium-sized processing ability in United Kingdom. The use of simulation techniques in conjunction with ML algorithms allows a low-cost, non-intrusive way to anticipating and optimizing building energy usage in the industrial business. Furthermore, application of neural networks provided anticipated building energy outlines for recognizing gaps in energy use to bridge the inappropriate and significant cost in the manufacture with manufacturing environment forecasting for condition monitoring with sensitive production environments (Mawson and Hughes 2020).

9.4.6 Sport Competitions

Sport has an influential impact by rain, lightning, etc.; thus, weather forecasters are employed. Lightning can be dangerous to a sport person, and wind greatly alter the outcome of an event. During rainfall, grounds needed to be covered, therefore weather condition forecasts indicate when this operation should be completed and planned. Thus, many contests throughout the world may be adequately organized such that little or no obstruction from weather parameters occurs and the competition is carried out smoothly. Prior to making projections, large contests such as the Olympics and World Cups may be planned correctly and successfully staged at appropriate sites with a proper management (Jain and Jain 2017). The indoor games not have any issue, but the outdoor games are highly influential; thus, it is very important for outdoor games arrangement and planning of schedule.

9.4.7 Transport Operations

Weather forecasting enables ships and workers all around the world be alert of threats of storms, allowing them to make accurate decisions about departure and routes. Additionally, flight plans may be delayed or cancelled owing to hazardous weather conditions; these issues may be resolved via weather forecasting (Zhang et al. 2017a, b). Transportation of products or other essential requirements, particularly in mountainous areas, is a difficult task, especially in adverse weather circumstances such as heavy rain, hailstorms, and so on. Visibility might be a difficulty at times, especially in the winter. Thus, previous predictions might be done using existing big data and its process to produce the desired result, allowing for effective planning and passengers, also think for alternative or may schedule their travelling according to forecast predicted. This will enhance the food grain transport and will provide proper supply to regions which are highly influenced by the weather especially remote and hilly regions.

9.4.8 Disaster Management

Natural disasters occur all around the world, but alert and forecasts can decrease unfortunate demise, save lives, and reduce economic losses. Yet, forecasting precision and time are heavily influenced by nature of disaster. Tornadoes and landslides are among of the natural hazards that may be predicted using weather prediction and BDA, potentially saving many lives (Braman et al. 2013). The key challenges include dealing with data gathering and management technologies in an effective manner, as well as establishing efficient methods for reducing the risks connected with natural disasters and managing catastrophes in ways that result in the least amount of human life and property damage. Other issues include developing algorithms through systems that may be used to solve operational issues and increase disaster prediction accuracy.

Recent research has improved resilience through the use of "big data" tactics improving the rapidity and efficiency of linkages between disaster information and systemic response. The adaptive, absorptive, and transformative capacity of sociopolitical systems to withstand and cope with the negative repercussions of tragedy is referred to as resilience. Big data technologies such as remote sensing photography, social media data, crowdsourcing data, geographic information system (GIS), and mobile metadata are examples of big data tactics that may be easily applied in the many stages of disaster management and resilience. The verdicts are significant for legislators, administrators, and other stakeholders who are responsible for preparing for, responding to, and recovering from disasters through training, early warning systems, emergency evacuation, relief distribution, and other vital infrastructure components.

More research on big data applications is needed to improve the efficiency of the public sector in developing tactics to mitigate the harmful effects of natural disasters. Overall, further progress is required to investigate Big Sensing Data challenges, predominantly in rising data volume of streaming videos, such as effective data management, quick data transit with its easy display (Sarker et al. 2020; Yu et al. 2018).

9.4.9 Power Management

Usage of renewable energy is fetching popularity. So, operative weather forecasting is mandatory to gather the maximum data from various sources of energy. Solar and wind energy well-organized usage is reliant on weather circumstances. Precision weather forecast allows for the development of future intelligent power systems to cope up with the scarcity of power. Weather forecasting is influenced by two major factors: technical advancements and human activity (Casas et al. 2009). The advancement of communication and information technologies results in the continuous progress of weather data in both structured and disorderly. Typically, organized data is processed through systems. These systems are incapable of processing all jumbled data. Big data analytics can be employed to overcome this challenge. The accurate and timely weather forecasting may benefit everyone in society (Reddy and Babu 2017). With the ongoing deployment of sensors, wireless transmission, network connectivity, and cloud computing services, massive volumes of data are being gathered in the power industry, to harness the promise of big data in power and get answers for smart energy management (Zhou et al. 2016).

9.5 Data Mining and BDA for Geographic Monitoring

Potential disease hazards to important food crops can be investigated using data mining tools. Archival data and available recent “big data” from various structured data sources such as scientific papers, printed documents, PubMed and CAB Abstracts, Google N gramme, and social media, image posts, and twitter are employed in conjunction with natural-language analysing strategies to monitor and map causes of plant diseases (Collier 2012).

Data mining technologies may be employed on a regular basis for plant disease management through forecasting. Big data provides decision makers with new information for effective agricultural disease management and to make food supply more robust to epidemics. It may able to forecast the pliability of agroecosystems in certain geographic zones and detect endemic areas. Natural language analysing technologies and geoparsing were used to monitor nineteenth-century dispersal of potato late blight in USA all around ports of New York and Philadelphia, revealing disease control tactics and theories about the origins of outbreaks. In another case, more than 12,500 diseases were reported and compiled in order to investigate the prevalence and pathogenicity of *Phytophthora* species. *Phytophthora* epidemics were anticipated with the aggressive dominating species. Data mining tools used to track and anticipate plant pathogen outbreaks and surge global food security vigilance, policy development, research, and phytosanitary service execution (Cheng et al. 2019; Cramer et al. 2016).

9.6 Challenges with Big Data Analytics

9.6.1 *Time Taking Process*

The predictive analytics work of forecasting is a continuous and time taking process since it involves several phases of development, testing, and adaptation (Blackburn et al. 2015). Taking together professionals from multiple functions with diverse sights would be a tough task as it might take months to years to see benefits in complex systems. BDA adoption requires ongoing support from senior management and key institutes. Obtaining access to various data possessed by various sectors of the corporation and to merge, verify, and purify will be cumbersome and may necessitate the project management team's undivided attention.

9.6.2 *Scarcity of Assets*

Data and analytics capabilities varies among the institutes in a system. Conflicts will arise as a result of a lack of IT resources and the ability to transfer data and information in real time. Big data deployment should priorities collaboration and cross-functional team formation across several institutes (Dutta and Bose 2015). However, barriers to building an interorganizational cross-functional team include network rivalry, principal conflicts, incentive systems, data sharing regulations, etc. A data-driven culture is crucial for BDA competences, and fact-based management should be encouraged across the network as a strategy for successful BDA utilization and delivering professional value. Leadership is also essential for the successful implementation of BDA (Seah et al. 2010).

9.6.3 *Confidentiality and Safety*

Big data complicates matters such as privacy, security, unethical data use, and incorrect processing leading to biased outcomes. Professionals highlighted worries regarding data privacy and security, claiming that outdated regulation is one of the most significant impediments to data sharing. Privacy, security, and data rules offer significant challenges for transnational networks that must adhere to the laws while transferring data across global networks. These challenges may be avoided if an effective data governance plan was implemented as part of the data integration and display concerns (Richey et al. 2016).

9.6.4 Social Acceptability

From a behavioural aspect, actual data and information utilization may be problematic in meantime decision makers might respond to even minute variations in the physical environment, raising risk, and inventory cost. There is a danger of identifying many statistically significant but irrelevant links with no causal link when using big data. Big data relies on correlation rather than cause, requiring human interpretation to address massive concerns (Radke and Tseng 2015).

9.6.5 Uncertainty of Returns on Investment

Uncertain advantages and uncertainty regarding investment return representatives takes sceptical approach of implementing BDA. Gaining monetary rewards from BDA is extremely challenging since it is heavily reliant on downstream workers who finish the activity and processing. The data-driven culture and individual workers' absorptive capacity are crucial in collecting and digesting knowledge (Davenport et al. 2001).

9.6.6 Absence of Multiskilled Experts

The main hurdles include inexperienced personnel, time restrictions, a poor degree of integration, a lack of an effective predictive analytics system, and issues with change management. Moreover, it was demonstrated that professionals who anticipate to use analytics in future but are now not doing so cite a lack of data and an inability to locate acceptable data as a major barrier. It was also suggested that a data scientist must have both analytical and subject expertise, which is hard to find because someone with analytical skills may be unwilling to gain domain knowledge. There is a critical shortage of BDA expertise (Waller and Fawcett 2013; Richey et al. 2016).

9.6.7 Scalability

Implementing BDA in forecasting systems on a huge scale is critical; however, supreme studies could not specify whether their tactics would work effectively on a large scale. Despite the fact, precise scalability of weather forecasting models is an important precept, most technique-based research have mostly focused on strategies in small-scale ways. Still, this is an essential and difficult topic.

In order to store recent processed data, organizations will store their data after a particular interval. Replace relational databases with more current infrastructure such as Hadoop distributed databases, distributed file systems, parallel computing, and cloud computing capability to notice scalability issues. A NoSQL database with a high level of scalability is looked-for to deal with unstructured data generated by IoT. However, employing cloud computing power to store BD may put more financial strain on organizations. To avoid this, organizations may employ techniques to optimize the data collection process and reduce unnecessary data output (Rehman et al. 2016; Kang et al. 2015).

9.6.8 Quality

There are quality issues related to data generating procedure linked to the product manufacturing course, as deprived data superiority impedes data analytics processes and influences decisions. The issues of noise, outliers, inconsistent incomplete data in old style data organization will be exacerbated for BDA. Data cleansing is a critical step in managing data, particularly in weather prediction. To capture and collect weather data through various sensors-based technologies make it easy to deliver more partial and inconsistent data. So, processing is necessary but challenging procedure since determining which numbers are reliable and applicable data must be extracted from a data is which is never straightforward. Accuracy of data was assigned through technique-based tactics, with negligence of reliability, time, scalability, etc., so contributing a multi-objective way for producing a trade among the parameters drive an open issue. Hence, some efforts, such as constructing grouping on changing data subsets, may be quite difficult, and providing more efficient approaches that to be employed in meteorological data processing is never straightforward. As a result, modern data infrastructures and finest practices in big data organization, such as Hadoop, MapReduce, and statistical process control, are crucial for efficient information flow (Saneja and Rani 2018).

Optimal model performance is vigorously trailed in agricultural image analysis in order to make visual identification such as image classification, segmentation, object detection, and localization of biological variability and unstructured surrounding. Large-scale, balanced, and ground-truthed picture datasets are incredibly important to create efficient models which is usually challenging. As artificial intelligence has impact on agricultural image analysis and modelling through deep learning with picture augmentation having a crucial role in improving model performance with lowering manual actions for image collecting and labelling.

The generative adversarial network (GAN), developed in computer vision community in 2014, offers a set of distinct techniques for learning appropriate data representations and producing extremely realistic cases. In agriculture, it had been used for image augmentation or synthesis since 2017 to enhanced model performance (Lu et al. 2022).

9.6.9 Unavailability and Unawareness About Techniques

Inability to use the data effectively is not only challenge but its quality also matters. Forecasting approaches have a strong emphasis on endogenous time-serious aspects for predicting, with minimal attention to exogenous variables and information sources. This obviously suggests to improve data management skills and forecasting methods. Furthermore, using expert verdict as a covariate in predictive models is problematic. Ordinal scales are also used to examine expert judgements and opinions (Meixell and Wu 2001). Many method-based strategies in BDA for weather forecasting such as assembly, regression, classification, and dimension reduction are needed and are regarded as a subset of the NP-hard or NP-complete class from the complex aspect.

9.6.10 Implementation Task

Discovering more impactful algorithms, participating BDA tactics and practices to predict weather has promoted precision for models. Weather models based on numerical one are linked with ML algorithms or other methodologies tend to produce better results. The integration prospect might be an interdisciplinary threat and big challenge. Various BDA methodologies are employed in forecasting within real-world scenarios. The majority of these methodologies have been validated through the analysis of datasets and simulated settings. For model's performance test, the aforementioned methodologies must be evaluated in a realistic world setting. As a consequence, realistic world implementation is quite appealing (Fig. 9.2).

9.7 Future Path

Enormous losses occur due to natural disasters such as earthquakes, hurricanes, and cyclones, which have a negative impact on people. Predicting environmental effect has become a difficult challenge in the modern world. A thorough rainfall forecast system is incredibly beneficial and necessary for most countries, such as India, because agriculture is their major source of income and raw material is provided from this sector. Prediction is critical for evaluating yield productivity, water resource utilization, and rainfall. In the literature, there are several statistical and predictive models for rainfall forecasting systems. Due to the dynamic temperament of climate circumstances, model will deliver less accurate prediction models in large-scale rainfall forecasting. Weather damage prediction is a critical research subject that involves both science and technology in order to forecast weather damage in a specific region. There are numerous good prediction systems, but there is no fair calculation for future

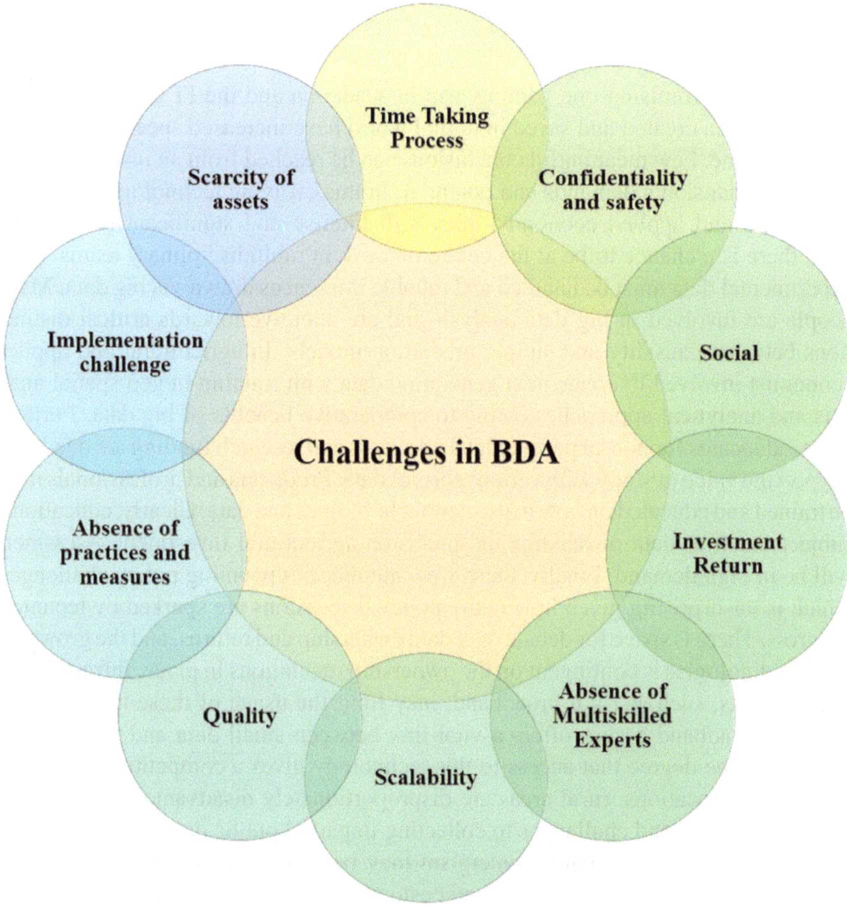

Fig. 9.2 Challenges in BDA

harm since the time span between now and when forecasting is done and disseminated varies. So, anticipation of the quantities of damage by gathering prior datasets from datacentres such as India Meteorological Departments in a given place and utilizing regression analysis can be done.

Because of the dynamic temperament of the climate, the model is unable to provide precise prediction models in rainfall forecasting. Weather damage forecast is an important one that combines science and technology to predict damage in a given location. There are many good prediction methods, but no fair accounting for future harm exists since the time period between now and when forecasting is done and broadcast changes. So, predicting the amounts of damage by gathering prior datasets from datacentres such as India Meteorological Departments at given location and using regression analysis is possible.

9.8 Conclusion

Big data is a promising one gaining grip in academia and the IT community. The enormous data created and saved in digital world have increased speedily in a short period of time. Few meaningful conclusions can be reached from an investigation of big data expansion's demands and potential. In this evolving technological context, agriculture and applied economics play a distinctive and significant role. In big data, there is a chance to be at the centre of work in multidisciplinary teams. Non-experimental data must be handled and reliable inferences drawn via big data. Many people are involved in big data analysis and are attentive towards critical distinctions between causality and simple prediction models. In agricultural and applied economist involved to create next generation data with training in geo-spatial analysis and analytical approaches owing to comparative benefits of big data. Furthermore, advocates for non-experimental data access and research funding are needed to support the value of study with various sorts of data. Producers and professionals must be trained and educated on how to use new technologies and data. Clearly, educational subjects such as data ownership and precision agricultural investment assessment will be in high demand. Finally, there are a number of upcoming policy challenges, which is unsurprising given how many policy discussions are sparked by technical progress. There is space for debate over data ownership and returns, and the growth of these technologies is contingent on the ownership regulations in place. Infrastructure requirements, such as rural broadband, may limit the usage of these technologies, as rural broadband access offers a vital link between small data and big data. As a result, to the degree that access to this technology gives a competitive advantage to particular locations, rural areas are disproportionately disadvantaged. There are both possibilities and challenges to collecting impartial public data and government programme data. The greatest synergism may be produced by combining private data, government programme data, and customized data gathering surveys designed to supplement other accessible methods, making big data a possible instrument for the future of not only agriculture but other industries.

References

Abkenar SB, Kashani MH, Mahdipour E, Jameii SM (2021) Big data analytics meets social media: a systematic review of techniques, open issues, and future directions. Telemat Inform 57:101517

Balti H, Abbes AB, Mellouli N, Farah IR, Sang Y, Lamolle M (2020) A review of drought monitoring with big data: issues, methods, challenges and research directions. Ecol Inform 60:101136

Bendre MR, Thool RC, Thool VR (2015) Big data in precision agriculture: Weather forecasting for future farming. In: 2015 1st international conference on next generation computing technologies (NGCT). IEEE, pp 744–750

Beresford RM, Mackay, AH (2012) Climate change impacts on plant diseases affecting New Zealand horticulture. Ministry for Primary Industries, Wellington

Bhuvaneswari V (2022) Role of big data analytics in industrial revolution 4.0. In: Big data applications in industry 4.0. Auerbach Publications, pp 85–105

Blackburn R, Lurz K, Priese B, Göb R, Darkow IL (2015) A predictive analytics approach for demand forecasting in the process industry. Int Trans Oper Res 22(3):407–428

Braman LM, van Aalst MK, Mason SJ, Suarez P, Ait-Chellouche Y, Tall A (2013) Climate forecasts in disaster management: red cross food operations in West Africa, 2008. Disasters 37(1):144–164

Casas DM, González JÁT, Rodríguez JEA, Pet JV (2009) Using data-mining for short-term rainfall forecasting. In: International work-conference on artificial neural networks. Springer, Berlin, pp 487–490

Cheng Y, Zheng Z, Wang J, Yang L, Wan S (2019) Attribute reduction based on genetic algorithm for the coevolution of meteorological data in the industrial internet of things. Wirel Commun Mob Comput

Collier N (2012) Uncovering text mining: a survey of current work on web-based epidemic intelligence. Glob Public Health 7(7):731–749

Cramer S, Kampouridis M, Freitas A (2016) A genetic decomposition algorithm for predicting rainfall within financial weather derivatives. In: Proceedings of the genetic and evolutionary computation conference, pp 885–892

Davenport TH, Harris JG, De Long DW, Jacobson AL (2001) Data to knowledge to results: building an analytic capability. Calif Manage Rev 43(2):117–138

Dutta D, Bose I (2015) Managing a big data project: the case of ramco cements limited. Int J Prod Econ 165:293–306

Fathi M, Haghi Kashani M, Jameii S M, Mahdipour E (2021) Big data analytics in weather forecasting: a systematic review. Arch Comput Methods Eng 1–29

Fenu G, Malloci FM (2019) An application of machine learning technique in forecasting crop disease. In: Proceedings of the 2019 3rd international conference on big data research, pp 76–82

Frechtling D (2012) Forecasting tourism demand. Routledge

Hazyuk I, Ghiaus C, Penhouet D (2012) Optimal temperature control of intermittently heated buildings using model predictive control: part I-building modeling. Build Environ 51:379–387

Jaber MM, Ali MH, Abd SK, Jassim MM, Alkhayyat A, Aziz HW, Alkhuwaylidee AR (2022) Predicting climate factors based on big data analytics based agricultural disaster management. Phys Chem Earth 128:103243

Jain H, Jain R (2017) Big data in weather forecasting: applications and challenges. In: 2017 International conference on big data analytics and computational intelligence (ICBDAC). IEEE, pp 138–142

Kang Y, Park I, Rhee J, Lee Y (2015) MongoDB-based repository design for IoT-generated RFID/sensor big data. IEEE Sens J 16(2):485–497

Laurenson MR, Manktelow DW (1992) Monitor: a computer-based tool for monitoring apple black spot infection periods. In: III International symposium on computer modelling in fruit research and orchard management, vol 313, pp 197–204

Leu JS, Su KW, Chen CT (2014) Ambient mesoscale weather forecasting system featuring mobile augmented reality. Multimed Tools Appl 72(2):1585–1609

Li J, Xu L, Tang L, Wang S, Li L (2018) Big data in tourism research: a literature review. Tour Manag 68:301–323

Li K, Liu YS (2005) A rough set based fuzzy neural network algorithm for weather prediction. In: 2005 international conference on machine learning and cybernetics, vol 3. IEEE, pp 1888–1892

Lu Y, Chen D, Olaniyi E, Huang Y (2022) Generative adversarial networks (GANs) for image augmentation in agriculture: a systematic review. Comput Electron Agric 200:107208

Mawson VJ, Hughes BR (2020) Deep learning techniques for energy forecasting and condition monitoring in the manufacturing sector. Energy Build 217:109966

Meixell MJ, Wu SD (2001) Scenario analysis of demand in a technology market using leading indicators. IEEE Trans Semicond Manuf 14(1):65–75

Mittal S, Sangwan OP (2018) Big data analytics using data mining techniques: a survey. In: International conference on advanced informatics for computing research. Springer, Singapore, pp 264–273

Pandey, K., Singh, K.G., and Singh, A (2023) Multi-Sensors based smart nutrient reuse management system for closed soilless culture under protected cultivation. Comput Electron Agric 107495
Pathak AR, Pandey M, Rautaray S (2018) Construing the big data based on taxonomy, analytics and approaches. Iran J Comput Sci 1(4):237–259
Poornima S, Pushpalatha M (2018) A survey of predictive analytics using big data with data mining. IJBRA 14(3):269–282
Radke AM, Tseng MM (2015) Design considerations for building distributed supply chain management systems based on cloud computing. J Manuf Sci Eng 137(4)
Rathore MMU, Paul A, Ahmad A, Chen BW, Huang B, Ji W (2015) Real-time big data analytical architecture for remote sensing application. IEEE J STARS 8(10):4610–4621
Reddy PC, Babu AS (2017) Survey on weather prediction using big data analytics. In: 2017 Second international conference on electrical, computer and communication technologies (ICECCT), pp 1–6
Richey RG, Morgan TR, Lindsey-Hall K, Adams FG (2016) A global exploration of big data in the supply chain. Int J Phys Distrib Logist Manag
Ristaino JB, Anderson PK, Bebber DP, Brauman KA, Cunniffe NJ, Fedoroff NV, Wei Q (2021) The persistent threat of emerging plant disease pandemics to global food security. Proc Natl Acad Sci 118(23):e2022239118
Saneja B, Rani R (2018) A hybrid approach for outlier detection in weather sensor data. In: 2018 IEEE 8th international advance computing conference (IACC), IEEE, pp 321–326
Sarker MNI, Peng Y, Yiran C, Shouse RC (2020) Disaster resilience through big data: way to environmental sustainability. IJDRR 51:101769
Seah M, Hsieh MH, Weng PD (2010) A case analysis of Savecom: the role of indigenous leadership in implementing a business intelligence system. IJIM 30(4):368–373
Tsai CW, Lai CF, Chao HC, Vasilakos AV (2015) Big data analytics: a survey. J Big Data 2(1):1–32
Rehman MH, Chang V, Batool A, Wah TY (2016) Big data reduction framework for value creation in sustainable enterprises. IJIM 36(6):917–928
Waller MA, Fawcett SE (2013) Data science, predictive analytics, and big data: a revolution that will transform supply chain design and management. J Bus Logist 34(2):77–84
Xiao Z, Liu B, Liu H, Zhang D (2012) Progress in climate prediction and weather forecast operations in China. Adv Atmos Sci 29(5):943–957
Xu XM, Butt DJ (1993) PC-based disease warning systems for use by apple growers 1. EPPO Bulletin 23(4):595–600
Yu M, Yang C, Li Y (2018) Big data in natural disaster management: a review. Geosci 8(5):165
Zhang B, Tang L, Roemer M (2017a) Probabilistic planning and risk evaluation based on ensemble weather forecasting. IEEE Trans Autom 15(2):556–566
Zhang P, Gao J, Thomas AG, Alagupackiam KP, Mannava K, Bosco PI, Chiao S (2017b) On building a big data analysis system for California drought. In: 2017 IEEE third international conference on big data computing service and applications (BigDataService), IEEE, pp 49–156
Zhou K, Fu C, Yang S (2016) Big data driven smart energy management: from big data to big insights. Renew Sustain Energy Rev 56:215–225

Chapter 10
Crop Monitoring System Integrating with Internet of Things and Artificial Intelligence

Asit Kumar Pradhan, Dilip Kumar, Himanshushekhar Chaurasia, Sneha Murmu, Ipsita Samal, Tanmaya Kumar Bhoi, G. A. K. Kumar, and Biswajit Mondal

Abstract The agricultural industry contributes significantly to the economy. As demand for food is rising due to burgeoning population pressure, improved technologies like artificial intelligence (AI) and the Internet of Things (IoT) in farming systems will help to reap more production per farm to meet the increasing food demand and the net income of farmers as well. To achieve the goal, it is essential to manage damage to crops by insects, pests, and diseases through primary plant protection measures. The major constraints faced by the farmer are lack of awareness and proper information. So, for providing real-time information to the farmers, a smart crop monitoring system integrating advanced technologies is needed. This comprises crop health monitoring, yield and field mapping, soil monitoring, weather forecasts, and other tools that can assist farmers in making smart informed decisions to prevent crop losses and enhance production quality. A smart crop monitoring system has

A. Kumar Pradhan (✉) · G. A. K. Kumar · B. Mondal
ICAR-National Rice Research Institute, Cuttack, Odisha 753006, India
e-mail: asitpradhan20672@gamil.com

D. Kumar
ICAR-National Institute of Agricultural Economics and Policy Research, New Delhi 110012, India

H. Chaurasia
ICAR-Central Institute for Research on Cotton Technology, Nagpur, Maharashtra 440023, India

S. Murmu
ICAR-Indian Agricultural Statistics Research Institute, New Delhi 110012, India

I. Samal
ICAR-National Academy of Agricultural Research Management, Hyderabad, Telangana 500030, India

T. Kumar Bhoi
ICFRE-Arid Forest Research Institute, Jodhpur, Rajasthan 342005, India

K. Pandey et al. (eds.), *Artificial Intelligence and Smart Agriculture*,
Advances in Geographical and Environmental Sciences,
https://doi.org/10.1007/978-981-97-0341-8_10

developed using technologies like the Internet of Things (IoT) and artificial intelligence (AI) to cater to numerous farming applications, such as detecting soil quality, meteorological conditions, and crop requirements.

Keywords Artificial intelligence · Internet of Things · Crop monitoring system · Yield mapping

10.1 Introduction

Agriculture is one of the prime economical activities in India, which along with the allied sectors provide 60–70% of India's livelihood. After the USA, it boasts the second-most arable land. This is a result of the enormous network of irrigation water sources and the fertile soil. Even when resources are readily available, they do not yield outcomes commensurate with accessibility. It is due to a lack of available technologies, ineffective application of technology, farmers' ignorance and unawareness, as well as the use of some antiquated techniques (Bhargava et al. 2014). Additionally, pests, insects, and diseases harm the majority of crops, lowering grain yields and quality. Since they can be hazardous to both animals and helpful living things as insecticides and pesticides are not always proven to be effective. Moreover, it harms food chains and the natural animal food web (Shinde and Kulkarni 2017). A 20–40% yield reduction in agriculture is brought on by insects, pests, viruses, animals, and weeds (Oerke 2006). Additionally, they have a variety of aspects, some of which affect global food security immediately and others in the long term (Savary et al. 2012). In India's semi-arid climate, crop output losses from pests and diseases are particularly significant (Shinde and Kulkarni 2017). The impact of weather on agricultural production is enormous. In general, weather-based frangible agriculture systems produce greater crops. According to surveys, if the population increases to 10 billion, we will undoubtedly reach a food catastrophe by the year 2050. It means that unless we develop and expand smart agricultural technologies, our ability to produce food will go bankrupt. Therefore, it is essential to provide a cost-effective technology for Indian farmers to properly manage the country's scarce resources. Advances in technology like artificial intelligence (AI) and the Internet of Things (IoT) in farming systems will assist farming communities to enhance their production to meet future food demand and their net income. Demand for food is rising due to burgeoning population pressure. Technology has altered farming practices over time, and it has a wide range of effects on the agriculture industry. Conventional methods, however, fall short of this massive need. Farmers and farming systems are being forced by this to create innovative methods for increasing output and lessening waste. As a result, AI in integration with IoT technology is progressively becoming a part of the technical advancement of the agriculture sector. For providing real-time information to the farmers, a smart crop monitoring system is essential. This includes soil monitoring, yield mapping, field mapping, weather forecasts, crop health monitoring, etc. The system should assist farmers in increasing food production and quality by enabling timely crop health monitoring. This is due to their ability to strategically deploy sensors remotely and monitor ambient parameters, soil parameters, and plant parameters. The producers can benefit from an early prediction of pests and diseases induced by numerous harmful organisms, whether by a lack of or excess in the normal values of the monitored parameters so that they can effectively take additional precautions against the attack of these pests and insects. This will stop the use of any chemicals and lower the number of diseases occurring in fauna. It would also unquestionably boost output, closing the gap between rising food consumption and population growth. Farmers will be able to work more productively while also improving crop quality and quantity and assuring a speedier time to market with the aid of AI-powered technologies. The Internet of Things (IoT) and artificial intelligence (AI) technologies have enabled the development of smart agriculture through crop monitoring systems (Fig. 10.1).

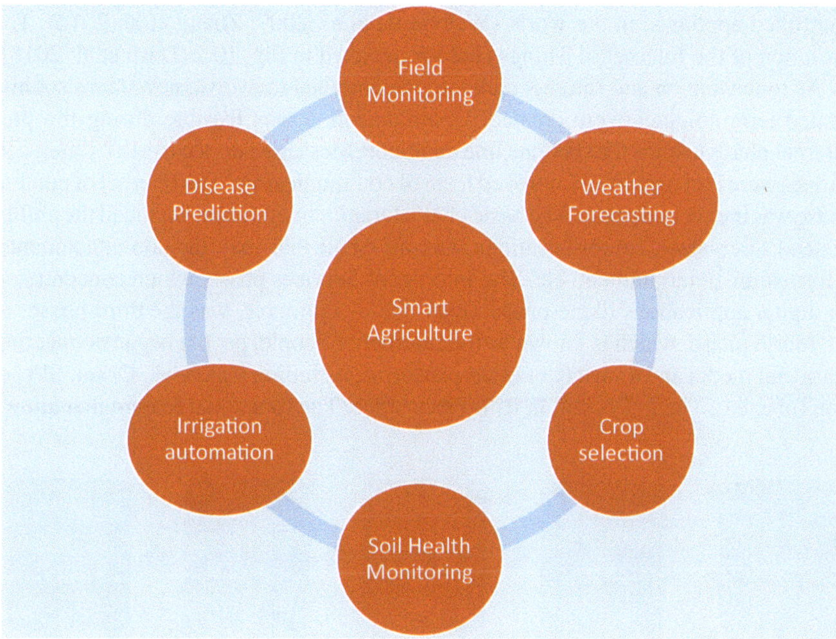

Fig. 10.1 Components of smart agriculture

10.2 Technology Evolution

10.2.1 Evolution of the Internet of Things (IoT)

In the middle of the 1980s, letters or voice via phone lines was the only means of contact. With time, the term "Internet" emerged, providing a new platform for communication. Over the years, Voice over Internet Protocol (VoIP) becomes a possibility. The idea of the Internet of Things (IoT) has swapped the concept of the Internet in the modern day. The IoT is a method for taking control of devices by connecting existing resources to the Internet. At the Massachusetts Institute of Technology (MIT) Auto-ID Labs in the early 1990s, the initial IoT concept was put forth. Trojan Room coffee pot was the first IoT application was developed in 1999 by University of Cambridge, England (Jia et al. 2012). Later in the same year, a toaster that could be remotely turned on and off over the Internet became the first Internet-controlled appliance in the world (Welbourne et al. 2009; Zhang et al. 2012). The evolution of the Internet of Things (IoT) is depicted in Fig. 10.2 (Dash et al. 2018).

As time went on and Internet technology continued to evolve, new, more sophisticated terminologies were created. Communication was feasible during the pre-Internet phase using a fixed phone line and Short Message Service (SMS). Later, cell phones were included as an improved form of communication. The Internet of content phase was the second phase. The basic characteristics of this phase included the ability to send large-sized communications, i.e., an email that may include attachments, information, entertainment, etc. The Internet of Services phase, which concentrated on digital applications like e-productivity and e-commerce, was the third phase. In the fourth phase, which is known as the Internet of People, people began connecting via social media and a variety of other platforms, including Facebook, Orkut, Skype, YouTube, etc. The IoT is still in its infancy stage. The functional feature that allows

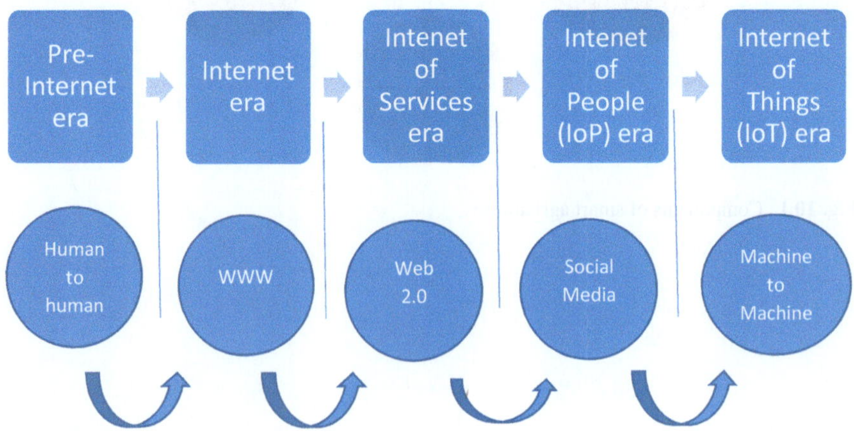

Fig. 10.2 Evolution of the Internet of Things (IoT)

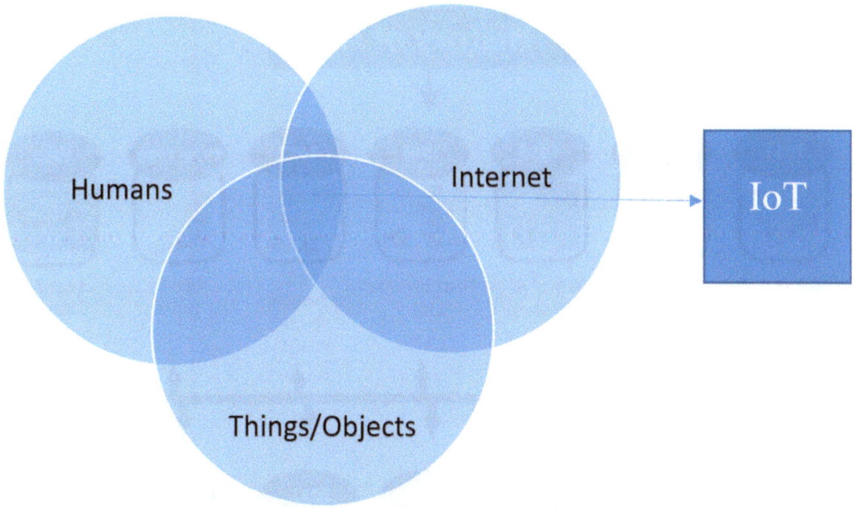

Fig. 10.3 Relationship among the three aspects of IoT

for Internet-based devices is connectivity. As a result, these devices can interact with one another and carry out a variety of tasks when programmed to do so following the functionality and design of diverse things. Although the Internet of Things is primarily about synchronizing things (Objects), this idea is only conceivable because of human intervention. The link between the three IoT aspects is shown in Fig. 10.3.

10.2.2 Evolution of Artificial Intelligence

The science of neural networks emerged in the history of artificial intelligence from the 1950s to the 1970s. From the 1980s to the 2010s, machine learning applications began to take off. Because of ongoing research, rising interest, and extensive application, machine learning has made way for the more complex concept of deep learning. Warren McCulloch and Walter Pits started work on AI in 1943. They put forth a model of synthetic neurons, and Donald Hebb showed how to change the strength of the connections between neurons using an updating mechanism. Hebbian learning is the modern name for his rule. Alan Turing an English mathematician invented machine learning in 1950. He defined a test in his book entitled "Computing Machinery and Intelligence". To assess whether a machine is capable of acting intellectually on par with a person, a Turing test can be employed. Then gradually this science has evolved into many areas like expert systems, machine learning, deep learning, artificial neural networks, etc. (Fig. 10.4), https://www.javatpoint.com/history-of-artificial-intelligence.

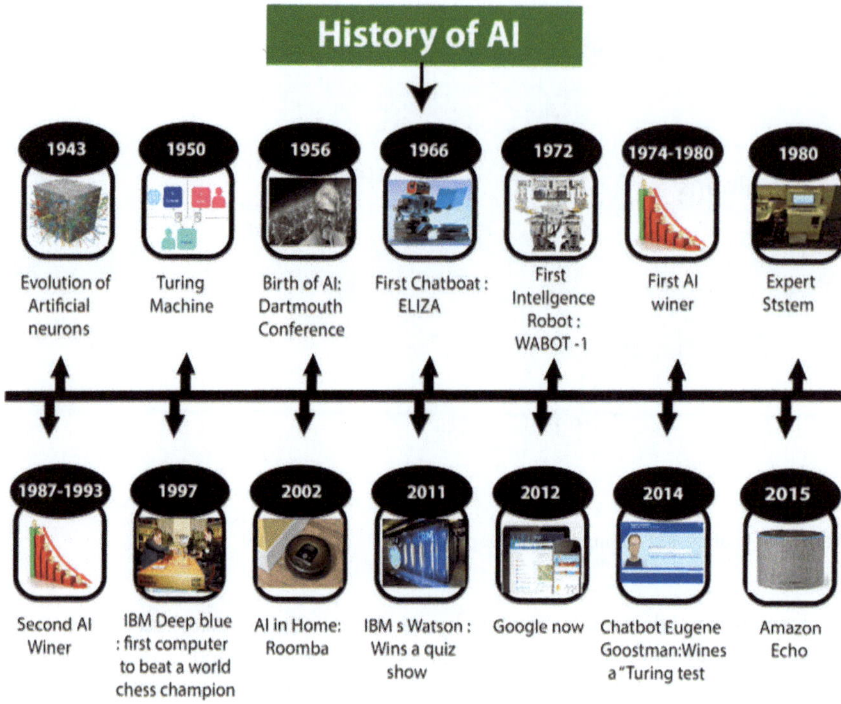

Fig. 10.4 Evolution of AI

10.3 Applications of IoT and AI in Agriculture

10.3.1 Applications of IoT

Smart farming is now possible due to the Internet of Things. A futuristic, capital-intensive approach to producing food that is both sustainable and clean is smart farming. The Internet of Things-based smart farming has benefits for environmental concerns as well. It can help farmers use water efficiently and get the most out of their treatments and inputs. The major applications of IoT-based smart farming are as follows.

10.3.1.1 Precision Farming

To lower its carbon footprint, the globe is turning to sustainable agriculture. Smart farming technologies play a crucial role in this context. It lessens the amount of trash produced, which lessens the harm to the environment while increasing production and financial gains. This method gathers data on every component of the agricultural ecosystem using IoT sensing and communication technology. For instance, information is gathered on the temperature, lighting, humidity, and state of the soil.

10.3.1.2 Agricultural Drones

Unmanned aerial vehicles (UAVs) or drones are used in IoT applications in agriculture for real-time data collection and processing. These drones could have a terrestrial or aerial base. These drones' main uses include agricultural health evaluation, the study of the soil and field type, specialized fertilizer and pesticide application to crops, crop planting, and irrigation.

10.3.1.3 Livestock Monitoring

IoT devices can now be used by farmers to monitor the health of their cattle in real time. Animals can be fitted with IoT sensors and smart gadgets to track their whereabouts and health. Typically, farmers can use drones that offer live cattle tracking capabilities to lower their costs. These IoT devices warn the owners of the animals' health state, allowing them to remove sick animals from the herd and stop the spread of disease. Drone technology allows farmers to monitor their live animals while spending less on labor charges.

10.3.1.4 Smart Greenhouses

The development of a crop depends heavily on environmental conditions. Crop productivity is enhanced by monitoring and controlling factors including humidity, temperature, pressure, and light levels in greenhouse farming. Since manual intervention results in significant labor expenses and energy loss, it is ineffective for managing greenhouses. It has developed a system that can design a smart greenhouse that can track and automatically manage the environmental conditions necessary for the growth of crops (Pandey et al. 2023).

10.3.1.5 Monitoring Climate Conditions

A key consideration for farming is the climate. Once farmers should have full knowledge of the weather conditions, then only they can take preventative actions. Through IoT weather station devices, this information can be made available. Farmers install these sensors in various locations, where they gather data about the local environment and store it on cloud servers. Farmers utilize this information to understand how the weather impacts the crops they select. Farmers can then take the appropriate actions to maximize output by adopting precision farming.

10.3.2 Applications of AI in Agriculture

Artificial intelligence has become one of the most important technological developments across many sectors, including banking, robotics, agriculture, and education. It is revolutionizing the agriculture industry and playing a very important part in the agricultural sector. AI protects the agriculture industry from several concerns also, including food safety, population increase, climate change, etc. Due to the use of AI, the agriculture system of today has upgraded to an advanced level. Different high-tech computer-based systems are developed to identify numerous essential criteria, such as weed detection, yield detection, crop quality, and many more (Banerjee et al. 2018). This agriculture sector is adopting artificial intelligence (AI) technologies to improve a variety of activities related to agriculture throughout the whole food supply chain. These technologies have the potential to assist farmers in various agricultural related tasks, including better crop growth, pest management, soil and growing conditions monitoring, data organization for farmers, and a host of other improvements. The use of AI for various crucial aspects has been described in the following sub-sections.

10.3.2.1 Weather Forecasting

Due to climate change and increased pollution, farmers are finding it difficult to determine the optimal time to plant seeds. Farmers can use artificial intelligence to analyze weather patterns and weather forecasts to help them plan the kinds of crops that can be grown and the best times to sow seeds.

10.3.2.2 Crop and Soil Health Monitoring

The type of soil and its nutrition has a substantial impact on the crops that are grown as well as their quality. Artificial intelligence (AI)-enabled applications may be able to identify soil nutrient deficits as well as plant diseases and pests, providing producers the information they need to use fertilizer to improve the quality of their produce. The farmer can utilize smartphones to snap images of the plants using picture recognition technology.

10.3.2.3 Agricultural Robotics

AI industries are creating robots that can do a range of tasks in farming fields with ease. These machines are programmed to control weeds and harvest crops more quickly and efficiently than people. These robots are programmed to pick and pack produce while inspecting the quality of the produce and looking for weeds. The problems faced by agricultural laborers can also be handled by these machines.

10.3.2.4 Pest and Disease Detection System

AI systems employ satellite photos and chronological data to determine whether any insects have descended and, if so, which species—such as locusts, grasshoppers, and others—have done so. AI can help farmers fight off diseases and pests by warning them on their mobile devices, so they can take required precautions and apply the suitable pest control.

10.3.2.5 Predictive Analytics

The practice of AI in agriculture has generated tools and applications that assist farmers in conducting precise and controlled farming by giving them the required instructions regarding the type of crops to be grown, the best time to plant, water management, crop rotation, pest attacks, timely harvesting, and nutrition management. AI-enabled systems use information about atmospheric parameters like temperature, precipitation, wind speed, and sun radiation along with images captured by satellites and drones to anticipate the weather, monitor agricultural sustainability, and inspect farms for the presence of pests, diseases, and undernourished plants.

10.4 Crop Monitoring System Using IoT and AI

Crops are continuously and carefully inspected throughout the growing season as part of crop monitoring. The monitoring system integrating these cutting-edge technologies will help farmers to manage multiple fields and take reliable and timely decisions at the farm level. A generalized crop monitoring system using IoT and AI is depicted in Fig. 10.5. Various sensors are installed on the sites and these constitute the source for data generation. To access the data gathered by these sensors, the processing unit—which might be any reasonably priced microcontroller like Arduino, Raspberry Pi, etc.—is employed. The processing unit's data are separated at the server or cloud, where the AI algorithm can be used to generate useful output. The actuator may be controlled based on the algorithm's output, and the right action can be taken depending on the various circumstances. Then, all these analytical results and sensor data can be displayed to the users on real-time basis.

The crop monitoring system integrating IoT and AI technology has various components, i.e., sensors, microcontroller/actuators, cloud/server, various AI models (machine learning models, deep learning models, ANN models, etc.), user interface, and users. These components have been described below.

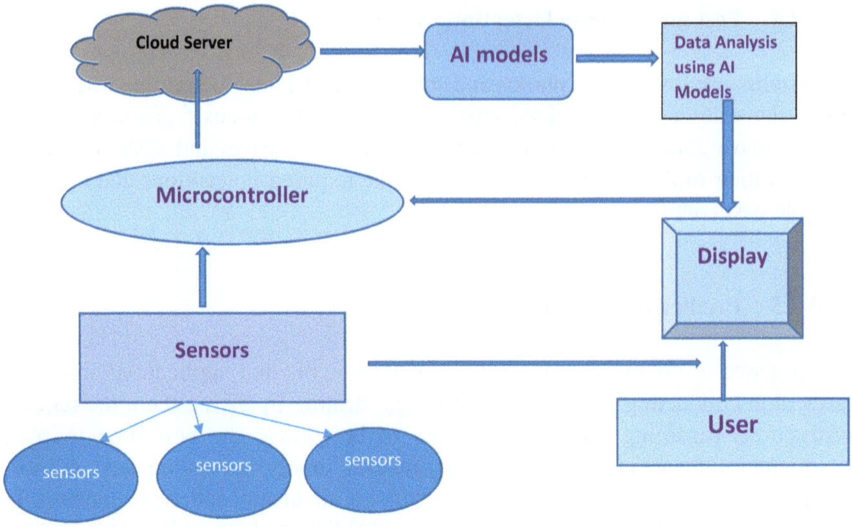

Fig. 10.5 Crop monitoring system integrating AI and IoT

10.4.1 Sensors

A sensor is a device that can identify particular inputs like light, motion, pressure, heat, or moisture and respond by translating those inputs into signals or representations that humans can understand. Different types of sensors are used in agriculture for various purposes as given below.

10.4.1.1 Optical Sensors

It uses multiple light frequencies to measure the qualities of the soil. These sensors, which are mounted on cars or drones, enable the collection and processing of information about plant color and soil reflectance. Clay, organic matter, and soil moisture content can all be measured by optical sensors.

10.4.1.2 Electrochemical Sensors

Data on the chemical composition of the soil are useful. The measurement of soil nutrients is made possible using electrochemical sensors.

10.4.1.3 Mechanical Soil Sensors

These sensors use a tool that pierces the earth to capture the force measured by load cells or pressure scales. It measures the moisture level of the soil. The farm makes use of numerous rain check stations in addition to moisture sensors.

10.4.1.4 Location Sensors

Any location within the required area can have its range, distance, and height measured by these sensors. They rely on GPS satellites for this.

10.4.1.5 Electronic Sensors

It is mounted on tractors and other outdoor machinery to inspect equipment performance. The information was then instantaneously transmitted to computers or sent via email to others using cellular and satellite connection networks. The information can then be retrieved by the field executive using their office computer or mobile device.

10.4.1.6 Airflow Sensors

Its measurements can be taken while moving to certain spots. The pressure required to force a predetermined volume of air into the ground at a fixed depth is the desired output.

10.4.2 Microcontroller

The microcontroller of an embedded system is a tiny integrated circuit that manages a single process. Input/output (I/O) peripherals, a CPU, memory, and other components are all found on a single chip in a conventional microcontroller. Microcontrollers, often referred to as embedded controllers or microcontroller units (MCU), are found in a wide range of products, such as office supplies, robotics, vending machines, and medical equipment. They lack a sophisticated front-end operating system. A microcontroller is built into a system to control a single device function. It achieves this by analyzing data that it gets from its I/O peripherals using its core CPU. The microcontroller's processor reads temporary data from its data memory and processes it using programmed memory instructions in order to understand and apply the incoming data. The essential action is subsequently carried out after it has communicated with its I/O peripherals.

10.4.3 Cloud/Server

A cloud computing provider, which can be accessible via the public Internet or a dedicated private network connection, provides cloud storage, a service that enables us to store data and files online. The provider securely stores, administers, and maintains the storage servers, infrastructure, and network to ensure that we can access the data at *essentially* any scale and with *flexible* capacity. Since we can access the data at any time and from any location without having to buy or manage our data storage infrastructure, cloud storage gives us flexibility, scalability, and durability.

10.4.4 AI Models

Artificial intelligence (AI) models help to automate intelligent decision-making (machine learning and deep learning). This methodology contributes to the faster and wiser development of analytics by having the capacity to expand along with ever-increasing data volumes. The third stage of the data pipeline involves the creation of clever machine learning models to support advanced analytics, which comes after data collection and data preparation. These models *help us* by finding patterns in the data and *inferring* conclusions using several *Machine learning and deep learning techniques*. Model development, model training, and inference drawing are the three fundam*ental processes in AI modeling.*

10.4.5 User Interface

User interfaces are the means by which people interact with computers, websites, and applications. A user interface is important for meeting clients' expectations and promoting our site's effective operation. A well-designed user interface enables successful interaction between the user and the program, app, or machine.

10.5 Future Perception and Constraints

Despite the enormous promise that AI and IoT hold for agricultural applications, farms all over the world still lack knowledge of cutting-edge high-tech machine learning solutions. The following list outlines significant impediments and restrictions to the deployment of this advanced monitoring system (Dharmaraj and Vijayanand 2018). These factors were grouped into four categories, viz. Technical (T), Social (S), Economic (E), and Organizational (O), based on their relevance (Narwane et al. 2022; Table 10.1).

Table 10.1 List of categorical critical factors/constraints

Sl. No.	Factors/constraints	Description	Category
1	High capital investment cost	To ensure the widespread adoption of these systems in the agri-food sector, the capital investment costs for their deployment must be kept to a minimum	E, O
2	Lack of security	The deployment of multiple embedded devices along the smart contract increases the risk of hacking, which could result in system manipulation or illicit data collecting. As a result, the main technological issue preventing the widespread adoption of this system is a lack of data privacy and security	T
3	Lack of interoperability	The integration of IoT and AI in this system in the agri-food sector necessitates seamless data flow between numerous companies, individuals, and systems. Additionally, IoT applications demand more system and stakeholder interoperability	T
4	Network challenges	IoT systems rely profoundly on network communication technology, and problems with its adoption, such as communication range, transmission losses, routing, have a big impact	T
5	Lack of standardization	The interpretation and distribution of data are hampered by the availability of numerous conflicting communication, frequency, and other standards and protocols. Thus, a key element impacting the acceptance of this system is the absence of standardization	T, O
6	Energy efficiency	IoT systems frequently consist of a large number of battery-operated sensor nodes that are scattered widely over remote locations. There are so many sensor nodes that energy-efficient solutions are required	T, E
7	Technical complexity	Each unique technology has its difficulties and limitations, such as complicated hardware and software. The technology selected will have a significant impact on whether this system is adopted in the agriculture industry	T
8	Huge volume of data	These solutions' broad range makes it difficult to gather, store, analyze, and visualize large amounts of diverse data	T

(continued)

Table 10.1 (continued)

Sl. No.	Factors/constraints	Description	Category
9	Lack of connectivity	To create a network over the Internet with this method, numerous sensor connections are necessary. However, a significant concern is the absence of Internet connectivity in rural areas. Emerging economies must have adequate 4G/5G coverage in rural areas	T
10	Scalability	In the future, the agricultural industry will require the deployment of a substantial number of IoT devices. The system's architecture must be created so that an increase in computational power can be accommodated without requiring a complete redesign of the system	T, E
11	Lack of data reliability	This system may be used to keep track of various environmental conditions, perishability, etc., and may need to rapidly notify users of important information. For this to be reliable, there must be little data loss during transmission. As a result, numerous researchers have been testing out different data compression approaches	T, O
12	Data management	The data gathered would be massive and prone to inaccuracies because the agriculture industry is so broad and dynamic. Additionally, to make the obtained raw data relevant, it might need to be filtered and processed. As a result, issues with data management affect the adoption of AI and IoT in the agro-food business	T, O
13	Technical skill requirement	One of the main challenges influencing the adoption of this monitoring system in the agriculture sector is the lack of information and expertise regarding these technologies as well as the absence of training facilities	T, S
14	User acceptance	This system's end user might not comprehend how to use and apply the solutions. Therefore, there is a need to raise user knowledge of the applications of AI and IoT in the cases and to educate them by providing training	T, S, E

(continued)

Table 10.1 (continued)

Sl. No.	Factors/constraints	Description	Category
15	Ease of use	The solutions must be made such that they are simple to install and manage because the end consumers of these products might not be technically savvy. Furthermore, they ought to require little upkeep throughout their existence	S
16	Risk assessment	It is necessary to do a risk assessment of these technical advancements and how they will be used in the agriculture industry. This would persuade everyone involved and speed up the adoption of this cutting-edge monitoring system	S, E

Source Narwane et al. (2022)

Technological considerations played a significant role in the development of this crop monitoring system using AI and IoT technology. Issues with compatibility, security, and technical complexity make implementation difficult. Security, complexity, and interoperability problems can be mostly resolved through recent technological advancements. This system generates a lot of data, and thus, issues with data security can be handled by giving multilayer protections to the data. When adopting this cutting-edge technology, the volume of data, data dependability, and data administration are top priorities. However, new developments in artificial intelligence (AI) and big data analytics can get around these data-related technical problems.

10.6 Conclusion

Farmers are under pressure to increase farm productivity despite deteriorating yield patterns in many staple crops because of concerns and pressures the growing global food demand has placed on the agriculture sector. Growers and the global agricultural industry are relying on technologies to convert modern farming into an efficient and automated crop production method to come over these challenges. Using AI and IoT to monitor crops, smart farming is a new trend that aims to boost farm productivity. IoT-based AI-powered solutions can help farmers to increase productivity while also enhancing crop quality and quantity and ensuring a quicker time to market. The real-time data are gathered using sensors, which are sent to the IoT Cloud Platform. In data analytics using AI models, several machine learning algorithms like Decision Trees, Logistic Regression, and Random Forest for weather prediction and crop monitoring modules can be used. This is very helpful for a variety of farming applications, including determining crop requirements, soil quality, and meteorological conditions.

References

Banerjee G, Sarkar U, Das S, Ghosh I (2018) Artificial intelligence in agriculture: a literature survey. Int J Sci Res Comput Sci Appl Manag Stud 7(3):1–6. ISSN: 2319-1953

Bhargava K, Kashyap A, Gonsalves TA (2014) Wireless sensor network based advisory system for Apple Scab prevention. In: Twentieth national conference on communications (NCC) in 2014. Institute of Electrical and Electronics Engineers, pp 1–6. https://doi.org/10.1109/NCC.2014.6811263

Dash A, Pal S, Hegde C (2018) Ransomware auto-detection in IoT devices using machine learning. no. December, pp 0–10

Dharmaraj V, Vijayanand C (2018) Artificial intelligence (AI) in agriculture. Int J Curr Microbiol Appl Sci 7(12):2122–2128. https://doi.org/10.20546/ijcmas.2018.712.241

Jia X, Feng Q, Fan T, Lei Q (2012) RFID technology and its applications in Internet of Things (IoT). In: 2012 2nd international conference on consumer electronics, communications and networks (CECNet). Institute of Electrical and Electronics Engineers, pp 1282–1285. https://doi.org/10.1109/CECNet.2012.6201508

Narwane VS, Gunasekaran A, Gardas BB (2022) Unlocking adoption challenges of IoT in Indian Agricultural and Food Supply Chain. Smart Agric Technol 2:100035. https://doi.org/10.1016/j.atech.2022.100035

Oerke EC (2006) Crop losses to pests. J Agric Sci 144(1):31–43. https://doi.org/10.1017/S0021859605005708

Pandey K, Singh KG, Singh A (2023) Multi-sensors based smart nutrient reuse management system for closed soilless culture under protected cultivation. Comput Electron Agric 107495

Savary S, Ficke A, Aubertot JN, Hollier C (2012) Crop losses due to diseases and their implications for global food production losses and food security. Food Secur 4(4):519–537. https://doi.org/10.1007/s12571-012-0200-5

Shinde SS, Kulkarni M (2017) Review paper on prediction of crop disease using IoT and machine learning. In: 2017 International conference on transforming engineering education (ICTEE). Institute of Electrical and Electronics Engineers, pp 1–4. https://doi.org/10.1109/ICTEED.2017.8586207

Welbourne E, Battle L, Cole G, Gould K, Rector K, Raymer S, Borriello G (2009) Building the internet of things using RFID: the RFID ecosystem experience. IEEE Internet Comput 13(3):48–55. https://doi.org/10.1109/MIC.2009.52

Zhang M, Sun F, Cheng X (2012) Architecture of internet of things and its key technology integration based-on RFID. In: 2012 fifth international symposium on computational intelligence and design. Institute of Electrical and Electronics Engineers, vol 1, pp 294–297. https://doi.org/10.1109/ISCID.2012.81

Chapter 11
Soil Moisture and Temperature Management Using IoT for Sustainable Farming

P. Kannan, M. Mohamed Roshan Abu Firnass, Jay Bose, S. Ponmani, and C. Swaminathan

Abstract Soil temperature and moisture are the two most essential indicators of soil health, since they influence numerous critical agricultural, environmental, and morphological activities inside the soil. The temperature and moisture content of the soil are essential determinants in agricultural, engineering, geological, biological, and hydrological research. Soil moisture, a critical component of hydro-meteorological systems, is characterized by the existence of a limited number of molecules of water within the pore spaces. Surface and subsurface soil temperatures and moisture should be regularly monitored and measured using high-resolution data. The Graphical User Interface (GUI) displays the soil's temperature and moisture trends. The most crucial aspect of this system is its data output, that can be viewed and handled remotely using Wi-Fi and an online web application called LabVIEW. The wireless soil temperature and moisture monitoring sensors aid in data collection, analyzing, presentation, and storage via an easy-to-use interface for irrigation water scheduling over the web. The predominant goal of this chapter is to offer a sophisticated wireless network of sensors for a farming scenario. This report provides farmers with a more efficient and cost-effective technique of measuring many environmental factors important for crop growth, such as temperature, moisture, and relative humidity. These insights might help with irrigation scheduling, soil management, fertilization, and drought monitoring, along with depth analysis of soil moisture and temperature.

P. Kannan (✉) · M. Mohamed Roshan Abu Firnass
Department of Soils and Environment, Agricultural College and Research Institue, Madurai 625104, India
e-mail: kannan.p@tnau.ac.in

J. Bose
School of Science, University of Western Sydney, Sydney, NSW 2753, Australia

S. Ponmani
Mother Terasa College of Agriculture, Pudukkottai 622102, India

C. Swaminathan
Centre for Water and Geospatial Studies, Tamil Nadu Agricultural University, Coimbatore 641003, India

K. Pandey et al. (eds.), *Artificial Intelligence and Smart Agriculture*, Advances in Geographical and Environmental Sciences,
https://doi.org/10.1007/978-981-97-0341-8_11

Keywords Soil temperature · Soil moisture · Sensor · IoT and sustainable agriculture

11.1 Introduction

Monitoring and regulating soil moisture and temperature are crucial choice-making criteria in easy and effective water irrigation and management. In reality, agricultural output is reliant on these two major characteristics, which are a vital and integral part of human health and well-being (Mujumdar et al. 2021; Xu et al. 2018). The expense of monitoring and managing soil moisture and temperature, and hence acquiring data, is a crucial factor in agricultural productivity (Raduła et al. 2018).

Soil moisture and temperature fluctuation is a significant aspect of farming, environmental disasters (landslides and debris flows), ecological, hydro-meteorological, and climatic research (Chakraborty et al. 2018; Datta et al. 2018; Kalita et al. 2016; Adeyemi et al. 2016). The amount of water in the soil fluctuates dramatically as time passes and across geographical sizes. Several characteristics, such as soil texture, plant diversity, bulk density, rainfall, and irrigation, vary from centimetres to decimetres (Zhang et al. 2018).

Due to the geographical heterogeneity of soil water in the field, evaluating soil water status using the thermogravimetric approach, which entails obtaining soil samples directly from the field and exposing them to heat to remove moisture before calculating heat loss, is problematic. It is also time intensive, laborious, and costly. It uses destructive sampling, which makes repeating measurements on a given sample impractical.

Soil biological and chemical activities, such as organic matter breakdown and mineralisation, are directly impacted by daily and yearly temperature changes (Shamshiri and Weltzien 2021). Temperature influences plant development both directly and indirectly by influencing crop physiological activity and soil nutrient levels. Soil temperature influences crop development during seed emergence and germination (Dias et al. 2016; Saeed et al. 2019).

Accurate and exact assessment of soil moisture and temperature is essential due to many applications in various research disciplines (Iwata et al. 2017; Placidi et al. 2020). Several ways for recording and analyzing it have been devised. One way to reduce the expenses associated with conventional, inconvenient, bulky, less effective, and cost-inefficient wired network is to utilize wireless soil moisture and temperature monitoring sensors, which enable connection between physically and or financially different to access (Kapilaratne and Lu 2017). It aids in the preservation of natural assets and water for favourable environmental sustainability. Numerous important agronomic activities, such as surface water run, leaching, seedling emergence, evapotranspiration, and organic matter mineralisation, cannot be assessed without knowing the moisture condition of the soil (Then et al. 2016).

Soil temperature and moisture content are the main factors affecting crop development and output (Kojima et al. 2016; Nagahage et al. 2019; Gao et al. 2018). At the moment, there is a significant advancement in the theory and application of stratified non-contact soil temperature and moisture measuring technology both domestically and internationally (Bachuwar et al. 2017). Wireless technology with numerous sensors for precision farming has grown widespread in greenhouse effect studies (Zhou et al. 2019). Individual people are incorporating the benefits of embedded systems into agricultural parameter control and monitoring systems. Remote access is an excellent way for avoiding environmental interference and increasing productivity (John et al. 2015).

To transfer signals in a remote monitoring system, Ethernet networks, RF modules, and ZigBee wireless networks are now employed (Goswami et al. 2018). The NImyDAQ system was utilized to collect, regulate, and monitor soil moisture and temperature data, which was then transmitted via the web to provide a data visualization analysis in the GUI format. The GUI interface analyzes the data, calibrates, displays, and saves it. This system has a function to set soil temperature and moisture thresholds, that helps in the irrigation system automation. The designed GUI holds variety of advances from soil temperature and moisture nodes put in farms.

Soil moisture and temperature are measured using a variety of approaches, including remote sensing and in situ platforms (Palaparthy et al. 2017). Traditional soil moisture and temperature measuring methods, such as thermogravimetric, sensor-based approaches, are categorized as follows: resistive sensor, inductive sensor, conductive sensor, electromagnetic induction, dielectric methods, neutron moisture meter, tensiometer method, and so on (Dias et al. 2016). This chapter also discusses the technique and safety precautions followed during system assembly and development, as well as the calibration and standardization of the results obtained.

11.2 Materials and Methods

11.2.1 Wireless Sensor Networks

A wireless sensor network (WSN) is a system that includes a radio frequency (RF) transmitter and receiver, sensors, microcontrollers, and power supplies. Latest advancements in wireless sensor network technologies have resulted in the creation of inexpensive and low-energy multi-purpose sensor nodes (Prasad and Madkar 2013). Sensor nodes offer environmental sensing as well as processing of data. Monitoring of various surroundings is possible because of the sensors including humidity, temperature, and volatile substance detection which are used (Cho et al. 2011). They can communicate with other sensing devices and transfer information to outside users.

Sensor networks are employed for a broad array of applications such as wireless data collection, machine maintenance and smart building monitoring and roadways, environmental sensing, security standards, automatic on-site tracking of expensive items, safety practices, and a variety of other services are available (Hwang and Yoe 2010). The typical WSN system consists of the application layer, data connection layer, transport layer, network layer, mobility management plane, physical layer, task management plane, and power control plane (Cheng et al. 2010).

WSN currently supports Bluetooth and ZigBee as standard features both operate at 2.4 GHz in the Industrial Scientific and Medical (ISM) band, which allows for licence-free operation, vast spectrum allotment, and global interoperability. Frequency increases typically result in bandwidth expansions that enable quicker data rates, but they also increase power requirements and dramatically shorten transmission distance (Wang et al. 2011).

Since multi-hop communication needs less electricity than typical single-hop communication, it is possible in WSN. Wi-Fi (IEEE 802.11) may also be employed to build a WSN; however, this protocol is commonly utilized in PC-based systems since it was meant to expand or replace a wired LAN. It consumes a lot of power, and the minimal flexibility of a battery supply is still a big disadvantage (Salunke et al. 2015).

11.2.2 ZigBee

The IEEE 802.15.4 system is the basis for the ZigBee technology. The IEEE 802.15.4 system provides the physical surface and the medium access control (MAC) surface for low-rate wireless personal area networks. The physical surface provides three unique frequency bands with varying gross data rates: 2450 MHz (250 kbs^{-1}), 915 MHz (40 kbs^{-1}), and 868 MHz (20 kbs^{-1}). It also includes capabilities such as channel selection, connectivity estimation, energy evaluation, and clear channel rating (Haefke et al. 2011). ZigBee regulates the network as well as the application layer.

The network component is in charge of organizing and routing communication over a multi-hop network, as well as specifying topologies for networks such as tree, star, and peer-to-peer mesh. The application layer provides a framework for developing and communicating apps across networks (Pang 2011). It is widely used in building management, automation, surveillance, consumer products, computer accessories, medical inspection, and toys, as well as agriculture and food. For such uses, a technology with long battery life, reliability, automated or semiautomatic installation, the ability to simply add or remove network nodes, data transfer through walls and ceilings, and a cheap system cost is required (Haefke et al. 2011) (Fig. 11.1).

11.2.2.1 System Architecture

The proposed gear for this system comprises an 8-bit AVR, a Bluetooth component, temperature, humidity, and soil moisture detectors, as well as an LCD. The system is inexpensive in cost and power usage, making it affordable to anybody. The monitored data is stored on the server. Precision farming can benefit from it. The system should be developed so that even illiterate farmers can use it (Prasad and Madkar 2013).

Fig. 11.1 ZigBee soil temperature and moisture sensor

They may examine various soil factors such as salinity, acidity, moisture, and so regularly.

Since the power supply is not constant during the irrigation season, they must track their remote pump house over the night (Cho et al. 2011). The system may be installed outside the village in the pump house; it is linked to the pump opener, and sensors are put in at various positions around the field for data gathering and recording. They can use this method to turn on their pump from their home whenever they need it (Hwang and Yoe 2010).

Temperature Sensor

The output voltage of the LM35 series of precision-integrated circuit temperature sensors reflects the temperature in Celsius. It is a cheap sensor with a tiny footprint. It operates at temperatures ranging from − 55° to + 150 °C (Fig. 11.2).

Humidity Sensor

When moisture gets absorbed, humidity monitoring equipment frequently relies on other values such as pressure, mass, temperature, or an electrical or physical shift in a substance. These observed values can be converted to humidity measurements by calibration and computation (Fig. 11.3).

Fig. 11.2 Temperature sensor and ZigBee module with wire antenna

Fig. 11.3 Humidity sensor

Fig. 11.4 Soil moisture sensor

Soil Moisture Sensor

The soil moisture detector used is a capacitive kind. At 100% moisture, the sensor provides an analog signal of no volts, and at 0% moisture, it produces a signal of five volts (Fig. 11.4).

11.2.2.2 Working

The system operates in two components, viz., transmitter and receiver.

Transmitter

After the system has been reset, the power is switched on for the first time. Different sensors read signals, and the output is sent to the microcontroller. Sensor output is sent to the microcontroller via eight-channel ADC pins. The microcontroller's output is delivered to ZigBee via the Rx and Tx pins (Fig. 11.5).

Receiver

At the receiver side of ZigBee, there are just one Tx and one Rx pin. The microcontroller receives a signal and monitor variables such as temperature, soil moisture, and humidity. These characteristics are tracked on a computer through an RS-232

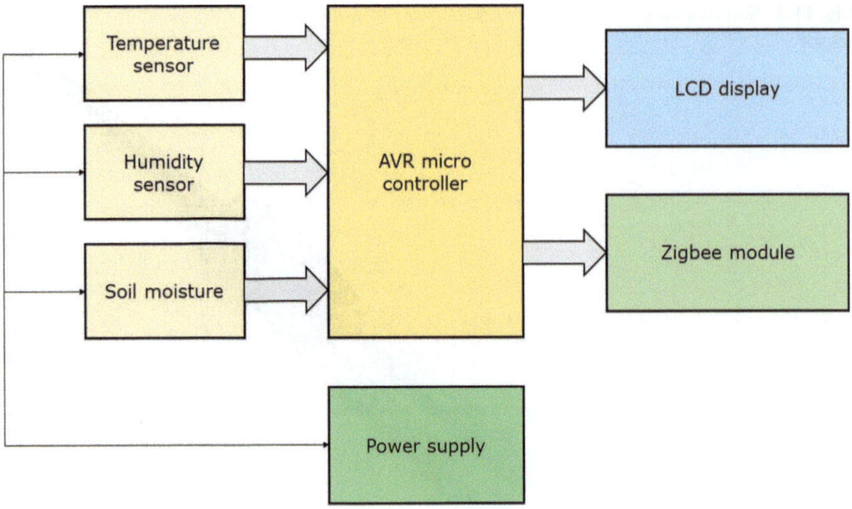

Fig. 11.5 Block diagram of a transmitter

interface. This information may be utilized to improve precision farming. Microcontroller data may be used to control the actuators. This is how the entire automatic irrigation system works.

Different sensors, such as temperature and moisture sensors, detect the needed temperature and moisture content. As a result, it contributes to providing a suitable environment for agricultural cultivation. The land was irrigated using a different irrigation technology. Water is initially stored in tanks and trench pipes before being used to water crops with drip irrigation and sprinklers systems. We must thoroughly irrigate the land in agricultural fields. We must irrigate based on the soil, the changes in elevation of the land, and where it is required.

Chavan and Karande (2014) created a ZigBee wireless sensor system to track agricultural field regions by putting moisture sensors in the field to identify locations where water levels are low. Based purely on the findings, we can irrigate that region. As a result, we can protect water resources and lower the chance of land flooding. Crops now contain more fertilizer than ever before, which has an impact on humans. We collect information about the soil and determine its acidity using pH sensors. This enables us to fertilize crops just when necessary and prevents overfertilization. In the surroundings of a paddy field, temperature might vary (Fig. 11.6).

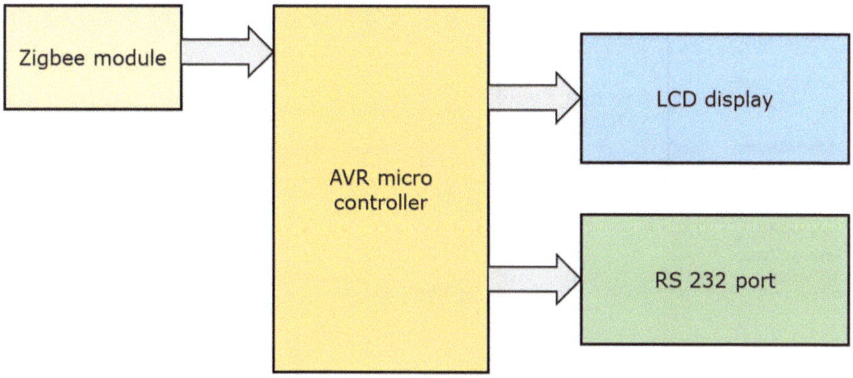

Fig. 11.6 Block diagram of receiver

11.2.3 System Based on IoT for Continual Assessment and Monitoring of Soil Temperature and Moisture

The suggested system is a cost-effective design based on farmers' needs for increased crop output (Kumar et al. 2016). It also provides data on weather variables and soil moisture levels without the need for human assistance.

11.2.3.1 System Block Diagram

The system consists of a humidity and temperature sensor named DHT11, a capacitive soil moisture sensor (Ayday and Safak 2009), a resistive soil moisture sensor, an ESP8266 Wi-Fi module, and an Arduino Uno Board. The variables detected by the sensors are passed into the ATMEGA328 microprocessor. It is then shown on the LCD together with remotely on the PC over the internetFigs. 11.7 and 11.8.

The DHT11 humidity and temperature sensor module delivers a digital value of relative humidity and temperature to the microcontroller. The capacitive sensor and soil moisture sensor detect moisture in the soil and generate an analog signal corresponding to the moisture content. The output voltage is then converted into soil moisture by the microcontroller (Kumar et al. 2016; Nath et al. 2018). The data is then sent to the Wi-Fi module (ESP8266) via serial port (Singh and Saikia 2016).

The Wi-Fi module is driven by an 80 MHz TensilicaXtensa 32-bit RISC CPU and delivers data to the Thingspeak site via the web. Once the information has been uploaded to the cloud system, it may be accessed remotely over the web. Temperature, humidity level, and soil moisture values are available on the Thingspeak streams. Thus, the data from the ATMEGA328 microcontroller is displayed on the LCD and also transmitted to the PC for remote viewing.

Two instruments monitor soil moisture. A resistive soil moisture sensor is one type, whereas a capacitive soil moisture sensor is another (Ayday and Safak 2009).

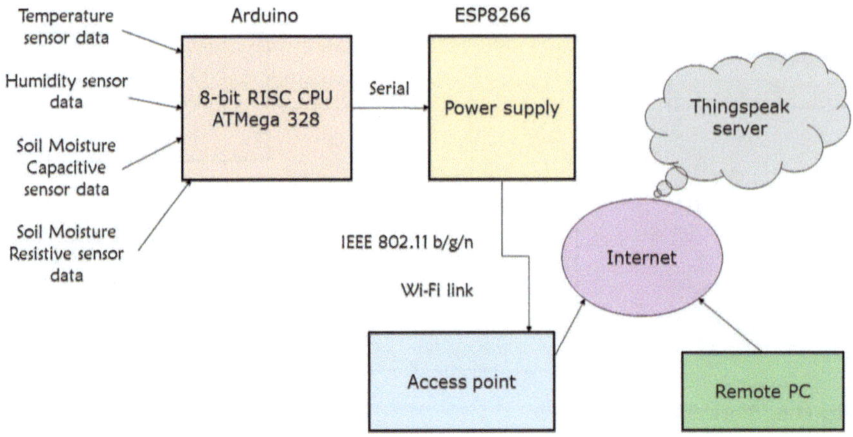

Fig. 11.7 Block diagram of the system

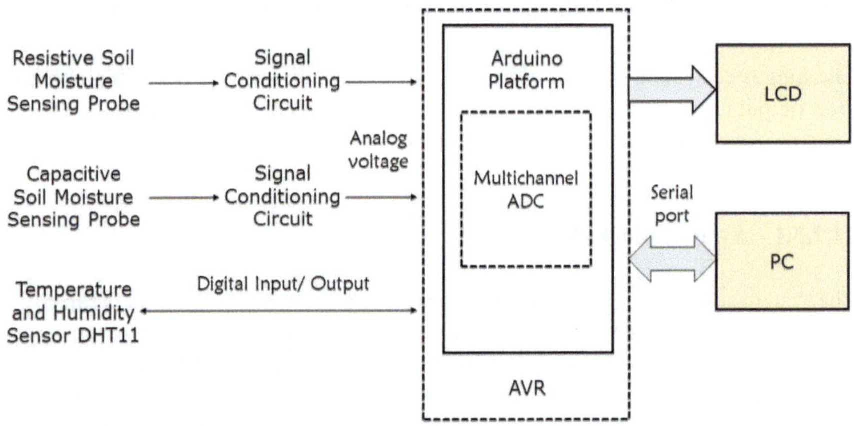

Fig. 11.8 Block diagram of the measurement unit

Soil moisture sensors detect the presence of water in the soil. It is an essential factor in agricultural environment study (Songhua et al. 2010). Soil moisture must be measured and monitored to determine when and how much to irrigate the crops. The soil moisture is estimated here using the gravimetric approach.

To begin, a soil sample is oven dried to eliminate moisture content and then weighed. The dry sample is then weighed after adding a small amount of water. The moisture content is then calculated based on the weights of the dry and moist samples. Both soil moisture sensors are put into a dry soil sample and the associated voltages are recorded. These voltages are subsequently fed into the ATMEGA328 microprocessor, which converts them to moisture levels and displays them (Kaur

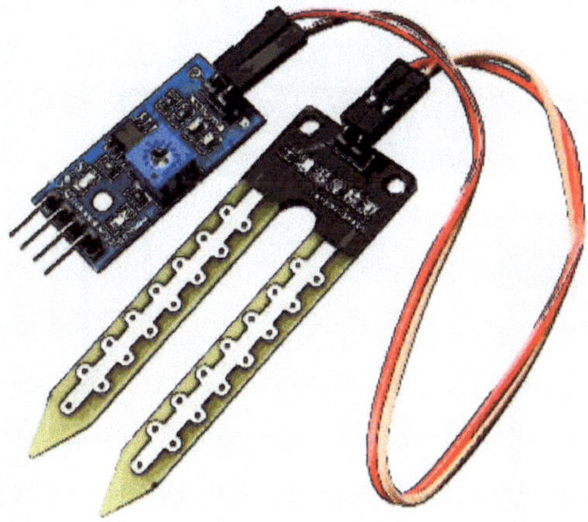

Fig. 11.9 Soil moisture resistive sensor

Fig. 11.10 Soil moisture capacitive sensor

et al. 2016). Figure 11.11 depicts the experimental setup for studying the features of resistive sensors (Fig. 11.9) and capacitive sensors (Fig. 11.10).

11.2.3.2 Hardware Components

The Arduino UNO is a microcontroller board with an ATmega328. It has fourteen digital I/O pins, six analog ports, a ceramic resonator with a frequency of 16 MHz

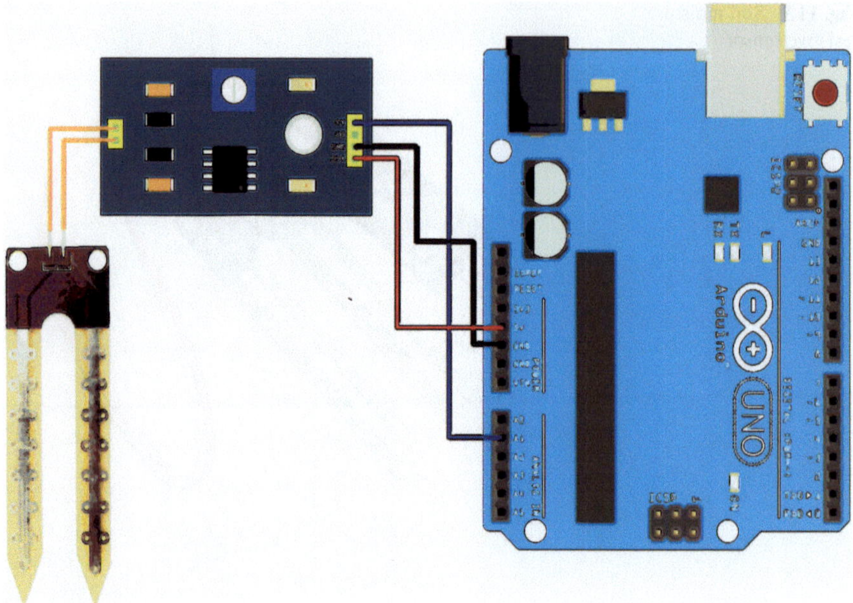

Fig. 11.11 Experimental setup for measuring soil moisture

a power connector, an USB port, an ICSP header, and a button to reset to support the microcontroller. It is connected to the PC via USB port and displays the data on the computer's screen. The Atmega16U2 has been programmed to function as a USB-to-serial converter. It is a free to use hardware and software architecture that is simple to use. It is easy to use and great for making interactive applications. The Arduino detects the environment by collecting data from multiple sensors and turning it into humidity, temperature, and soil moisture measurements. The results are then shown on the LCD as well as the PC.

ESP8266

The ESP8266 is a cheap Wi-Fi chip with a complete TCP/IP stack and microcontroller unit (MCU) functionality manufactured by Shanghai-based Chinese firm Espressif Systems (Fig. 11.12).

11.2.4 Flowchart

See Figs. 11.13 and 11.14.

Fig. 11.12 ESP8266 Wi-Fi module

11.3 Results and Discussion

11.3.1 Wireless Soil Temperature and Moisture Tracking System Based on LabVIEW for Irrigation Water Planning

11.3.1.1 NImyDAQ System Design and Development

Bachuwar et al. (2017) utilized NImyDAQ to connect the soil temperature and moisture sensors. The soil moisture sensor was connected to channels + 0 and − 0, whereas the soil temperature sensor was connected to channels + 1 and − 1. The water pump was wired to the NImyDAQ's analog output. Figure 11.15 displays a schematic representation of our system.

NImyDAQ includes two analog input networks, each of which may be configured as either versatile high-impedance voltage input or audio input. The analog inputs are combined, which means that a single analog-to-digital converter (ADC) samples both channels. In broad-application mode, voltage signals of up to 10 V may be generated at rates of up to 200 kS/s per channel. The two channels imitate left and right stereo line level inputs in audio mode. NImyDAQ has two channels for inputs and one output channel in our example.

Both channels have a digital-to-analog converter (DAC) that may be updated simultaneously. NImyDAQ also contains eight DIO lines, each of which is a programmable function interface (PFI). As a result, it may act as an universal software-timed digital input or output, as well as an unique input or output for a digital meter.

Bachuwar et al. (2017) initialized NImyDAQ utilizing a measurement I/O tool kit (a measurement I/ONI-DAQmxDAQ Assistant). Channels of NImyDAQ were set up utilizing an acquired signal → analog input → voltage → select channel ai0 and ai1 → finish. Furthermore, samples were chosen continuously while keeping a read rate

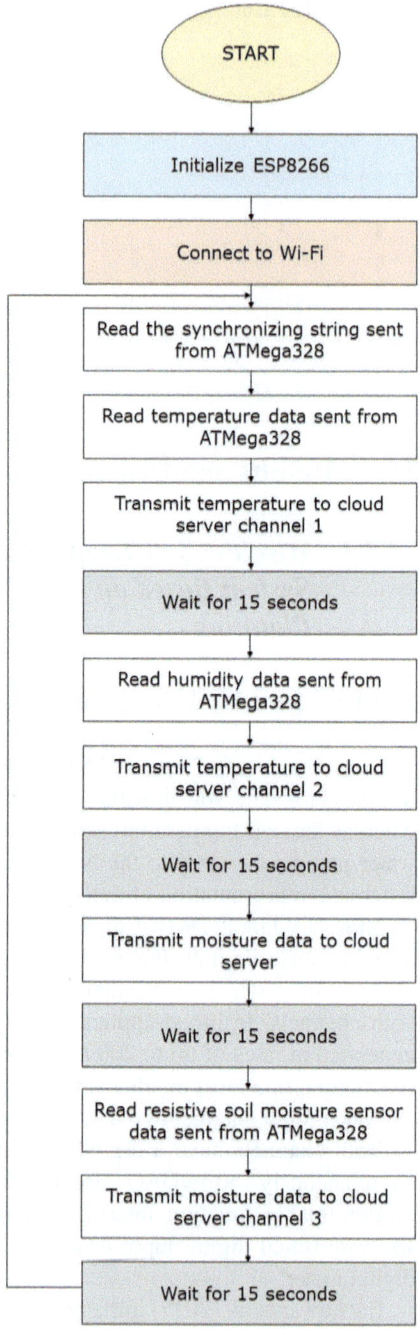

Fig. 11.13 Working operation of ATMega328 microcontroller

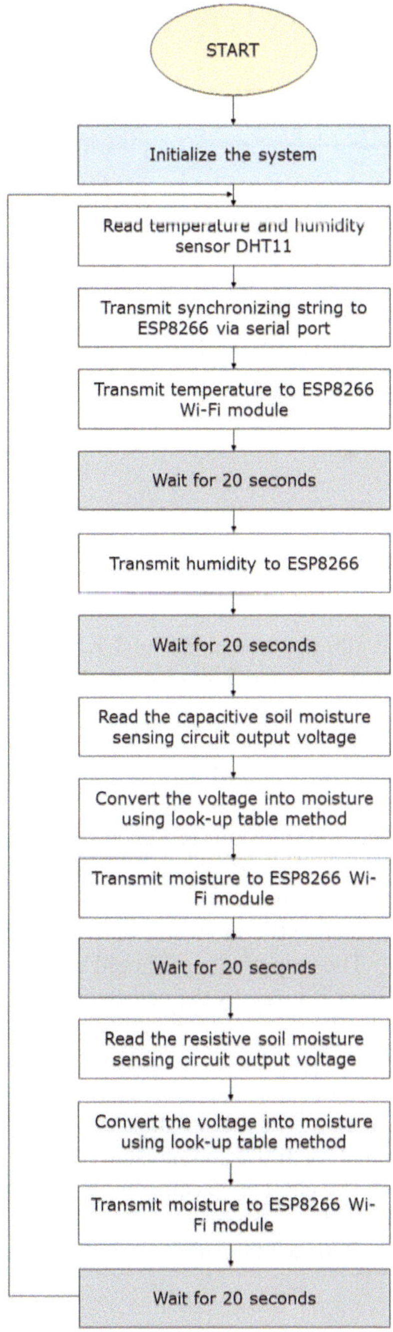

Fig. 11.14 Working operation of ESP8266 microcontroller

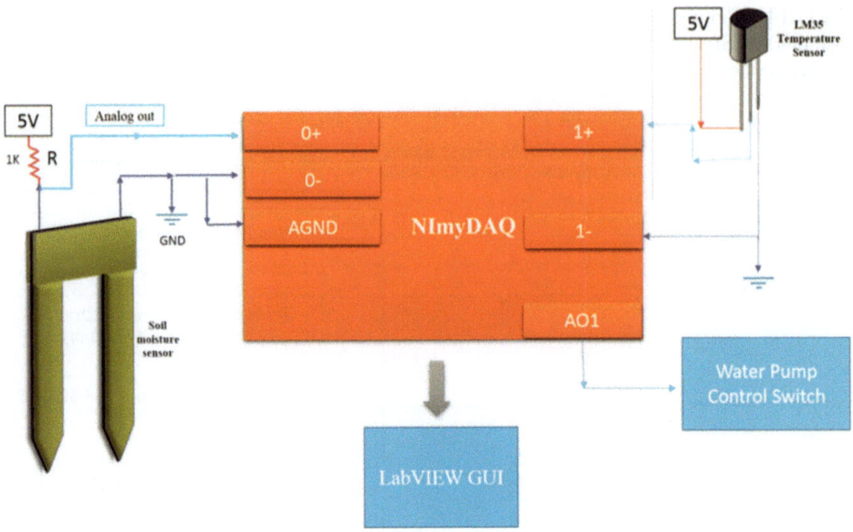

Fig. 11.15 Soil moisture and temperature monitoring using NImyDAQ

of 100 and a rate (Hz) of 1 K in setup. The voltage input for channel ai0 was kept between 0 and 5 V, whereas channel ai1 was kept between 0 and 1 V.

11.3.1.2 Soil Moisture Calibration

The soil moisture sensor has been calibrated across two conditions: 0 and 100% soil wetness. To obtain soil moisture data, voltage-type soil moisture sensors with voltage outputs of 5 V for moist soil, indicating 100% moisture, and 0 V for dry soil, indicating 0% moisture, were used.

The output voltage should be proportional (linear) to soil moisture. A calibrating factor is given by the percentage conversion of soil moisture.

$$\text{Soil moisture in}\% = (V_{\text{out}}/5) * 100, \tag{11.1}$$

where V_{out} is the soil moisture sensor's output voltage in volts, while 5 is the output value for 100% soil moisture.

The alternative equation for determining soil moisture is:

$$\text{Soil moisture in } \%(y) = 20 * \text{output voltage of soil moisture sensor}(x). \tag{11.2}$$

11.3.1.3 Calibration of Soil Temperature (LM35)

The soil temperature was determined using an LM35 accurate digital temperature sensor, the output of which represents a linear function of temperature. The sensor has a temperature rating of – 550–1500 °C. The calibration factor needed to convert sensor output to temperature is given as follows:

$$\text{Soil temperature(in °C)} = (\text{Output voltage of LM35}/10) * 1000. \tag{11.3}$$

The alternative equation for determining the soil temperature is:

$$\text{Soil temperature(in °C)}(y) = (\text{Output voltage of LM35}(x) * 10. \tag{11.4}$$

11.3.2 Remote Access Web Publishing Tool

The web publishing tool in LabVIEW 2014 was used to provide a Wi-Fi internet connection, allowing remote access to actual time soil temperature and moisture data. The following are the steps to accessing the GUI:

Front panel → web publishing tool → browse the GUI.

VI → view mode → monitor → seconds between updates is 1 s → next as shown in Fig. 11.16.

The created technology aids in the acquisition of continuous data on soil temperature and moisture via remotely operated sensors. With a nominal inaccuracy of 0.5%

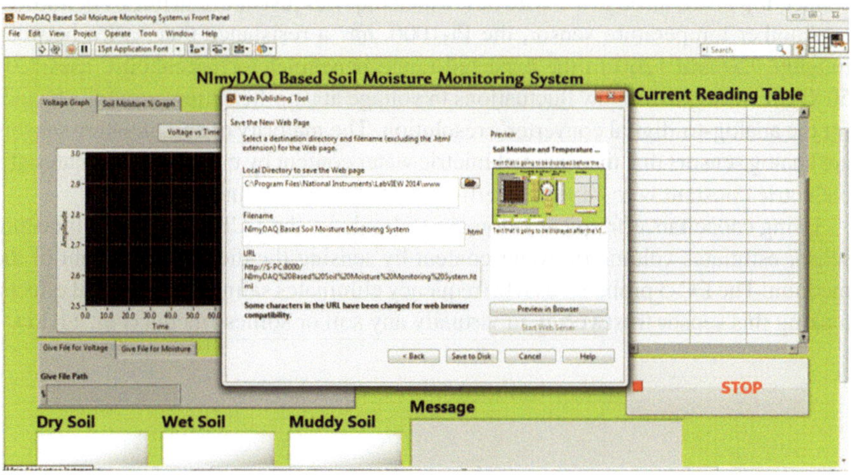

Fig. 11.16 Web publishing tool

± 0.2 mV, the data collection system is precise. The online publishing tool enables end users to connect wirelessly. GUI allows for greater flexibility in determining threshold values for sensor characteristics in diverse crop-growing environments.

Furthermore, the GUI offers a user-friendly platform for evaluating the efficacy of sample time delay and sampling rate. The GUI has the extra benefit of displaying data on a spreadsheet as well as parameter trends. The scale of the sensing network may be increased by using the NIUSB6009 DAQ or the NImyRIO card.

11.4 Examples with Real-World Applications

The majority of soil moisture sensors, particularly the 10HS, use capacitance technology to detect the dielectric characteristics of the soil and calculate its volumetric water content for scientific examination. These sensors generally operate at a frequency of 70 MHz, which eliminates salt and textural impacts while providing detailed information that allows hourly or daily monitoring of soil moisture levels through analog voltage proportionate to water content. These sensors offer low salt and temperature sensitivity as well as low power consumption. They can be attached directly or connected via interfaces to IoT boards for real-time monitoring.

A light level sensor, also known as a light-dependent resistor (LDR), is a live sensor made of a light-sensitive semiconductor that has high accuracy, quick response, and resistance. It minimizes resistance to absorbing brightness (light) on the component's sensitive surface.

The DS18b20 waterproof sensor is a long-lasting and anti-corrosive sensor that may be used to assess soil temperature. This sensor has wire lengths of 1.8 and 3 m and provides 9-bit Celsius temperature data. The DS18B20 communicates with the connecting board through a 1-wire bus, which requires just one data line by definition.

Another temperature sensor, the Pt-1000, has a resistance that varies between roughly 920 and 1200 in the expected range beneficial for greenhouse uses (20–50°C), resulting in too low fluctuations in voltage at significant temperature changes for the analog-to-digital converter's resolution. The majority of soil moisture sensors are analog sensors that measure volumetric water content by measuring the material's dielectric constant with capacitance/frequency-domain technology.

Using capacitance/frequency-domain technology, the ECHO EC-5 soil moisture sensor estimates volumetric water content by sensing the dielectric constant of the medium. The EC-5 probe's 70 GHz frequency eliminates salinity and textural effects, making this sensor trustworthy in virtually any soil or soilless media (Fig. 11.17).

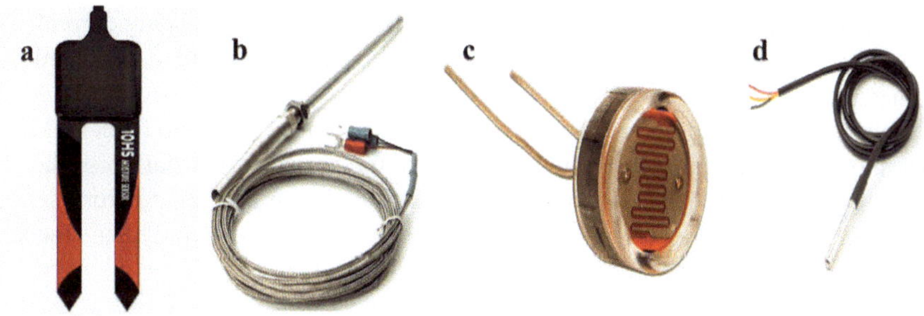

Fig. 11.17 Some of the most common soil moisture sensor probes used in digital agricultural applications; **a** 10HS soil moisture sensors, **b** Pt-1000 temperature sensor, **c** light-dependent resistor, **d** waterproof DS18b20 soil temperature sensor

11.5 Future Prospects

For hydro-meteorological, agricultural modelling, and other applications, real-time data on soil temperature and moisture profile fluctuation are essential. Researchers can examine various crops and moisture fluctuations with the use of this monitoring system in order to improve irrigation methods. Additionally, a variety of soil types can be accommodated by soil moisture sensors. Using this surveillance system, researchers may investigate different crops and moisture fluctuations to enhance irrigation techniques.

Soil moisture and temperature sensors can also be tailored to a number of soil types. According to the software, Bluetooth and LoRA may be utilized to wirelessly deliver output data to the user without requiring internet access. To detect the quantity of moisture in a given location, a GPS module can be installed. A camera may be attached to this monitoring system to capture sky and ground parameters on a regular basis in order to monitor crops and their growth. Wireless field monitoring not only helps users to decrease human power, but it also enables users to view exact changes in the field. It is less expensive and uses less energy.

11.6 Conclusion

At the moment, the area of agricultural cultivation is expanding, which expands the area of control of agro-based intelligent systems. As a result of this circumstance, the sensor, which acts as the agricultural IoT's front end, requires network connection capacity. Sensors with transmission capabilities serve as the foundation for real-time continuous data collection. The application sensor may be placed in a remote node for IoT applications. For the progress of digital farming, more sensors in agricultural IoT must be created and installed.

The soil moisture and temperature sensors were critical in acquiring current and precise soil moisture and temperature status and providing correct data for microclimate surveillance. Some researchers took a more specific approach to building soil monitoring systems, creating a low-cost, computerized model to track and alter moisture levels in soil samples in order to meet the needs of rural farmers who are technologically illiterate. In conclusion, the soil moisture and temperature sensor network assessed in this chapter has the potential for in situ deployment for soil moisture and temperature management and monitoring.

Acknowledgements The authors acknowledge the ICAR-NAHEP-IDP for funding to visit University of Western Sydney for doing part of literature survey in digital library at School of Science.

References

Adeyemi O, Norton T, Grove I, Peets S (2016) Performance evaluation of three newly developed soil moisture sensors. In: Proceedings of the CIGR-AgEng conference, Aarhus, Denmark, pp 26–29

Ayday C, Safak S (2009) Application of wireless sensor networks with GIS on the soil moisture distribution mapping. In: Symposium GIS Ostrava, pp 1–6

Bachuwar VD, Shligram AD, Prabhakar P, Deshmukh LP (2017) Lab view based wireless soil moisture and temperature monitoring system for water irrigation management. Indian J Sci Res 15(2):186–190

Chakraborty M, Kalita A, Biswas K (2018) PMMA-coated capacitive type soil moisture sensor: design, fabrication, and testing. IEEE Trans Instrum Meas 68(1):189–196. https://doi.org/10.1109/TIM.2018.2838758

Chavan CH, Karande PV (2014) Wireless monitoring of soil moisture, temperature & humidity using zigbee in agriculture. Int J Eng Trends Technol (IJETT) 11(10):493–497. https://doi.org/10.14445/22315381%2FIJETT-V11P296

Cheng W, Qiao X, Liu Y, Yu C (2010) Low power research and design in plant eco-physiological monitoring system based on ZigBee. In: 2010 world automation congress. IEEE, pp 67–71

Cho H, Park D, Park CY, Kim HG, Shin CS, Cho YY, Park JW (2011) A study on localization based Zigbee and Monitoring system in greenhouse environment. In: The 3rd international conference on data mining and intelligent information technology applications. IEEE, pp 190–195

Datta S, Taghvaeian S, Ochsner TE, Moriasi D, Gowda P, Steiner JL (2018) Performance assessment of five different soil moisture sensors under irrigated field conditions in Oklahoma. Sensors 18(11):3786. https://doi.org/10.3390/s18113786

Dias PC, Cadavid D, Ortega S, Ruiz A, França MBM, Morais FJ, Ferreira EC, Cabot A (2016) Autonomous soil moisture sensor based on nanostructured thermosensitive resistors powered by an integrated thermoelectric generator. Sens Actuators A Phys 239:1–7. https://doi.org/10.1016/j.sna.2016.01.022

Gao Z, Zhu Y, Liu C, Qian H, Cao W, Ni J (2018) Design and test of a soil profile moisture sensor based on sensitive soil layers. Sensors 18(5):1648. https://doi.org/10.3390/s18051648

Goswami MP, Montazer B, Sarma U (2018) Design and characterization of a fringing field capacitive soil moisture sensor. IEEE Trans Instrum Meas 68(3):913–922. https://doi.org/10.1109/TIM.2018.2855538

Haefke M, Mukhopadhyay SC, Ewald H (2011) A Zigbee based smart sensing platform for monitoring environmental parameters. In: 2011 IEEE international instrumentation and measurement technology conference. IEEE, pp 1–8. https://doi.org/10.1109/IMTC.2011.5944154

Hwang JH, Yoe H (2010) Paprika greenhouse management system for ubiquitous agriculture. In: 2010 international conference on information and communication technology convergence (ICTC). IEEE, pp 555–556. https://doi.org/10.1109/ICTC.2010.5674751

Iwata Y, Miyamoto T, Kameyama K, Nishiya M (2017) Effect of sensor installation on the accurate measurement of soil water content. Eur J Soil Sci 68(6):817–828. https://doi.org/10.1111/ejss.12493

John J, Palaparthy VS, Sarik S, Baghini MS, Kasbekar GS (2015) Design and implementation of a soil moisture wireless sensor network. In: 2015 twenty first national conference on communications (NCC). IEEE, pp 1–6. https://doi.org/10.1109/NCC.2015.7084901

Kalita H, Palaparthy VS, Baghini MS, Aslam M (2016) Graphene quantum dot soil moisture sensor. Sens Actuators B Chem 233:582–590. https://doi.org/10.1016/j.snb.2016.04.131

Kapilaratne RJ, Lu M (2017) Automated general temperature correction method for dielectric soil moisture sensors. J Hydrol 551:203–216. https://doi.org/10.1016/j.jhydrol.2017.05.050

Kaur K, Mahajan R, Bagai D (2016) Zigbee based soil moisture monitoring system. Int J Innov Res Comput Commun Eng 4(6)

Kojima Y, Shigeta R, Miyamoto N, Shirahama Y, Nishioka K, Mizoguchi M, Kawahara Y (2016) Low-cost soil moisture profile probe using thin-film capacitors and a capacitive touch sensor. Sensors 16(8):1292. https://doi.org/10.3390/s16081292

Kumar MS, Chandra TR, Kumar DP, Manikandan MS (2016) Monitoring moisture of soil using low cost homemade soil moisture sensor and Arduino UNO. In: 2016 3rd international conference on advanced computing and communication systems (ICACCS), vol 1. IEEE, pp 1–4. https://doi.org/10.1109/ICACCS.2016.7586312

Mujumdar M, Goswami MM, Morrison R, Evans JG, Ganeshi N, Sabade SS, Krishnan R, Patil SN (2021) A study of field-scale soil moisture variability using the cosmic-ray soil moisture observing system (COSMOS) at IITM Pune site. J Hydrol 597:126102. https://doi.org/10.1016/j.jhydrol.2021.126102

Nagahage EAAD, Nagahage ISP, Fujino T (2019) Calibration and validation of a low-cost capacitive moisture sensor to integrate the automated soil moisture monitoring system. Agriculture 9(7):141. https://doi.org/10.3390/agriculture9070141

Nath S, Kumar Nath J, Sarma KC (2018) IoT based system for continuous measurement and monitoring of temperature, soil moisture and relative humidity. Technology 9(3):106–113

Palaparthy VS, Singh DN, Baghini MS (2017) Compensation of temperature effects for in-situ soil moisture measurement by DPHP sensors. Comput Electron Agric 141:73–80. https://doi.org/10.1016/j.compag.2017.07.010

Pang N (2011) ZigBee mesh network for greenhouse monitoring. In: 2011 international conference on mechatronic science, electric engineering and computer (MEC). IEEE, pp 266–269. https://doi.org/10.1109/MEC.2011.6025452

Placidi P, Gasperini L, Grassi A, Cecconi M, Scorzoni A (2020) Characterization of low-cost capacitive soil moisture sensors for IoT networks. Sensors 20(12):3585. https://doi.org/10.3390/s20123585

Prasad RK, Madkar SR (2013) Design & implementation of wireless transceiver for data acquisition in wireless sensor network. Int J Adv Res Comput Sci Softw Eng 3(7):521–526

Raduła MW, Szymura TH, Szymura M (2018) Topographic wetness index explains soil moisture better than bioindication with Ellenberg's indicator values. Ecol Ind 85:172–179. https://doi.org/10.1016/j.ecolind.2017.10.011

Saeed IA, Wang M, Ren Y, Shi Q, Malik MH, Tao S, Cai Q, Gao W (2019) Performance analysis of dielectric soil moisture sensor. Soil Water Res 14(4):195–199. https://doi.org/10.17221/74/2018-SWR

Salunke SA, Chincholikar SY, Kharde SP (2015) An overview on wireless sensor technologies for the development of agriculture. Int J Comput Sci Mob Comput 4(6):416–418

Shamshiri RR, Weltzien C (2021) Development and field evaluation of a multichannel LoRa sensor for IoT monitoring in berry orchards. 41. GIL-Jahrestagung, Informations-und Kommunikationstechnologie in kritischen Zeiten

Singh P, Saikia S (2016) Arduino-based smart irrigation using water flow sensor, soil moisture sensor, temperature sensor and ESP8266 WiFi module. In: 2016 IEEE region 10 humanitarian technology conference (R10-HTC). IEEE, pp 1–4. https://doi.org/10.1109/R10-HTC.2016.7906792

Songhua Y, Jianya G, Hanwu L (2010) Research on soil moisture sensor nodes and their placement in distributed sensor networks. In: 2010 ninth international symposium on distributed computing and applications to business, engineering and science. IEEE, pp 165–168. https://doi.org/10.1109/DCABES.2010.38

Then YL, You KY, Dimon MN, Lee CY (2016) A modified microstrip ring resonator sensor with lumped element modeling for soil moisture and dielectric predictions measurement. Measurement 94:119–125. https://doi.org/10.1016/j.measurement.2016.07.046

Wang Z, Zhao C, Zhang H, Fan H (2011) Real-time remote monitoring and warning system in general agriculture environment. In: 2011 international conference of information technology, computer engineering and management sciences, vol 3. IEEE, pp 160–163. https://doi.org/10.1109/ICM.2011.228

Xu Z, Zhou W, Zhang H, Shen M, Liu Y, Cai D, Li Y, Lei Y, Wang G, Bagtzoglou AC, Li B (2018) Flat thin mm-sized soil moisture sensor (MSMS) fabricated by gold compact discs etching for real-time in situ profiling. Sens Actuators B Chem 255:1166–1172. https://doi.org/10.1016/j.snb.2017.05.154

Zhang X, Zuo W, Zhao S, Jiang L, Chen L, Zhu Y (2018) Uncertainty in upscaling in situ soil moisture observations to multiscale pixel estimations with kriging at the field level. ISPRS Int J Geo Inf 7(1):33. https://doi.org/10.3390/ijgi7010033

Zhou W, Xu Z, Ross D, Dignan J, Fan Y, Huang Y, Wang G, Bagtzoglou AC, Lei Y, Li B (2019) Towards water-saving irrigation methodology: field test of soil moisture profiling using flat thin mm-sized soil moisture sensors (MSMSs). Sens Actuators B Chem 298:126857. https://doi.org/10.1016/j.snb.2019.126857

Chapter 12
Internet of Things-Enabled Irrigation Management System for Precision Agriculture

Siddharam, L. Aiswarya, Venkatesh Gaddikeri, G. M. Rajesh, Malkhan Singh Jatav, and Jitendra Rajput

Abstract Agriculture is crucial for economies worldwide, contributing to GDP and food security. The burgeoning population necessitates increased demand for nutritious food and water resources. Declining land availability and water allocations require effective water management systems for agricultural production. Precision agriculture and smart irrigation techniques enable precise water application for the "per drop more crop" objective. Top technologies for agriculture water management include IoT, sensors, real-time monitoring, and Unmanned Arial Vehicles (UAVs). This chapter focuses on implementing an IoT-based smart irrigation system. IoT comprises hardware, software tools, and Internet-triggered events. Connectivity methods like Wi-Fi and Ethernet link devices to the Internet. The IoT architecture consists of application, sensing, network, and data processing layers. Studies show that IoT-based irrigation systems are feasible, cost-effective, and optimize water resources for agriculture. This technology offers user-friendly features, saves time, and improves efficiency compared to existing systems, leading to enhanced crop yields and agricultural sustainability.

Keywords Internet of Things · Automatic irrigation · Precision and smart irrigation · Sensors agricultural water management · Automation

Siddharam (✉) · L. Aiswarya · G. M. Rajesh
Kelappaji College of Agricultural Engineering and Technology, Kerala Agricultural University, Tavanur 679573, India
e-mail: siddharampatil360@gmail.com

V. Gaddikeri (✉) · M. S. Jatav · J. Rajput
ICAR-Indian Agricultural Research Institute, Pusa, New Delhi 110012, India
e-mail: venkatg303@gmail.com

K. Pandey et al. (eds.), *Artificial Intelligence and Smart Agriculture*,
Advances in Geographical and Environmental Sciences,
https://doi.org/10.1007/978-981-97-0341-8_12

12.1 Introduction

Agriculture has been viewed as a significant water use sector in the recent era as it uses 70% of the world's freshwater resources to irrigate 25% of the world's cropland (Bwambale et al. 2022). Nearly 90% of freshwater is used in India is for irrigation, and it is recorded that only 5% of water is consumed for domestic purposes in India (Ritchie and Roser 2017). In recent times, there has been increasing competition among various water use sectors, which may result in reduced water availability for irrigation in the future. Additionally, the combined effects of climate change and population growth are exerting additional stress on natural resources, including water, land, and forests (Ungureanu et al. 2020).

By 2050, the global population anticipated to reach 9.7 billion, resulting in an increased need for nutritious food and water resources. According to the World Resources Institute's Aqueduct Water Risk Atlas, India was placed thirteenth out of the world's 17 'extremely water-stressed' countries. The report defines "extremely high levels of water stress" as the annual average water usage by irrigation, industries, and municipalities exceeding 80% of the available water in the country. In India, Punjab is identified as the most water-stressed state, followed by Haryana, Rajasthan, and Uttar Pradesh. This type of water scarcity in the country as well as in the world can be minimized by the adoption of water saving technologies, improving water use efficiency and integration of high-level technologies with agriculture for water management. This type of adoption strategies may help in sustainable development, energy saving, cost reduction and increasing productivity per unit input (Bwambale et al. 2022). Recent advancements in Information and Communication Technology (ICT) can be used effectively in agriculture for resource management. Because of ICT concepts like smart, precision, and technology-driven irrigation system have gained prominence. Precision and smart irrigation technologies ensure higher water use efficiency, which is the fusion of various technologies, including real-time statistics, processing of sensor data, machine learning, and decision support systems (Evans and Sadler 2008).

Precision and smart irrigation enable farmers to save valuable water resources; this can be accomplished by applying water at the appropriate time, quantity, quality, and location/site. Efficient irrigation scheduling can be achieved by considering spatiotemporal variation weather parameters, soil moisture condition, soil and crop information, and plant physiological parameters. So, irrigation application is not same in all conditions and it varies with time and space as a result, the actual depth of irrigation water received by plants varies (Vories et al. 2021).

Development in ICT, labors shortage, cost of production leads to shift from manually operated irrigation to automated control irrigation systems, which can be timer-based, sensor-based, weather–based, and Internet of Things (IoT)-based smart irrigation technology. IoT is a collection of independent objects linked together and exchanging information remotely via the Internet. The current book chapter concentrates on IoT for irrigation management, including its components, architecture, critical requirements, and limitations.

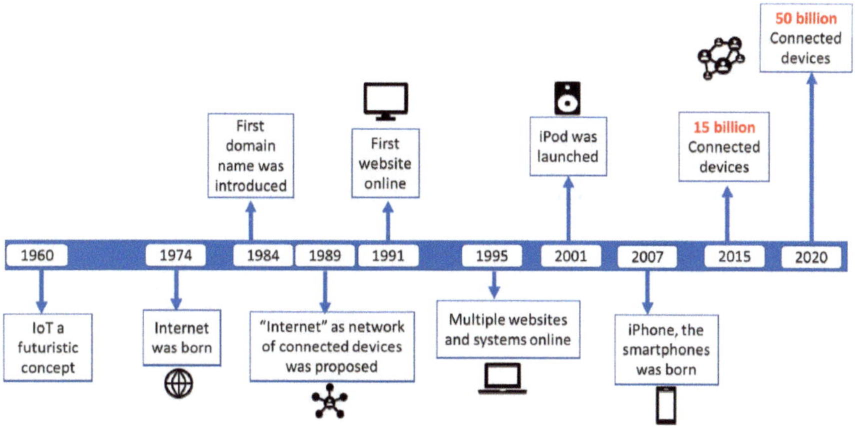

Fig. 12.1 Brief history of how Internet of Things become a thing

12.2 Smart Irrigation Management System Powered by IoT

12.2.1 Internet of Things (IoT)

The term "Internet of Things" was invented by Kevin Ashton (1999). In his vision, computers would be used to organize and manage things in a world where everything has its own digital identity. A system of networked computing devices, mechanical and digital machinery, items, animals, or people, known as the IoT, can transfer data over a network without requiring human-to-human or human-to-computer interaction (Muangprathub et al. 2019). IoT refer to a network of physical objects that communicate with one another via Internet. Objects or things can allow the transfer information wirelessly without interaction of human. Any object that has the capacity to send data over a network and can be given an IP address is referred to as a "thing" (Tao et al. 2021). Our devices are getting smarter each day from age old technologies to smart phones, smart TV, car, and kitchen. Everything is now being linked to Internet. There are numerous similar devices on the market now. Examples include washers and dryers that use Wi-Fi for remote monitoring and smart thermostat systems (Ismail et al. 2019). The brief history of Internet and IoT is presented in Fig. 12.1.

12.2.2 IoT Architecture

IoT is an architecture that includes specialized hardware boards, software systems, web triggered event that occurs over the Internet. Devices are connecting to the Internet via a various means including Wi-Fi, Ethernet, and other technologies.

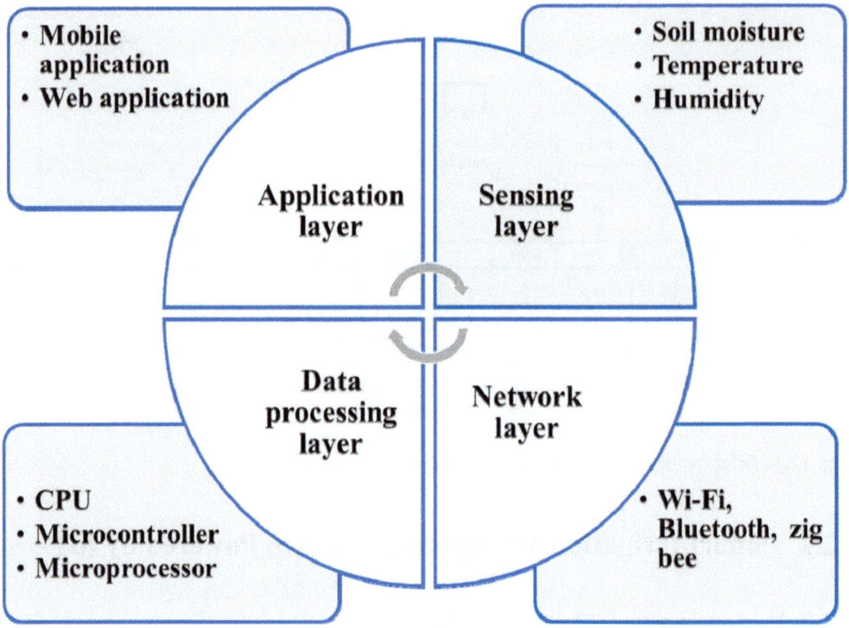

Fig. 12.2 IoT architecture layers

Furthermore, it might not be necessary for all devices to be connected to Internet independently (Seethalakshmi et al. 2021). The architecture layers of IoT are discussed and shown in Fig. 12.2.

12.2.2.1 Sensing Layer

The sensing layer, which consists of a number of sensors, is primarily responsible for detecting any phenomena in the device's periphery and collecting information from the outside environment. One of the key characteristics of IoT devices is the use of multiple sensors to sense the moisture, light, motion, pressure, heat, or any other environmental phenomena.

12.2.2.2 Network Layer

Data collected in the sensing area can be transferred to other network devices through the network layer. To enable data transfer between other devices on the same network, the network layer in IoT devices uses a variety of communication technologies (such as Wi-Fi, Bluetooth, Zigbee, LoRa, Z-Wave, cellular network, and so on).

12.2.2.3 Data Processing Layer

The data processing layer is a fundamental component of IoT devices, responsible for processing the primary data collected by the sensing layer. It analyzes the gathered information and makes decisions based on it. In certain IoT devices like smartphones, smartwatches, and smart home hubs, the data processing layer also stores previous analysis results to enhance the user experience. The network layer enables other connected devices to access the outcomes of the data processing, facilitating seamless communication and data sharing among devices.

12.2.2.4 Application Layer

To enable the implementation of various IoT device applications, the application layer utilizes the outputs of the data processing layer and presents them to the user. The application layer serves as a user-centric component that handles a range of user-related functionalities.

12.2.3 Applications of IoT

The IoT can be used in a variety of technological applications/sectors, including transportation, smart home appliances, smart farming, smart cities, power efficient building, smart water supply, traffic cramming, infrastructure management, manufacturing units, health care, media and entertainment, security and energy management. The several applications are presented in Fig. 12.3.

12.2.4 IoT in Agriculture

IoT devices may provide significant assistance in improving agricultural productivity and yield since they can be used to monitor numerous inputs, climate factors, and data gathering facilities. In smart farming, technology integrated with IoT helps to monitor the crop growth, keep record on the climatic parameters (temperature, soil moisture, humidity, etc.), and computerized decision making for yield prediction. Besides that, farmers can get information about pest infestation and soil nutrient status. In precision farming, IoT can help for the precise management of resources such as fertilizer, nutrient, water, labor, and money. When IoT integrated with drones can help in collection of spatiotemporal variation in field moisture, nutrient status, plant physical condition indices, plant counting, harvest prediction, plant altitude dimension, chlorophyll quantity, and livestock monitoring. The Radio Frequency Identification (RFID) sensors are utilized in livestock farming operations to measure body temperature, heart rate, and blood oxygen levels as well as the temperature and

Fig. 12.3 Various application of IoT

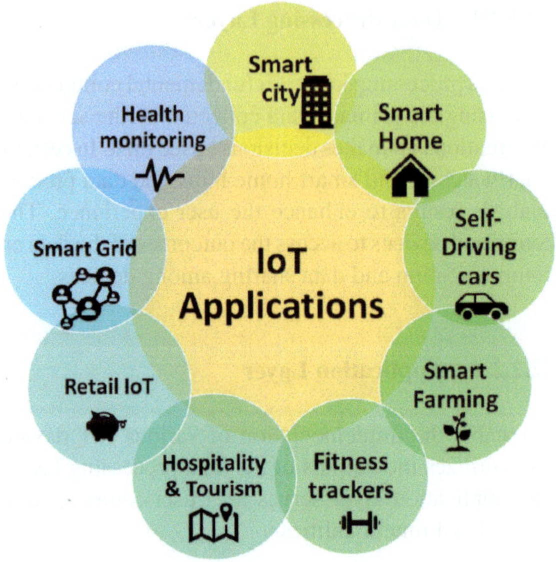

pH level inside the rumen of the cow. In greenhouse cultivation, IoT which can help to monitor the climatic parameters, maintain suitable crop environment, irrigation management, and ventilation system can be achieved by this technology. Because of its real-time, accurate, and shareable features, IoT will fundamentally change the agricultural distribution network and provide design objective for creating a seamless flow of agricultural logistics (Thakur et al. 2020) (Fig. 12.4).

Advantages of IoT in enhancing resource use efficiency in agriculture (Kumar et al. 2021) are as follows:

(1) Efficient water management using IoT by reducing wastage of water.

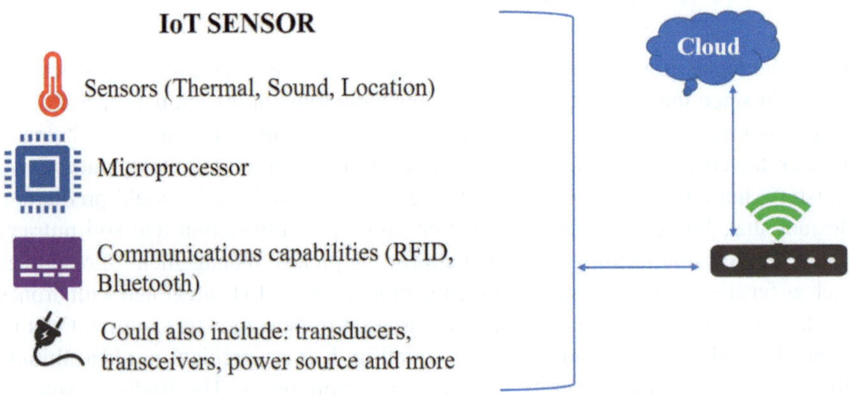

Fig. 12.4 Basic components of IoT

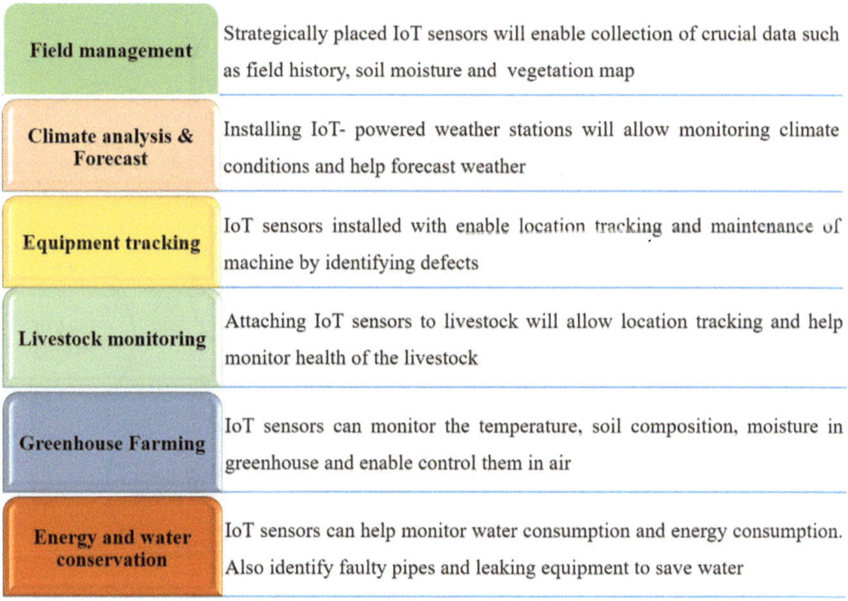

Fig. 12.5 IoT in agriculture Ratnaparkhi et al. (2020)

(2) Continuous monitoring of land and pest management so that early precautions can be performed.
(3) It improves the resources use efficiency, reduce the work and time, and makes efficient farming.
(4) Crop monitoring is a simple way to track crop progress.
(5) It is easy to identify soil management factors like pH level, moisture content, and nutrient status.
(6) Without being limited by geographical constraints, farmers can easily connect to the global market (Fig. 12.5).

12.2.5 Challenges of IoT in Agriculture

Even though implementing IoT in the agricultural sector has many advantages and there are certain difficulties. The usage of IoT in agriculture is mostly unknown to farmers. The fundamental problem is that some of them are unwilling to adopt new concepts, even when they provide significant advantages. The main challenges faced by the farmers to adopting IoT technologies in agriculture are lack of infrastructure, high hardware cost, lack of security, poor Internet connectivity in farm, lack of stakeholder interest, untrained manpower, and disrupted connectivity to the cloud (Punitha and Geetha 2018).

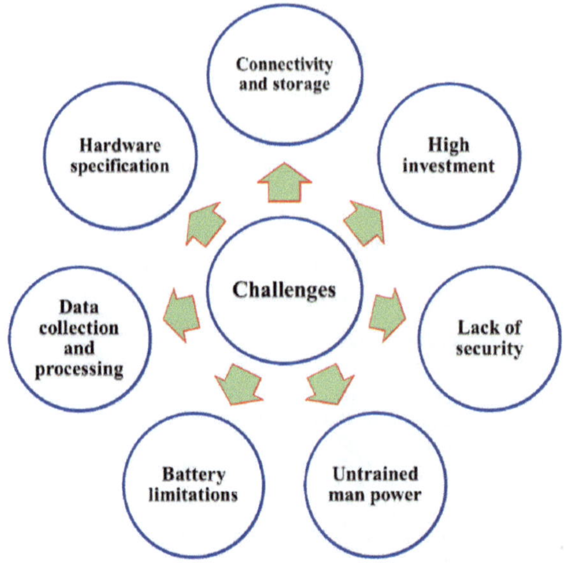

Fig. 12.6 Challenges of IoT in agriculture

To overcome these challenges, it is essential to increase awareness among farmers about the benefits of IoT and educate them on how to effectively use IoT technologies such as drones, sensors, and other relevant tools to improve their agricultural practices. Real-world examples can be shared to demonstrate the practical benefits of IoT in farming, making it easier for farmers to understand and embrace these technologies, ultimately enhancing their productivity and efficiency (Fig. 12.6).

12.3 Application of IoT in Irrigation Management

Due to the lack of efficient and innovative irrigation systems, the agriculture sector, particularly irrigation, consumes a significant amount of water. The application of IoT in irrigation water management has revolutionized the agricultural sector. By integrating IoT technology, farmers can monitor and control irrigation systems remotely, optimizing water usage and improving efficiency. IoT-enabled sensors gather real-time data on soil moisture levels, weather conditions, and crop water requirements (Fig. 12.7).

The various components using smart irrigation with IoT are depicted in Fig. 12.8. It primarily consists of a microcontroller, sensors, and an application for data display. Sensors are used as sensing devices to measure soil moisture, temperature, pH level, and electrical conductivity (EC), and they are installed in the soil. The sensed data is transmitted to the microcontroller via Wi-Fi or Ethernet. The microcontroller, a small device capable of network connectivity and running programs, can be programmed to receive data from sensors and transmit it to the Internet for storage and analysis.

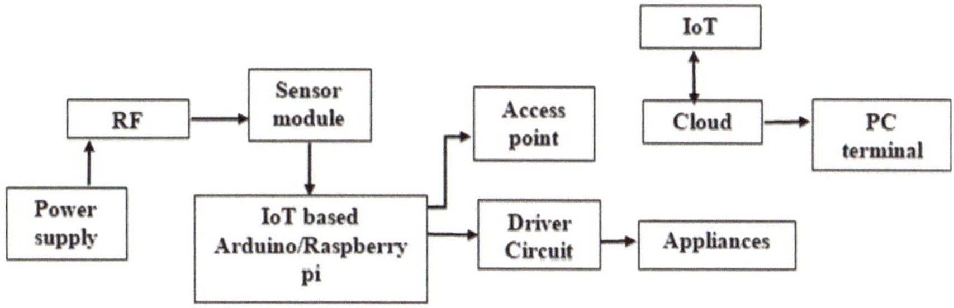

Fig. 12.7 Hardware components of IoT

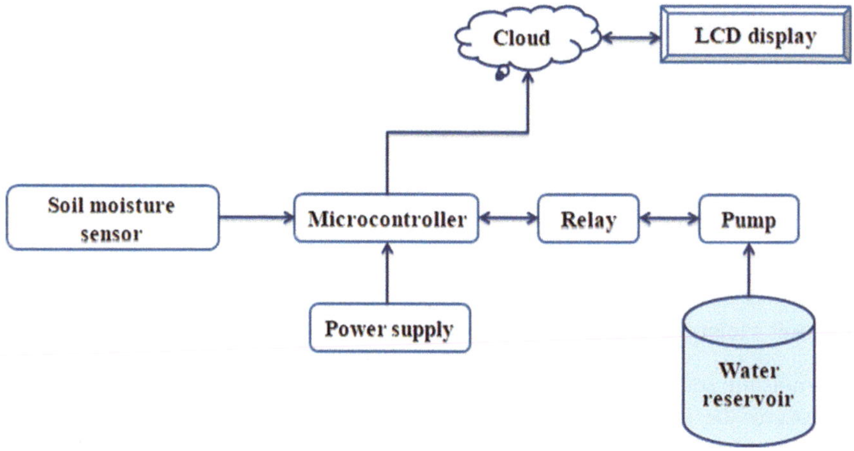

Fig. 12.8 Basic component of IoT-based irrigation

An application, which is essentially a web-based program, receives data from the controller and presents it in a user interface (Singh and Ahmed 2021; Pandey et al. 2023).

12.3.1 IoT-Based Irrigation System Components

12.3.1.1 Hardware Components

The hardware system consists of microprocessor, microcontroller, sensors, Wi-Fi module, GSM module, solenoid valve, water pump, and relay switch.

Microprocessor

A microprocessor is a Central Processing Unit (CPU) that performs arithmetic and logical operations, including those of the Arithmetic Logical Unit (ALU). It serves as the control unit of a microcontroller. Additionally, a microprocessor can establish communication with other devices connected to it. Examples of microprocessors include Raspberry Pi (R-Pi), Orange Pi, Banana Pi, and Arduino.

Microcontroller

A single metal-oxide semiconductor integrated circuit chip serves as the physical home for a microcontroller, which is a tiny computer. It has memory, programmable input/output peripherals, and one or more CPUs. This microcontroller can connect to the Internet and serve as a server. They are utilized in automatically controlled goods and gadgets like autos, implanted medical equipment, remote controls, workplace equipment, and embedded systems (Sudharshan et al. 2019).

Sensors

(a) **Soil Moisture Sensor**

Soil moisture sensor is the very vital part in irrigation scheduling using IoT system. It measures the soil moisture present in the soil in terms of volumetric water content or in terms of soil water potential. Standard gravimetric measurement of free-soil moisture is accurate one but laborious and time consuming. In order to overcome this issue, various soil properties are used to indirectly estimate the soil moisture. These properties are dielectric constant, electrical resistance, or interaction with neutrons. Once these properties have been detected by the sensor, they will be converted into respective soil moisture content in gravimetric content or volumetric content. Then, irrigation managers or programmed algorithm will then determine how much amount of water need to be applied to bring the soil moisture to field capacity. Among various sensors, electrical resistivity sensor is widely used due to its nature of operation and availability. It consists of two electrodes through which (Fig. 12.9) current passes in soil and then it is connected to Wheatstone bridge to measure the resistance applied by the soil due to absence or presence of soil moisture. When there is enough soil moisture, electrical resistance will be less and vice versa. These resistances value need to be calibrated and need to develop soil electrical resistance vs. soil moisture content curve. From this curve, we can determine the soil moisture present with respect to soil resistance. Generally, the sensors are platinum and anti-rust coated to increase efficiency and longevity (Thakur et al. 2020). The sensors position, depth of installation, soil characteristic, and salt content affect the electrical resistance which needs to be taken care while installing the sensor for irrigation scheduling.

Fig. 12.9 Soil moisture sensor

Fig. 12.10 Temperature and humidity sensor

(b) **Temperature and Humidity Sensor**

The temperature of its surroundings is sensed by a temperature sensor, a sort of electronic equipment, which converts the quantitative measurements into electronic data to collect, monitor, or notify temperature changes. The market has a variety of temperature sensors, which are primarily categorized into contact type and non-contact type. Infrared sensors are non-contact sensors, whereas contact type sensors are thermocouples and thermistors. This type of sensors can be used to temperature-based irrigation schedule. Once crop canopy temperature sensed by the sensor, then their value can be used to irrigate the crop. One of the sensors called DHT 11 sensor, by detecting the electrical resistance across two electrodes, it may identify water vapor (Ismail et al. 2019) and used to assess the humidity and temperature value via single sensor from surrounding air (Fig. 12.10). It is embedded with capacitive humidity sensor and thermistor sensor. It has various benefits, including low cost, low power consumption, and signal transmission distances of up to 20 m.

Wi-Fi Module

Wi-Fi (Wireless Fidelity) module (Fig. 12.11) is used to connect the different device and communicate between them through the Wi-Fi. Besides that, they can also receive commands via Wi-Fi. Such type of modules is commonly used in the IoT devices. There are various types Wi-Fi modules available in the markets, namely ESP8266, ESP-01, ESP-02, ESP-03, ESP-14, etc. Any microcontroller may connect to any Wi-Fi network to the ESP8266 Wi-Fi module, a self-contained System on Chip (SoC) with an embedded Transmission Control Protocol/Internet Protocol (TCP/IP) stack.

Fig. 12.11 Wi-Fi module

Fig. 12.12 GSM module

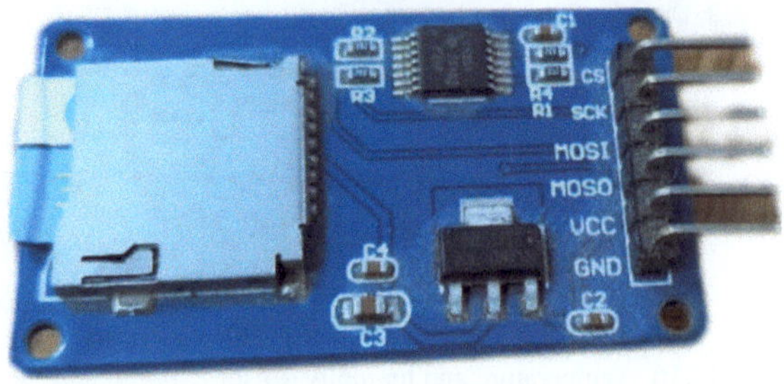

Each preprogrammed ESP8266 module can be readily linked to a microcontroller device to add Wi-Fi functionality. This module can be linked with sensors and other application-specific devices to its robust on-boarding processing and large storage capacity (Anitha et al. 2020).

GSM Module

The Global System for Mobile communication (GSM) module standard was created by the European Telecommunication Standards Institute (ESTI) to specify the standards for second-generation (2G) digital cellular networks used among smartphones. A digitized, circuit-switched network known as GSM is one that has been enhanced to provide data communications and packet data transmission via General Packet Radio Services (GPRS) (Krishnan et al. 2020). It is a chip or circuit that joins a computing device or mobile device to a GSM network. GSM modules enable microcontrollers to communicate wirelessly with other devices and instruments. Because of its simplicity, affordability, and accessibility, GSM is currently the most widely used network technology in IoT applications (Fig. 12.12).

Solenoid Valve

A solenoid valve is type of electromechanical valve and made up of two separate components such as solenoid and valve. The combined action of the plunger and the magnetic force created by passing current through the coil will help in controlling

Fig. 12.13 Solenoid valve

Fig. 12.14 Water pump

the flow of fluids through a pipe line. In order to open or close the valve, the solenoid valve transfers electrical energy into mechanical energy. There are two types of solenoid valve; they are namely open and closed solenoid valve. Their design is based on the flow rate, operating pressure, and type of fluid to be handled. With their rapid and secure switching, good reliability, long service lives, good medium adaptability of their materials, low control power, and compact size, solenoids are a popular switching technology. (Sudharshan et al. 2019). This type of device is connected to Wi-Fi and other microprocessor devise for supply irrigation water to the various location. When the controller receives signal, it will open the valve. Further integration with flow meter will help in fluid flow monitoring. Once the required quantity of water delivered, it will turn off the devises (Fig. 12.13).

Water Pump

The water from source is pumped under pressure to the irrigation system. The motor of the pump is connected to the microcontroller (Sharma and Kumar 2021). It is used to sprinkle water on the land based on environmental conditions such as moisture, temperature, and humidity. This will be activated when the soil moisture sensor detects a low moisture level and will automatically adjust the moisture parameter to its optimum value (Fig. 12.14).

Fig. 12.15 Relay module

Relay Module

Relay is an electromechanical component that functions as a switch. This kind of relay unit is frequently used in automatic control units and is used to control high current by applying low current. The electromechanical relay unit consists of electromagnet, mechanically movable contact, switching point, and spring. A temporary magnetic field is created when the electromagnetic field is activated by the relay circuit upon detection of a fault current. The relay armature moves as a result of the magnetic field, enabling the interconnections to be opened or closed. The high-power relay has two contacts for opening the switch, compared to the small power relay's single contact. They are typically utilized to join an electrical circuit to an electronic circuit, which runs at a low voltage (which operates at a very high voltage). For instance, a relay can switch a 230 V AC mains circuit using a 5 V DC battery circuit. When it receives a signal from the operators or from programmed algorithm, this type of relay unit can be used to turn on and off the irrigation pump. To save and deliver water, an IoT-enabled irrigation system, irrigation water pump, relay unit, and IoT devices work in tandem.

Relays that are frequently used come in a variety of operating voltage configurations, including 6, 9, 12, and 24 V. A basic relay unit has three contactors: a normally open (NO), normally close (NC), and common contactor (COM). In the absence of input, the COM is connected to NC. When the operational voltage is delivered, the relay coil activates and the COM changes connection to NO (Kumbhar and Ghatule 2016) (Fig. 12.15).

12.4 Application of Machine Learning in Smart Irrigation

Machine learning can be used to predict water usage in a water management system. It is a data analysis technique that extracts information directly from the data using a variety of algorithms and models. The machine learning application facilitates in the proper maintenance of irrigation systems. For instance, it maximizes water use and supplies the field with enough water and fertilizer, increasing yield production while lowering manual labor and crop disease outbreaks (Janani and Jebakumar 2019).

It learns agriculture data using a computational method. The goal of the learning process is to learn how to perform a given task from training data. A set of data examples is described by a set of features. This feature can be numeric, binary, or nominal. Machine learning is used in the irrigation system to regulate water, crop management, plant development, and soil management, yield prediction, disease detection, weed identification, species recognition, and crop quality are all applications in agriculture (Janani and Jebakumar 2019).

Machine learning applications facilitate the maintenance of proper irrigation systems, efficient water usage, water conservation, and the improvement of production quality and quantity. It will predict soil moisture using a support vector regression model or the k-means clustering algorithm based on data from field sensors and weather forecasting (Liakos et al. 2018).

12.4.1 Machine Learning Techniques for Smart Irrigation

12.4.1.1 Support Vector Regression (SVR)

A nonparametric method called support vector regression (SVR) makes it possible to build nonlinear models. The dependent variable is numerical rather than categorical in this revised method of the Support Vector Machine (SVM). The kernel functions are used in the SVR technique to build the model. The most used kernel functions include sigmodal, radial basis, linear, and polynomial functions (Hearst et al. 1998).

12.4.1.2 k-means Clustering

K-means clustering uses a clear-cut and basic process of clustering to split the given dataset into a certain number of clusters. It works in tandem with SVR to increase the precision of the soil moisture difference measurement. One per each bunch, *k* centroids have to be found. Before calculating the cluster centroids (mean points) of each subgroup or cluster, it separates n items into *k* non-empty subgroups or clusters. Then, it determines the separations between each point and the centroids and places each object in the cluster with the shortest separation from the centroid. The points are repeatedly reassigned, and the centroid of the new clusters is determined.

Sharma et al. (2022) suggested two-hybrid DL models, namely Convolution Neural Network-eXtreme Gradient Boosting (CNN-XGB) and Convolution Neural Network-Support Vector Regression (CNN-SVR), to evaluate daily ET_c values of wheat and rice crops employing scant climate data. In order to manage circumstances where there is a lack of data, the prediction of ET_c values uses limited climate data, including minimum temperature (T_{min}), mean temperature (T_{mean}), maximum temperature (T_{max}), and solar radiation (Rs). The outcomes show that the suggested hybrid models work satisfactorily, with Nash–Sutcliffe Efficiency (NSE) values of 0.95 for rice and 0.976 for wheat ET_c, respectively.

In order to estimate ET_0 in various sites situated in the diverse climatic region of Andalusia, Bellido et al. (2021) evaluated various temperature-based Multilayer Perceptron (MLP), Support Vector Machines (SVM), Generalized Regression Neural Network (GRNN), XGBoost, Extreme Learning Machine (ELM), and Random Forest (RF) models (Southern Spain). Due to its high efficiency of NSE performance across all configurations and regions, it was concluded that the usage of ELM is advised.

12.5 IoT-Based Smart Irrigation System in Greenhouse Condition

In greenhouse cultivation, the climatic conditions have a significant impact on crop production. Temperature, humidity, light intensity, moisture, and CO_2 concentration are the various parameter need to be monitor in greenhouse (Hakkim et al. 2014; Bai et al. 2019; Carpenter et al. 2019; Pandey et al. 2023; Sharda et al. 2023). The relevant information related to the individual effects of climatic parameters in green house is obtained by continuous monitoring. Sensors are key components for the monitoring various climatic parameters in greenhouse systems. As the name implies, a sensor is any electronic instrument or device that measures the characteristics of the environment, further it sends it data through the network layers such as Wi-Fi, Zigbee, GSM, etc., to other devices. After these data are acquired, that will be further processed so that an useful interpretation can be done.

Four components, including a water pump, fogger, fan, and LED lights, are controlled by this system using input parameters such as soil moisture, temperature, humidity, and light. This system's brain is the microcontroller. The microcontroller begins functioning based on the data received from the sensors after the various sensors perceive the various parameters inside the greenhouse. When the soil moisture sensor detects a moisture level that is lower than the predetermined value, the microcontroller activates the water pump and notifies the farmer by text message that the motor is ON. Similar to this, when the temperature and humidity sensor detects any deviation from the required level, the microcontroller is engaged and turns on or off the fan or foggers depending on the situation, resulting in the cooling action. Like this, the microcontroller will switch on the lights when it detects that the light intensity is below the predetermined value. Figure 12.16 displays a block diagram of an IoT-based monitoring and control system for greenhouses.

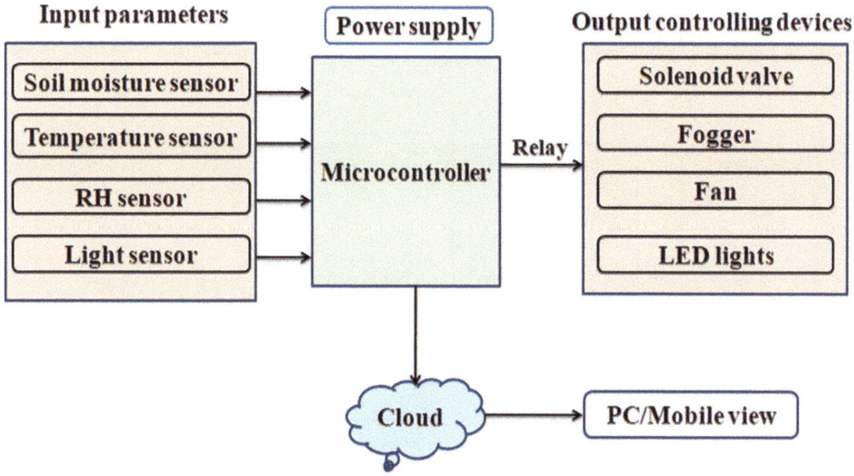

Fig. 12.16 Block diagram of greenhouse monitoring and controlling using IoT

12.6 IoT-Based Aeroponic System

Each Internet of Things device collects temperature, humidity, and pH balance sensor data during the cropping process, as seen in the aforementioned Fig. 12.17. If the nutrient level is below a threshold value after verifying it, the microcontroller will activate a relay. The attached dosing pumps start supplying the aeroponics system with nutrients. The aeroponics system will stop delivering nutrients when the nutrition level reaches a specified threshold. The operation of the dosing pump and the submersible pump is identical.

12.7 IoT-Based Hydroponic System

Hydroponic farming is a method of growing plants that does not require sunlight or soil. Plants are grown in this method their roots exposed to a mineral–water mixture rather than soil as a medium. This technique is a form of weather-independent indoor agriculture that reduces the need for labor and ploughing. Using Internet of Things (IoT) devices connected to wireless sensor networks and Internet connectivity, irrigation, and microclimate control is carried out. The authorized person can track the status of the plant's growth in real time from a distance with the help of IoT technology.

During the cropping period, an IoT device collects sensor data on temperature, pH balance, and humidity. If the nutrient level is below a predefined threshold after verifying it, the microcontroller will activate a relay. The linked dosing pumps start feeding the hydroponic system with nutrients. It will stop supplying nutrients to the

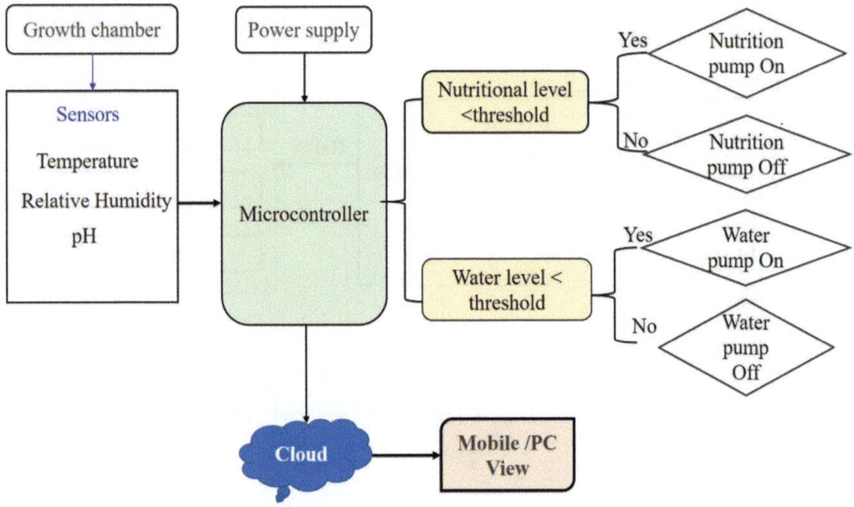

Fig. 12.17 Block diagram of IoT-based aeroponic system

hydroponic system once the nutrition level reaches a particular threshold. To maintain the ideal water level in the system, the submersible pump process functions similarly to the dosing pump process. Similar to this, the microcontroller will turn ON the LED lights when it detects that the light intensity is below the threshold value. The block diagram of IoT-based hydroponic system is shown in Fig. 12.18.

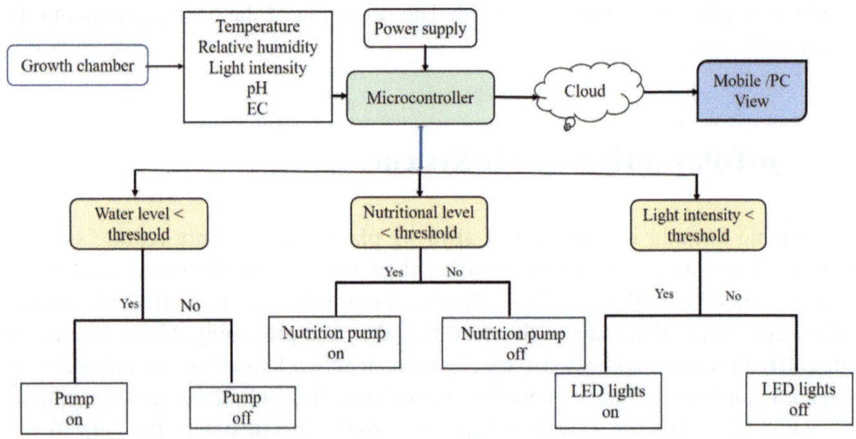

Fig. 12.18 Block diagram of IoT-based hydroponic system

12.8 Conclusion

Agriculture sector consumes more than 70% of fresh water resource to irrigate 25% world's cropland in order to meet food security. As a result, water is a precious resource that must be protected using the most advanced technologies available. IoT and other cutting-edge technologies have just started to play a big part in our daily lives, boosting our ability to perceive and change our environment.

It has been demonstrated that the automated irrigation system is practical and economical for maximizing the use of water resources for agricultural production. The main goal of the smart irrigation system is to improve upon the current system by making it more creative, user-friendly, time-saving, and efficient. IoT plays an important role in agriculture to enhancing crop yields, improving quality, and lowering costs, allowing cultivation in water-stressed areas, and thus improving sustainability. Also helps in reduces water consumption to a great extent. The most attractive feature of this system is the visit free monitoring for people far away from their farms.

IOT-based smart irrigation systems are an economical way to maximize agricultural production while conserving water resources. This automated and intelligent system benefits the farmer. Water can only be provided in the required quantity at the required time and location by burying several sensors in the soil. Machine learning can be used in IoT-based smart irrigation system for prediction of soil moisture and irrigation scheduling. To maximize the use of water resources in agriculture, smart irrigation systems combining IoT and machine learning techniques are therefore practical and affordable.

References

Anitha A, Sampath N, Jerlin MA (2020) Smart irrigation system using IoT. In: International conference on emerging trends in information technology and engineering, pp 1–7

Bai X, Wang Z, Sheng L, Zhen W (2019) Sensor networks with asynchronous measurement for greenhouse monitoring. Trans Control Syst Technol 27(3):1036–1046

Bellido-Jiménez JA, Estévez J, García-Marín AP (2021) New machine learning approaches to improve reference evapotranspiration estimates using intra-daily temperature-based variables in a semi-arid region of Spain. Agric Water Manag 245:1–16

Bwambale E, Abagale FK, Anornu GK (2022) Smart irrigation monitoring and control strategies for improving water use efficiency in precision agriculture: a review. Agric Water Manag 260:1–12

Carpenter N, Ma D, Maki H, Rehman UT, Tuinstra RM, Jin J (2019) Greenhouse environment modeling and simulation for microclimate control. Comput Electron Agric 162:134–142. https://doi.org/10.1016/j.compag.2019.04.013

Evans RG, Sadler EJ (2008) Methods and technologies to improve efficiency of water use. Water Resources Res 44:1–15

Hakkim AVM, Barin P, Hareesh MN (2014) Greenhouse management and operations, Farm Information Bureau. Department of Agriculture, Government of Kerala, 208p

Hearst MA, Dumais ST, Osuna E, Platt J, Scholkopf B (1998) Support vector machines. Intell Syst Appl 13(4):18–28

Ismail N, Rajendran S, Tak WC, Xin TK, Anuar NSS, Zakaria FA, Al Quhaif YMS, Karakhan HAMH, Rahim HA (2019) Smart irrigation system based on IoT (IOT). J Phys Conf Ser 1339:1–8

Janani M, Jebakumar R (2019) A study on smart irrigation using machine learning. Cell Cell Life Sci J 4(2):1–8

Krishnan RS, Julie EG, Robinson YH, Raja S, Kumar R, Thong PH (2020) Fuzzy logic based smart irrigation system using internet of things. J Clean Prod 252:119902

Kumar SV, Suresh V, Kasiviswanath N, Reddy YJ (2021) Improvisation of irrigation system in a smart way using IoT and cloud. Int J Adv Res Eng Technol Sci 12:78–85

Kumbhar SR, Ghatule AP (2016) Microcontroller based controlled irrigation system for plantation. In: Proceedings of the international multi conference of engineers and computer scientists, vol 2

Liakos KG, Busato P, Moshou D, Pearson S, Bochtis D (2018) Machine learning in agriculture: a review. Sensors 18(8):2674p

Muangprathub J, Boonnam N, Kajornkasirat S, Lekbangpong N, Wanichsombat A, Nillaor P (2019) IoT and agriculture data analysis for smart farm. Comput Electron Agric 156:467–474

Pandey K, Singh KG, Singh A (2023) Multi-sensors based smart nutrient reuse management system for closed soilless culture under protected cultivation. Comput Electron Agric 107495

Punitha A, Geetha V (2018) Review on challenges & opportunities of IOT in agriculture. J Adv Res Appl Sci 5(11):25–31

Ratnaparkhi S, Khan S, Arya C, Khapre S, Singh P, Diwakar M, Shankar A (2020) Smart agriculture sensors in IOT: a review. Mater Today Proc 1–8

Ritchie H, Roser M (2017) Water use and stress. Our World in Data

Seethalakshmi E, Shunmugam M, Pavaiyarkarasi R, Joseph S (2021) An automated irrigation system for optimized greenhouse using IoT. Mater Today Proceed 1–6

Sharda R, Singh A, Pandey K (2023) Paddy and maize straw-based media as an alternative for cocopeat in soilless cultivation. J Plant Nutrition 1–19

Sharma BB, Kumar N (2021) IoT-based intelligent irrigation system for paddy crop using an internet-controlled water pump. Int J Agric Environ Inf Syst 12(1):21–36

Sharma G, Singh A, Jain S (2022) Hybrid deep learning techniques for estimation of daily crop evapotranspiration using limited climate data. Comput Electron Agric 202:1–16

Singh M. & Ahmed, S. (2021). IoT based smart water management systems: A systematic review. Materials Today: Proceedings. pp. 5211–5218.

Sudharshan N, Karthik AK, Kiran JS, Geetha S (2019) Renewable energy based smart irrigation system. Procedia Comput Sci 165:615–623

Tao W, Zhao L, Wang G, Liang R (2021) Review of the IoT communication technologies in smart agriculture and challenges. Comput Electron Agric 189:1–16

Thakur D, Kumar Y, Vijendra S (2020) Smart irrigation and intrusions detection in agricultural fields using IoT. Procedia Comput Sci 167:154–162

Ungureanu N, Vladuț V, Voicu G (2020) Water scarcity and wastewater reuse in crop irrigation. Sustainability 12(21):9055p

Vories E, Oshaughnessy S, Sudduth K, Evett S, Andrade M, Drummond S (2021) Comparison of precision and conventional irrigation management of cotton and impact of soil texture. Precision Agric 22(2):414–431

Chapter 13
Unmanned Aerial Vehicles (UAV) for Smart Agriculture

R. Pungavi and C. Praveenkumar

Abstract Global food and environmental security would be threatened by current socioeconomic and climatic concerns. Agriculture systems are being stressed by the emergence of population-feeding issues, and increased use of pesticides, fertilisers, uneven irrigation scheduling, the appearance of pests and diseases, etc., are leading to a sharp decline in overall production as well as disturbed soil microbial communities and soil quality maintenance. Smart farming involves utilising drones, to change conventional farming operations in order to reduce the current damaging issues and to achieve the objective of sustainable agriculture. Drone technology is used in smart agriculture for a variety of tasks, crop health surveillance, planting, spraying process, field scrutiny, evaluation of soil, and weed control which are all part of the process. UAV are equipped with a variety of sensors, to map field water, schedule irrigation for crop plants, detect water shortages, and detect impurities in water. Drones use various spectral images in combination with cameras (3D cameras, thermal, multispectral, and optical imaging cameras), global positioning system (GPS), and other technologies to track down and locate plant diseases as well as to prevent pests from attacking healthy crop plants. Complex agricultural environments are also examined. A promising strategy to assist effective, lucrative, and sustainable agricultural production is smart farming (SF), which is based on improvements in drone technology outfitted with cameras, sensors, and GPS systems.

Keywords Smart farming · Drone technology · Imaging cameras · Sensors · GPS · Crop monitoring · Detect water shortage · Sustainability · Agriculture

---

R. Pungavi (✉)
Department of Entomology, Faculty of Agriculture, Annamalai University, Chidambaram, Tamil Nadu 608002, India
e-mail: r.pungavientomology@gmail.com

C. Praveenkumar
Department of Agricultural Entomology, Pandit Jawaharlal Nehru College of Agriculture and Research Institute, Karaikal 609603, India

K. Pandey et al. (eds.), *Artificial Intelligence and Smart Agriculture*,
Advances in Geographical and Environmental Sciences,
https://doi.org/10.1007/978-981-97-0341-8_13

13.1 Introduction

Two of humanity's most basic needs are met by agriculture, which is a significant source of wealth for many nations: food and fibre. The Green Revolution, sometimes referred to as the third invention of agriculture, took place between the 1960s and the 1980s and involved irrigation, synthetic fertilisers, pesticides, and crop modification varieties that increased crop food output and nutritional security, particularly in underdeveloped countries. The issue of wise resource use is brought on by the world population's continued rise and the declining amount of resources available. The consumption of agricultural and food products is anticipated to increase by more than 75% by 2050 (Singh et al. 2022). Rapid population results in expanding agriculture, leading to usage more fertilisers, insecticides, freshwater, and other resources. The ecology, however, is negatively impacted by increased agricultural input use, including eutrophication, groundwater extraction, and diminished surface streams. Improper or inefficient use of natural resources, fertilisers, and pesticides in agricultural output generates long-term degradation of soil microorganism and environment. It is vital to develop techniques that can enhance agricultural output through higher production quality while decreasing environmental waste in order to create manufacturing systems that are both economical and environmentally friendly (Rahman et al. 2021).

There has been a rise in interest in deploying autonomous ways to check on the health of farmland. Technologies has entered this industry, offering creative and practical solutions to various processes including harvesting and ploughs (Bac et al. 2014). Drones are significantly more profitable in agriculture and innovative farming than satellite imagery because unmanned aerial vehicles (UAV) can provide farmers with an up-close view of their fields while remaining close to the landscape and so delivering more precise assessments. In farming, drones could be used to maximise crop selection, water accessibility, fertilisation, and insecticide use. Drones will help reduce the cost of agricultural production overall while also maintaining a good yield and crop quality (Gao et al. 2020). Due to increased agricultural productivity and population, new problems are created. Drones are being utilised to encourage accurate measurement, current information collecting, profitable farming growth, and improved crop production, accuracy, and problem-solving abilities. Additionally, the agriculture industry must become more efficient and competitive due to the recent globalisation of the food market. In these conditions, smart farming which methodically makes use of cutting-edge technology like robotics, monitoring, information, and communication technology would be critical in promoting productive, labour-saving, profitable, and agricultural sustainability. This chapter's objective is to present the use of drone technology to better understand farming challenges and profitable, safe innovations that increase the profitability of agricultural products in the future.

13.2 A New Paradigm Shifts in Agriculture

The agricultural sector is quite intricate. Weather has always been a factor in harvesting success. For a successful harvest, there must be an appropriate ratio of sunny and wet days, pleasant temperatures, and no weather-related catastrophes. Historically, farmers have made weather predictions using their knowledge of the land and natural indications. A career in farming is not one that many young people choose (Figs. 13.1 and 13.2).

They instead opt for more promising careers. Ageing agricultural workers and decreased production follow from this. The actual amount of food harvested is unpredictable and difficult to forecast because it depends on so many outside factors. Regular and manual procedures are used in many agricultural processes. Water scarcity is a further issue. Water is necessary for harvesting, but due to a global shortage brought on by climate change, agricultural firms must either pay more to get it or use less water, which results in a lesser harvest. Even though the agriculture sector generates enough food to feed everyone on the earth, about 10% of the

Fig. 13.1 Field scouting done by UAVs

Fig. 13.2 Spraying done by UAVs

world's population still faces hunger. Even still, supply chain flaws and unconscious shopping habits are the main causes of the wasteful consumption of one-third of the food produced.

Precision agriculture (PA), which is our present agricultural practice going towards smart farming, has been nicknamed "the farmer of the century" and is quickly gaining popularity in today's technologically sophisticated world. This computerised farming management strategy combines cutting-edge technologies to track and enhance agricultural manufacturing activities. PA employs cutting-edge technology and concepts to reduce temporal and spatial variability in many sectors of agricultural output in order to increase farm productivity. Spatial and temporal distribution is a variance that significantly affects agriculture. Examples of spatial and temporal distribution include changes in soil, yield, crop, etc. Unmanned aerial vehicles (UAVs) and sensor technologies are frequently used to achieve these goals.

13.3 Different Types of UAV Models Used in Agriculture

In comparison to multi-rotor UAVs, fixed wing UAVs have a completely different design, and the aerodynamic geometry of their two wings makes gliding much easier. Based on the number of rotors fitted to the main body, rotary UAVs are categorised. The terms "single copter", "tricopter", "quadcopter", "hexacopter", and "octocopter" refer to multi-rotor aircraft with one, three, four, six, or eight rotors. A quadcopter is a special type of unmanned aerial vehicle (UAV) with four rotors in its model. When there are four rotors, the two opposite rotors rotate clockwise (CW), while the other two rotate counterclockwise (CCW). Pitch, roll, and yaw are the three directions that a quadcopter can rotate around its axis (clockwise and counterclockwise). Basically, there are two ways to configure a quadcopter: the plus (+) model and the cross configuration (×). Compared to the plus model, the cross model is highly common and more reliable (Kedari et al. 2016).

13.4 Camera and Sensor Types Used in Drone

Many well-known consumer drones have built-in cameras that work well for photography and scouting. However, drones marketed for gauging the health of agricultural crops may be equipped with specialist cameras, frequently at a price significantly greater than that of consumer drones. The final user will determine the required camera for agriculture. The various camera types used for agricultural purposes are listed below (Figs. 13.3, 13.4, 13.5, 13.6 and 13.7).

Seeing Cameras A visual camera, often known as an red–green–blue (RGB) camera, may be included with less expensive consumer drones. These cameras will catch the entire visual spectrum, just like the human eye, making them ideal for field scouting. Spectral cameras. Due to intrinsic leaf refraction, plant reflection is

Fig. 13.3 Single copter

Fig. 13.4 Tricopter

Fig. 13.5 Quadcopter

greatest in the near-infrared (NIR) range. The NIR waveband is among the several that the multispectral camera employs. For crop monitoring, it is also utilised to obtain various vegetation indicators (Figs. 13.8, 13.9 and 13.10).

Fig. 13.6 Hexacopter

Fig. 13.7 Octocopter

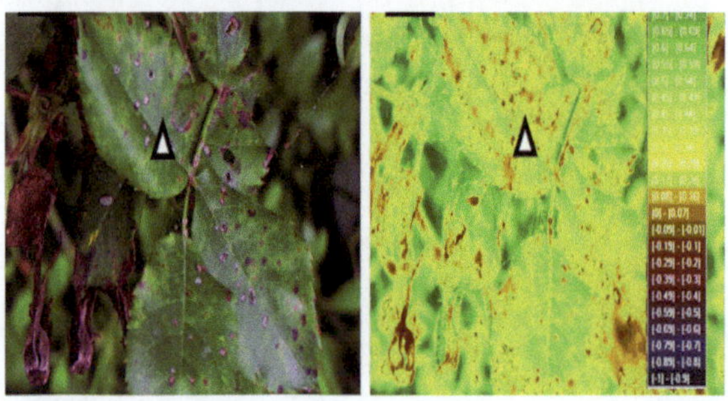

Fig. 13.8 RGB images

Fig. 13.9 Hyperspectral images

Fig. 13.10 Multispectral images

Infrared Cameras—It is possible to measure thermal radiation or temperature using longer infrared wavelengths. Thermal cameras are primarily used to assess biotic, thermal, and water stressors on crops. Hyperspectral photography with small wave bands, hyperspectral imaging gathers and analyses data from all areas of the electromagnetic spectrum. It may give additional specific information about the object. Of all the sensors, it is the most expensive. This kind of camera is used for weed detection, agricultural monitoring, and disease monitoring, among other things. LiDAR Lasers are used by light detection and ranging (LiDAR) to measure object distance. These sensors provide extremely accurate elevation maps of the landscape.

13.5 Applications of UAV in Smart Farming

1. **Crop Health Status Monitoring**

Visual examination and manually taking ground samples from a random selection of areas were the traditional techniques for crop monitoring. Colour and infrared photography taken on various platforms for more than 50 years has been used to track crop progress. In recent years, drone technology is one of these inventions that has grown in popularity, and precision agriculture has made extensive use of it. Regulation requires that a number of optical sensors be linked to a GPS so that the location of the readings can be recorded. Farmers can use these sensors to actively monitor the development of growing crops, and depending on the crop analysis to be done on a farm, cameras fitted with drones along with sensors with functions are listed in Table 13.1 (Haghighattalab et al. 2016).

Vegetation indices like normalised difference vegetation index (NDVI) imaging equipment with detailed colour information are available on drones with hyperspectral and multispectral cameras to display the health of the plants. This enables farmers to keep an eye on crops as they grow and deal with any problems as quickly as possible to save the plants (Fig. 13.11).

According to Adak et al. (2021), the normalised difference red edge index (NDREI) at the booting stage can be used to predict wheat production in advance. Drones equipped with "conventional" cameras are also used to monitor crop health. Many scientists and farmers typically use information from satellite photos to monitor crop growth, density, and colour.

2. **Detecting Weed Using UAVs**

In order to find weeds in a field, drones can be employed. To prevent them from vying for resources with the main crop, these weeds can be swiftly eliminated from the field. Nowadays the issue of examining weeds were solved using UAV imagery technology which aid in to determine a field's current weed situation and crop and weed distribution in space through camera-fitted drones. By concentrating a detection

Table 13.1 List of drones' technology used for monitoring different agricultural crops

Sl. No.	Techniques	Target	Reference
1	Multispectral camera on UAV—normalised difference vegetation index (NDVI)	Monitoring the crop	
2	RGB imaging with UAV to detect disease in the vineyard field	Vegetation indices	
3	UVIRCO camera on the drone	For monitoring the crop health and crop harvest size	
4	UAV imaging for different rice cultivars	Leaf area index (LAI)	Gong et al. (2021)
5	UAV-derived normalised difference red edge index (NDRE) in corn	Mapping chlorophyll content	Simic Milas et al. (2018)
6	UAVs helicopter	Vegetation indices—differentiating healthy, unhealthy plants and weeds	Swain et al. (2010)
7	Multispectral imaging with UAV	Track stress levels on crops and analyse their health	Paredes et al. (2017)
Monitoring on forest using UAVs			
8	UAV imagery—normalised difference vegetation index (NDVI), digital surface model (DSM), topographic and digital aerial photogrammetry	Monitoring health status in priority riparian forests	
9	UAV-based multispectral indices—in uneven-aged mixed conifer-broadleaf forest	Seasonal monitoring and careful estimation of photosynthetic activity	Vivar-Vivar et al. (2022)

on a per-plant basis, it is possible to assess the quantity of crops and different weed species and give this information to farmers (Lottes et al. 2017).

3. **Planting Seeds and Seedlings Using UAVs**

Sowing and planting was well efficiently done using drone with any doubt, as an illustration, Salaan et al.'s (2019) work demonstrates the effective use of UAVs over a sizable area of uneven rice paddies. They employed a UAV-based system for the quick and effective distribution of seeds, fertiliser, and plant nutrients. Plant seeds and seedlings are still in its early stages; researchers are aiming to build UAVs with image recognition technology and an enhanced planting strategy. Startups have developed drone-based planting systems with a 75% adoption rate and 85% cost saving. These

Fig. 13.11 Crop monitoring using drones

systems inject soil with plant nutrients, providing the plant with all the nutrition it requires to survive (Kim et al. 2019).

4. **Soil Analysis Using UAVs**

Drones can be used to mount sensors that can analyse the soil conditions, terrain conditions, soil pH (Jack et al. 2019), moisture content (Duangsuwan et al. 2020), soil tillage quality parameters (Fanigliulo et al. 2020), nutrients content, and fertility levels (Paredes et al. 2017).

5. **Crop Protection Using UAVs**

Arthropod pest and disease outbreaks frequently exhibit non-uniform geographical distributions in field crops and orchards. First, reflectance-based crop monitoring can be used to locate pest and disease hotspots using remote sensing technology. This is the first step in precision pest control. Second, localised remedies may be offered by precision control devices like pesticide spray rigs and distributors of natural enemies. Both technologies can be attached to machinery that is being transported through fields by manned or unmanned vehicles, as well as by aerial drones. Remote sensing technologies have been successfully used to detect stress caused by a number of arthropod pests and illnesses on a wide range of agricultural crops (Nansen 2016; Nansen and Elliott 2016) (Table 13.2).

Drones could be used to deliver biocontrol agents in the same way that pesticides are sprayed (Berner and Chojnacki 2017a, b). Predatory mites, *P. persimilis*, and the egg parasite *Trichogramma* spp. have received the most interest in the development of drone-mounted dispensers. In Australia, a company is also utilising drones to

Table 13.2 Use of drones for detecting arthropod-induced stress and disorders

Pest	Assessment	Technique	Author
Melanaphis sacchari	Arthropod counts	Multispectral	Stanton et al. (2017)
Two-spotted spider mite, *Tetranychus urticae* in cotton	Assessments of damage	Multispectral	Huang et al. (2018)
Wheat fall armyworm, *Spodoptera frugiperda*	Outbreak reported by grower	Red green blue, multispectral	Zhang et al. (2014)
Grape phylloxera, *Daktulosphaira vitifoliae*	Acting as traps and visual assessments of plants	Red green blue, Hyperspectral, multispectral,	Vanegas et al. (2018)
Onion thrips	Damage assessment	Multispectral	Nebiker et al. (2016)
Leptinotarsa decemlineata	Damage assessment	Multispectral	Hunt and Rondon (2017)
Monitoring in rice field	Insect sampling	Rotary drone	Kim et al. (2019)

spread predatory mites on strawberry and maize (Drone Agriculture 2018; Pearl 2015). *Trichogramma* and *Cotesia flavipes* are also used through drone discharge to combat sugarcane pests (*Saccharum* spp., *Diatraea saccharalis*) (Yang et al. 2018; Rangel, 2016) (Table 13.3).

Table 13.3 Drone-based detection of plant diseases

Pathogen	UAV with number of rotors	Technique	Assessments	Reference
Xanthomonas citri in	Multi-rotor with 4 rotors	Hyperspectral	Visual inspection of images	Abdulridha et al. (2019)
Banana yellow sigatoka	Multi-rotor-quadcopter with 4 rotor	RGB	Visual images, degree of severity	Calou et al. (2020)
Potato *Phytophthora infestans*	Multi-rotor with 6 rotor	Multispectral	Visual images, degree of severity	
Gemini virus in maize	Multi-rotor with 6 rotor	Multispectral	Visual images	Chivasa et al. (2021)
Wheat *Fusarium graminearum*	Multi-rotor with 4 rotors	Hyperspectral	Visual images	
Wheat *Puccinia striiformis*	Multi-rotor with 4 rotors	Hyperspectral	Diseases monitoring	Guo et al. (2021)
Ustilaginoidea virens in paddy	Multi-rotor with 4 rotors	Hyperspectral	Damage assessment	An et al. (2021)
Grapevine vein-clearing virus	Multi-rotor with 4 rotors	Hyperspectral	Visual images	Nguyen et al. (2021)

6. Spray Application

Manual pesticide application affects humans, causing problems such as cancer, hypersensitivity, asthma, and other disorders. To solve these issues, a drone-mounted sprayer is being employed. The capacity to cover more ground, increase chemical effectiveness, and streamline and speed up the spraying process have all been improved by the use of drone-mounted sprayers in the field. These days, drones can spray fields as needed by flying following pre-mapped lines while carrying up to a 40-L pesticide tank. Drones are capable of scanning the ground, properly dispensing liquid, and spraying in real time for even coverage while adjusting their distance from the surface. Improved effectiveness was the ultimate result, with less chemical leaking into groundwater. Drones can actually do aerial spraying up to five times more swiftly than traditional equipment, according to experts (Figs. 13.12 and 13.13; Table 13.4).

Trichogramma parasitoids are also used through drone discharge in China to combat sugarcane (*Saccharum* spp.) pests (Yang et al. 2018). Other lepidopteran pests that affect other crops are also controlled by *Trichogramma* varieties. The use of *Trichogramma* via drone is a cost-effective and efficient method of controlling the European maize borer. *T. ostriniae* released 500 k parasitoids/ha at 125 release points per ha in 11 ml of vermiculite-*Trichogramma* mix and 4000 eggs per release point with flying speed of 3–5 m/s and height of 10 m (Figs. 13.14 and 13.15; Table 13.5).

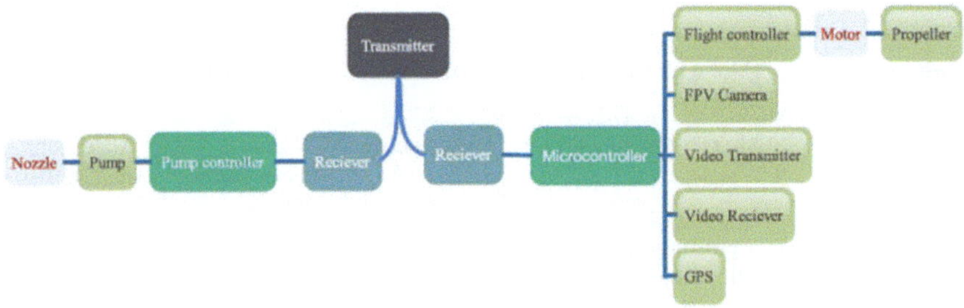

Fig. 13.12 Spraying procedure in UAVs

Fig. 13.13 Spraying using UAV

Table 13.4 Different spraying system

Controller	Sprinkling speed	Nozzle type	Reference
FC	1.15 ha/h	Flat fan	Yallappa et al. (2017)
Rotomotion's SR200	0.4 ha/min (1 acre/min), 2.2 m/s	Micron air ULV–A+	Huang et al. (2015)
Yamaha RMAX	47 l/ha	Flat fan	Giles and Billing (2015)
Arduino	1 l/min	Universal	Meivel et al. (2016)
MSP430 single-chip	0.6–1 l/min	Centrifugal	Xue et al. (2016)
-	13.61 g/30 s	Micron air–A+	Zhu et al. (2010)
DJIs 900 model	0.2 MPa (pressure)	Flat fan	Berner and Chojnacki (2017a, b)
N-3 type	850 ml/min	Rotary atomizer	Qin et al. (2018)
UAV ZHKU-0404-01	0.2–1.0 MPa (pressure)	Flat fan	Tang et al. (2018)

Fig. 13.14 Biocontrol agents sprayed using UAVs

Fig. 13.15 Tricho capsules

Table 13.5 Analysis of different spraying speeds of UAV different crops

UAV and number of nozzles	Speed (max load with max speed)	Reference
WSZ-0610 (2)	0.72 (10 L, 4 m/s)	Shilin et al. (2017)
HY-B-15L (5)	0.38 (15 L, 4.5 m/s)	
Six-rotor UAV (4)	3.15 (10 kg, 16 m/s)	Yanliang et al. (2017)
N-3 UAV (2)	15 L/m^2 (25 L, 4 m/s)	Qin et al. (2018)
Knapsack-type Electric fog sprayer 3WBD-16L (1)	300 L/hm^2 (20 L, 0.8–1 m/s)	
Hexacopter (4)	2.5 L/min (5 L, 3.6 kmh/1)	Yallappa et al. (2017)
Six rotors (6)	680 mL/min (30 L, 5 m/s)	Xin et al. (2018)
DJI AGRAS MG 1S (6)	0.43 L/min (10 L, 8 m/s)	Sarghini et al., 2019

13.6 Drone Pollination

A New York-based startup has created a pollen-dumping drone that assists in the pollination of apples, cherries, and other fruits like almonds. The corporation asserts that its drone rate might rise from 25 to 65%, despite the fact that an external audit revealed that these numbers have not been finalised. Despite this, some growers of fruits and vegetables are upbeat about the potential of drones in orchards. Some of the most current drone farming applications are still in the development stages. One of the most common uses of drone technology is pollination. Scientists from Japan and the Netherlands are developing a tiny drone that can safely pollinate plants (Croptracker). The following phase is to develop autonomous pollinating drones that can keep an eye on the health of crops without constant human monitoring (Adaka et al. 2022).

13.7 Climate Smart Agriculture

The goal of climate-smart agriculture (CSA) is to increase the adaptability of agricultural systems to climate change (Campbell et al. 2014). According to the FAO, CSA is agriculture that increases production, strengthens resilience, lowers GHG emissions, and makes it easier to accomplish national food security and development goals (Lipper et al. 2014). A CSA-based system built on three main pillars—raising agricultural productivity, boosting adaptive capacity at multiple scales, and reducing greenhouse gas emissions—can help enhance food supply in the face of climate change and increased climate variability (Bai et al. 2019; Stein and Steinmann 2018; Sharda et al. 2023). These systems, which are based on intercropping and crop rotation, have the potential to retain long-term soil fertility while also maximising biological nutrient cycling between plants and animals (Pervaiz et al. 2020). Mixed farming is defined by the FAO as farming operations managed by households or enterprises in which crop production and livestock husbandry are merged into a single farming system. There are numerous advantages to mixed farming, including effective resource utilisation by feeding livestock on crops and grassland and using animal waste to fertiliser the fields, crops, or livestock needs. Sustainability in global food production is achieved through the integration of crops and animals in a single system, which acts as an ecological and environmental safeguard. Advanced smart technologies will improve crop and animal management and boost food security. These technologies include agricultural drones, drones with sensors, and remote sensing (Barasa et al. 2021; Pandey et al. 2023).

13.8 Water Management Using UAVs

Drones fitted with imaging sensors can pinpoint the precise location of crop fields in water and heat-stressed areas. It can be used to supply irrigation based on the needs of the crops. Water waste will be reduced as a result, and efficient irrigation water use will be ensured. Because drought conditions are more impacted by climate change, it is imperative to supply more efficient irrigation systems. Drones can measure the soil's moisture content with extreme precision using microwave detection without upsetting the plants in the most effective way to resources conservation (Adaka et al. 2022).

Koparan et al. (2018) describe a method for evaluating on-site water quality with a hexacopter fitted with electronic sensors that can measure temperature, EC, dissolved oxygen, and pH in water. The results proved to be accurate in measuring all the parameters, and sensors fitted capture maps depicting the spatial distribution of the measured paramecium across the pond and also measures water level in the field using imagery-based technology (Gao et al. 2019). In the state of Maharashtra, the area irrigated is measured using drones. Drone technology has been used successfully in the water and irrigation industries. The work was done in Pune, and it resulted in a significant increase in overall irrigated area from 10.71 lakh ha last year to 13.884 lakh ha this year. As a result, the valuation was raised by Rs. 50 crore. Drone technology increased departmental efficiency and transparency, reducing the need for human resources. The rapid regulation of the irrigation schedule and identification of cultivars with improved water use efficiency under field conditions will be made possible by UAVs fitted with remote sensors (Gago et al. 2014).

13.9 Conclusion

Drones could totally transform Indian agriculture. Future technological advancements are anticipated to make drone manufacturing economically feasible. Due to the arduous labour and tedium required in farming, modern adolescents are not drawn to it. Drones are the most common varieties employed in agricultural crop production and crop protection activities like pest detection etc. The size and cost of these drones are shrinking by the day. Drone helicopters are typically used for spraying fertiliser or pesticides; however, the utilisation of multi-copters for pesticide spraying is increasing all the time because of better solutions for spot spraying and their greater flight stability. The weight, size, and resolution of drone cameras have all changed significantly over time. Multispectral cameras are replacing RGB cameras as feature extraction requirements increase.

The application of machine learning in drones has made it feasible to create a system that is beneficial to farmers. However, there are other issues with deploying drone technology in agriculture that must be addressed in order to raise the acceptance rate of drones. Cost of technology, limitations on drone battery life, vision loss, end-user technological literacy, and flaws in image processing and data analysis are the main issues. Young people may be enthralled by the potential usage of drones and be motivated to pursue careers in agriculture. The next agricultural revolution will be data-driven, and drones can be a big part of that. Without harming the environment, the proper use of data can increase farmer productivity and quality of life. Drones may thus play a crucial role in agriculture in the future by helping farmers manage their lands and resources more effectively and sustainably.

References

Abdulridha J, Batuman O, Ampatzidis Y (2019) UAV-based remote sensing technique to detect citrus canker disease utilizing hyperspectral imaging and machine learning. Remote Sens 11(11):1373

Adak S, Bandyopadhyay KK, Sahoo RN, Mridha N, Shrivastava M, Purakayastha TJ (2021) Prediction of wheat yield using spectral reflectance indices under different tillage, residue and nitrogen management practices. Curr Sci 121(3):402

Adaka S, Mandala N, Maitya PP, Mukhopadhyaya A (2022) Drones: a modern breakthrough for smart farming. Food Sci Rep 3(9):54–57

An G, Xing M, He B, Kang H, Shang J, Liao C et al (2021) Extraction of areas of rice false smut infection using UAV hyperspectral data. Remote Sens 13(16):3185. https://doi.org/10.3390/rs13163185

Bac CW, Van Henten EJ, Hemming J, Edan Y (2014) Harvesting robots for high-value crops: state-of-the-art review and challenges ahead. J Field Robot 31(6):888–911. https://doi.org/10.1002/rob.21525

Bai X, Huang Y, Ren W, Coyne M, Jacinthe PA, Tao B et al (2019) Responses of soil carbon sequestration to climate-smart agriculture practices: a meta-analysis. Glob Change Biol 25(8):2591–2606

Barasa PM, Botai CM, Botai JO, Mabhaudhi T (2021) A review of climate-smart agriculture research and applications in Africa. Agronomy 11(6):1255

Berner B, Chojnacki J (2017a) Use of drones in crop protection. In: IX International scientific symposium, Lublin, Poland. https://doi.org/10.24326/fmpmsa.2017.9

Berner B, Chojnacki J (2017b) Influence of the air stream produced by the drone on the sedimentation of the sprayed liquid that contains entomopathogenic nematodes. J Res Appl Agric Eng 62(3). https://doi.org/10.31838/jcr.07.06.117

Calou VBC, dos Santos Teixeira A, Moreira LCJ, Lima CS, de Oliveira JB, de Oliveira MRR (2020) The use of UAVs in monitoring yellow sigatoka in banana. Biosys Eng 193:115–125

Campbell BM, Thornton P, Zougmoré R, Van Asten P, Lipper L (2014) Sustainable intensification: what is its role in climate smart agriculture? Curr Opin Environ Sustain 8:39–43

Chivasa W, Mutanga O, Burgueño J (2021) UAV-based high-throughput phenotyping to increase prediction and selection accuracy in maize varieties under artificial MSV inoculation. Comput Electron Agric 184:106128. https://doi.org/10.1016/j.compag.2021.106128

Drone Agriculture (2018) Formerly aerobugs. Available from https://www.droneagriculture.com.au/

Duangsuwan S, Teekapakvisit C, Maw MM (2020) Development of soil moisture monitoring by using IoT and UAV-SC for smart farming application. Adv Sci Technol Eng Syst J 5(4):381–387. https://doi.org/10.25046/aj0504244

Fanigliulo R, Antonucci F, Figorilli S, Pochi D, Pallottino F, Fornaciari L, Grilli R, Costa C (2020) Light drone-based application to assess soil tillage quality parameters. Sensors 20(3):728. https://doi.org/10.3390/s20030728

Gago J, Douthe C, Florez-Sarasa I, Escalona JM, Galmes J, Fernie AR, Flexas J, Medrano H (2014) Opportunities for improving leaf water use efficiency under climate change conditions. Plant Sci 226:108–119

Gao A, Wu S, Wang F, Wu X, Xu P, Yu L, Zhu S (2019) A newly developed unmanned aerial vehicle (UAV) imagery based technology for field measurement of water level. Water 11(1):124

Gao D, Sun Q, Hu B, Zhang S (2020) A framework for agricultural pest and disease monitoring based on internet-of-things and unmanned aerial vehicles. Sensors 20(5):1487. https://doi.org/10.3390/s20051487

Giles D, Billing R (2015) Deployment and performance of a UAV for crop spraying. Chem Eng Trans 44:307–312

Gong Y, Yang K, Lin Z, Fang S, Wu X, Zhu R, Peng Y (2021) Remote estimation of leaf area index (LAI) with unmanned aerial vehicle (UAV) imaging for different rice cultivars throughout the entire growing season. Plant Methods 17(1):1–16

Guo A, Huang W, Dong Y, Ye H, Ma H, Liu B et al (2021) Wheat yellow rust detection using UAV-based hyperspectral technology. Remote Sens 13(1):123. https://doi.org/10.3390/rs13010123

Haghighattalab A, González Pérez L, Mondal S, Singh D, Schinstock D, Rutkoski J, Ortiz-Monasterio I, Singh RP, Goodin D, Poland J (2016) Application of unmanned aerial systems for high throughput phenotyping of large wheat breeding nurseries. Plant Methods 12(1):1–15

Huang Y, Hoffman WC, Lan Y, Fritz BK, Thomson SJ (2015) Development of a low-volume sprayer for an unmanned helicopter. J Agric Sci 7(1):148

Huang H, Deng J, Lan Y, Yang A, Deng X, Zhang L et al (2018) A two-stage classification approach for the detection of spider mite-infested cotton using UAV multispectral imagery. Remote Sens Lett 9(10):933–941. https://doi.org/10.1080/2150704X.2018.1498600

Hunt ER, Rondon SI (2017) Detection of potato beetle damage using remote sensing from small unmanned aircraft systems. J Appl Remote Sens 11(2):026013. https://doi.org/10.1080/01431161.2018.1455244

Jack R, Mohidin H, Tamrin KF, Banchit A, Khan MYMA, Narawi A, Dollah N (2019, December) Soil pH mapping of pineapple crop: a feasibility study using aerial photo. In: 2019 International conference on computer and drone applications (IConDA). IEEE, pp 5–8. https://doi.org/10.1109/IConDA47345.2019.9034909

Kedari S, Lohagaonkar P, Nimbokar M, Palve G, Yevale P (2016) Quadcopter-a smarter way of pesticide spraying. Imp J Interdiscip Res 2(6):1257–1260

Kim J, Kim S, Ju C, Son HI (2019) Unmanned aerial vehicles in agriculture: a review of perspective of platform, control, and applications. IEEE Access 7:105100–105115

Koparan C, Koc AB, Privette CV, Sawyer CB (2018) In situ water quality measurements using an unmanned aerial vehicle (UAV) system. Water 10(3):264

Lipper L, Thornton P, Campbell BM, Baedeker T, Braimoh A, Bwalya M, Caron P, Cattaneo A, Garrity D, Henry K, Torquebiau EF (2014) Climate-smart agriculture for food security. Nat Clim Chang 4(12):1068–1072

Lottes P, Khanna R, Pfeifer J, Siegwart R, Stachniss C (2017, May) UAV-based crop and weed classification for smart farming. In: 2017 IEEE international conference on robotics and automation (ICRA). IEEE, pp 3024–3031

Meivel S, Maguteeswaran R, Gandhiraj N, Srinivasan G (2016) Quadcopter UAV based fertilizer and pesticide spraying system. Int Acad Res J Eng Sci 1:8–12

Nansen C (2016) The potential and prospects of proximal remote sensing of arthropod pests. Pest Manag Sci 72(4):653–659. https://doi.org/10.1002/ps.4209

Nansen C, Elliott N (2016) Remote sensing and reflectance profiling in entomology. Annu Rev Entomol 61(1):139–158. https://doi.org/10.1146/annurev-ento-010715-023834

Nebiker S, Lack N, Abächerli M, Läderach S (2016) Light-weight multispectral UAV senors and their capabilities for predicting grain yield and detecting plant diseases; International archives of the photogrammetry. Remote Sens Spat Inf Sci Prague Czech Repub 1

Nguyen C, Sagan V, Maimaitiyiming M, Maimaitijiang M, Bhadra S, Kwasniewski MT (2021) Early detection of plant viral disease using hyperspectral imaging and deep learning. Sensors 21(3):742

Pandey K, Singh KG, Singh A (2023) Multi-sensors based smart nutrient reuse management system for closed soilless culture under protected cultivation. Comput Electron Agric 107495

Paredes JA, González J, Saito C, Flores A (2017, June) Multispectral imaging system with UAV integration capabilities for crop analysis. In: 2017 First IEEE international symposium of geoscience and remote sensing (GRSS-CHILE). IEEE, pp 1–4. https://doi.org/10.1109/GRSS-CHILE.2017.7996009

Pearl E (2015) Drone used to drop beneficial bugs on corn crop. The University of Queensland, Australia, News (UQ News). Available from https://www.uq.edu.au/news/article/2015/04/drone-used-drop-beneficial-bugs-corn-crop

Pervaiz ZH, Iqbal J, Zhang Q, Chen D, Wei H, Saleem M (2020) Continuous cropping alters multiple biotic and abiotic indicators of soil health. Soil Syst 4(4):59

Qin W, Xue X, Zhang S, Gu W, Wang B (2018) Droplet deposition and efficiency of fungicides sprayed with small UAV against wheat powdery mildew. International Journal of Agricultural and Biological Engineering 11(2):27–32

Rahman MFF, Fan S, Zhang Y, Chen L (2021) A comparative study on application of unmanned aerial vehicle systems in agriculture. Agriculture 11(1):22. https://doi.org/10.1002/rob.21525

Rangel RK (2016) Development of an UAVS distribution tools for pest's biological control "Bug Bombs!". In: IEEE aerospace conference, 5–12 Mar 2016, Big Sky, MT. 1, pp 1–8

Salaan CJ, Tadakuma K, Okada Y, Sakai Y, Ohno K, Tadokoro S (2019) Development and experimental validation of aerial vehicle with passive rotating shell on each rotor. IEEE Robot Autom Lett 4(3):2568–2575. https://doi.org/10.1109/LRA.2019.2894903

Sarghini F, Visacki V, Sedlar A, Crimaldi M, Cristiano V, de Vivo A (2019, October) First measurements of spray deposition obtained from UAV spray application technique. In: 2019 IEEE international workshop on metrology for agriculture and forestry (MetroAgriFor). IEEE, pp 58–61

Sharda R, Singh A, Pandey K (2023) Paddy and maize straw-based media as an alternative for cocopeat in soilless cultivation. J Plant Nutr 1–19

Shilin W, Jianli S, Xiongkui H, Le S, Xiaonan W, Changling W, Zhichong W, Yun L (2017) Performances evaluation of four typical unmanned aerial vehicles used for pesticide application in China. Int J Agric Biol Eng 10(4):22–31

Simic Milas A, Romanko M, Reil P, Abeysinghe T, Marambe A (2018) The importance of leaf area index in mapping chlorophyll content of corn under different agricultural treatments using UAV images. Int J Remote Sens 39(15–16):5415–5431. https://doi.org/10.1080/01431161.2018.1455244

Singh AP, Yerudkar A, Mariani V, Iannelli L, Glielmo L (2022) A bibliometric review of the use of unmanned aerial vehicles in precision agriculture and precision viticulture for sensing applications. Remote Sens 14(7):1604. https://doi.org/10.3390/rs14071604

Stanton C, Starek MJ, Elliott N, Brewer M, Maeda MM, Chu T (2017) Unmanned aircraft system-derived crop height and normalized difference vegetation index metrics for sorghum yield and aphid stress assessment. J Appl Remote Sens 11(2):026035. https://doi.org/10.1117/1.JRS.11.026035

Stein S, Steinmann HH (2018) Identifying crop rotation practice by the typification of crop sequence patterns for arable farming systems—a case study from Central Europe. Eur J Agron 92:30–40

Swain KC, Thomson SJ, Jayasuriya HP (2010) Adoption of an unmanned helicopter for low-altitude remote sensing to estimate yield and total biomass of a rice crop. Trans ASABE 53(1):21–27

Tang Y, Hou CJ, Luo SM, Lin JT, Yang Z, Huang WF (2018) Effects of operation height and tree shape on droplet deposition in citrus trees using an unmanned aerial vehicle. Comput Electron Agric 148:1–7

Vanegas F, Bratanov D, Powell K, Weiss J, Gonzalez F (2018) A novel methodology for improving plant pest surveillance in vineyards and crops using UAV-based hyperspectral and spatial data. Sensors 18(1):260. https://doi.org/10.3390/s18010260

Vivar-Vivar ED, Pompa-García M, Martínez-Rivas JA, Mora-Tembre LA (2022) UAV-based characterization of tree-attributes and multispectral indices in an uneven-aged mixed conifer-broadleaf forest. Remote Sens 14(12):2775. https://doi.org/10.3390/rs14122775

Xin F, Zhao J, Zhou Y, Wang G, Han X, Fu W, Jizhong D, Lan Y (2018) Effects of dosage and spraying volume on cotton defoliants efficacy: a case study based on application of unmanned aerial vehicles. Agronomy 8(6):85

Xue X, Lan Y, Sun Z, Chang C, Hoffmann WC (2016) Develop an unmanned aerial vehicle based automatic aerial spraying system. Comput Electron Agric 128:58–66

Yallappa D, Veerangouda M, Maski D, Palled V, Bheemanna M (2017, October) Development and evaluation of drone mounted sprayer for pesticide applications to crops. In: 2017 IEEE global humanitarian technology conference (GHTC). IEEE, pp 1–7

Yang S, Yang X, Mo J (2018) The application of unmanned aircraft systems to plant protection in China. Precision Agric 19(2):278–292. https://doi.org/10.1007/s11119-017-9516-7

Yanliang Z, Qi L, Wei Z (2017) Design and test of a six-rotor unmanned aerial vehicle (UAV) electrostatic spraying system for crop protection. Int J Agric Biol Eng 10(6):68–76

Zhang C, Walters D, Kovacs JM (2014) Applications of low altitude remote sensing in agriculture upon farmers' requests—a case study in northeastern Ontario, Canada. Plos One 9(11):e112894. https://doi.org/10.1371/journal.pone.0112894

Zhu H, Lan Y, Wu W, Hoffmann WC, Huang Y, Xue X, Liang J, Fritz B (2010) Development of a PWM precision spraying controller for unmanned aerial vehicles. J Bionic Eng 7(3):276–283

Chapter 14
IoT-Enabled Unmanned Aerial Vehicle: An Emerging Trend in Precision Farming

Manish Yadav, B. B. Vashisht, Niharika Vullaganti, S. K. Jalota, S. L. Yadav, Gurjeet Singh, Arun Kumar, and Shivani Kothiyal

Abstract Over the last decade, smart farming (SF) and climate smart agriculture (CSA) have benefited greatly from the integration of cutting-edge technologies into unmanned aerial vehicles (UAVs) platforms. In the coming years, smart farming will be practised all over the world. These smart farming technologies provide the ability to simultaneously reduce operational expenses while increasing agricultural productivity and the effectiveness of inputs. Unmanned aerial vehicles (UAVs) have an immense potential for SF and CSA. Drones are the most used UAV platform in agriculture for low-altitude remote sensing. This chapter presents a useful guide on technical characterizations of popular UAV types and on-board sensors that are best appropriate for various agricultural applications utilized in SF and CSA. An effort has been made in this chapter to examine the categories of sensors appropriate for SF and CSA, as well as prospective needs and difficulties in using UAVs in smart agriculture. The UAV technology can be used for spraying the pesticides, crop monitoring, weed mapping, water stress detection, irrigation scheduling, and precision input application in agriculture. However, the affordability and simplicity of operating UAVs for smart

M. Yadav (✉) · B. B. Vashisht · N. Vullaganti · S. K. Jalota
Department of Soil Science, Punjab Agricultural University, Ludhiana, Punjab 141004, India
e-mail: manish.soil017@gmail.com

S. L. Yadav
Anand Agricultural University, Anand, Gujarat 388110, India

G. Singh
AgriLife Research Center, Texas A&M University, Beaumont, TX 77713, USA

A. Kumar
Livestock Research Station, Rajasthan University of Veterinary and Animal Sciences, Beechwal, Bikaner 334001, Rajasthan, India

N. Vullaganti
Agricultural and Biosystems Engineering, North Dakota State University, Fargo 58102, USA

S. Kothiyal
Department of Climate Change & Agricultural Meteorology, Punjab Agricultural University, Ludhiana, Punjab 141004, India

K. Pandey et al. (eds.), *Artificial Intelligence and Smart Agriculture*,
Advances in Geographical and Environmental Sciences,
https://doi.org/10.1007/978-981-97-0341-8_14

farming may be significant factors in encouraging farmers to employ them. UAVs are typically operated using radio waves by remote controllers.

Keywords Climate-smart agriculture (CSA) · Crop monitoring · Drones · Remote sensing · Sensors · Smart farming (SF) · Unmanned aerial vehicles (UAVs)

14.1 Introduction

Just in last 20 years, technology has advanced in numerous ways and significantly transformed many aspects of our society. Use of satellites via remote sensing tools has been adopted for smart farming and climate resilient agriculture to supervise large-scale operations. The use of remote sensing technology has its own limitations of spatial and temporal resolution trade-offs for ground observations. The limitations of high-altitude satellite-based remote sensing technology can be overcome by use of unmanned aerial vehicles (UAVs). The most common mean of UAVs is drones and shows the potential to use of drones in climate-smart agriculture and precision agriculture. In the course of time, growers may originate untold uses from drones based on the farm needs like 'mosquito-sized drones' with sensors. By using these types of drones, it will be possible to measure the leaf water and nutrient status of all plants. While it may appear awkward to have so many drones working just to minimize accuracy, a high level of accuracy may not be required for on-farm management plans. Drones are cited as the most recent gadget that could be used for implementing smart farming and climate-smart agriculture (Caldwell 2015). A useful feature of drone services is the ability to provide accurate images of crops. Growers can use drone technology because it is relatively quick and easy to use. The farmers only need a brief tutorial on flight control and image processing. Drones are economically efficient when compared to farm workers. Furthermore, drones are highly economical as they reduce the huge need for labour. With the capacity of drones cover thousands of acres in one hour while spraying chemicals, farm workers may be replaced largely. Meanwhile, drones are creating large numbers of jobs requiring computer skills. Let us now pay particular attention to agricultural drones and how they help with smart framing and climate-smart agriculture. Grassi (2014) cites five different ways' drones which can be used in smart framing and climate-smart agriculture. They are as follows:

a. In midseason, farmers can inspect their crops for growth traits, height, chlorophyll content, and leaf area with drones equipped with multi-spectral sensors. Drones can cover 50 acres of area in 15–30 min, while eliminating human error and fatigue.
b. Agricultural drones can be used to monitor smart farming aspects, which is based on soil moisture variations and crop needs. Smart farming involves moving centre-pivot systems in a field at prescribed rates. Maize, sugarcane, and tall sorghum crops need meticulous scrutiny of the centre pivot and nozzles from

above. North American cereal belt farmers are increasingly using drones to inspect nozzles and sprinklers.

c. The first step in smart weed eradication involves a thorough survey (on land) or aerial imagery using drones. In order to assess the spread and intensity of weeds, drones with multi-spectral imaging are employed. We can read drone images properly based on the spectral signature of weeds and cropping areas. In order for variable-rate herbicide applicators to receive the necessary electronic signals, the NDVI data and postflight image processing are required (Grassi 2014).
d. Drone imagery can help to identify diseased/pest-attacked areas, as well as indicate the intensity of infection. Drone imagery is becoming increasingly popular during variable-rate pesticide and fungicide supply. A few models of drones are capable of carrying chemicals as variable-rate nozzles, which are highly profitable economically. They accomplish the task at a rapid pace.
e. Fertilizer application to crops is often based on nutrient availability study of the soil and creation of fertility maps that show variances. The fertilizer needs of a crop are also influenced by residual effects of previous season's inputs and cropping systems. The maps generated using drones for crop growth are efficient for fertilizer distribution.

As stated by Grassi (2014), farmers could use drones during smart farming to application of N, P, and K. The practice of crop phenotyping with drones is a relatively new one. Precision farming relies heavily on yield data and maps of each grid cell that showing variations in productivity. The yield map of several years is layered together, and then fertilizer and water schedules are determined based on this information.

Unmanned aerial vehicles are miniature, electronically controlled machines that are launched from the edge of a field. Recently, as the cost of implementing aerospacing engineering and sensor technology is falling, their application is becoming more prevalent (Abdullahi et al. 2015). UAVs can be used by farmers, consultants, and crop insurance agents to examine fields for weather-related damage, harmful insects, diseases, and chemical drift (Maes and Steppe 2019). UAVs are directed by a radio controller, and they are now programmable to work with smartphone applications.

14.2 Opportunities and Challenges in Smart Farming and Climate-Smart Agriculture

Smart farming and climate-smart agriculture require a series of services including meteorological data monitoring and high-resolution drone images of the crop environment (Trimble 2015; Precision Hawk 2014; Pandey et al. 2023). The knowledge of spatial variability of soil nutrient status and ground features and crop productivity within the field is essential for site-specific input applications or precision farming. Drones with the appropriate sensors may be used to measure the spatial variability within a field, and they can also be used to monitor changes in the temporal input

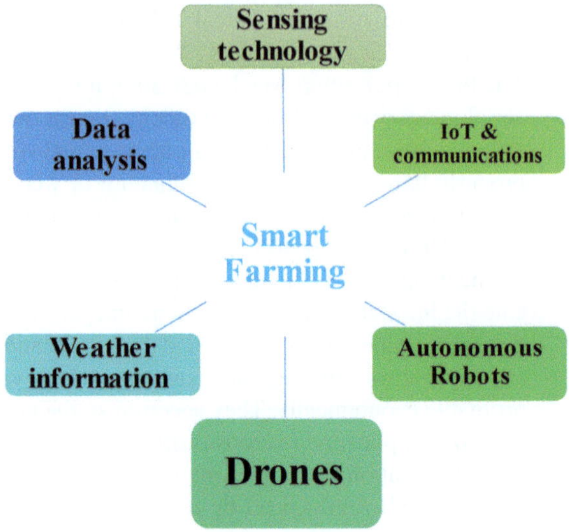

Fig. 14.1 Components of smart farming (https://www.smlease.com/entries/technology/what-is-smart-farming-future-of-agriculture/)

demand and the diagnosis of crop diseases by routine flight of these sensor-equipped drones. For visual examination, the drones can be fitted with RGB sensors, and multi-spectral sensors can be used to track crop vegetative development. Estimation of crop ET demands can be done through combined use multi-spectral and thermal sensors, and it is further used for irrigation scheduling of the crops. It is possible to use drones with kinematic base stations to collect elevation data (Hashem 2020) (Fig. 14.1).

Agricultural drones equipped with sophisticated sensors and imaging technologies are enabling farmers to boost yields and prevent crop loss. These drones are capable of capturing images and alerting farmers to any unseen issues, such as fungal infections. Growers use the drones as an essential aerial surveillance tool that enables them to carefully monitor their crops, locate pests, and make sure that their water management is efficient. With the low-altitude view, farmers can gain a previously unseen perspective.

It is far less expensive and has a greater resolution when compared to satellite imaging. It is unhindered and always available since it is taken under the clouds. Additionally, it is far less expensive than crop imaging from a manned aircraft (Fig. 14.2).

Major challenges for SF and CSA (Djojodihardjo et al. 2022)

- Food security is a significant concern as a consequence of climate change. Based on optimistic lower-end temperature rise forecasts, agricultural yields may decline by 10–20% by 2050 as a result of climate change.
- A concerted effort should be made to optimize synergies and ameliorate trade-offs between productivity and mitigation to achieve the necessary levels of growth. Furthermore, adequate financial resources, institutions, and incentives must be established to facilitate climate-smart transitions.

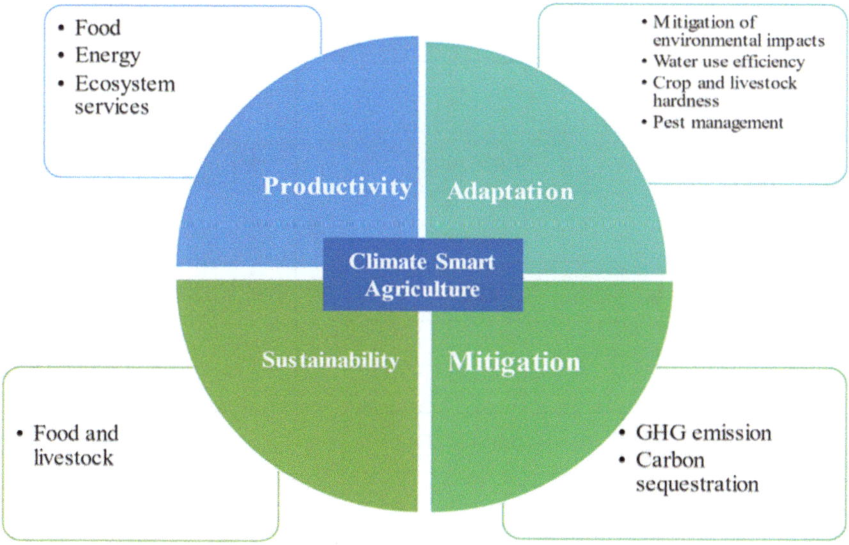

Fig. 14.2 Components of climate-smart agriculture (CSA) (Jariwala et al. 2022)

- Over the next forty years, food production must rise by 70% due to population expansion and economic prosperity.
- Climate change will pose a significant threat to rising global agricultural consumption and decreasing arable land per person.
- Activities that are climate-smart have a great chance of addressing environmental issues and cutting CO_2 emissions.
- The requirement for automated surveillance, trimming, thinning, and even harvesting crops as a substitute for a reducing human labour force.
- Promoting of climate resilient agriculture tactics and encouragements.
- Early intervention of climate resilient agriculture policy.
- Provision of finance for climate resilient agriculture.

14.3 Technological Specifications of the Unmanned Aerial Vehicle (UAV) Platforms

The key features that set UAVs apart from other types of small drones (such MAVs and NAVs) are the operational purpose of the system, the materials utilized in its construction, and the cost of the control system. UAVs come in a broad variety of sizes and designs. Unmanned aerial vehicles can be categorized into a number of groups based on their aerodynamic properties. Multi-rotor, fixed-wing, single-rotor,

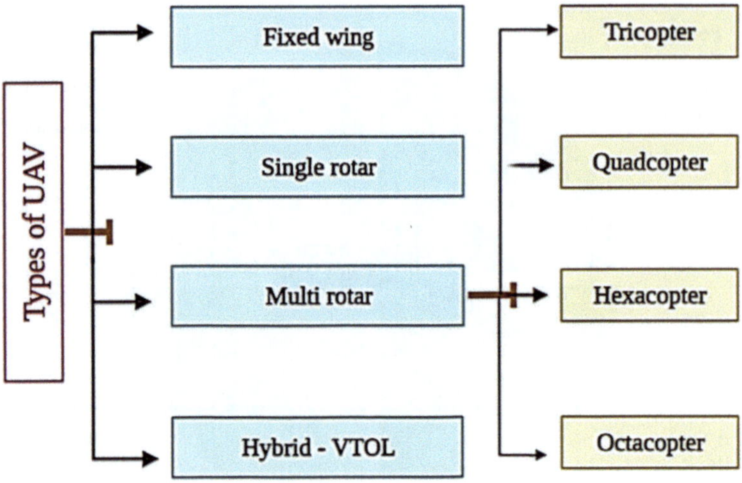

Fig. 14.3 Types of UAV platforms

and hybrid vertical take-off and landing (VTOL) are the four main types (Velusamy et al. 2021) (Fig. 14.3).

1. Fixed-wing drones

Fixed-wing drones lack propellers in favour of wings. They are constructed similar to regular aeroplanes. They exhibit excellent aerodynamics. High-resolution cameras are frequently included in these drones, allowing for improved aerial mapping and observation.

2. Single rotor drone

These have a construction and design that resemble helicopters. They feature a large rotor on top and a small rotor on the tail for direction control. Compared to multi-rotor drones, they are most difficult to fly and have a longer flight time as they are powered by gas engines. A flight control system, heading sensor, GPS receiver, gyroscope, image and telemetry transmitter, and spraying components are included in the helicopter.

3. Multi-rotor drone

A multi-rotor drone uses a number of propellers to fly and navigate. These are the UAVs that professionals use most frequently. They are employed in aerial mapping, video surveillance, and photography. The least expensive UAVs are those of this category. They are further divided into four classes:

a. Tricopter

A drone with three propellers (rotor) which will assist in stabilizing the centre of gravity of the tricopter during flight (Velusamy et al. 2021).

b. Quadcopter

Unmanned aerial vehicles (UAVs) with quadcopters have an advantageous design in that they have four rotors. These rotors provide this kind of UAV its lift.

c. Hexacopter

A hexacopter is an unmanned aerial vehicle with six rotors, each driven by an independent brushless direct current (BLDC) motor. This airborne vehicle may be used to apply pesticides for agricultural purposes, and the tank included beneath its airframe has a maximum capacity of 5 L of liquid. The liquid is then forced through a spray motor, which subsequently channels it to the spray nozzles (Table 14.1).

d. Octocopter

Similar to a hexacopter, an octocopter has eight rotors and is used for agricultural spraying. Typically, as the number of rotors of the drone increases, so does its payload.

4. Fixed-wing hybrid vertical take-off and landing (VTOL)

They exhibit the combined abilities of fixed-wing and multi-rotor drones since they feature both wings and propellers. Wings are used to fly, and propellers are utilized to take off. They fly with the efficiency of fixed-wing systems and are effective at taking off like multi-rotor systems. They are still in developmental stages (Table 14.2).

Unmanned aerial vehicle low-altitude remote sensing (UAV-LARS) is a novel approach to aerial remote sensing that makes use of unmanned aerial vehicles (UAVs)

Table 14.1 Comparison of rotor and fixed-wing UAVs (Abdullahi et al. 2015; Olson and Anderson 2021)

Particulars	Rotor UAV	Fixed-wing UAV
Ease of operation	Easy to operate	Difficult to fly
Flight duration	Shorter (20–40 min)	Longer (1–2 h)
Stability	Greater stability and hovering in one place	Lower stability and the inability to hover in a place
Area covered per battery	Limited	High
Space required	Little open space is required for vertical take-off and landing	Horizontal take-off and landing require more open space
Wind pressure	Able to fly in winds from 20 up to 50 mph	Rather of flying across the wind, it is possible to fly in and out of it
Change in direction	Can move in all four directions	Can only move forward
resolution	High resolution ± 1 cm RGB, 5 cm multi-spectral	High resolution ± 8 cm RGB and 15 cm multi-spectral
Payload capacity	Higher	Lower
Cost	Lower cost compared to fixed-wing UAV	Higher
Area coverage	1–8 ha	10–40 ha

Table 14.2 Comparison of types of UAVs (Velusamy et al. 2021)

Parameters	Types of UAV			
	Fixed wing	Single rotor	Multi-rotor	Hybrid VTOL
Number of rotors	01	02	03–08	01
Construction and upkeep	Simple	Complicated	Complicated	Complicated
Cost	High	High	Low	High
Time of flight	2 h on battery power and 16 h on a gas engine	Higher	20–40 min	Capacity of covering longer distances
Tolerance	High	High	Limited	High
Energy	Battery: they never use energy to maintain aerial motion	Gas Power	Battery: they utilize energy to maintain aerial motion	Battery
Speed of flight	High	Low	Low	High
Applications	Long-distance aerial mapping and surveillance	Aerial scanning	Aerial photography, short distance aerial mapping and surveillance	Mapping and land surveying, mining, surveillance, and security
Drawbacks	Aerial photography is not feasible due to the need for prolonged airborne suspension	Dangerous to handle and harder to fly	Limited payload	Imperfect hovering and a little payload
Requirement of training in flying	A runway or a catapult launcher is required for the take-off of a fixed-wing aircraft, while a parachute or a landing net is required for the landing	Not required	Not required	Not required

as aerial platforms for carrying sensor payloads like digital, multi-spectral, and hyper-spectral cameras. UAV-LARS sensors are a versatile, effective instrument for remote sensing since they may be frequently deployed and can gather high-resolution images from low altitudes with reduced cloud interference. The fundamental UAV-LARS system consists of a flight system, a mission load system, and a ground control system (Fig. 14.4).

A route planning programme, an image sensor, an inertial navigation system, and other UAV components are also regarded as essential. There are many different sizes and configurations of drones (Zhang et al 2021) (Table 14.3).

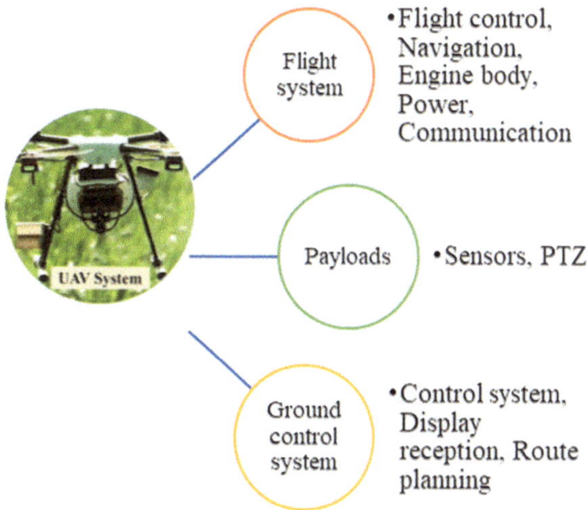

Fig. 14.4 Components of the unmanned aerial vehicle (UAV) monitoring system

14.4 UAV Applications in Smart Farming and Climate-Smart Agriculture

(i) **Crop type classification and planting**

Crop type maps are one of the most essential components of agricultural operations, such as predicting crop yields. Accurate crop type identification is also essential for managing subsidies, monitoring biodiversity, protecting the soil, and applying fertilizer sustainably. Similar to land cover/land use classification, the objective of the crop type classification is to map out different types of crops using the data supplied by low-altitude remote sensing (Liu et al. 2021). Depending on the requirements of certain tasks, it can be applied at various spatial scales. SAR sensors are used for classification on a larger scale, and RGB images from UAVs or SAR data combined can be used for classification on a smaller scale. Start-ups have has developed drone planting devices that have 75% adoption rate and 85% reduction in planting expenses. These devices release nutrient- and seed-filled pods into the ground, providing the plant with everything it needs to survive.

(ii) **Crop monitoring**

UAVs can help to gather phenotypic information of the crop like plant height (Han et al. 2018), lodging (Zhao et al. 2019), canopy cover (Lee and Lee 2011), leaf colour (Zhao et al. 2018), stand count, flower count, and fruit count in a non-destructive, quick, and efficient way that helps for making management and breeding decisions to optimize agricultural production (Olson and Anderson 2021). Chen et al. (2020) studied the use of imagery from UAV for agriculture land use classification. As crops progress through the crop season, drones can monitor the conditions so that timely

Table 14.3 Main characteristics of UAVs used in agriculture (del Cerro et al. 2021)

Type	UAV	Payload (kg)	Endurance (min)
Eight rotors	MK Okto XL 2	4	46
	Okto-XL	2.5	25
	DJI Agras MG-1	10	–
	Spreading Wings S1000	6–11	15
	ARF-MikroKopter	2.5	28
	AT8 UAV	2	30
Six rotors	EM6-800	1.2	25
	Matrice 100/S1000	1	40
	DJI M600	4.5	16
	HexacopterP-Y6/A2500_WH	3	21
	HEXA-PRO™ UAV	2	40
	AIR-200	3	40
Four rotors	Parrot AR/2.0	–	12
	Parrot Anaf	–	25
	Phantom 2/3 Pro/4 Pro	0.3	30
	DJI Mavic Pro/inspire 1/inspire 2	0.2	21
	Jifei P20 UAV	12	20
	MD4-100/1000	1.2	45
	3DR Iris/Solo	0.4	22
Helicopters	Yamaha Fazer R	32	25
	Rotomotion SR200	20	240
	SKeldar V200	40	300
Fixed wing	Gatewing X100	0.2	15
	Zangaouav	1.1	60
	Trinity F90 +	–	90
	eBee SQ/Plus	0.3	59
	M23 UAV	14.6	25
	Tuffwing Mapper	2	40
	Airborne XT	200	3900

and need-based action can be taken. The reflection pattern at different wavelengths may be used to calculate a variety of multi-spectral indices using different visible, near-infrared, and thermal infrared sensor types. Along with assessing the effects of water stress, nutritional stress, insect–pest attack, and diseases, these indicators may also be used to assess the condition of the crops. Drones are equipped with sensors that can detect diseases or deficiencies even before symptoms appear. Therefore, they are useful for detecting diseases at an early stage. Thus, drones can provide early warning systems so that remedial measures can be applied in a timely manner depending on the degree of stress. There are several indices by which UAVs (drones) can observe a crop (Simelli and Tsangaris 2015). The quad copter is equipped with thermal and multi-spectral cameras (Colomina and Molina 2014) which can record vegetation canopy reflectance. The camera captures and retains one image every second using telemetry. Data obtained from the multi-spectral camera through telemetry was analyzed using the geographic indicator Normalized Difference Vegetation Index (NDVI) (Reinecke and Prinsloo 2017).

$$\text{NDVI} = (R_{\text{NIR}} - R_{\text{RED}})/(R_{\text{NIR}} + R_{\text{RED}}).$$

R_{NIR} = Reflectance of the near-infrared band.
R_{RED} = Reflectance of the red band.

This metric measures the health of green vegetation based on the normalization difference vegetation index. Values range from − 1 to + 1, with near to zero indicating no vegetation on the crop and close to one (0.8–0.9) denoting the maximum density of green leaves. These findings make it easier for farmers to monitor and diagnose crop conditions. As a result of these results, farmers can identify the field where pesticides should be sprayed easily. By acting quickly and appropriately, yield loss can be prevented. By using this technology, farmers will no longer have to inspect their crops visually. They are able to maintain a close watch on plants growing in isolated locations, such hilly terrain. They may also keep a watch on tall crops and trees that are challenging for farmers to physically scout.

(iii) **UAV imaging in health, nutrient, and biomass estimate**

Visually assessing nutritional status is currently the most common method, using plant colour guides that do not allow for quantitatively rigorous evaluations (Graeff et al. 2008). Laboratorial leaf analyses are needed for a more accurate evaluation, which requires the use of particular techniques for a proper the analysis of the data (Dezordi et al. 2016). Some nutrients can be predicted indirectly, such as nitrogen, using chlorophyll meters (soil–plant analyses development-SPAD) (Balasubramaniam and Ananthi 2016), but this causes delays and is not always accurate (Jia et al. 2004). By using cameras to capture photos and sensors to gather the dataset, UAVs can help with decision-making on farms. They collect data at higher spatial resolutions showing the comparison of fields by centimetres unlike satellite imagery which shows differences in metres. Drones provide immediate visualization of the crop covering larger areas, that help farmers with fast decision-making. UAVs have a great potential in site-specific crop management. Drone can be used to obtain crop imagery at different bands at various spatio-temporal resolutions to calculate NDVI and study its relationship with other indices that determine the crop status. NDVI, biomass, nitrogen content in the crop, and yield can be obtained and correlated to apply it in precision farming. The variability of field can be identified from these parameters, and accordingly, variable application of fertilizer and irrigation can be taken up.

A Tetracam ADC Lite digital camera was installed on a microdrone MD4-200 by Vega et al. (2015) to capture photographs of the field in several spectral bands and calculate the NDVI, aerial biomass, yield, and nitrogen content of the sunflower crop. Agüera et al. (2012) collected images from a quadcopter flying at 70 m above a sunflower field and found a significant correlation between applied nitrogen and NDVI. Using an unmanned aircraft equipped with an image collecting system, Hunt et al. (2005) calculated the biomass and nitrogen status for crops of maize, alfalfa, and soybeans. In order to evaluate biophysical parameters, Berni et al. (2009) used thermal and narrow band multi-spectral photos from an unmanned aircraft. These images showed a strong correlation between water stress, chlorophyll concentration, and leaf area index. Swain et al. (2010) employed a radio-controlled UAV platform to take high-quality spatial and temporal resolution images of a rice field to assess grain production and total aerial biomass. A UAV equipped with a multi-spectral sensor and thermal camera was utilized by Baluja et al. (2012) to evaluate the water condition of a Tempranillo cultivar of *Vitis vinifera* L. Zarco-Tejada et al. (2012) used thermal and hyperspectral imaging from a UAV to demonstrate the capacity to monitor stress levels in citrus crops. Citrus greening was identified by Garcia-Ruiz et al. (2013) using a UAV equipped with a multi-spectral camera. To assess leaf area, plant cover, and chlorophyll content, Hunt et al. (2014) employed sensor-mounted UAV to capture NIR, red, and green images and calculate NDVI and GNDVI. UAVs are aircraft that fly at low altitudes, allowing for frequent acquisition/analysis in almost real-time as well as ultra-high spatial resolution photos at minimal cost (Agüera et al. 2012; Hardin and Hardin 2010; Xiang and Tian 2011). It is possible to determine precisely which regions of the crops are stressed or suffering and take immediate measures on them by employing mosaics of the NDVI. Multi-spectral/NIR images delivered by UAVs can be used to detect the crop deficiencies or stress earlier, thus identifying the zones where management practices are required. It is therefore possible to target these management zones before they negatively impact the development and yield of crops.

(iv) **Abiotic and biotic stress detection**

UAVs can be used in abiotic and biotic stress detection (Olson and Anderson 2021). Abiotic stress like water stress, nutrient stress, and to determine soil characteristics and properties can be assessed with UAVs. Due to the distinctive spectral signatures of soil, remote sensors mounted on UAVs can be used to assess its parameters. The heterogeneity in the properties can be identified and mapped. Biotic stress caused due to weeds, pests, and disease can be detected using UAVs. There are three main types of crop diseases: fungal, bacterial, and viral. An infrared camera on a drone can provide a clear view within plants, revealing their status (Klemas 2015). Farming can be done in a way that prevents the infection from spreading to neighbouring plants, such as removing the plant when an infection is detected before it spreads. Image-based methods can be useful when human evaluation is not possible, reliable, or readily available, especially with the expanded coverage offered by UAVs (Altas et al. 2018). Several techniques have been used for gathering data on the examined areas, including RGB (Altas et al. 2018), multi-spectral (Dash et al. 2018), hyperspectral (Calderón et al. 2013), and thermal pictures. The latter technique is mostly utilized to identify water stress symptoms brought on by the targeted condition.

(v) **Weed detection and mapping**

In agricultural crops, weeds are undesirable plants that cause a variety of problems. Crop growth and yields suffer as a result of competition for limited resources like water or even available space. In India, weeds have impacted a variety of crops, including: rice (10–100%), wheat (10–60%), maize (30–40%), sugarcane (25–50%), vegetables (30–40%), jute (30–70%), potatoes (20–30%), etc. Weeds are typically controlled with herbicides. In conventional farming, it is typical for farmers to pull up weeds as soon as they appear and to evenly apply herbicides across the whole field, even in weed-free areas. UAVs help in weed scouting. Using herbicides is essential for crop development since weeds are the primary cause of losses in agricultural output. However, using pesticides excessively will lead to a lot of environmental problems. According to Liu et al. (2021), UAV-RS can help with weed location, weed type and density analysis, quantitative herbicide administration at fixed sites, and improved and targeted mechanical soil tillage. The goal of weeds' detection and mapping is to locate and identify weeds in the UAV-RS photos that have been collected. This is often done by comparing the geographical distribution, shape, spectral signatures, or combinations of these characteristics of weeds to those of typical crops. As a result, RGB, multi-spectral, and hyperspectral sensors are often the most important sensors for UAV payload. Herbicide overuse can lead to herbicide-resistant weeds, affecting crop growth and yield. Differentiating weeds with different resistance to glyphosate-based on their spectral signatures (Li et al. 2016). Different weed species can be classified using RGB sensors (Huang et al. 2018).

A drone equipped with hyperspectral sensors monitored weeds in relation to plant canopy chlorophyll content and leaf density (Malenovský et al. 2017). To solve the aforementioned issues, it is required to create an accurate weed cover map for the correct application of herbicide. Drones may be used to collect data and photographs throughout the whole field to determine where the chemicals are required. Preemergence and postemergence weed control can also be done by agro-drone application. The spraying process can be carried out in any field condition in any field circumstance (muddy, weeds, insects, etc.) as well as in sunny or drizzly conditions. Using drones to apply weedicides is efficient and optimizes their use. Easy to use, transport, and maintain. Remotely operate in a very safe manner.

(vi) **Crop pest and disease detection**

In order to effectively apply pesticides and respond quickly to production costs and environmental impacts, early detection of pests and diseases from UAV-RS images is essential for pest and disease management. Crop pest and disease detection using an observed UAV to locate potentially infested or diseased areas on leaves via observed UAV-RS pictures, and detection based on their spectrum differences (Liu et al. 2021). For high spatial or spectral resolution studies to learn more about pests and diseases on leaves, UAVs often fly at low altitudes. According to the information gathered from UAVs, variable rates of fertilizer and chemical application can be done to improve the input use efficiency in agriculture. UAVs can position the applicators and apply particular amounts of pesticides to specified places, giving farmers GPS direction to allow variable-rate technology (Abdullahi et al. 2015).

Fig. 14.5 Drone-based spray at farmer's field (*Photo source* Prakash bhai Rangani, Agri Drone Spray, Kanakpur, Gujarat)

(vii) **UAVs in spraying**

Pesticides sprayed by humans is the potential cause for nervous system disorders. In order to avoid pesticide poisoning, remote-controlled unmanned aerial vehicles (UAVs) are used to spray pesticides and fertilizers. Drone-based spray can spray precisely and quick as compared to conventional prayers. Additionally, drone-based spray saves the farmers money and safe for pesticide induced health hazards (Fig. 14.5).

Spray systems when coupled with UAVs can lead to accurate site-specific crop protection. UAVs have been employed in agriculture for pest management and remote sensing. Chen et al. (2021) reviewed various UAVs used in spraying technologies. The Yamaha model helicopter was created and utilized mostly for agricultural purposes, such as the control of insect pests in rice fields, soybeans, and wheat. (Miller 2005) documented a study into the efficacy of employing a UAV to disperse insecticides to minimize insect-related human sickness. Using an off-the-shelf Yamaha RMAX UAV outfitted with both liquid and granular pesticide dispersion devices, a number of trials were carried out to evaluate the effectiveness of the UAV for aerial pesticide distribution. Results showed that the UAV pesticide-dispersal system performed consistently throughout. Huang et al. (2009) studied various nozzles, controlled the range of spray pump speed, and successfully developed a spray system for a UAV application platform. This has demonstrated an effective pest and vector management in areas that are difficult to reach and access by human labour or large equipment. Xiongkui et al. (2017) showed various types of drones being used for plant protection in East Asia.

High productivity requires production and protection materials in Indian agriculture. It is frequently necessary to use fertilizer and chemicals to kill insects and promote crop growth. Drones may spray chemicals like fertilizer and insecticides based on the spatial variability of the crops and land. Insect–pest attacks and crop conditions can affect the amount of chemicals to be sprayed. A platform for vector control and pest management can be offered by combining UAVs with sprayers. A large crop field requires precise site-specific application. To spray large areas, heavy lift UAVs (Sarghini and Vivo 2017) are required. Quad copters (QCs), also known as unmanned aerial vehicles (UAVs), are lightweight and low cost (Kedari et al. 2016). It is a small quadcopter that may be used for both indoor and outdoor farmings. Using an Android device, quadcopters spray pesticides and fertilizer autonomously. Real-time communication between the quadcopter and the Android device is accomplished via a Bluetooth connection. This system reduces agricultural field-related problems and increases agricultural yield. The PWM controller (Zhu et al. 2010) improves the effectiveness of the spraying system installed to the UAV for applying pesticides. For spraying pesticides at lower altitudes, an aerial automated pesticide sprayer (AAPS) based on GPS coordinates was created (Vardhan et al. 2014).

To overcome this low-cost problem, an Android app called Freyr was developed to allow users to spray pesticides using a flexible drone. In order to reduce pesticide wastage, an electrostatic sprayer is designed with hexarotor UAV technology (Spoorthi et al. 2017). The use of drone pesticide application is suitable for all situations, especially those where labour is hard to find. By spraying from a lower altitude, environmental pollution is reduced. Furthermore, it has the potential to enhance pest management for small and large crops by entail highly accurate site-specific application. This eventually improves the effectiveness of chemicals used, lowering their negative environmental consequences by reducing soil and water contamination. In this way, sustainable agriculture can be achieved. Chemicals are sprayed more quickly by drones than by other methods. As a result, input costs can be reduced as well as the amount of chemicals used (Table 14.4).

14.5 UAV Applications in Water Management

Managing water efficiently and sustainably is a complex challenge throughout the world (Yadav et al. 2022). The existence of life on earth depends on water. It preserves organic life, provides for livestock, promotes food security, and meets residential and industrial needs. Today, there are political, legal, environmental, social, economic, and even religious considerations in addition to the technical and scientific components of water. In addition, because of the widening difference between supply and demand, water separates people into urban and rural, wealthy and poor groups. If water management and optimal use of this priceless, necessary, and rare resource are not put into place, the environment will experience a water crisis in the future. The problem is anticipated to get worse due to climate change. More than one billion Asians are projected to experience negative impacts as a result of Climate Change by 2050, according to the Intergovernmental Panel on Climate Change. As many as

Table 14.4 Application of drones in agriculture

Application	Sensor/model used	References
Water stress monitoring	Multi-spectral/hyperspectral sensors, NDVI, GNDVI, PRI model, thermal IR sensor, canopy temperature ($T_c - T_a$), RGB sensor, and WDI	Zúñiga et al. (2017), Zarco-Tejada et al. (2012), Dezordi et al. (2016), Balasubramaniam and Ananthi (2016), Jia et al. (2004)
Nutrient disorders	RGB, multi-spectral, and hyperspectral sensors	Maimaitijiang et al. (2017)
Diseases monitoring	RGB sensor, multi-spectral images sensor, hyperspectral and thermal sensor	Dash et al. (2018), Calderón et al. (2013)
Weeding	Hyperspectral sensor and RGB sensors	Li et al. (2016), Malenovský et al. (2017) Huang et al. (2018)
Evapotranspiration	Multi-spectral and thermal sensor	Hoffmann et al. (2015), Xia et al. (2016)
Spraying	GPS sensor, accelerometer, and gyroscope sensors	Vardhan et al. (2014), Plant et al. (2000)

three out of four people in the world could be affected by water scarcity by 2050, according to the United Nations.

Agriculture and irrigation management are increasingly being solved by drones. The utilization of unmanned aerial vehicles (UAVs), commonly known as drones, and linked analytics might significantly help and resolve some of the most serious concerns confronting agriculture in terms of access to actionable real-time quality data. In the next five years, Goldman Sachs predicts that agricultural will deploy drones in second-largest numbers. A crucial role for advanced Information and Communication Technology (ICT) is playing in improving water usage efficiency. Drones are now being widely used in water management to address vital issues in agriculture and irrigation. As a result of climate change, several agricultural management issues emerge, and advanced technologies such as artificial intelligence (Bakala et al. 2020) and image processing are urgently needed. A pilot experiment using drones is necessary to curb the issues faced by this sector in order to achieve the Sustainable Development Goals (SDGs) by 2030.

According to Shi et al. (2019) these are the steps for variable rate irrigation (VRI) application using UAV technology. They formed a Decision Support System, called DSS-VRI. The steps are follows:

1. The first part of system requires the basic data for irrigation scheduling, i.e. unmanned aerial vehicle (UAV) outputs as multi-spectral high-resolution image and ground meteorological parameters for as input for the model.
2. Second step is the data processing of selected bands like red, near-infrared, red edge, and green bands and estimation of crop ET (ETc), crop water stress index (CWSI) and precipitation.

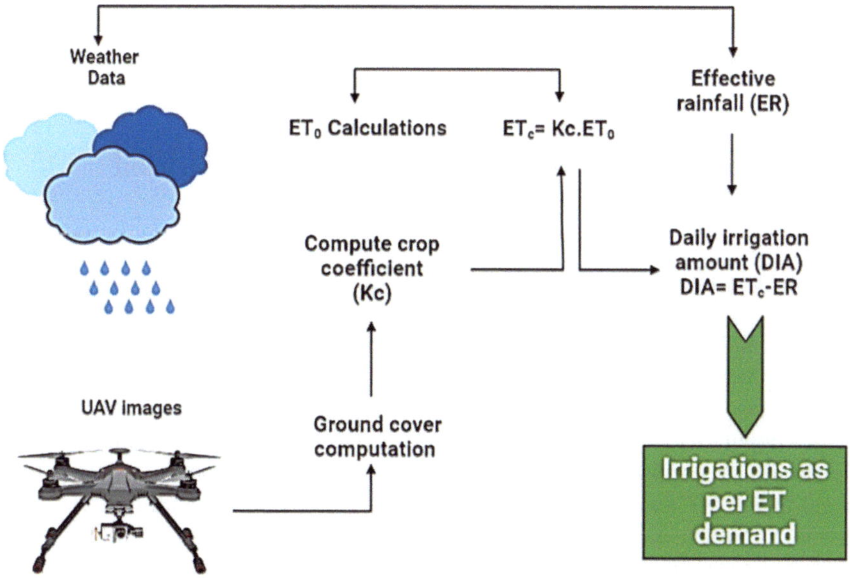

Fig. 14.6 UAV-based climatological approach for irrigation scheduling (Lin 2019)

3. The third step involves the data input to the fuzzy system. It performs the standardizing inputs and blurring inputs to fuzzy interface and de-fuzzy.
4. The last step involves the duty-cycle control map of different irrigation levels for a partial management zone (Fig. 14.6).

Nowadays use of integrated drone tools along with weather sensors is preferred for the accurate simultaneously real-time crop water needs (Alexandris et al. 2021). As a substitute for or addition to satellite observations, UAVs are now used to monitor vegetation. Due to the ability of UAVs to collect images at a resolution of 1 m above the ground, the resolution of the images is far better than that of high-altitude other open satellites. Nowadays, in USA, many companies are offering the commercial NDVI and thermal spectrum high-resolution images on weekly basis. The grower can access images and NDVI from company website within few hours of survey. This drone-based survey gives the details of spatial variability of field and water stress status of crop (Cahn and Johnson 2017). Drones equipped with infrared sensors can measure the water content of natural plants and crops using thermal images (Krishna 2018). Drones are frequently used to detect soil moisture and depletion pattern of soil moisture over the time, crop water status, and experimental observation of crops to response to different irrigation schedules. Estimation of CWSI is the main aspect for irrigation scheduling through UAV's tools (Krishna 2018). CWSI is usually computed based on the crop canopy temperature and ambient temperatures (Bellvert et al. 2014).

$$\text{CWSI} = \frac{(T_\text{c} - T_\text{a}) - (T_\text{c} - T_\text{a})\text{LL}}{(T_\text{c} - T_\text{a})\text{UL} - (T_\text{c} - T_\text{a})},$$

where T_c is canopy temperature; T_a is air temperature; LL is lower limit of ($T_c - T_a$) values, and UL is upper limit of ($T_c - T_a$) values.

14.6 Future Prospectus of UAV Platforms for Precision Agriculture

As drones can be used in a variety of industries, they are advanced, intelligent, and versatile. Industrial drones are becoming increasingly useful to farmers as technology advances and manufacturers invest in regulatory changes. Modern agriculture relies heavily on industrial drones to keep up with technological advancements. While drones have been used in the private sector for decades, they have only recently gained popularity in agriculture. Agriculture is one of the most promising industries for drones. According to projections, the world's population will reach 9.7 billion people by 2050, and agricultural consumption will rise along with it (Duro et al. 2020). Simultaneously labour shortage and wage for agriculture operations have also expected to increase in the future. So, UAVs may helpful in the following ways:

- Increased productivity and reduced costs.
- Improve the accuracy and efficiency of operations.
- Drones ensure effective crop surveillance.
- Drones simplify the use of other cutting-edge technologies in agriculture.
- Drones—a universal data collection tool.
- Additionally, this technology is eco-friendly to environment.

14.7 Conclusion

In this chapter, the architecture, their potential ability, and application of UAVs in smart farming, climate-smart agriculture, and water management have been explored and presented. We have also discussed different types of drones that may useful in agriculture such as fixed rotor, single-rotor, multi-rotor, and hybrid VTOL rotor drones which are used for precision farming and water management. The technological capabilities of UAVs used for field-based agricultural remote sensing are summarized in this article. The use of UAVs sowed the beneficial remarks for spraying the pesticides, monitoring crop health, weed mapping, and precision input application in agriculture. This chapter helps growers and interested researchers in a preliminary understanding of how to choose the most suitable UAVs for smart farming operations. The difficulties and drawbacks of using UAVs in climate-smart agriculture and smart farming were explored, and several solutions were put forth to maximize their utilization. The significance of application of low-altitude UAVs in irrigation scheduling and monitoring the water stress is the precious gift of the decade for smart farming. Finally, the main research issues and future directions of UAVs have now been emphasized and explored.

References

Abdullahi HS, Mahieddine F, Sheriff RE (2015) Technology impact on agricultural productivity: a review of precision agriculture using unmanned aerial vehicles. In: Wireless and Satellite Systems: 7th International Conference, WiSATS 2015, Bradford, UK, July 6–7, 2015. Revised Selected Papers 7. Springer International Publishing, pp 388–400

Agüera F, Carvajal F, Pérez M (2012) Measuring sunflower nitrogen status from an unmanned aerial vehicle-based system and an on the ground device. Int Arch Photogramm Remote Sens Spat Inf Sci 38:33–37

Alexandris S, Psomiadis E, Proutsos N, Philippopoulos P, Charalampopoulos I, Kakaletris G, Paraskevopoulos A (2021) Integrating drone technology into an innovative agrometeorological methodology for the precise and real-time estimation of crop water requirements. Hydrology 8(3):131

Altas Z, Ozguven MM, Yanar Y (2018) Determination of sugar beet leaf spot disease level (*Cercospora beticola* Sacc.) with image processing technique by using drone. Curr Investig Agric Curr Res 5:621–631

Bakala HS, Singh G, Srivastava P (2020) Smart breeding for climate resilient agriculture. In: Plant breeding-current and future views. IntechOpen

Balasubramaniam P, Ananthi VP (2016) Segmentation of nutrient deficiency in incomplete crop images using intuitionistic fuzzy C-means clustering algorithm. Nonlinear Dyn 83:849–866

Baluja J, Diago MP, Balda P, Zorer R, Meggio F, Morales F, Tardaguila J (2012) Assessment of vineyard water status variability by thermal and multispectral imagery using an unmanned aerial vehicle (UAV). Irrig Sci 30(6):511–522

Bellvert J, Zarco-Tejada PJ, Girona J, Fereres EJPA (2014) Mapping crop water stress index in a 'Pinot-noir'vineyard: comparing ground measurements with thermal remote sensing imagery from an unmanned aerial vehicle. Precision Agric 15:361–376

Berni JA, Zarco-Tejada PJ, Suárez L, Fereres E (2009) Thermal and narrowband multispectral remote sensing for vegetation monitoring from an unmanned aerial vehicle. IEEE Trans Geosci Remote Sens 47(3):722–738

Cahn MD, Johnson LF (2017) New approaches to irrigation scheduling of vegetables. Horticulturae 3(2):28

Calderón R, Navas-Cortés J, Lucena C, Zarco-Tejada P (2013) High-resolution hyperspectral and thermal imagery acquired from UAV platforms for early detection of verticillium wilt using fluorescence, temperature and narrow-band indices. In: Proceedings of the workshop on UAV-based remote sensing methods for monitoring vegetation, Cologne, Germany, 2013, pp 7–14

Caldwell L (2015) Unmanned aerial vehicles—how can they help agronomists? Lachlan Fertilizers

Chen PC, Chiang YC, Weng PY (2020) Imaging using unmanned aerial vehicles for agriculture land use classification. Agriculture 10(9):416

Chen H, Lan Y, Fritz BK, Hoffmann WC, Liu S (2021) Review of agricultural spraying technologies for plant protection using unmanned aerial vehicle (UAV). Int J Agric Biol Eng 14(1):38–49

Colomina I, Molina P (2014) Unmanned aerial systems for photogrammetry and remote sensing: a review. ISPRS J Photogramm Remote Sens 92:79–97

Cristinel F, Daniela BL, Vitali Z, Silvius S (2022) The future of agriculture drone technology trends and prospects. In: Economic and Social Development: Book of Proceedings, pp 68–77

Dash JP, Pearse GD, Watt MS (2018) UAV multispectral imagery can complement satellite data for monitoring forest health. Remote Sens 10(8):1216 del Cerro J, Cruz Ulloa C, Barrientos A, de León Rivas J (2021) Unmanned aerial vehicles in agriculture: a survey. Agronomy 11(2):203

Dezordi LR, Aquino LAD, Aquino RFBDA, Clemente JM, Assunção NS (2016) Diagnostic methods to assess the nutritional status of the carrot crop. Rev Bras Ciência Solo. https://doi.org/10.1590/18069657rbcs20140813

Djojodihardjo H, Ahmad D (2015) Opportunities and challenges for climate-smart agriculture. In: Proceedings of the 3rd International Conference on Natural Resource Management (NRM'15), New Delhi, India, pp 10–13

Duro JA, Lauk C, Kastner T, Erb KH, Haberl H (2020) Global inequalities in food consumption, cropland demand and land-use efficiency: a decomposition analysis. Glob Environ Chang 64:102124

Garcia-Ruiz F, Sankaran S, Maja JM, Lee WS, Rasmussen J, Ehsani R (2013) Comparison of two aerial imaging platforms for identification of Huanglongbing-infected citrus trees. Comput Electron Agric 91:106–115

Graeff S, Pfenning J, Claupein W, Liebig HP (2008) Evaluation of image analysis to determine the N-fertilizer demand of broccoli plants (*Brassica oleracea* convar. botrytis var. italica). Adv Optic Technol

Grassi, M. (2014). Five actual uses for drones in precision agriculture today. Drone Life, pp 1–3. http://dronelife.com/2014/12/30/5-actual-uses-drones-precision-agriculturetoday/. Accessed 18 May 2015.

Han X, Thomasson JA, Bagnall GC, Pugh NA, Horne DW, Rooney WL, Cope DA (2018) Measurement and calibration of plant-height from fixed-wing UAV images. Sensors 18(12):4092

Hardin PJ, Hardin TJ (2010) Small-scale remotely piloted vehicles in environmental research. Geogr Compass 4(9):1297–1311

Hashem (2020) UAV imagery offers multiple irrigation applications. https://irrigationtoday.org/features/uav-imagery-offers-multiple-irrigation-applications/

Hoffmann H, Nieto H, Jensen R, Guzinski R, Zarco-Tejada PJ, Friborg T (2015) Estimating evapotranspiration with thermal UAV data and two source energy balance models. Hydrol Earth Syst Sci Discuss 12(8):7469–7502

Huang Y, Hoffmann WC, Lan Y, Wu W, Fritz BK (2009) Development of a spray system for an unmanned aerial vehicle platform. Appl Eng Agric 25(6):803–809

Huang Y, Reddy KN, Fletcher RS, Pennington D (2018) UAV low-altitude remote sensing for precision weed management. Weed Technol 32(1):2–6

Hunt ER, Cavigelli M, Daughtry CS, Mcmurtrey JE, Walthall CL (2005) Evaluation of digital photography from model aircraft for remote sensing of crop biomass and nitrogen status. Precision Agric 6:359–378

Hunt Jr E, Horneck D, Hamm P, Gadler D, Bruce A, Turner R, Brungardt J (2014) Detection of nitrogen deficiency in potatoes using small unmanned aircraft systems. In: 12th International Conference on Precision Agriculture. California, USA Sacramento

Jariwala H, Santos RM, Lauzon JD, Dutta A, Wai Chiang Y (2022) Controlled release fertilizers (CRFs) for climate-smart agriculture practices: a comprehensive review on release mechanism, materials, methods of preparation, and effect on environmental parameters. Environ Sci Pollut Res 29(36):53967–53995

Jia L, Chen X, Zhang F, Buerkert A, Römheld V (2004) Use of digital camera to assess nitrogen status of winter wheat in the northern China plain. J Plant Nutr 27(3):441–450

Kedari S, Lohagaonkar P, Nimbokar M, Palve G, Yevale P (2016) Quadcopter-a smarter way of pesticide spraying. Imp J Interdiscip Res 2(6):1257–1260

Klemas VV (2015) Coastal and environmental remote sensing from unmanned aerial vehicles: an overview. J Coastal Res 31(5):1260–1267

Krishna KR (2018) Agricultural drones: a peaceful pursuit. Apple Academic Press, Waretown

Lee KJ, Lee BW (2011) Estimating canopy cover from color digital camera image of rice field. J Crop Sci Biotechnol 14(2):151–155

Li L, Fan Y, Huang X, Tian L (2016) Real-time UAV weed scout for selective weed control by adaptive robust control and machine learning algorithm. In: 2016 ASABE Annual International Meeting. American Society of Agricultural and Biological Engineers, p 1

Lin Z (2019) Unmanned aerial systems and crop modeling for irrigation scheduling in the Southern High Plains. Doctoral dissertation

Liu J, Xiang J, Jin Y, Liu R, Yan J, Wang L (2021) Boost precision agriculture with unmanned aerial vehicle remote sensing and edge intelligence: a survey. Remote Sensing 13(21):4387

Maes WH, Steppe K (2019) Perspectives for remote sensing with unmanned aerial vehicles in precision agriculture. Trends Plant Sci 24(2):152–164

Maimaitijiang M, Ghulam A, Sidike P, Hartling S, Maimaitiyiming M, Peterson K et al (2017) Unmanned Aerial System (UAS)-based phenotyping of soybean using multi-sensor data fusion and extreme learning machine. ISPRS J Photogramm Remote Sens 134:43–58

Malenovský Z, Lucieer A, King DH, Turnbull JD, Robinson SA (2017) Unmanned aircraft system advances health mapping of fragile polar vegetation. Methods Ecol Evol 8(12):1842–1857

Mayeed MS, Darveau G (2016) Designing an unmanned aerial vehicle for specific aerial applications of insecticides and herbicides. ASME Int Mech Eng Cong Expos 50688:V014T07A005

Miller JW (2005) Report on the development and operation of an UAV for an experiment on unmanned application of pesticides. AFRL, USAF

Olson D, Anderson J (2021) Review on unmanned aerial vehicles, remote sensors, imagery processing, and their applications in agriculture. Agron J 113(2):971–992

Pandey K, Singh KG, Singh A (2023) Multi-sensors based smart nutrient reuse management system for closed soilless culture under protected cultivation. Comput Electron Agric 204:107495

Plant R, Pettygrove G, Reinert W (2000) Precision agriculture can increase profits and limit environmental impacts. Calif Agric 54(4):66–71

Precision Hawk (2014) Lancaster platform, pp 1–4. http://www.precisionhawk.com/index.html#industries. Accessed 5 Aug 2014

Reinecke M, Prinsloo T (2017) The influence of drone monitoring on crop health and harvest size. In: 2017 1st International conference on next generation computing applications (NextComp). IEEE, pp 5–10

Sarghini F, De Vivo A (2017) Analysis of preliminary design requirements of a heavy lift multirotor drone for agricultural use. Chem Eng Trans 58:625–630

Shi X, Han W, Zhao T, Tang J (2019) Decision support system for variable rate irrigation based on UAV multispectral remote sensing. Sensors 19(13):2880

Simelli I, Tsangaris A (2015) The use of unmanned aerial systems (UAS) in agriculture. In: HAICTA, pp 730–736

Spoorthi S, Shadaksharappa B, Suraj S, Manasa VK (2017) Freyr drone: pesticide/fertilizers spraying drone—an agricultural approach. In: 2017 2nd International Conference on Computing and Communications Technologies (ICCCT). IEEE, pp 252–255

Swain KC, Thomson SJ, Jayasuriya HP (2010) Adoption of an unmanned helicopter for low-altitude remote sensing to estimate yield and total biomass of a rice crop. Trans ASABE 53(1):21–27

Trimble (2015) Trimble UX5 aerial imaging solution for agriculture, pp 1–3. http://www.trimble.com/Agriculture/UX5.aspx. Accessed 20 May 2015

Vardhan PH, Dheepak S, Aditya PT, Arul S (2014) Development of automated aerial pesticide sprayer. Int J Eng Sci Res Technol 3(4):458–462

Vega FA, Ramirez FC, Saiz MP, Rosua FO (2015) Multi-temporal imaging using an unmanned aerial vehicle for monitoring a sunflower crop. Biosys Eng 132:19–27

Velusamy P, Rajendran S, Mahendran RK, Naseer S, Shafiq M, Choi JG (2021) Unmanned aerial vehicles (UAV) in precision agriculture: applications and challenges. Energies 15(1):217

Xia T, Kustas WP, Anderson MC, Alfieri JG, Gao F, McKee L et al (2016) Mapping evapotranspiration with high-resolution aircraft imagery over vineyards using one-and two-source modelling schemes. Hydrol Earth Syst Sci 20(4):1523–1545

Xiang H, Tian L (2011) Method for automatic georeferencing aerial remote sensing (RS) images from an unmanned aerial vehicle (UAV) platform. Biosys Eng 108(2):104–113

Xiongkui H, Bonds J, Herbst A, Langenakens J (2017) Recent development of unmanned aerial vehicle for plant protection in East Asia. Int J Agric Biol Eng 10(3):18–30

Yadav M, Vashisht BB, Jalota SK, Kumar A, Kumar D (2022) Sustainable water management practices for intensified agriculture. In: Soil-water, agriculture, and climate change: exploring linkages. Springer International Publishing, Cham, pp 131–161. https://doi.org/10.1007/978-3-031-12059-6_8

Zarco-Tejada PJ, Ustin SL, Whiting ML (2005) Temporal and spatial relationships between within-field yield variability in cotton and high-spatial hyperspectral remote sensing imagery. Agron J 97(3):641–653

Zarco-Tejada PJ, González-Dugo V, Berni JA (2012) Fluorescence, temperature and narrow-band indices acquired from a UAV platform for water stress detection using a micro-hyperspectral imager and a thermal camera. Remote Sens Environ 117:322–337

Zhang H, Wang L, Tian T, Yin J (2021) A review of unmanned aerial vehicle low-altitude remote sensing (UAV-LARS) use in agricultural monitoring in China. Remote Sensing 13(6):1221

Zhao B, Zhang J, Yang C, Zhou G, Ding Y, Shi Y, Liao Q (2018) Rapeseed seedling stand counting and seeding performance evaluation at two early growth stages based on unmanned aerial vehicle imagery. Front Plant Sci 9:1362

Zhao X, Yuan Y, Song M, Ding Y, Lin F, Liang D, Zhang D (2019) Use of unmanned aerial vehicle imagery and deep learning unet to extract rice lodging. Sensors 19(18):3859

Zhu H, Lan Y, Wu W, Hoffmann WC, Huang Y, Xue X et al (2010) Development of a PWM precision spraying controller for unmanned aerial vehicles. J Bionic Eng 7(3):276–283

Zúñiga EC, Khot LR, Sankaran S, Jacoby PW (2017) High resolution multispectral and thermal remote sensing-based water stress assessment in subsurface irrigated grapevines. Remote Sensing 9(9):961

Chapter 15
Smart Irrigation Management Through Unmanned Aerial Vehicles (UAVs)

L. Aiswarya, Siddharam, G. M. Rajesh, Venkatesh Gaddikeri, Malkhan Singh Jatav, Jitendra Rajput, and K. R. Asha

Abstract With the ever-growing global population, meeting the escalating demand for food has become a pressing challenge. To address this issue, the adoption of smart agriculture techniques has gained significant momentum. Leveraging the latest advancements in Information and Communications Technology (ICT), smart agriculture integrates climate change mitigation and other environmental considerations to ensure sustainable and efficient food production. Among the array of innovative tools available, Unmanned Aerial Vehicles (UAVs) emerge as a promising solution due to their sensor-equipped capabilities and remote controllability. By harnessing the potential of UAVs, the realm of smart farming can be revolutionized. UAVs offer a multitude of applications, including crop quality assessment, surveillance of pest, weed, and animal threats, irrigation planning, and identification of bio-processing errors. These applications not only empower agricultural experts to make informed decisions but also lead to cost savings and knowledge advancement. This chapter aims to present numerous examples of smart agriculture and elucidate the benefits and applications of deploying UAVs in the agricultural domain. Specifically, the utilization of UAVs for pest control, intelligent irrigation, agricultural monitoring, and livestock surveillance will be explored. The integration of UAVs in agriculture holds immense potential in enhancing sustainability and achieving the Sustainable Development Goals (SDGs).

Keywords Smart agriculture · Unmanned Aerial Vehicles (UAVs) · Crop monitoring · Pest management · Smart irrigation · Livestock monitoring

---

L. Aiswarya (✉) · Siddharam · G. M. Rajesh
Kelappaji College of Agricultural Engineering and Technology, Kerala Agricultural University, Tavanur, India
e-mail: aiswaryasona1234@gmail.com

V. Gaddikeri (✉) · M. S. Jatav · J. Rajput · K. R. Asha
ICAR-Indian Agricultural Research Institute, Pusa, New Delhi 110012, India
e-mail: venkatg303@gmail.com

K. Pandey et al. (eds.), *Artificial Intelligence and Smart Agriculture*,
Advances in Geographical and Environmental Sciences,
https://doi.org/10.1007/978-981-97-0341-8_15

15.1 Introduction

Feeding the growing global population is a paramount concern, as agricultural production takes center stage. The responsibility of ensuring a secure food supply for the entire world places immense pressure on decision-makers (Norasma et al. 2019). Projections indicate that by 2050, the population will reach a staggering 9.8 billion, signifying a 25% increase from the current figures (Kc and Lutz 2017). In India, the Food and Agriculture Organization (FAO) estimates that 82% of farmers are categorized as small and marginal, with 70% of rural households relying predominantly on agriculture for their livelihood. As we look ahead, it is anticipated that agriculture and food consumption will surge by 75% by 2050 (Singh et al. 2022; Refaai et al. 2022).

In the face of a rapidly expanding global population, ensuring an adequate food supply becomes increasingly challenging due to the potential impacts of climate change and the depletion of natural resources. These factors can significantly affect food production and further strain the availability of natural resources. To address these challenges, it is crucial to manage the existing resources in a scientifically informed manner, aiming to minimize losses and maximize productivity. This gave rise to the concept of precision agriculture, which emphasizes the minimization of inputs while achieving optimal production levels. By employing advanced technologies and data-driven approaches, precision agriculture offers the potential to mitigate the adverse effects of climate change and resource depletion, enabling sustainable and efficient food production for the growing population.

Precision agriculture (PA) has emerged as a vital farming approach in the technologically advanced world, often referred to as the "farmer of the century." Using advanced technologies, this farming administration approach monitors and optimizes agricultural manufacturing processes using computerized farming administration. This approach regulates agricultural output spatially and temporally, leveraging current technology to enhance farm productivity. Various aspects such as soil characteristics, field conditions, crop growth, yield variations, and management factors are observed through diverse techniques. UAVs, sensor technology, positioning systems, satellite tracking, and the Internet of Things (IoT) are just a few of the techniques which are employed to achieve these goals (Pandey et al. 2023). Synergy between IoT, sensor technology, and UAVs has facilitated sustainable smart farming practices. With the integration of IoT, real-time identification of infestations such as weeds, pests, and diseases, monitoring of weather conditions to mitigate extremes, and measurement of soil conditions to prevent detrimental weed growth have become possible. The availability of UAVs and IoT technology empowers users to optimize the utilization of valuable resources like water, nutrients, and agrochemicals, thereby promoting efficient and sustainable agricultural practices.

The utilization of UAVs opens up new avenues for intelligent farming, particularly in terms of agricultural surveillance and imaging analysis (Kim et al. 2016). UAVs enable comprehensive field monitoring by conducting patrols and performing image analysis and agricultural data processing (Mogili and Deepak 2018). These aerial platforms can transmit collected data, providing valuable insights to ground tracking stations. UAVs have proven effective in various applications, including prospecting and targeted application of pesticides and fertilizers, detection of crop stress and pest infestations, fertility assessment, field mapping, and precise seed planting. Their integration into agricultural practices enhances efficiency and facilitates informed decision-making for farmers and agricultural experts.

15.2 Unmanned Aerial Vehicles (UAVs)

In recent years, UAVs or drones have emerged as a cost-effective alternative to sensing technologies and data analytics methods. The advantages of utilizing electromagnetic energy-based remote sensing techniques for assessing objects from a distance include thoroughness, non-intrusiveness, timeliness, and flexibility. However, due to the complexity of remote sensing, agricultural production, and soil characteristics, remote sensing data obtained from conventional methods often deviate from actual data observed in the field (Eisenbeiss and Sauerbier 2011). Experts suggest that UAVs can serve as a viable alternative for collecting precise field data. UAVs have been recognized as a potential technology capable of providing high-resolution images (around 1 m) and frequent updates, allowing for tactical awareness and generating relevant information on crop and field conditions (Elarab et al. 2015). The use of mini-UAVs, which are less expensive than traditional manned aircraft, is particularly suitable for low-altitude remote sensing (LARS). This cost advantage may contribute to the increasing market demand for UAVs, surpassing production capabilities (Zhang et al. 2016).

An UAV is autonomous aircraft that serves various functions. Initially developed for military applications such as air defense training and data collection in hazardous environments, UAV technology has expanded its scope and found relevance in diverse fields. UAVs operate either autonomously based on pre-programmed flight plans or under remote control from a ground station, eliminating the need for an on-board pilot. Agriculture presents numerous opportunities for UAV utilization (Cano et al. 2017). One such application involves leveraging geographical data gathered by UAVs to generate precise and evidence-based forecasts of agricultural harvests. Farmers can obtain an aerial view of their fields through UAV imaging, providing valuable insights into irrigation issues, soil conditions, fungal and insect infestations, and other challenges. UAVs are also instrumental in monitoring soil conditions, weed and insect presence, variable growing seasons, weather patterns, water availability, and more. In livestock management, UAVs aid in animal counting, observing feeding patterns, and assessing overall health. By leveraging the data collected, farmers can swiftly identify issues, make informed management decisions, boost agricultural productivity, and ultimately increase their revenue (Mukherjee et al. 2020; Bacco et al. 2018).

One of the advantages of UAVs is their ability to capture high-resolution photographs of crops using various camera lenses, thereby providing farmers with access to images across different spectral ranges. These images can be processed and analyzed, enhancing the understanding of crop health and identifying areas that require additional attention. The mini-UAV, in particular, presents a valuable option for farmers looking to integrate farming practices with remote sensing. Its ease of operation and low maintenance requirements make it an accessible choice, requiring minimal training (Allred et al. 2020). By harnessing the capabilities of UAVs, farmers can enhance their agricultural practices and make informed decisions for improved crop management.

15.2.1 Components of UAVs

Typically, an UAV for precision agriculture consists of the following essential components (Fig. 15.1):

I. *One or more UAVs*: Flying machines that can operate remotely or autonomously, without the need for a pilot.
II. *A Ground Control Station (GCS)*: It is a computer that either interacts with the UAV control system or directly manages and supervises the UAV. The GCS keeps track of data pertaining to the UAV's flight. The user can receive information pertinent to the aircraft's flight as well as information captured by the sensors that assist with that flying. The GCS also includes the software necessary for processing the data obtained by the UAV and extracting the data required by the operator for crop monitoring.
III. *UAV Control System (UAV-CS):* It is employed to pilot the UAV. It could be a built-in computer or a two-way data link, like a remote control (usually with inbuilt GPS). The flight management system and/or the autonomous

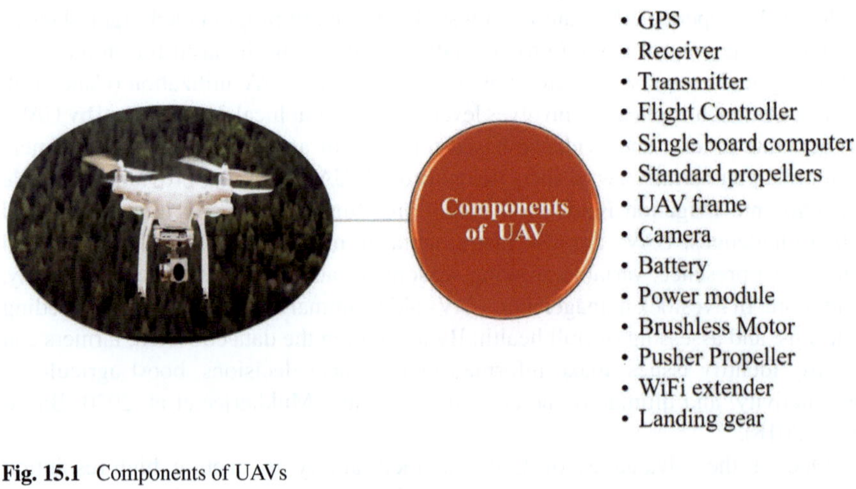

Fig. 15.1 Components of UAVs

system, which manage the UAV's operation, are both included in the UAV-CS. For the proper operation of the UAV, this system gathers and analyzes data from the autopilot or flight management system. It typically has sensors to track many aspects of flight, including air force, distance from the ground, and other parameters. The controlling system has the capacity to process sensor data in order to address any issues that may develop, as well as to wirelessly and instantly exchange the required data with the GCC.

IV. *Sensors for data acquisition*: These are cameras designed to gather the necessary data. The potential methods of information gathering and the technologies used are presented in depth in the next section. The appropriate parts are substituted for the sensors when UAVs are employed for activities other than information gathering, such as spraying.

Based on their design characteristics, the many UAVs used for PA can be categorized into various categories (Fig. 15.2).

1. *Fixed-wing*: This unmanned aircraft has wings and can only launch themselves into the air using a catapult or a runway. This class of UAVs can fly at great speeds and has a long endurance. In particular, fixed-wing UAVs can transport greater payload and cover larger areas per flight. They do cost more than the other kinds, though.
2. *Rotary-wing*: Rotor-wing aircraft UAVS, also known as rotorcrafts or vertical take-off and landing (VTOL) aircraft, provide the benefits of stable flying at one

Fig. 15.2 UAVs' platform types

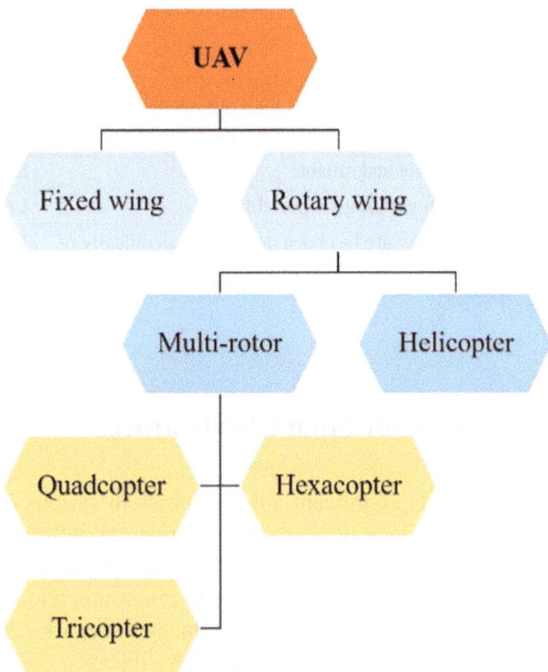

location while maintaining the attribute of agility. These capabilities come in handy for a wide range of missions. They are unable to fly very quickly or hover in the air for an extended period of time. This type UAV may be:

(a) An unmanned helicopter: Includes main and tail rotors such as conventional helicopters (overall, usage is 4% of the works).

3. *Multi-rotor*: The rotary-wing UAVs with four or more rotors fall under this category (quadcopter, hexacopter, octocopter, etc.). In general, these planes fly more steadily than unmanned helicopters (overall usage is 72% of the works).
4. *Blimps:* This kind of UAV is larger in size compared to the others, is lighter than the air, has high endurance, and flies at low speeds. Despite being comparatively safe in the event of a collision, their production characteristics allow them to stay in the air even in the event of a complete loss of power. They are not frequently utilized in precision agriculture applications.
5. *Flapping wing:* These UAVs are very small, flexible, shaped little wings inspired by the way birds and insects fly and high consumption of energy, not used in PA.
6. *Parafoil-wing*: These kinds of aircraft often have a propeller or several propellers at the back to regulate their direction of flight and to harness the power of the air to fly without using a lot of energy. They can also move bigger loads. They are not typically employed for PA applications.

15.2.2 Advantages and Disadvantages of UAV Applications

Advantages	Disadvantages
Helps in examining inaccessible areas	Limited payload, flight route control
More affordable and reliable	Limited battery life
Saves time, night usage is possible	Legal restrictions
Processing software has been developed specifically of both snapshot imaging and line scanning systems	Expensive

15.3 UAV for Smart Agriculture

IoT has made significant advancements in various domains, including agriculture. However, the lack of adequate communication infrastructure, such as base stations or Wi-Fi, poses a hindrance to the widespread adoption of IoT in the agricultural industry. In the absence of reliable communication facilities, wireless sensor data cannot be effectively transmitted. In such scenarios, UAVs offer an alternative solution. UAV systems can establish long-distance connections with wireless sensors, facilitating data collection for further analysis and research. Among the diverse applications of smart agriculture, UAVs, categorized as robotic systems, have emerged as crucial tools. Farmers frequently employ UAVs for field observation, inspection, and agricultural development. In challenging terrains where human mobility is limited and crops exhibit varying heights, specific UAVs are utilized for tasks such as water spraying and pesticide application. The integration of UAV technology in agriculture enhances operational efficiency and supports precision-based practices.

In the realm of precision agriculture, UAVs equipped with precise sensors and high-resolution cameras have emerged as valuable tools for comprehensive agricultural monitoring. With the ability to cover vast expanses of agricultural land, UAVs play a crucial role in various agricultural applications, particularly in forestry and crop monitoring where extensive areas need to be surveyed (Uddin et al. 2018). The collection and transmission of precise data, coupled with efficient and thorough monitoring capabilities, are vital for optimizing agricultural output. Currently, two primary methods are employed to obtain aerial images of fields or agricultural zones: satellites and manned aircraft. While both options offer the advantage of monitoring large areas, they often fall short in terms of image quality. These images lack the necessary spectral and spatial resolution required for detailed analysis. Furthermore, capturing images at regular intervals is challenging due to limited temporal resolution. Another significant challenge is the potential obstruction of imagery by cloud cover when inspections are conducted above cloud level, particularly in unfavorable weather conditions. Addressing these limitations and ensuring reliable and high-quality image acquisition are essential for leveraging UAVs effectively in precision agriculture.

UAVs, serving as an "eye in the sky," provide a solution to address the previously mentioned challenges with a micro-scale perspective. The imagery captured by UAVs holds significant value, tailored to the specific needs of the application, and often surpasses the resolution capabilities of satellite imaging. Particularly in terms of evaluating crop conditions, UAVs enable the rapid acquisition of high-quality Normalized Difference Vegetation Index (NDVI) values, facilitating tasks such as weed identification and leaf assessment. This timely information empowers farmers to make informed decisions and take immediate action. It is worth noting that UAV operations are subjected to weather conditions, except during rainfall, as they are frequently deployed multiple times throughout the day. Nonetheless, UAVs remain a reliable and effective tool in precision agriculture, providing invaluable insights for optimizing farming practices (Fig. 15.3).

15.3.1 Smart Crop Monitoring Using UAV

Traditionally, farmers have relied on visual observation and manual soil sampling to monitor their crops for potential hazards such as disease, pests, and growth issues. However, with the advent of advanced technology, a more sophisticated approach has emerged. Utilizing UAVs equipped with cameras and employing sophisticated image

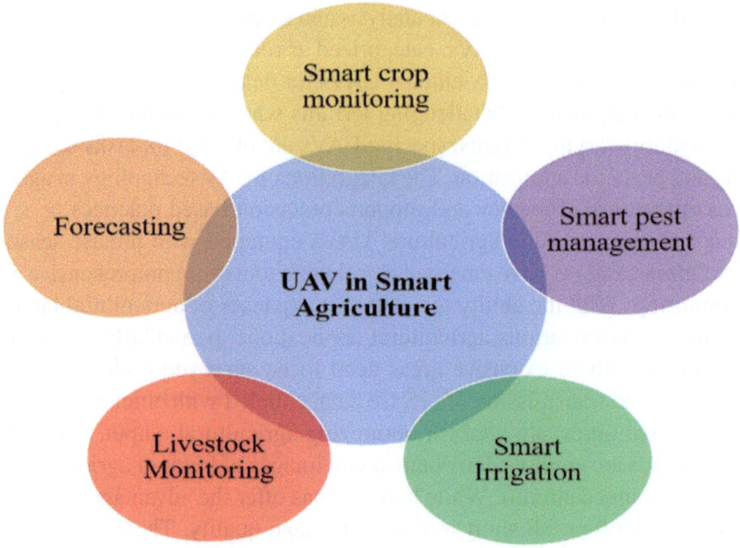

Fig. 15.3 Application of UAVs in smart agriculture

data analysis tools, farmers can now locate plants with infections or flaws more effectively. The UAV's photos can be used to generate maps of vegetation indices, which provide valuable insights into plant ailments, fertilizer requirements, and water stress. These vegetation indices play a crucial role in identifying weeds, diseased plants, and thriving plants, as they are derived from the crop's visual spectrum, directly linked to its health status (Cuaran and Leon 2021). Moreover, there exist strong correlations between crop yield and vegetation indices calculated at specific stages of crop growth. These relationships are vital for evaluating agricultural productivity and yield. By leveraging these connections, farmers can make informed decisions and enhance crop management practices to maximize productivity.

In a study conducted by Saha et al. (2018), a cutting-edge real-time UAV system integrated with IoT technology was proposed for crop data monitoring. Real-time data analysis was conducted using intelligent sensors and modules. The system incorporated intelligent sensors and modules to enable real-time data analysis. Additionally, a Raspberry Pi 3B module was employed to integrate the suggested solution with the UAV. Machine learning techniques were utilized to train the system in forecasting crop health. Comparative analysis revealed that this system outperformed existing UAV systems in terms of user-friendliness and accuracy in assessing crop health status. The innovative combination of UAV and IoT technology showcased the potential for advanced and precise crop monitoring, offering significant benefits to agricultural practitioners and stakeholders.

15.3.1.1 Vegetation Growth Monitoring and Yield Estimation

Regular utilization of UAVs for data collection and visual representation of crops enhances the potential for monitoring crop growth and documenting fluctuations in field parameters. The information obtained through UAVs can be employed to generate three-dimensional digital representations of the field, as well as measure various other factors such as crop height, plant spacing, and the Leaf Area Index (LAI). The systematic gathering of crop data by UAVs empowers farmers to effectively plan their crop management strategies, optimize the use of resources like nutrients, determine the harvest time, identify soil and yield diseases, and proactively address potential management errors in advance (Jung et al. 2018).

15.3.2 UAV for Smart Pest Management

The utilization of UAVs in agriculture finds its primary application in pesticide use. In 1983, Yamaha Motor Co. Ltd. introduced the first UAV designed for pesticide application in Shizuoka, Japan. Extensive research has been conducted to enhance various aspects of pest management, including spraying volume, droplet size, droplet dispersion and penetration, and pesticide effectiveness (Lou et al. 2018). These optimization factors have predominantly been applied to agricultural pesticide use, with a particular focus on crops that require substantial pesticide application. Rice (Qin et al. 2016), maize (Songchao et al. 2017), wheat (Wang et al. 2019), cotton (Lou et al. 2018), and bell pepper (Xiao et al. 2020) have been the subject of significant research involving UAVs for pesticide spraying, owing to their high pesticide usage.

In both industrialized and developing nations, the practical utilization of UAVs has led to the production of crops in larger contiguous blocks. To enhance the efficacy of pesticide application on crops, UAV operating characteristics, such as flight speed, altitude, nozzle type, payload, and UAV configuration, must be tailored to specific situations. Research has shown that a UAV equipped with four rotors, two fan nozzles, a flight altitude of 2–3 m, a flying speed of 3–5 m per second, and a payload capacity of 15 L is ideal for effective pesticide spraying in agricultural fields (Zhang et al. 2011). The deployment of UAV-mounted sprayers has improved the capacity to cover more ground, increased the effectiveness of chemicals used, and enhanced the speed and convenience of pesticide spraying. UAVs can now autonomously follow premapped routes while carrying pesticide containers with capacities of up to 40 L. UAVs offer exceptional coverage for fields that are difficult to access for tractors and conventional aircraft. The advantages of using UAVs in agriculture include extensive area coverage, reduced pesticide usage, labor savings, rapid response times, and timely deployment before pest infestations reach economic thresholds (Lv et al. 2019).

The flowchart in Fig. 15.4 illustrates the operational process of a UAV's unified spraying system. The system incorporates both a receiver and a transmitter to facilitate signal transmission and reception. The received signal is then relayed to the

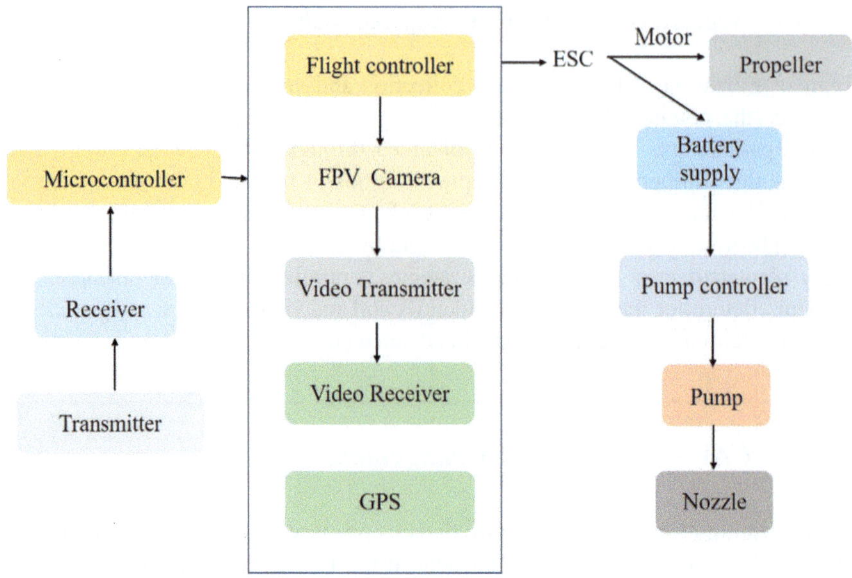

Fig. 15.4 Schematic chart of semi-controlled unmanned vehicles for pesticide application

flight controller, where it undergoes processing through specialized sensors, namely an accelerometer and a gyroscope. Following analysis, the signal is forwarded to the Electronic Speed Controller (ESC), which utilizes the signal to regulate the current supplied to the motor. The motors and propellers, mechanically connected, generate rotation and thrust. The flight controller also powers and captures footage using a first-person-view (FPV) camera. The transmitter processes the video signals, which are subsequently received by the ground-based receiver. Meanwhile, the fluid from the storage tank is pressurized by a battery-powered pump. Once inside the nozzle, the pressurized liquid is sprayed. The transmitter has the ability to modulate the input current, enabling adjustment of the pump's flow rate (Shaw and Vimalkumar 2020).

15.3.2.1 Vegetation Health Monitoring and Diseases Detection

Crops should be regularly monitored so that diseases can be detected early and prevented from spreading. Traditionally, this task has been performed on-site by human experts. The time it takes to inspect an entire crop, which can take months, limits the possibility of "continuous" monitoring. Tools for data processing based on UAVs use crop imaging data to monitor changes in plant biomass and health. Early detection of infections allows farmers to implement measures to minimize losses. UAVs have two applications in disease control. Firstly, during the initial phase of infection, UAVs gather crop health data, detecting potential infections before visible symptoms manifest and mapping the extent of the infestation across different areas of the crops. Secondly, throughout the treatment phase, farmers can utilize UAVs for targeted spraying and precise monitoring of their intervention's progress (Kerkech et al. 2018).

15.3.2.2 Crops Spraying

The primary spraying equipment used in conventional farming is the manual air-pressure and battery-powered knapsack sprayers. These conventional sprayers, however, take a lot of time and can cause significant pesticide losses. Using sophisticated distance-measuring technologies, UAVs can follow the contours of the ground and keep a consistent height. In order to spray the appropriate amount of herbicide spatially, an aircraft can modify its height and the amount of herbicide sprayed based on the crop site it is in. Crop spraying is especially important when illnesses have been identified and it is necessary to limit pesticide use without affecting crop output. UAV-based solutions can considerably enhance agricultural spray control, in conclusion (Xue et al. 2016).

15.3.3 UAVs for Smart Irrigation

Efficient irrigation management is of paramount importance for achieving optimal crop production while minimizing negative environmental impacts. In the context of precision agriculture (PA), field-level irrigation and moisture management play a crucial role. Camera-equipped agricultural UAVs offer valuable insights into specific problem areas. By utilizing UAV cameras, farmers can identify regions with low soil moisture, signs of plant dehydration, and waterlogging issues and assess overall plant health in the field. The UAV is deployed to fly over designated farm areas, collecting data, while electronic sensors gather environmental information such as temperature, humidity, and soil moisture. This data is transmitted from the UAV's Wi-Fi module to the cloud for processing using selected algorithms. Specialists or farmers can then access information regarding the farm's water supply status. Subsequently, irrigation planning can be implemented manually or through automatic irrigation systems, thus contributing to optimized resource management and crop productivity (Fig. 15.5).

By locating regions that need intensive irrigation, farmers can save time and water resources. Higher crop quality and yield may also arise from such precision farming techniques. In the context of precision agriculture, the field is divided into multiple irrigation zones to efficiently manage the resources. Using UAVs with the appropriate sensor types, it is possible to identify crop areas that need more water. The aforementioned technologies also facilitate the construction of particular maps that depict the soil's structure, which facilitates more efficient irrigation planning for each crop separately (Quebrajo et al. 2018).

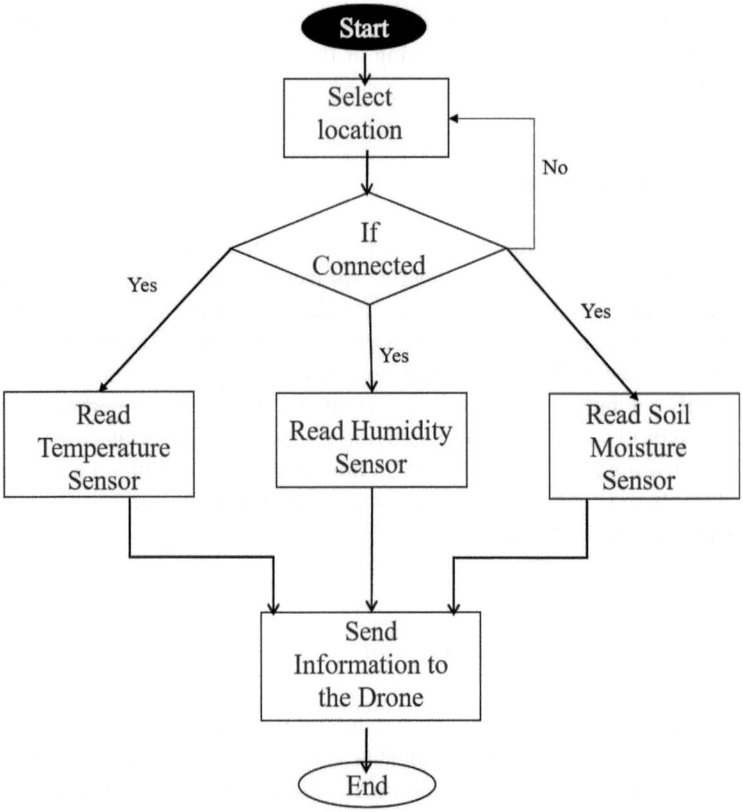

Fig. 15.5 Monitoring the environmental data using IoT and UAV system

15.3.4 UAVs for Livestock Monitoring

UAVs have emerged as a valuable tool for livestock monitoring, offering innovative solutions to enhance animal welfare, health assessment, and overall management practices in the livestock industry. UAVs offer the ability to observe and track the movements of animals within a herd, overcoming ground-based traffic restrictions and maintaining a close proximity to the livestock. By utilizing a system comprising switchable-battery UAVs and ground-based UAV bases for battery replacement, continuous animal tracking and monitoring missions can be conducted by seamlessly switching between multiple UAVs. The integration of GPS collars with UAV-based data sinks holds the potential to transmit a greater volume of messages and collect data at higher frequencies. Furthermore, wireless backhaul links to satellites or nearby base stations enable the upload of collected data from UAVs to a central server for further analysis (Fig. 15.6). UAV-based livestock monitoring extends beyond location tracking, with the possibility of incorporating real-time video monitoring by connecting cameras to the UAVs (Li and Xing 2019).

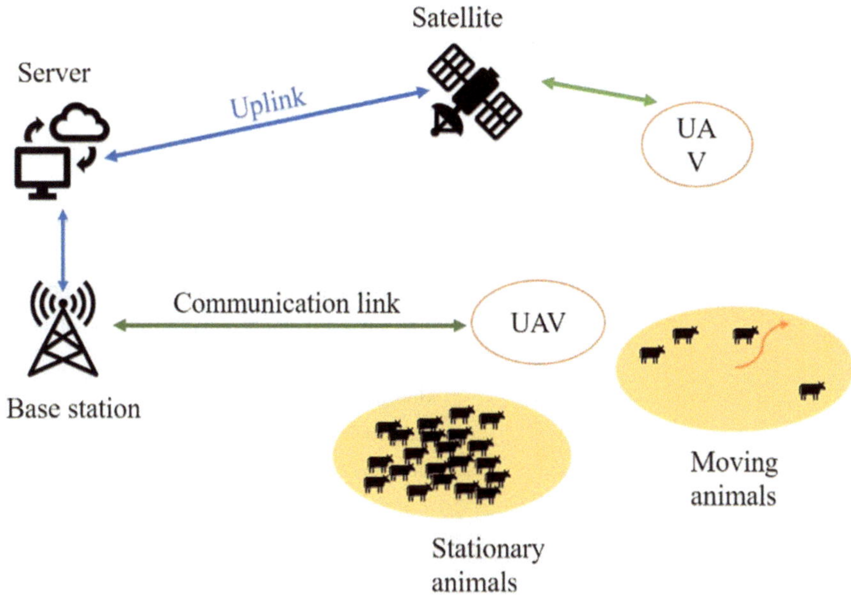

Fig. 15.6 Surveillance and monitoring using UAV

15.3.5 Weed Mapping and Management

The most common practice in conventional farming for managing weeds is to uniformly spray herbicides to the whole field, even the areas where there are no weeds. Excessive pesticide use can lead to the development of herbicide-resistant weeds, which can also affect the growth and yield of crops. Additionally, it seriously endangers the ecosystem through pollution. The method of administering herbicides in spatially diverse patterns as opposed to equally dusting them across an entire area is known as Site-Specific Weed Management (SSWM). In this particular scenario, the field is partitioned into distinct zones, each with its own customized approach to managing weed growth. To achieve this, it is necessary to generate an accurate map illustrating the distribution of weeds, enabling the strategic application of herbicides. To create such a precise weed cover map, which highlights the areas necessitating the highest, lowest, or no pesticide treatment, aerial drones can be employed to gather comprehensive field data and capture photographs (Bah et al. 2017).

15.3.6 Crops Monitored by UAV-Based Systems

UAVs have proven to be valuable tools for monitoring various crop species, with over 30 distinct species identified for tracking purposes. Some of the most frequently

Table 15.1 Crop features monitored with UAVs (Tsouros et al. 2019)

	Crop features
Vegetation	Biomass
	Nitrogen status
	Moisture content
	Vegetation color
	Spectral behavior of chlorophyll
	Temperature
	Spatial position of different elements and plants
	Size and shape of different elements and plants
	Vegetation indices
Soil	Moisture content
	Temperature
	Electrical conductivity

monitored crops using UAV technologies include maize, vineyards, wheat, soya, cotton, and rice. Interestingly, UAV techniques have been successfully employed to monitor crops with diverse features, such as rice and olive trees. One important aspect is that UAV monitoring can be conducted throughout different growth phases, including the initial stages, even before conclusions can be drawn based on soil features. This continuous monitoring enables the tracking of variables that may impact the crop's performance and allows for the evaluation of corrective measures taken in response to earlier issues during the development process.

Key characteristics observed using UAV-based remote sensing for precision agriculture (Table 15.1).

Considering the scale of the crops being monitored, UAVs offer feasibility in the following scenarios:

- Monitoring small farms or specific portions of a field: UAVs can efficiently gather data from smaller agricultural areas, providing valuable insights and assisting in decision-making processes.
- Monitoring large fields (>10 ha): UAVs can cover vast areas and collect data from all parts of the field, ensuring comprehensive monitoring and analysis.
- Monitoring highly heterogeneous areas: UAVs equipped with automatic pilot systems and ground-level sensors can effectively monitor areas with significant variations in crop types or conditions, providing detailed information for targeted interventions.

By utilizing UAVs with automatic pilot systems and ground-level sensors, crop monitoring becomes more efficient and accurate. This combination allows for precise data collection, analysis, and decision-making, ultimately improving crop management practices and maximizing agricultural productivity

15.4 Steps Involved in the Processing of UAVs Imageries

Despite the advanced capabilities of newer UAV models, which include kinematic global real-time navigation satellite systems (RTK-GNSS) that can achieve planimetric precision of 1 cm, the installation of Ground Control Points (GCPs) within the surveyed area is generally necessary to ensure high position accuracy. In the process of creating final orthomosaics, the acquired images must undergo geometric alignment (orthorectification), calibration, and adjustment, accounting for atmospheric factors. These steps are crucial in processing UAV-based thermal imagery and ensuring accurate and reliable results (Tsouros et al. 2019). The following section outlines the essential stages involved in the processing of UAV-based thermal imagery:

- Image preprocessing, which entails eliminating blurry photos and transforming all images into TIFF files with 16-bit dynamic range to make sure that each image's temperature value equates to an identical Digital Number (DN) value.
- Image alignment, where the timestamp of each image and the on-board log file GPS are used to create early estimations of image location.
- By manually inserting GCPs with known positions, either via real time kinematics (RTK) GPS or from the decoded RGB images, images are spatially co-registered to RGB (or other) orthophotos.

15.5 UAV Data Acquisition System

Depending on the various crop metrics that need to be monitored, an agricultural UAV can use various steps for data acquisition. The system begins with mission planning, where the flight area and objectives are defined. And then UAVs are equipped with various types of sensors. The sensors must primarily satisfy the requirements of lightweight, low energy consumption, and tiny size. All of the aforementioned factors must, of course, be matched with the capacity to take good-quality pictures. Most contemporary commercial on-board sensors used for PA that abide by the aforementioned constraints fall into one of the following four categories:

A. Visible light sensors (RGB).
B. Multispectral sensors.
C. Hyperspectral sensors.
D. Thermal sensors.

Every sort of sensor can keep an eye on many aspects of the plant, including its color, texture, and geometric shape. Moreover, certain sensors are capable of detecting specific wavelengths of radiation. Once the flight is completed, the acquired data is downloaded and processed. This includes data preprocessing steps such as georeferencing, image stitching, and sensor calibration. Georeferencing ensures that the collected data is accurately aligned with a geographic coordinate system. Image stitching combines multiple images to create seamless orthomosaics or 3D models. Sensor calibration accounts for any distortions or biases in the sensor measurements.

Post-processing techniques may involve data enhancement, feature extraction, or analysis, depending on the specific application. Advanced algorithms and software tools are utilized to extract valuable information from the collected data, aiding decision-making processes in various industries. This technique is used to monitor vital agricultural characteristics such as soil moisture, biomass from plants, and vegetation health throughout the various growth stages.

15.6 UAV Data Processing Techniques

UAVs have become instrumental in collecting data related to various aspects of cultivated areas through the utilization of specialized sensors. This section highlights the most common image analysis techniques employed for leveraging UAV footage in precision agriculture.

15.6.1 Photogrammetry Techniques

Photogrammetry is the exact rebuilding of an area or an object using multiple overlapping images. Photogrammetric algorithms may analyze the 2D data and identify the geometric relationships between the several pictures and the object(s) to produce 3D models. In order to generate 3D models, photogrammetry relies on the use of at least two images that overlap and depict the same area or objects from different angles (Sa et al. 2018). In comparison to other remote sensing methods, the collection of low-altitude data using UAVs allows for the creation of 3D models with much higher spatial resolution, surpassing that of satellites. Yet, in order to get data for the entire field being studied, a large collection of photographs is necessary. In order to produce Digital Elevation Models (DEMs) of the crops or orthophotos, it is typically essential to gather a large number of overlapping photographs (also referred to as orthomosaics). These techniques provide insights into crop health, vegetation indices, and topography. They support informed decision-making for resource allocation and crop management in precision agriculture.

15.6.2 Machine Learning (ML) Methods

In PA, machine learning techniques are frequently used to extract information from the vast amounts of data collected by UAVs. With the use of machine learning, it is possible to estimate certain crop growth rate metrics, identify diseases, and even distinguish between different objects in photos. Due to the rapid developments occurring, particularly in the deep learning field, the use of ML has significantly expanded recently. ML techniques are utilized for image segmentation, allowing the partitioning of images into meaningful regions based on pixel-level analysis. This enables precise delineation of fields, identification of boundaries, and measurement of area and crop coverage. Some of the mostly used ML techniques such as convolutional neural networks (CNNs), Random Forests (RF), Support Vector Machines (SVM), Recurrent Neural Networks (RNNs), and Long Short-Term Memory (LSTM) networks.

15.6.3 Vegetation Indices' Calculation

Vegetation indices (VIs) are widely recognized outcomes of remote sensing methods in precision agriculture. These indices utilize different mathematical operations involving at least two electromagnetic spectral bands to enhance the distinctive characteristics of vegetation, while minimizing the impact of unrelated variables that may cause confusion. They can provide accurate temporal and spatial information on the crops being monitored in agriculture. Most frequently, a large number of VIs are retrieved and used to make judgments. They can be found using information from either each individual image or after orthophotos showing that the complete crop is created. The computation of vegetation indices may aid in the detection of useful agricultural qualities, which include the biological and physical characteristics of the vegetation. The common vegetation indices used are NDVI, Green Normalized Difference Vegetation Index (GNDVI), Enhanced Vegetation Index (EVI), Normalized Difference Red Edge (NDRE).

15.7 Tools for UAVs' Data Processing

Numerous software tools and methods have been created to speed up data processing because it can sometimes be time-consuming. Table 15.2 lists the software programs that have been used the most frequently to support and expedite the data analysis process in the works reviewed.

Table 15.2 Software tools for image processing (Tsouros et al. 2019)

Software tool	Description
Adobe Photoshop	Applied to correct distortions
Agisoft Photoscan, MATLAB	Allows the calculation of VI
QGIS	Calculation of VI from multispectral data
Pix4D	The tool that is widely utilized for VI calculations, 3D model creation, and orthomosaic construction

15.8 Challenges Involved in UAV Application in Precision Agriculture

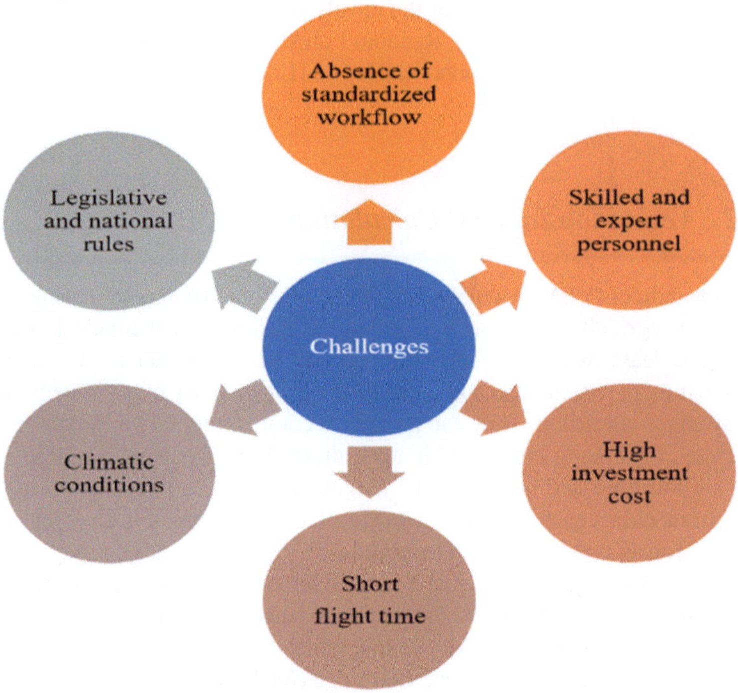

15.9 Directions for Future Research

Precision agriculture's utilization of UAVs is still in its infancy and may still have opportunity for technological and agricultural applications growth. The payload, sensors employed in the UAV, cost, flying time, data analytics, environmental factors, and needs are defined as the obstacles. The primary barrier to UAV utilization is cost, which is compounded by the need for several sensors, mounting components, innovation programs, and the software required for data analyses. Currently, businesses provide solutions for renting out different UAVs with all necessary remote sensing equipment. Once the data has been gathered from the numerous sensors put on the UAVs, data analytics is a crucial obstacle to overcome in order to achieve outcomes at regular intervals of time. It generates countless terabytes of data that must be properly handled, stored, and evaluated using the right tools. Similar to this, as payload and flight time are restricted for UAV use in fields, it is challenging to create a UAV that can both identify hotspots of the pest and the remedies used for them. Another aspect that restricts UAV activities and the sensing process is the weather, such as rain, snow, clouds, and fog. Farmers may readily adapt to this technique because it is suitable with their needs for agriculture and offers affordable solutions. It is probable that as UAV technology advances, so will low volume sprayers, enhanced image processing techniques, reduced costs, flight times, batteries, new camera designs, and nozzle kinds. Numerous experimental investigations have been conducted on UAV-based remote sensing in agriculture. These systems' advantages for precision farming and environmental monitoring are becoming more and more obvious.

UAV can also be used to quickly move harvested commodities from the fields to the storehouse for packing and transportation while utilizing less fuel and labor. UAV can be used to convey goods to markets and reduce product costs. More robust and powerful UAV systems with cutting-edge camera technology like NIR and automated georeferencing methods are projected to enter the market in the upcoming years.

15.10 Conclusion

The utilization of smart IoT and UAVs in agriculture has become a promising approach toward achieving the goals set by the "2030 Sustainable Development Agenda" in eradicating hunger. Recent advancements in data processing, transmission, and embedded systems have transformed UAV technologies into fully automated systems, offering great benefits to farmers. By incorporating machine learning, UAVs have become valuable tools for smart agriculture, enabling efficient crop monitoring, pest management, and animal monitoring. However, the widespread adoption of UAVs in agriculture still faces challenges that need to be addressed. The high cost of UAVs suitable for agricultural use, along with operational and maintenance expenses, pose obstacles in convincing farmers and other stakeholders to integrate UAVs into their practices. Furthermore, limitations such as battery capacity, safety concerns, and legal considerations must be overcome to fully leverage the potential of UAVs in agriculture.

To overcome these barriers, continued research and innovation are necessary to enhance UAV technology, making it more affordable, reliable, and safe for agricultural applications. Collaboration among stakeholders, including governments, researchers, and industry players, is essential to establish supportive policies, address regulatory issues, and provide necessary training and resources for effective UAV implementation in agriculture. With concerted efforts, the integration of UAVs in agriculture holds tremendous promise for increasing productivity, reducing resource wastage, and contributing to sustainable food production. It is crucial to overcome the challenges and capitalize on the potential of UAVs to unlock the benefits they offer to farmers, ultimately leading to a more resilient and food-secure future.

References

Allred B, Martinez L, Fessehazion MK, Rouse G, Williamson TN, Wishart D, Koganti T, Freeland R, Eash N, Batschelet A, Featheringill R (2020) Overall results and key findings on the use of UAV visible-color, multispectral, and thermal infrared imagery to map agricultural drainage pipes. Agric Water Manag 232:1–19

Bacco M, Berton A, Ferro E, Gennaro C, Gotta A, Matteoli S, Paonessa F, Ruggeri M, Virone G, Zanella A (2018) Smart farming: opportunities, challenges and technology enablers. In: 2018 IoT Vertical and Topical Summit on Agriculture-Tuscany (IOT Tuscany), pp 1–6

Bah MD, Hafiane A, Canals R (2017) Weeds detection in UAV imagery using SLIC and the Hough transform. In: Proceedings of the 2017 Seventh International Conference on Image Processing Theory, Tools and Applications (IPTA), Montreal, QC, Canada, 28 November–1 December 2017, 1–6.

Cano E, Horton R, Liljegren C, Bulanon DM (2017) Comparison of small unmanned aerial vehicles performance using image processing. J Imag 3(1):4

Cuaran J, Leon J (2021) Crop monitoring using unmanned aerial vehicles: a review. Agric Rev 42(2):121–132

Eisenbeiss H, Sauerbier M (2011) Investigation of UAV systems and flight modes for photogrammetric applications. Photogram Rec 26(136):400–421

Elarab M, Ticlavilca AM, Torres-Rua AF, Maslova I, McKee M (2015) Estimating chlorophyll with thermal and broadband multispectral high resolution imagery from an unmanned aerial system using relevance vector machines for precision agriculture. Int J Appl Earth Obs Geoinf 43:32–42

Ge Y, Thomasson JA, Sui R (2011) Remote sensing of soil properties in precision agriculture: a review. Earth Science 5(3):229–238

Jung J, Maeda M, Chang A, Landivar J, Yeom J, McGinty J (2018) Unmanned aerial system assisted framework for the selection of high yielding cotton genotypes. Comput Electron Agric 152:74–81

Kc S, Lutz W (2017) The human core of the shared socioeconomic pathways: population scenarios by age, sex and level of education for all countries to 2100. Glob Environ Change 42:181–192

Kerkech M, Hafiane A, Canals R (2018) Deep leaning approach with colorimetric spaces and vegetation indices for vine diseases detection in UAV images. Comput Electron Agric 155:237–243

Kim W, Khan GF, Wood J, Mahmood MT (2016) Employee engagement for sustainable organizations: keyword analysis using social network analysis and burst detection approach. Sustainability 8(7):1–11

Li X, Xing L (2019) Unmanned aerial vehicles for livestock monitoring based on streaming livestock k-means mon. IFAC PapersOnLine 52(30):324–329

Lou Z, Xin F, Han X, Lan Y, Duan T, Fu W (2018) Effect of unmanned aerial vehicle flight height on droplet distribution, drift and control of cotton aphids and spider mites. Agronomy 8(9):187

Lv M, Xiao S, Yu T, He Y (2019) Influence of UAV flight speed on droplet deposition characteristics with the application of infrared thermal imaging. Int J Agric Biol Eng 12(3):10–17

Mogili UR, Deepak BBVL (2018) Review on application of drone systems in precision agriculture. Proc Comput Sci 133:502–509

Mukherjee A, Misra S, Sukrutha A, Raghuwanshi NS (2020) Distributed aerial processing for IoT-based edge UAV swarms in smart farming. Comput Netw 167:1–14

Norasma CYN, Fadzilah MA, Roslin NA, Zanariah ZWN, Tarmidi Z, Candra FS (2019) Unmanned aerial vehicle applications in agriculture. Conf Ser Mater Sci Eng 506(1):1–11

Pandey K, Singh KG, Singh A (2023) Multi-sensors based smart nutrient reuse management system for closed soilless culture under protected cultivation. Comput Electron Agric 204:107495

Qin WC, Qiu BJ, Xue XY, Chen C, Xu ZF, Zhou QQ (2016) Droplet deposition and control effect of insecticides sprayed with an unmanned aerial vehicle against plant hoppers. Crop Prot 85:79–88

Quebrajo L, Perez-Ruiz M, Pérez-Urrestarazu L, Martínez G, Egea G (2018) Linking thermal imaging and soil remote sensing to enhance irrigation management of sugar beet. Biosyst Eng 165:77–87

Refaai MR, Dattu VS, Gireesh N, Dixit E, Sandeep CH, Christopher D (2022) Application of IoT-based drones in precision agriculture for pest control. Adv Mater Sci Eng 2022:1–10

Sa I, Popović M, Khanna R, Chen Z, Lottes P, Liebisch F, Nieto J, Stachniss C, Walter A, Siegwart R (2018) WeedMap: a large-scale semantic weed mapping framework using aerial multispectral imaging and deep neural network for precision farming. Remote Sensing 10(9):1423

Saha AK, Saha J, Ray R, Sircar S, Dutta S, Chattopadhyay SP, Saha HN (2018) IOT-based drone for improvement of crop quality in agricultural field. In: 2018 IEEE 8th Annual Computing and Communication Workshop and Conference (CCWC), pp 612–615

Shaw KK, Vimalkumar R (2020) Design and development of a drone for spraying pesticides, fertilizers and disinfectants. Eng Res Technol

Singh AP, Yerudkar A, Mariani V, Iannelli L, Glielmo L (2022) A bibliometric review of the use of unmanned aerial vehicles in precision agriculture and precision viticulture for sensing applications. Remote Sensing 14(7):1604

Songchao Z, Xinyu X, Zhu S, Lixin Z, Yongkui J (2017) Downwash distribution of single-rotor unmanned agricultural helicopter on hovering state. Int J Agric Biol Eng 10(5):14–24

Tsouros DC, Bibi S, Sarigiannidis PG (2019) A review on UAV-based applications for precision agriculture. Information 10(11):1–26

Uddin MA, Mansour A, Denis LJ, Ayaz M, Aggoune M (2018) UAV-assisted dynamic clustering of wireless sensor networks for crop health monitoring. Sensors 18:555

Wang G, Lan Y, Qi H, Chen P, Hewitt A, Han Y (2019) Field evaluation of an unmanned aerial vehicle (UAV) sprayer: effect of spray volume on deposition and the control of pests and disease in wheat. Pest Manag Sci 75(6):1546–1555

Xiao Q, Du R, Yang L, Han X, Zhao S, Zhang G, Fu W, Wang G, Lan Y (2020) Comparison of droplet deposition control efficacy on *Phytophthora capsica* and aphids in the processing pepper field of the unmanned aerial vehicle and knapsack sprayer. Agronomy 10(2):215

Xue X, Lan Y, Sun Z, Chang C, Hoffmann WC (2016) Develop an unmanned aerial vehicle based automatic aerial spraying system. Comput Electron Agric 128:58–66

Zhang W, Jiang F, Ou J (2011) Global pesticide consumption and pollution: with China as a focus. Proc Int Acad Ecol Environ Sci 1(2):125–144

Zhang Y, Wang L, Duan Y (2016) Agricultural information dissemination using ICTs: a review and analysis of information dissemination models in China. Inform Process Agric 3(1):17–29

Chapter 16
Biotic Stress Management in Field Crops Using Artificial Intelligence Technologies

Shubham Anand and Sarabjot Kaur Sandhu

Abstract India is an agriculture dependent country and due to changing climatic scenarios, nation is facing challenges in ensuring food security to its ever-increasing population. Abiotic and biotic stresses have a big impact on crop productivity. Due to climate change, biotic stresses, such as insects and diseases in various field crops are responding differently. The Food and Agriculture Organisation of the United Nations (FAO) estimates that crop insects and diseases are responsible for 20–40% of the annual losses in global food production. Inexperienced pesticide users run the risk of causing infections and insects to become resistant, which greatly reduces the host plant's capacity for self-defence. Various scientists are studying the intricate relationship between plants, pests, pathogens, and environment using a range of computer modelling and simulation tools, including artificial neural networks and conventional multiple regression methods, which can be used to forecast the occurrence of insects and diseases. Artificial neural network (ANN) techniques have received a lot of interest recently, partly due to their broad range of applicability and ease of use for treating complex issues even with noisy and erroneous data. The models enable research into the factors influencing pest outbreaks and the development of yield loss mitigation techniques. By utilising need-based chemicals, they also guarantee the safety of the operator, the consumer, and the environment. By focusing on the problems encountered with the methodology, procedures, and data employed, the main aim of this chapter is to describe in depth the artificial intelligence and machine learning algorithms used to anticipate plant insects and diseases.

Keywords Artificial intelligence · Artificial neural network · Biotic stress · Yield losses · Pest prediction

---

S. Anand (✉) · S. K. Sandhu
Department of Climate Change and Agricultural Meteorology, Punjab Agricultural University, Ludhiana, Punjab 141004, India
e-mail: 1904001@pau.edu

K. Pandey et al. (eds.), *Artificial Intelligence and Smart Agriculture*,
Advances in Geographical and Environmental Sciences,
https://doi.org/10.1007/978-981-97-0341-8_16

16.1 Introduction

Numerous biotic and abiotic elements have an impact on agricultural productivity throughout the world. Insect, pest, disease, and weed infestations are said to harm 40% of agricultural production (Roser 2013). The inability to make an accurate diagnosis at the appropriate time prevents the targeted degree of disease and pest management from being reached. The incorrect and excessive use of agrochemicals can result in financial and environmental issues. The Food and Agriculture Organisation (FAO) assessed that each year, pests cause 20–40% of global production to be lost (Frona et al. 2019; Ray et al. 2013). Similarly, each year plant diseases cost $220 billion in lost revenue, while pests cost $70 billion (FAO 2017). To increase the quality and lifespan of their harvests, producers use a variety of insecticides. In addition to potentially harmful conditions like severe respiratory infections, cancer, hereditary infections, and foetal mortality, continued application of pesticides can harm the environment (Deutsch et al. 2018). Agriculture considerably benefits from the application of new technical approaches in order to identify pests at their earliest stages and prevent the widespread use of harmful pesticides. To propagate the sooty mould disease, insects feed on the plants' leaves, fruits, stems, and roots. By hindering photosynthesis and producing tissue infections, the disease harms plants, lowering the quality and quantity of harvested items' market value. In order to identify pest infestations, farmers rely on their own knowledge and experience. Due to a lack of experience, application of pesticides is most common method of pest control because it is quick-acting and scalable (FAOSTAT 2022). Nevertheless, there is a need for less pesticide use as a result of growing environmental and health concerns. Spraying only where necessary is one of the most essential strategies for reducing the usage of pesticides. Spot spraying is reported to lessen the environmental contamination and cut the cost of applying pesticides by 90% (Kamilaris et al. 2017; Kartikeyan and Shrivastava 2021). Prior to spot application of pesticide, the primary loci of pest must be monitored and defined. Typically, pest detection is carried out physically, which is laborious and extremely prone to inaccuracy (Rajpurkar et al. 2017).

The latest improvements in computer science in relation to agriculture, the growth and health of crops now also include understanding insect pest and disease identification. For vast farms and orchards, managing autonomous pesticide spraying robots and turning on intelligent spraying systems depend on the ability to detect things at various phases of agricultural growth. However, due to their similar shapes, complex backgrounds, overlap caused by dense distribution, variable light in the vast geography of orchards, and a number of other issues, it can be challenging to distinguish target items with sufficient accuracy. But, because of advances in technology, it's now possible to identify insect pests through picture processing techniques. People started to show a growing amount of interest in precision agriculture to address these issues. Agricultural management became more accurate and controlled due to precision agriculture. These technologies include robots, remote sensing, data analysis, drones, and land vehicles (Wang et al. 2015). They also include the Global Positioning System (GPS) for tractor navigation. The precise recognition of insect/pest is the foundation key of precision agriculture.

By maximising the time and frequency of administration of control measures, weather-based forewarning systems lower production costs and maintain operator, consumer, and environmental safety by using fewer chemicals. Reduced fungicide use is a primary goal of many forewarning systems, and timely application of disease control methods is essential to prevent crop losses. Management choices could be influenced by a forecasting model developed on the basis of correlation between prevailing weather parameters at the time of crop management and the severity of late season diseases. Therefore, the violent behaviour of the disease can be avoided by judicious application of pesticide, if a reliable forecasting system is developed. Artificial neural networks (ANN), a machine learning technique, and more traditional multiple regression methods are used to support scientists synthesise and develop their understanding of this complex plant-pathogen-environment relationship. Recently, artificial neural network (ANN) techniques have received significant consideration, in large part because of their broad applicability and ease of use, especially when dealing with noisy and inaccurate data (Pandey et al. 2020; Pandey and Malik 2021). These methods are being successfully used in a staggering variety of problems occurring in agriculture, geology, finance, engineering, medicine, physics, biology, and as well as other fields. Because of its potential for application in prediction, neural networks are fascinating from a statistical perspective. Because the relationships between host, pathogen, and weather are so intricate, the ANNs technique can be used to forecast the arrival of pests and diseases. This chapter presents the importance of AI techniques in management of biotic stresses in different crops by including the work of different researchers on diagnosis, detection, and management of biotic stresses in crops using AI techniques.

16.2 Artificial Intelligence in Agriculture

Machine learning is utilised in different fields, including agriculture and horticulture, as a result of the advancement of artificial intelligence technology (Rehman et al. 2019). Machine learning algorithms can be used to create mathematical models using train data to make forecasting deprived of any specific knowledge to conduct analysis (Bishop 2006). According to Lee and Xing (2018), machine learning may be beneficial in recognising pest species. Building machine learning models for image classification in the past required significant expertise in selecting model architectures, as well as understanding of mathematics, image analysis, and coding (Feurer et al. 2015; Kotthoff et al. 2017). Due to the fact that bug species vary across different crops, geographical areas, and locales, it would be excellent if a special machine learning model could be created for each situation.

Different machine learning platforms are at present provided by various information technology companies like Google, it may be useful to identify pest species.

They offer an intuitive user interface and train models with minimal machine learning expertise. For instance, Google Cloud Auto ML Vision automatically trains models based on Neural Architecture Search (NAS) using pictures given through a straightforward graphical interface. Without technical knowledge of coding, machine learning, and image analysis, users can design a model.

16.3 Use of AI in Biotic Stress Diagnosis, Detection, and Classification

Extreme weather events have been a result of climate change, which is currently a major challenge for food security of the nation. By 2030, the temperature could rise by 1–2.5 °C, which could have a substantial impact on food output due to changes in photosynthesis, an increase in plant respiration, and changes in disease and pest populations (Bhanumathi et al. 2019). Diseases that affect plants and crops are significantly influenced by the weather. The conceptual model of diseases triangle is a fundamental idea of the elements involved in the development of disease. When the susceptible host and a virulent pathogen get together in a favourable environment, a disease is caused. Stevens published the first disease triangle illustration (Stevens 1960). The disease triangle includes components, such as humans, vectors, and time (Francl 2001). Numerous plant biologists are conducting more extensive research on the association and effects of different climate change components on spread of epidemics of crop diseases. Newbery et al. (2016) created a graphical approach to understand the potential connection of climate, growth of crop, and models for disease forecasting to provide estimates of crop development phases and disease development for various climate changing scenarios. Artificial intelligence (AI) emphasises the potential value of pattern and trend recognition in vast amounts of data utilising mathematical algorithms with the aim of completing a specific task. The task could be general, such as computer vision, natural language processing (NLP), or predictive modelling, or it could be more specialised and tied to a particular topic that would ordinarily require a subject matter expert. Machine learning (ML) and deep learning (DL) are fields within AI. While artificial intelligence (AI) is the general term used to classify any work that enables a machine or system to mimic human behaviour and intelligence, machine learning and deep learning are the specific methods used to accomplish this. Machine learning employs algorithms that learn from data, generalise, and construct rules in order to predict one or more target variables from a set of input variables (Goodfellow et al. 2016). Fortunately, machine learning accounts for understanding extremely complex relationships to specify with mathematical models, as well as aids in understanding and constructing new models. Verma et al. (2018) have investigated the use of electric nose (EN), deep learning, and image processing-based prediction models for the diagnosis of tomato plant diseases. Different models have been employed to forecast, recognise, and categorise the condition. For the prediction and, ideally, early forecast of plant diseases, they have utilised a variety of modelling techniques, including extreme learning machine (ELM) classifier, artificial neural networks (ANN), deep learning, support vector machines (SVM), etc. To detect the disease, there are various image processing techniques available, such as thermal imaging, RGB imaging, multispectral imaging, hyperspectral imaging, fluorescence spectroscopy, etc. These methods can anticipate the start of plant diseases, allowing for early treatment and the stoppage of the transmission of the disease to other possible hosts. Using a variety of neural network techniques, Muthukannan et al. (2015) identified spot ailments in leaves and categorised them according to the sorts of diseased leaves. The technique of classifying diseased plant leaves by processing the set of shape and texture features from the afflicted leaf picture utilising radial basis function networks (RBF), learning vector quantisation (LVQ), and feed forward neural networks (FFNN). The outcomes of the simulation demonstrated how well the suggested plan worked. A machine learning-based method for enhancing crop quality in the Indian economy can be developed with the aid of this effort.

16.4 Methods for Disease Diagnosis, Detection, and Classification

The various methods for disease diagnosis, detection, and classification (Fig. 16.1) are as discussed below:

16.4.1 Support Vector Machines (SVMs)

The separating hyper plane is a non-probabilistic algorithm that is used to officially describe SVM. It is available in linear and nonlinear SVM forms. When data can be split into linear groups, linear SVM is utilised. In essence, it divides the data into two categories. When there are just two output classes viz., healthy and diseased plant, linear SVM uses a hyper plane, which is a straight line. Data that cannot be categorised using a straight line but can nonetheless be converted into high-dimensional linear separable sparse data is utilised in nonlinear SVM. Multi-class SVM is used to categorise data that has more than two classes. Support vector machines, generalised regression neural networks, and back propagation neural networks were all tested by Kaundal et al. in 2006.

The association between environmental factors and disease level has been better described by the SVM-based regression approach, which may be helpful for disease management. Regression and SVM methods were evaluated by Wang and Ma (2011) to forecast yellow rust disease of wheat, i.e. serious disease of crop in China. The experiment's findings indicated that the SVM method's prediction had outstanding processing speed in addition to high fitting and predictive precision. The viability of

Methods for disease diagnosis, detection and classification	**Support vector machine (SVM)**
	SVM-based Multiple Classifier System
	Artificial neural networks
	Back Propagation Neural Network
	Feed-Forward Neural Network (FFNN)
	Multi-Layer Perceptron and Convolutional neural networks (CNNs)
	Deep Neural Network
	Bayesian network model
	Long Short-Term Memory network (LSTM)
	Logistic Regression (LR) algorithm
	K-Nearest-Neighbor
	Histogram of an Oriented Gradient (HoG)
	Progressive discriminant analysis
	Snake segmentation
	Communication Access Real Time analysis
	Computer Vision Technology & Fuzzy Logic
	Automated Vision-Based Diagnosis
	Internet of Things (IoT)
	Smart phone applications

Fig. 16.1 Methods used for disease diagnosis, detection, and classification

foliar sugar beet disease pre-symptomatic identification was established by Rumpf et al. (2010). This study combined many vegetative measures with the SVM algorithm to distinguish between diseased and healthy sugar beetroot leaves, identify, and diagnose different diseases prior the development and appearance of symptoms. With regards to the third objective, the classifier produced noteworthy outcomes. The authors proposed selecting wavelengths or wavelength combinations that were more specific for the intended task in order to further improve detection. By utilising support vector regression (SVR) and multilinear regression (MLR) to analyse numerous plant, soil, and environmental data, Badnakhe et al. (2018) developed a forecasting model for citrus gummosis. These included soil moisture (SM), leaf area index (LAI), soil temperature (ST), and chlorophyll (Cab), the values of which were obtained using the Inverse PROSAIL model and later verified using field data. The outcomes demonstrated that, in comparison with MLR, the SVR model performed reasonably well. This study not only identified the months with the highest disease occurrence rates, but also the variables that are most important for making predictions. They discovered that as humidity and soil moisture rise and soil temperature falls, disease incidence rises. The soil parameter was crucial in quantifying the models' ability to forecast disease.

16.4.2 SVM-Based Multiple Classifier System

Tian et al. (2010) employed an SVM-based multiple classifier system for the identification of wheat leaf diseases. Colour features are encoded in RGB to HIS by using GLCM, and seven invariant moments are taken as form parameters. For the purpose of identifying diseases in wheat plants offline, they used an SVM classifier with MCS.

The various types of classifiers are as follows:

(i) *Naive Bayesian (NB) classifier*: The continuous values of different attributes are calculated by Gaussian Naive Bayesian, and their distribution is determined by the Gaussian distribution, also known as the normal distribution.

(ii) *K-Nearest Neighbour (KNN) classifier*: KNN classifier utilised for regression and classification issues, but typically, KNN aids with classification concerns. KNN is a diagnostic spanning tree algorithm. Nonparametric, or the structure derived from the attributes of the data collection, is used when there is no imagination for the distribution of the data. KNN is used for forecasting when datasets don't follow pretended mathematical theories.

(iii) *Decision Tree (DT) classifier*: The most common and preferred method for classifying and making predictions is the DT classifier. A feature test is described by every internal node in DT, a test result is described by every branch and a data label is held by every terminal node. Decision trees can swiftly produce understandable rules. Because of its simple flowchart estimation, a decision tree is an accessible and useful value-based strategy. Zhao et al. (2018) carried out an experiment at the provincial level to anticipate the four severity levels of wheat powdery mildew infection (healthy, mild, moderate, and severe). The analysis of the NDVI, LST, rainfall, and temperature components resulted in an overall predicting accuracy of 83.33%. The spatial resolution restriction, the imaging quality of optical remote sensing imagery, and the difficulty in differentiating between healthy and moderate situations, particularly at the early infection stage, were the constraints that had an impact on the outcomes.

(iv) *Support Vector Machine (SVM) classifier*: SVM analysis for classification and regression can be helpful. Within the two classes of data, SVM determines the hyper plane with the increased margin. Support vectors are the name given to the hyper plane vectors. By taking into account crucial circumstances, SVM may create a margin of hyper plane that totally splits the hyper plane vector in two classes that don't intersect. However, this is not always the case, so this classifier will look for support vector hyper planes that have larger margins and fewer classification errors.

(v) *Random Forest (RF) classifier*: It is a supervised learning model which consists of multiple decision trees with the same nodes. It works by building different decision trees and then merging the decisions of other decision trees for getting a solution which is the average of decision trees. The factor that affects the performance of random forest is the strength of individual decision trees. An increase in the strength of trees results in lower error rate.

(vi) *Multilayer Perceptron (MLP) classifier*: The idea that is based on regression is known as a multilayer perceptron (MLP). By using this technique, nonlinear-based learners are converted, changing the input dataset. Changes from the input data is a property that is linearly distinct. A hidden layer is formed from the input data layer. The MLP classifier only forms only one hidden layer, failing which it functions as an artificial neural network.

16.4.3 Artificial Neural Networks

The human brain serves as the inspiration for artificial neural networks, or ANNs. It is made up of feed-forward networks and works by applying learning algorithms to change the biases and weight parameters to obtain the desired network output. To select which neuron to fire an output when, an additional activation function is present in ANN. Kulkarni and Patil (2012) employed artificial neural network (ANN) technology and other image processing approaches to identify plant diseases early and precisely. Since the proposed method was based on an ANN classifier for classification, it produced good outcomes with a recognition rate of 91%. A classifier built on ANN technology employs a combination of textures, colours, and traits to categorise and diagnose different plant diseases. Android (2016) concentrated on classifying and identifying plant diseases. The plant part where the symptoms can be noticed are likely to be affected. The leaf displays the symptoms by altering colour and making dots on it visible. There were other crucial stages, such as picture processing instruction and identification. The first step was resizing the acquired image and converting it to HIS colour space format using segmentation technology. Features like the major axis, minor axis, and abnormalities were extracted from the image. In the last step (ANN), the artificial neural network classified these features. The classifier was used to categorise the leaf disease. Ahmadi et al. (2017) used the artificial neural network analysis method to distinguish as well as classify fungal diseases in palm trees at an initial stage. The outcomes demonstrated that ANN can properly recognise weak and diseased palms that are not yet exhibiting obvious symptoms and can discover dynamic correlations between the components examined in the study.

16.4.4 Back Propagation Neural Network

Back propagation, sometimes referred to as backward propagation of errors, is an algorithm designed to detect faults as they move backwards from input nodes to output nodes. It is an essential mathematical tool for data mining and machine learning to boost the accuracy of forecasts. The "classification of pomegranate diseases based on back propagation neural network" proposed by Sannakki and Rajpurohit (2015) primarily makes use of the technique of segmenting the defective area and uses colour and texture as the features. In this case, the data was classified using a neural network classifier. The main advantage is that the image's chromaticity layers are retrieved by utilising a conversion to L*a*b (L*: Lightness. a*: Red/Green Value. b*: Blue/Yellow Value.). The categorisation is found to be 97.30% accurate. The fact that it is only used for a few crops is its main downside.

16.4.5 Feed-Forward Neural Network (FFNN)

In this method of ANN, feedback loops are not formed. As the name indicates, in this method information is passed only in one direction, i.e. forward. These networks have the reputation of being simple networks that link inputs and outputs. Fenu and Malloci (2019) investigated whether machine learning can be utilised to forecast the threat of potato late blight in southern Sardinia using meteorological data gathered by neighbourhood weather stations. A four-year period (2016–2019) hourly historical meteorological data, including temperature, humidity, rainfall, wind speed, and sun radiation were analysed to construct the potato late blight prediction model. A stratified k-fold cross approach, a FFNN and SVM classifier were implemented and validated. According to the results, SVM performed better than ANN did at differentiating medium risk levels. Temperature, humidity, and wind speed were all significant components in the prediction, according to Sunil et al.'s analysis from 2021. The models were later incorporated into a decision support system known as Laore Architecture Network Development for Sardinia (LANDS) to aid farmers in decision-making (Bhojani and Bhatt 2020).

16.4.6 Multilayer Perceptron (MLP) and Convolutional Neural Networks (CNNs)

Both MLP and CNN can be used for image classification, but MLP requires a vector as input, but CNN requires a tensor, allowing CNN to better comprehend the spatial relation (the relationship between adjacent pixels of the picture), which is important for complex images. In order to evaluate an early warning model to predict the signs of potato late blight at an initial stage, 126 multispectral photos obtained by unmanned aerial vehicles (UAV) were employed, according to Duarte-Garvajalino et al. (2018). In the study, support vector regression and random forest, two machine learning models, were used to assess the performance of CNN and MLP artificial neural networks. Random forests and MLP employing band differences both performed amazingly well.

16.4.7 Deep Neural Network (DNN)

It is an ANN method which uses number of layers between input and output layers. Different types of neural networks are there but they always use same components. Kaur and Kaur (2018) used k-means clustering and DNN to identify different citrus plant diseases and forecast diseases using picture data and meteorological conditions. Finally, four investigations predicted the occurrence of diseases using data from the weather, abiotic factors, and biotic factors.

16.4.8 Bayesian Network Model

A Bayesian model is a statistical model in which all of the model's uncertainty is represented by probability. Both the uncertainty of the model's input and its output are included in the category of uncertainties. According to disease triangle factors, i.e. host, pathogen, and favourable environment, Lu et al. (2020) formulated a distinctive Bayesian network model to anticipate the incidence of powdery mildew in vineyards. The model was evaluated by using thirteen years field data of three vulnerable varieties.

16.4.9 Long Short-Term Memory Network (LSTM)

Many recurrent neural networks (RNNs) are able to learn long-term dependencies, particularly in problems involving sequence prediction. Kim et al. (2018) used diverse approach to form a location specific model to forecast the rice blast incidence in seventeen varieties. The temporal trends of the climatic variables were modelled in this study. The scientists analysed historical blast onset data and meteorological data using an LSTM, in contrast to previous study that required climatic data from the same year as the predicting time. Additionally, when the input variables were combined, the effects of the LSTM input factors like scores for the rice blast disease, temperature, RH, and sunshine duration on the forecasting accuracy were investigated. Some varieties gave more accurate results than others but overall outcome of model was not up to mark. The accuracy of model outcomes varied in range from 40% to 79% across different locations.

16.4.10 Logistic Regression (LR) Algorithm

Based on some dependent variables, the machine learning classification process known as logistic regression is used to forecast the likelihood of a given class. Bhatia et al. (2020) utilised the SVM and logistic regression (LR) approaches to forecast the powdery mildew disease of tomato plants. The SVM was used to remove anomy from data and LR classifier was used to formulate the classification model after the SVM-ANR technique generated a training set. The recommended hybrid SVM-LR technique offers a high precision in forecasting this particular disease than the use of both these algorithms separately. No feature selection method was employed in the effort to identify the most important qualities. For additional research, it was advised to conduct a larger inquiry using an extreme learning machine (ELM) technique.

16.4.11 K-Nearest-Neighbour

The k-nearest neighbour algorithm, mostly denoted as KNN or k-NN, is a nonparametric supervised learning method used for classification and regression. T is one of the easiest and simplest method as it classifies the data point on the basis of classification of neighbour point. Using a modified K-nearest neighbour validation, Sannakki et al. (2013) combined the crop forecasting regression model with weather prediction using an ANN. The authors developed a method that could predict grape powdery mildew based on forecasted meteorological data, including temperature and humidity.

16.4.12 Histogram of an Oriented Gradient (HoG)

This algorithm divides image into smaller parts, i.e. cells and calculate each cells magnitude and orientation. After this the histogram is developed from collection of gradient information. The histograms developed explain the image features and define the objects within image.

16.4.13 Progressive Discriminant Analysis

Yeh et al. (2016) investigated the progression of leaf anthracnose in strawberry and obtained results of 88% by utilising progressive discriminant analysis through looking at hyper-spectral images to distinguish between a vigorous and symptomless leaf.

16.4.14 Snake Segmentation

This method can be used to address a variety of segmentation issues. The main job of the model is to locate and define the segmentation target item. It takes some prior understanding of the shape of the target object, particularly for complex objects. The method developed by Rothe and Kshirsagar (2015) for "Cotton Leaf Disease Identification using Pattern Recognition Techniques" makes use of Hu's moments and snake segmentation.

16.4.15 Communication Access Real-Time Analysis

Using communication access real-time analysis, Rodriguez-Moreno et al. (2020) combined hourly meteorological data with three-year wheat field disease data to forecast rust. They found a correlation between nighttime temperatures below 14.3 °C and total radiation levels below 521.7 Wm^{-2} and severity of the leaf rust disease. Relative humidity less than 88.7%, total radiation less than 597.4 Wm^{-2}, and dew point temperatures less than 16.1 °C were all associated with a greater severity of stripe rust.

16.4.16 Computer Vision Technology and Fuzzy Logic

This method involves type of thinking that reflects human reasoning and is known as fuzzy logic (FL). This is based on truthfulness of data. This approach explains about how humans make decisions. It includes "YES" and "NO" options to develop an algorithm. Rastogi et al. (2015) conducted a study related to disease diagnosis in crops. They used ANN, fuzzy logic and defective area is divided by using K-means clustering, and the severity of the condition is graded by using fuzzy logic.

16.4.17 Internet of Things (IoT)

Internet of Things refers to the complete network of interrelated devices as well as the technology that aids communication between them as well as with the cloud. Zhang et al. (2014) presented an Internet of Things-based system that can track and control weeds, pests, and diseases that affect wheat. This includes of a system for monitoring by professionals and cameras as well as a system for gathering, analysing, and transmitting data. The expert system gathers historical data on wheat-harming weeds, pests, and diseases in addition to a sizable number of digital images of these problems in wheat fields. Truong et al. (2017) have presented an IoT system that aids in fungal diagnosis and inhibition by gathering environmental data and anticipating weather. The microcontroller is automated to upload the data to cloud at defined intervals. The WiFi module also sends data to the free cloud storage service. Additionally, the website's design has received a GPS coordinate. Google Maps are used to show the locations of devices. The support vector machine regression algorithm was used to generate forecasts for the daily values of different weather parameters. These measures might be used to predict when harmful fungi infections will occur. Infections may be recognised by this gadget, and it can also predict their spread in agricultural fields. A sensor that can recognise fungal spores on the leaves could be added to the system in the future as an upgrade.

16.4.18 Smartphone Applications

In order to effectively control plant diseases and reduce their financial cost and environmental impact, early detection is essential. Due to their powerful computers, sharp displays, and comprehensive built-in accessory sets, including high-definition cameras, smartphones are a very new way to help diagnose diseases. For the purpose of diagnosing diseases, this method makes use of image processing technology, which is typically integrated with artificial intelligence such as neural networks. Depending on the tissue type and other factors, the accuracy ranges from 60 to 100%. In order to help farmers manage pests and diseases that affect fruit trees and safeguard their crops, Lee et al. (2017) developed a system that delivers information about these issues. Using the available weather information, the pest prediction tool assessed the risk of infection and the time it would occur. The prediction module inserted these anticipated dates into the database. The server notified the end user about the date of occurrence of insect/disease which was calculated. Two user apps were produced. Using one of two web applications and one android application, farmers could learn about pest forecasting, frost, and reporting services.

Some of the apps also provide direct communication between the user and the subject specialist. Some of the apps for plant disease diagnosis are listed below:

(a) **Agrio**: Artificial intelligence (AI)-based.
(b) **Crop Doctor**: This app is built by the Indian Government. The app supports both Hindi and English.
(c) **Crops AI**: AI-based, currently supports only a few crops.
(d) **Leaf doctor**: This app was produced by Cornell University and the University of Hawaii at Manoa, College of Tropical Agriculture and Human Resources.
(e) **Leaf Plant Tech**: A drone with an AI-based camera takes pictures of crops and sends information to the grower's mobile device and an app database.
(f) **Plantix**: This app is available in 18 regional languages of India.
(g) **Plants Disease Identification**: AI-based, presently provides information for some crops viz., apple, grape, squash, blueberry, cherry, pepper, soybean, corn, peach, orange, potato, tomato, and strawberry.

(h) **Plant Village Nuru**: It is AI-based technology created by the International Institute of Tropical Agriculture (IITA) and the Food and Agriculture Organisation of the United Nations (FAO).
(i) **Purdue Plant Doctor** identifies diseases of the tree, shrub, perennial and annual flowers, tomato, turf. No AI included in the app based on matching 1000 high-resolution photos of plant diseases from the database.
(j) **Tumaini**: Identification of diseases of banana, like Fusarium wilt.

16.5 Methods for Insect Detection and Classification

The various methods for insect detection and classification (Fig. 16.2) are discussed below

16.5.1 Knowledge-Based System (KBS)

It is a type of computer programme which includes both logic and knowledge to solve different problems. This can be applied to a different system. All KBS algorithms share two features: an effort to formally represent knowledge and an intellectual mechanism that permits it to conclude new knowledge. In initial experiments, two knowledge-based systems (KBS) were associated to recognise and count whiteflies in traps (Boissard et al. 2008). The sites for the specimens in the first KBS were chosen using a variety of image processing techniques. In order to completely define

Methods used for insect detection and classification	**Knowledge based system (KBS)**
	Boundary tracking algorithm
	Support vector machines (SVM)
	Zernike moments
	Artificial neural networks (anns)
	Convolutional neural networks (cnns)
	Multifractal analysis
	K-means clustering
	Multilayer feed-forward neural network
	RetinaNet

Fig. 16.2 Methods used for insect detection and classification

the samples of interest, another KBS used a range of numerical signifiers; these, along with few rules, assessed whether the pest attacked or not in specific areas.

16.5.2 Boundary Tracking Algorithm

A boundary tracking method is used to extract object (region) outlines from images. When using this method, it is assumed that the image has regions and either the image is binary or the regions have already been labelled. Qiao et al. (2008) created a technique for detecting whiteflies in traps that combined straightforward thresholding with a boundary tracking algorithm. Whiteflies were discovered in crops using a variety of image processing approaches by Wang et al. (2013). They presented a method that, after thresholding, splits clustered items by applying a median filter and an opening operation. By counting the linked objects, the final estimate is obtained.

16.5.3 Support Vector Machines (SVM)

SVM is used for both classification and regression. Maharlooei et al. (2017) used colour modifications and a few image processing approaches to detect aphids on soybean crop. The leaves were manually segmented, and the bugs were separated using the hue-saturation-intensity (HSI) colour space. SVMs along with the HoG feature were utilised by Liu et al. (2016) to identify wheat c aphids in photos taken in farms. SVM is most common machine learning classifier. They are particularly well-adapted to binary classifications since they seek out the hyper plane that divides two classes. A three-layer detection method was utilised by Yao et al. (2014) to find plant hoppers in rice fields. AdaBoost classifiers were employed in the first layer, SVM classifiers in the second layer, and colour and shape characteristics in the third layer were used to filter out any false positives discovered in the first two levels.

16.5.4 Zernike Moments

The set of Zernike complete orthogonal polynomials that make up its kernel are defined over the interior of the unit disc in polar coordinate space. Zernike moments are a subclass of orthogonal complex moments. Al-Saqer and Hassan (2011) described the morphologies of the target pest (Red Palm Weevil) using Zernike moments and regional features. It was used to determine if the pest was present or missing among other insect species by using traits discovered in two distinct ways.

16.5.5 Artificial Neural Networks (ANNs)

A computational model known as an artificial neural network (ANN) includes many processing components that accept inputs and produce results in agreement with predetermined activation functions. Vakilian and Massah (2013) used ANNs in conjunction with image processing approaches to recognise beetroot armyworms (*Spodoptera exigua*) in photos taken in controlled environments. Following the Canny Edge detector's initial segmentation of the pests, seven morphological and texture features were gathered and fed into three-layer neural networks. Following that, measurements and normalisations were made for four morphological and three textural features from the obtained images. Using back propagation supervised learning and inspection data, the performance of the ANN model was successfully tested for beet armyworm recognition in pictures. Results demonstrated that the suggested approach could recognise *S. exigua* in photos of other species.

16.5.6 Convolutional Neural Networks (CNNs)

Convolutional neural networks (CNN or ConvNet) are a type of deep learning network design that learns only from data. By analysing patterns in the images, CNNs may identify items, classes, and categories in photographs. They can be very helpful for classifying audio, time-series, and signal data.

16.5.7 Multifractal Analysis

Multiple factor analysis (MFA), a factorial approach, is used to analyse tables where a group of individuals is defined by a collection of variables (qualitative and/or quantitative) grouped in groups. To identify and count whiteflies on soybean leaves (hole filling), Barbedo (2014) used mathematical morphological techniques, thresholding, and colour space transformations. The early stages of nymph development were the focus of the detection. The remaining associated parts were tallied after choosing specific-sized objects to symbolise aphids. The degree of complexity and irregularity of an object can be determined via multifractal analysis. Since it is robust to changes in image size and rotation while maintaining the majority of the image's information, this method can be effective for pest detection. This methodology was applied in two investigations. Xia et al. (2012) used two distinct thresholding methods to identify candidate blobs that most likely matched the positions of whitefly traps. The features that define the items of interest were then recovered using multifractal analysis. The background of the image was initially eliminated by Li et al. (2015) using a Mahalanobis distance, and a few candidate blobs were then chosen using multifractal analysis. For the most accurate whitefly detection in the photos, features gathered for each blob were merged with size and form rules.

16.5.8 K-means Clustering

It is algorithm for unsupervised learning. The unlabelled dataset is divided into several clusters using K-means clustering. It allows us to categorise the data into different groups and offers a workable technique for quickly and accurately determining the groups in the unlabelled dataset without the need for any training. Yao et al. (2013) looked into the effectiveness of watershed, K-means clustering, and normalised cuts (NCuts) for sorting pest specimens in traps. The most precise outcomes were obtained using NCuts with the optical flow angle as the weight function. One of the classifiers explored by Liu et al. in 2016 was K-means clustering, which has also been used to improve the output of deep learning models (Sun et al. 2018).

16.5.9 Multilayer Feed-Forward Neural Network

The multilayer feed-forward neural network (MFFNN) is a weighted, multilayered, and interconnected artificial neural network that uses activation functions to compute outcomes. It is one of the neural network types where the signal moves forward, without loops, feedback, or reverse signal movement, from the output to the hidden and input layers. A three-step procedure was utilised by Espinoza et al. (2016) to identify thrips and whiteflies in traps. 14 morphological and colour attributes were gathered using a threshold from potential locations for the objects of interest. A multilayer feed-forward neural network was used to process these features and perform the final detection. The Squeezenet architecture was used to assess the model's capability to distinguish psyllids from other objects (such as insects and detritus) with accuracy.

16.5.10 RetinaNet

RetinaNet, one of the finest one-stage object detection models, has demonstrated success with both dense and small objects. Because of this, it has become a popular model for finding objects in aerial and satellite images. RetinaNet was downscaled by Sun et al. (2018) to locate beetles in traps. RetinaNet, a single-stage deep learning detector, was lowered via feature pyramid tailoring and depth-wise separable convolution. A residual classification subnet and k-means anchor optimisation were added to further improve the detector. The technique made it possible to distinguish red turpentine beetles from five other kinds of beetles.

16.6 Future Scope of the Study

Future potential exists for biotic stress control in field crops employing artificial intelligence (AI) technologies. The use of AI technology can help in the development of sustainable agricultural practices in light of the rising global demand for food production, the difficulties caused by climate change, and the appearance of new pests and diseases. Overall, there is a great potential for the development of novel and creative solutions that can assist farmers in enhancing crop yields, cutting costs, and enhancing sustainability. Biotic stress management in field crops utilising AI technology is expected to have a bright future.

16.7 Conclusion

One of the most important elements influencing plant fitness and, consequently, food production is plant stress. There have been significant technology advancements that help plant stress science get beyond these restrictions, producing large datasets that come from the many levels of the plant defence response. Because of this, machine learning (ML) technologies are now essential for processing and interpreting data in order to precisely model plant stress responses like genetic variation, protein expression, and metabolite production. Since the various biotic stresses result in significant yield losses, the farming community can get benefit greatly from biotic stress prediction in advance. It is clear from the current analysis that efficient and precise pest forecasting models may be created for a variety of crops in order to reduce pesticide loads, economic loads, environmental contamination, and increase crop yield.

References

Ahmadi P, Muharam FM, Ahmad K, Mansor S, Seman IA (2017) Early detection of ganoderma basal stem rot of oil palms using artificial neural network spectral analysis. Plant Dis 101:1009–1016. https://doi.org/10.1094/PDIS-12-16-1699-RE

Al-Saqer SM, Hassan GM (2011) Red palm weevil (*Rynchophorus Ferrugineous, Olivier*) recognition by image processing techniques. Am J Agric Biol Sci 6:365–376

Android OS (2016) Detection of plant leaf disease using image processing approach. Int J Sci Res Publ 6:73–76

Badnakhe MR, Durbha SS, Jagarlapudi A, Gade RM (2018) Evaluation of citrus gummosis disease dynamics and predictions with weather and inversion based leaf optical model. Comput Electron Agric 155:130–141. https://doi.org/10.1016/j.compag.2018.10.009

Barbedo JGA (2014) Using digital image processing for counting whiteflies on soybean leaves. J Asia Pac Entomol 17:685–694. https://doi.org/10.1016/j.aspen.2014.06.014

Bhanumathi B, Vineeth M, Rohit N (2019) Crop yield prediction and efficient use of fertilizers. In: IEEE international conference on communication and signal processing (ICCSP), pp 769–773

Bhatia A, Chug A, Singh AP (2020) Application of extreme learning machine in plant disease prediction for highly imbalanced dataset. J Stat Manag Syst 23:1059–1068

Bhojani SH, Bhatt NJNC (2020) Wheat crop yield prediction using new activation functions in neural network. Neural Comput Appl 32(17):13941–13951

Bishop CM (2006) Pattern recognition and machine learning. Springer, Berlin

Boissard P, Martin V, Moisan SA (2008) Cognitive vision approach to early pest detection in greenhouse crops. Comput Electron Agric 62:81–93. https://doi.org/10.1016/j.compag.2007.11.009

Deutsch CA, Tewksbury JJ, Tigchelaar M, Battisti DS, Merrill SC, Huey RB, Naylor RL (2018) Increase in crop losses to insect pests in a warming climate. Science 361:916–991. https://doi.org/10.1126/science.aat3466

Duarte-Garvajalino JM, Alzate DF, Ramirez AA, Santa-Sepulveda JD, Fajardo-Rojas AE, Soto-Suárez M (2018) Evaluating late blight severity in potato crops using unmanned aerial vehicles and machine learning algorithms. Remote Sens 10:1513. https://doi.org/10.3390/rs10101513

Espinoza K, Valera DL, Torres JA, Lopez A, Molina-Aiz FD (2016) Combination of image processing and artificial neural networks as a novel approach for the identification of *Bemisia tabaci* and *Frankliniella occidentalis* on sticky traps in greenhouse agriculture. Comput Electron Agric 127:495–505. https://doi.org/10.1016/j.compag.2016.07.008

FAO (2017) The future of food and agriculture: trends and challenges. FAO, Rome

Fenu G, Malloci FM (2019) An application of machine learning technique in forecasting crop disease. In: Proceedings of the 2019 3rd international conference on big data research. Paris, pp 76–82

Feurer M, Klein A, Eggensperger K, Springenberg J, Blum M, Hutter F (2015) Efficient and robust automated machine learning. Adv Neural Inf Process Syst 28:2962–2970

Food and Agriculture Organization of the United Nations FAOSTAT Pesticides Use (2022) https://www.fao.org/faostat/en/#data/RP/visualize. Accessed 10 Jan 2022

Francl L (2001) The disease triangle: a plant pathological paradigm revisited. Plant Health Instr

Frona D, Szenderak J, Harangi-Rakos M (2019) The challenge of feeding the world. Sustainability 11:5816. https://doi.org/10.3390/su11205816

Goodfellow I, Bengio Y, Courville A (2016) Deep learning. MIT Press

Kamilaris A, Kartakoullis A, Prenafeta-Boldú FX (2017) A review on the practice of big data analysis in agriculture. Comput Electron Agric 143:23–37. https://doi.org/10.1007/978-981-15-2414-1_66

Kartikeyan P, Shrivastava G (2021) Review on emerging trends in detection of plant diseases using image processing with machine learning. Int J Comput Appl 174:39–48

Kaundal R, Kapoor AS, Raghava GPS (2006) Machine learning technique in disease forecasting: a case study on rice blast prediction. BMC Bioinform 7:1–16. https://doi.org/10.1186/1471-2105-7-485

Kaur K, Kaur M (2018) Prediction of plant disease from weather forecasting using data mining. Int J Future Revolut Comput Sci Commun Eng 4:685–688

Kim Y, Roh JH, Kim HY (2018) Early forecasting of rice blast disease using long short-term memory recurrent neural networks. Sustainability 10:1–20. https://doi.org/10.3390/su10010034

Kotthoff L, Thornton C, Hoos HH, Hutter F, Leyton-Brown K (2017) Auto-WEKA 2.0: automatic model selection and hyperparameter optimization in WEKA. J Mach Learn Res 18:81–95. https://doi.org/10.1007/978-3-030-05318-5_4

Kulkarni AH, Patil ARK (2012) Applying image processing technique to detect plant diseases. Int J Mod Eng Res 2:3661–3664

Lee M, Xing SA (2018) Study of tangerine pest recognition using advanced deep learning methods (preprints). https://doi.org/10.20944/preprints2 01811 .0161.v1

Lee SH, Chan CS, Wilkin P, Remagnino P (2017) Deep-plant: plant identification with convolutional neural networks. In: Proceedings of IEEE international conference on image processing (ICIP). Quebec City, QC, pp 452–456

Li Y, Xia C, Lee J (2015) Detection of small-sized insect pest in greenhouses based on multifractal analysis. Optik Int J Light Electron Opt 126:2138–2143. https://doi.org/10.1016/j.ijleo.2015.05.096

Liu T, Chen W, Wu W, Sun C, Guo W, Zhu X (2016) Detection of aphids in wheat fields using a computer vision technique. Biosyst Eng 141:82–93. https://doi.org/10.1016/j.biosystemseng.2015.11.005

Lu W, Newlandsm NK, Carisse O, Atkinson DE, Cannon AJ (2020) Disease risk forecasting with bayesian learning networks: application to grape powdery mildew (*Erysiphe necator*) in vineyards. Agronomy 10:2–29. https://doi.org/10.3390/agronomy10050622

Maharlooei M, Sivarajan S, Bajwa SG, Harmon JP, Nowatzki J (2017) Detection of soybean aphids in a greenhouse using an image processing technique. Comput Electron Agric 132:63–70. https://doi.org/10.1016/j.compag.2016.11.019

Muthukannan K, Latha P, Selvi P, Nisha P (2015) Diseased plant leaves using neural network algorithms. J Eng Appl Sci 10:6619–6628

Newbery F, Qi A, Fitt BD (2016) Modelling impacts of climate change on arable crop diseases: progress, challenges and applications. Curr Opin Plant Biol 32:101–109. https://doi.org/10.1016/j.pbi.2016.07.002

Pandey K, Kumar S, Malik A, Kuriqi A (2020) Artificial neural network optimized with a genetic algorithm for seasonal groundwater table depth prediction in Uttar Pradesh, India. Sustainability 12(21):8932

Pandey K, Malik A (2021) Seasonal groundwater table depth prediction using fuzzy logic and artificial neural network in Gangetic plain, India. In: Advanced modelling and innovations in water resources engineering, lecture notes in civil engineering, 1st edn, Chap 37. Springer, Singapore

Qiao M, Lim J, Ji CW, Chung BK, Kim HY, Uhm KB, Myung CS, Cho J, Chon TS (2008) Density estimation of *Bemisia tabaci* (Hemiptera: Aleyrodidae) in a greenhouse using sticky traps in conjunction with an image processing system. J Asia Pac Entomol 11:25–29. https://doi.org/10.1016/j.aspen.2008.03.002

Rajpurkar P, Irvin J, Zhu K, Yang B, Mehta H, Duan T, Ding D, Bagul A, Langlotz C, Shpanskaya K, Lungren MP, Ng AY (2017) Chexnet: radiologist-level pneumonia detection on chest X-rays with deep learning. arXiv:1711.05225

Rastogi A, Arora R, Sharma S (2015) Leaf disease detection and grading using computer vision technology & fuzzy logic 2nd international conference on signal processing and integrated networks (SPIN)

Ray DK, Mueller ND, West PC, Foley JA (2013) Yield trends are insufficient to double global crop production by 2050. PLoS ONE 8:1–8. https://doi.org/10.1371/journal.pone.0066428

Rehman TU, Mahmud MS, Chang YK, Jin J, Shin J (2019) Current and future applications of statistical machine learning algorithms for agricultural machine vision systems. Comput Electron Agric 156:585–605

Rodriguez-Moreno VM, Jimenez-Lagunes A, Estrada-Avalos J, Mauricio-Ruvalcaba JE, Padilla-Ramirez JS (2020) Weather data based model: an approach for forecasting leaf and stripe rust on winter wheat. Meteorol Appl 27:e1896. https://doi.org/10.1002/met.1896

Roser M (2013) Future population growth. In: Our world in data. University of Oxford, Oxford

Rothe PR, Kshirsagar RV (2015) Cotton leaf disease identification using pattern recognition techniques. In: International conference on pervasive computing (ICPC)

Rumpf T, Mahlein AK, Steiner U, Oerke EC, Dehne HW, Plumer L (2010) Early detection and classification of plant diseases with support vector machines based on hyperspectral reflectance. Comput Electron Agric 74:91–99. https://doi.org/10.1016/j.compag.2010.06.009

Sannakki SS, Rajpurohit VS (2015) Classification of pomegranate diseases based on back propagation neural network. In: International research journal of engineering and technology (IRJET), vol 2, issue 02

Sannakki S, Rajpurohit V, Sumira F, Venkatesh H (2013) A neural network approach for disease forecasting in grapes using weather parameters. In: Proceedings of the 2013 fourth international conference on computing, communications and networking technologies (ICCCNT), Tiruchengode, India, pp 1–5

Stevens R (1960) An advanced treatise. Plant Pathol 3:357–429

Sun Y, Liu X, Yuan M, Ren L, Wang J, Chen Z (2018) Automatic in-trap pest detection using deep learning for pheromone-based *Dendroctonus valens* monitoring. Biosyst Eng 176:140–150. https://doi.org/10.1016/j.biosystemseng.2018.10.012

Tian U, Chunjiang Z, henglian L, Xinyu G (2010) SVM-based multiple classifier system for recognition of wheat leaf diseases. In: Proceedings of 2010 conference on dependable computing (CDC'2010)

Truong T, Dinh A, Khan W (2017) An IoT environmental data collection system for fungal detection in crop fields. In: Proceedings of IEEE 30th Canadian conference on electrical and computer engineering (CCECE)

Vakilian KA, Massah J (2013) Performance evaluation of a machine vision system for insect pests identification of field crops using artificial neural networks. Arch Phytopathol Plant Prot 46:1262–1269

Verma S, Chug A, Singh AP (2018) Prediction models for identification and diagnosis of tomato plant diseases, pp 1557–1563

Wang H, Ma Z (2011) Prediction of wheat stripe rust based on support vector machine. In: Proceedings of 7th international conference natural computation, pp 378–382

Wang K, Zhang S, Wang Z, Liu Z, Yang F (2013) Mobile smart device-based vegetable disease and insect pest recognition method. Intell Autom Soft Comput 19:263–273. https://doi.org/10.1080/10798587.2013.823783

Wang L, Leem CY, Tu Z, Lazebnik S (2015) Training deeper convolutional networks with deep supervision. arXiv 2015, arXiv:1505.02496

Xia C, Lee JM, Li Y, Chung BK, Chon TS (2012) In situ detection of small-size insect pests sampled on traps using multifractal analysis. Opt Eng 51:1–12

Yao Q, Liu Q, Dietterich TG, Todorovic S, Lin J, Diao G, Yang B, Tang J (2013) Segmentation of touching insects based on optical flow and NCuts. Biosyst Eng 114:67–77. https://doi.org/10.1016/j.biosystemseng.2012.11.008

Yao Q, Xian DX, Liu QJ, Yang BJ, Diao GQ, Tang J (2014) Automated counting of rice planthoppers in paddy fields based on image processing. J Integr Agric 13:1736–1745

Yeh YH, Chung WC, Liao JY, Chung CL, Kuo YF, Lin TT (2016) Strawberry foliar anthracnose assessment by hyperspectral imaging. Comput Electron Agric 122:1–9. https://doi.org/10.1016/j.compag.2016.01.012

Zhang S, Chen X, Wang S (2014) Research on the monitoring system of wheat diseases, pests and weeds based on lOT. In: Proceedings of 9th international conference on computer science & education (ICCSE 2014), pp 981–985

Zhao J, Xu C, Xu J, Huang L, Zhang D, Liang D (2018) Forecasting the wheat powdery mildew (*Blumeria graminis* f. Sp. tritici) using a remote sensing-based decision-tree classification at a provincial scale. Australas Plant Pathol 47:53–61

Chapter 17
Soilless Cultivation: A Distinct Vision for Sustainable Agriculture

Alka Chaudhary and Shubham Anand

Abstract In agriculture, the declining availability of cultivated land combined with an increase adverse climatic conditions has resulted in a considerable number of crops, particularly vegetables, being grown using protected soilless farming techniques. Any method that allows plants to be grown without using soil as a rooting medium and supply of vital nutrients through irrigation water is said to be soilless cultivation. Soilless farming provides advanced approaches for sustainable agricultural production, soil protection, and other high-tech industrial techniques. Furthermore, in order to supply people with sufficient and high-quality food, the modern plant cultivation practices are now being used which include hydroponics, aeroponics, and aquaponics. These contemporary methods of cultivating crops usually combine nutrient solutions with inert organic or inorganic growing substrates to escalate the yield and production of crops. Therefore, in case of hydroponics, all the essential nutrients are provided via various mediums such as rockwool, hydroton, and perlite for various crops like leaf lettuce, tomatoes, peppers, cucumbers, etc. While aeroponics is a variation of hydroponics, plants are suspended, with roots being intermittently sprayed with a nutrient solution by an automated sprinkler system connected to the main nutrient reservoir, with efficiency and precision. Lot of vegetables and fruits like beets, broccoli, cabbage, carrots, potato, tomatoes, and watermelon are cultivated by this method. This approach is used to cultivate a variety of vegetables and fruits, including beets, broccoli, cabbage, carrots, potatoes, tomatoes, and watermelon. Moreover, the merging of fish farming and hydroponics is known as aquaponics, which is regarded as an effective and sustainable technique of growing cannabis for tomatoes, cucumbers, capsicum, red salad onions, Chinese cabbage, lettuce, bell peppers, and snow peas. Thus, in this chapter, we will discuss distinct soilless culture methods that offer advanced method of farming crops in comparison to conventional soil culture.

A. Chaudhary
Department of Plant Pathology, Punjab Agricultural University, Ludhiana, Punjab 141004, India

S. Anand (✉)
Department of Climate Change and Agricultural Meteorology, Punjab Agricultural University, Ludhiana, Punjab 141004, India
e-mail: shubham-1904001@pau.edu

K. Pandey et al. (eds.), *Artificial Intelligence and Smart Agriculture*,
Advances in Geographical and Environmental Sciences,
https://doi.org/10.1007/978-981-97-0341-8_17

Keywords Soilless farming · Hydroponics · Aeroponics · Aquaponics · Nutrient medium · Crops

17.1 Introduction

Soil, a natural resource, has been scorched by human civilization. Because of the rapid increase in human population, per capita land available for agricultural uses has reduced. Less rainfall due to climate change and excessive use of groundwater have alleviated the problem. In addition to this, soil limits plant growth by harbouring nematodes and disease-causing organisms, as well as reducing crop productivity due to poor drainage and erosion (Singh et al. 2017; Pandey et al. 2023; Sharda et al. 2023). Therefore, soilless plant cultivation appears to be a viable alternative to address these challenges. Soilless cultivation, where plants are grown without soil, is an important thrust area in agricultural sciences that is becoming more relevant in the current environment. Due to its effective resource management, hydroponics, aeroponics, and aquaponics have emerged a viable solution that is currently gaining popularity throughout the world (Sundari et al. 2022). Hydroponics plantation is an advanced form of agriculture which enables the option of having complete control over the distribution and delivery of nutrition among the plants, whereas aeroponics is a kind of hydroponics in which water is employed as a nutrient-dissolving media. Aquaponics, on the other hand, is a sustainable farming technique that blends hydroponic growing with a recirculating aquaculture system (RAS) (AlShrouf 2017). Compared to conventional agriculture, soilless farming produces the majority of successful vegetable crops with the highest productivity. It can be operated in a variety of locations, including rooftops, greenhouses, and inaccessible farmland. Besides this, even while the constituent unit cost of soilless cultivation increased, the problem of power interruption, high risk of water-borne diseases and proper monitoring and maintenance etc., are some of the downsides of soilless system (Serio et al. 2022). Thus, seeing towards these prospects, there is a need to popularize soilless method by providing cutting-edge alternatives for large-scale food production to farmers and spread awareness in potential areas globally (Fussy and papenbrock 2022).

17.2 Soilless Cultivation System

The concept of soilless culture, which is man-made, is to give plants support and a place to store water and nutrients. According to Savvas et al. (2013), "any method of growing plants that does not need soil as a rooting medium, provides the inorganic nutrients absorbed by the roots through irrigation water". The earliest form of soilless farming, which is frequently referred to as solution culture or water culture, involves melting inorganic substances in a water-filled container (nutrient solution) to deliver the plants with all the nutrients they require. The fundamental objective of soilless cultivation is to encourage plant development while controlling dissolved oxygen, mineral salts, and water levels. Likewise, when roots are suspended in circulating water, they quickly assimilate nutrients and oxygen. Plant growth will be delayed if there is inadequate oxygen. On the other hand, plant development will accelerate, if the fluid is saturated with oxygen. Thus, in order to maximize output and quality, the grower must balance the needs with the availability of water, nutrients, and oxygen. For the optimum results, a few key factors must be considered, including temperature, light intensity, ventilation, CO_2 and humidity levels, pH, and the genetic makeup of the plant. Agriculture carried out without the use of soil includes hydroponics, aeroponics, aquaponics, as well as agriculture employing supporting mediators. However, plants do not require soil in soilless culture; instead, they require a supply of the minerals: Nitrogen (N), Calcium (Ca), Potassium (K), Magnesium (Mg), Copper (Cu), Phosphorous (P), Iron (Fe), Sulphur (S), Zinc (Zn), Manganese (Mn), Molybdenum (Mo), and others.

17.3 Classification of Soilless Cultivation System

There are primarily two sorts of soilless culture methods, i.e. open soilless culture and closed soilless culture.

17.3.1 Open Soilless Culture System

In open soilless frameworks, a fresh dissolved supplements are added for each watering cycle (Pradhan and Deo 2019). The dripping system is typically used to deliver the dissolved supplements to the plants. To maintain nutrient balance in the root zone in open soilless structures, an adequate amount of nutrients is synchronized in the root zone. All soilless farming only uses substrates and drip frameworks, and they can all be found in open soilless cultures. A few open soilless culture techniques are root dipping, hanging bag, and trench system, which are illustrated below (Fig. 17.1).

17.3.1.1 Root Dipping System

Under this process, plants are cultured in pots that have tiny holes at the bottom. These are placed in a nutritional solution-filled container containing substrate material, such as coconut fibre. A minimum of 1–3 cm of the pots' lower portion remains in direct contact with the nutritional media. Only a few roots are partially immersed in the growing medium, while others float in the air. Hence, this is a simple and economical method of growing small herbs or blooming plants.

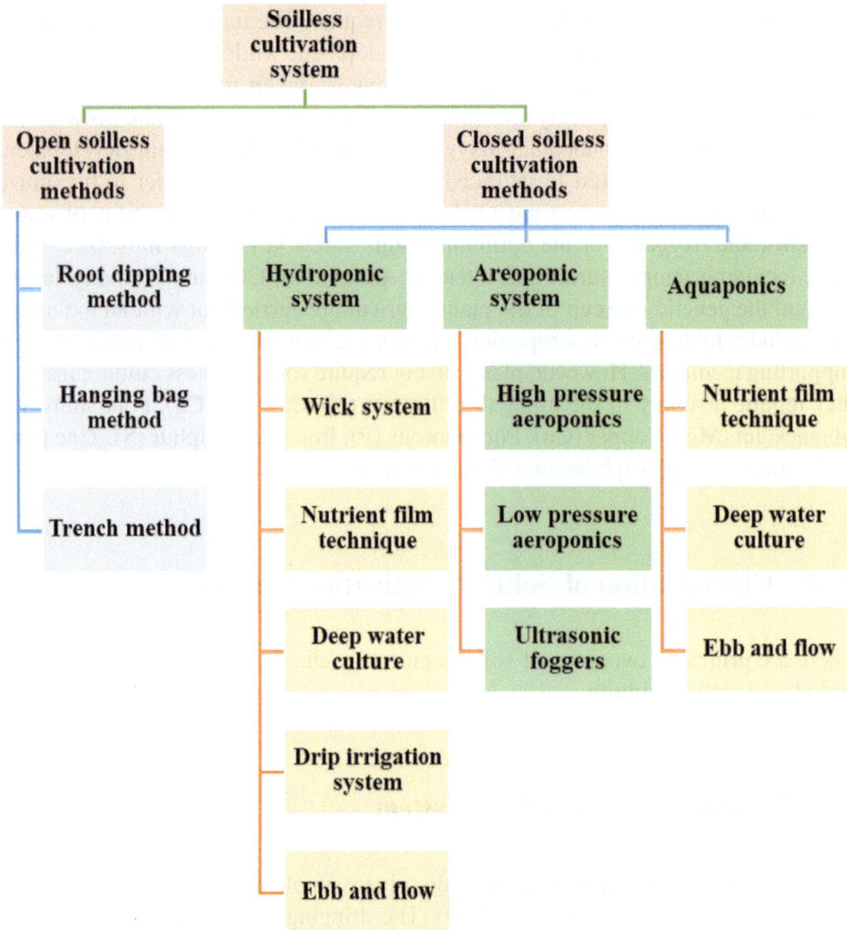

Fig. 17.1 Different categories of soilless cultivation methods

17.3.1.2 Hanging Bag System

This method makes use of polythene bags in the form of long cylinders. These have polyvinyl chloride (PVC) pipes attached to them at the upper end and are closed at the lower end. These are suspended vertically over a tank that holds vitamin supplements. Planting materials, such as seeds and fruits, are firmly pressed into the hanging bags hole after being acclimatized in netted pots. Apart from this, a microsprinkler is used to circulate the nutritional medium to the top of each hanging bag. The sprinkler evenly distributes the fertilizer solution throughout the hanging bag. Excess nutrient solution is collected in the solution stock tank, which is located at the bag's bottom. To prevent mould growth inside, nutrient solution tubes should be black in colour.

Using this approach, veggies like climbers, lettuce, small flowering plants, etc., can be cultivated.

17.3.1.3 Trench Method

Small herbs and shrubs are grown in trenches made of concrete blocks or bricks that are either above ground or below ground. Trenches' interior linings are lined with thick polythene sheets to keep the growth media from coming into direct touch with the earth. The dripping procedures deliver all nutrient additives as well as water. As a result, this technique is ideal for producing vine plants and herbs.

17.3.2 Closed Soilless Cultivation System

The dissolved supplements are recycled in closed soilless framework, and the supplement concentrations are similarly monitored and balanced. It is challenging to maintain supplement balance in such systems, and the dissolved supplements should be checked and dissected on a frequent basis. Moreover, results must balance the dissolved supplements. The dissolved supplements may escape the balance if not properly controlled. Both fundamental and sophisticated soilless culture frameworks are involved in closed soilless frameworks. Therefore, the diluted nutrient concentrations are identified and balanced for reuse using this procedure. Closed agricultural concepts involve both fundamental as well as contemporary frameworks. Few approaches of closed soilless culturing are mentioned underneath.

17.3.2.1 Hydro Agriculture (Hydroponics)

Hydroponics also known as "water working" is the process of growing plants without soil. Although soil is not necessary for plants, but they do require the vitamins and minerals that soil may offer. At the root zone, plants also require light, water, light, oxygen, and carbon dioxide. Under hydroponics, plants are cultivated on an inert substrate like pebbles or coco coir fibre and fed a prescribed solution containing a formulated combination of primary, secondary, and micronutrients (El-Kazzaz and El-Kazzaz 2017). As a result of its effective resource management, high-quality food output, and ability to grow almost any kind of plant, including fruits, vegetables, herbs, and flowers, hydroponic farming is also gaining popularity. Today's challenges for soil-based agriculture include urbanization, natural disasters, climate change, and the widespread use of pesticides and herbicides, which are reducing soil fertility and crop performance in cucumber, tomato, pepper, leafy greens, etc. As a consequence, wick, deep water culture (DWC), nutritional film technique (NFT) system, drip irrigation, and ebb and flow have all been covered in hydroponic practices (Sharma et al. 2018a).

Furthermore, hydroponic farming offers several advantages over traditional farming, such as shorter crop development cycles, year-round output, minimal disease and insect incidence, and the absence of chores like weeding, spraying, and watering. The NFT method has been used effectively in the commercial production of leafy and other vegetables all over the world, resulting in water reductions of 70–90%. The Netherlands, France, Canada, England, Australia, Israel, and the USA are among the world leaders in hydroponic technology. Commercial hydroponic technology must be effectively deployed, and thus, low-cost solutions that are easy to use and maintain, require less labour overall, and have reduced setup and maintenance expenses are vital.

17.3.2.2 Hydroponic Structures and Their Operation

In accordance with the recycling and reuse of nutrient solution and supporting medium, hydroponic systems are customized and altered. Thus, following are the most common types of hydroponics which are wick, nutrient film technology (NFT), drip, deep water culture, and ebb–flow.

Wick System

The wick, on the other hand, connects the container plant to the food solution in the reservoir. A nylon wick connects the plant roots to a reservoir of nutritional solution that is deposited in absorbent medium like coco coir, vermiculite, or vermiculite. Water or fertilizer solutions are transported to plants by capillary action. This approach works well for herbs and spices, but it is ineffective for plants that require a lot of water.

Nutrient Film Technique (NFT)

A recirculated design is used in the nutrition film technique to continuously distribute highly oxygenated dissolved nutrients over plant roots via a network of tubes. Plants used in this technique are commonly grown in baskets suspended in polyvinyl chloride (PVC) pipes. The nutrition solution is recycled because it is fed from a holding tank at the top of each sloping pipe and the runoff from the bottom of the channels is pumped back into the tank by irrigators. It is feasible to reduce the pipe's angle and include an overflow pipe, much like in an ebb and flow system. This would function as a nutrient reserve in the case of a power outage or pump failure. The nutrition film strategy is ideally suited to plants with small root balls, such as herbs, strawberries, and lettuce, due to the restricted space of a PVC pipe and necessity for nutrients to flow continually over their roots.

Water Culture or Deep Water Culture (DWC)

The most basic form of hydroponics system is water culture, often known as deep water culture. In deep water cultivation, an air stone delivers air directly to the plant roots, which are suspended in nutrient-rich water. Using a float platform, roots are floating in a hydroponic mineral solution and an air pump that works constantly provides oxygen directly to the plant roots. Glass basins (fish ponds), concrete basins, ice and plastic boxes coated in polypropylene sheets can all be used to build up a water culture system. There is no risk of plant injury if an air pump fails since the plants are floating and constantly in contact with the nourishing fluid. Seaweeds and fungus are well quickly thrive in the reservoir, and thus, nutrient concentrations, oxygen, salinity, and pH must all be monitored (Domingues et al. 2012). The bigger fruit-producing plants, notably the cucumber and tomato, flourish in this type of arrangement.

Drip System

Flow irrigation, often known as trickling irrigation, is a sort of microirrigation system that allows water to flow gently to plant roots, either above or below the soil surface. In this, at least two containers are used in a drip hydroponic system with one container being higher than the other. The fertilizer solution is in the bottom container, while the plants are in the top container. To carry the nutritional solution up to drips on each plant's stem, a water pump is employed, and an aquarium air stone is used to oxygenate the water. The nutrients pass via the plant roots and return to the container's bottom. With this kind of system, the water and air pumps often run continuously; thus, several crops can be cultivated like cotton, maize, tomatoes, citrus, sugarcane, eggplant, strawberries, etc., due to less water scarcity. The drip hydroponic method is often used by both commercial and home growers.

Ebb and Flow System

The ebb and flow system was the first commercial, low-cost hydroponic system to employ the flood and drain approach. The setup is similar to a drip system in that there are two containers: one at the bottom carries the nutritional solution, while the other holds the plants in pots with substrate on top. The nutrients are injected in enormous quantities into the top container, drowning it, rather than being given gradually to drippers at the stems of each plant. By pumping surplus liquid through the overflow pipe back to the bottom container, an overflow pipe defines the height of the nutrients, generally at the base of the stem. The pump in an ebb and flow system is switched on and off intermittently (perhaps for 30 min on, 15 min off), periodically flooding the grow tray (Kumar et al. 2018). When the pump is switched off, all of the nutrients are drained out from the grow tray via the pump line. An air stone is not strictly necessary for ebb and flow systems because the emptying phase lets oxygen to reach the roots. Additionally, it is feasible to grow any type of plant; however, root rot, algae, and mould problems are frequent, and thus, a modified system with a filtering unit is needed. Large root balls in plants make them excellent candidates for off-and-flow systems (Table 17.1).

17.3.2.3 Aeroponics

Aeroponics is a relatively newer mode of producing plants that is acquiring popularity among people due to its velocity, low price, and innovation. It is a promising soilless farming approach for overcoming future food crises. Vertical farming and hydroponics are both used in aeroponic farming. With this farming technique, the producer has entire access to the roots throughout the duration of the crop's life and had precise control over the environmental, nutritional, and hydration conditions in the root zone. An aeroponic farming seems to be commercially superior to conventional approaches of propagation, in terms of good aeration, water usage effectiveness, time and space requirements, seasonal independence, disease-free plant propagation, large-scale plant production of seed potato and tomato, etc. (Kumari et al. 2019).

Working Mechanism of an Aeroponic System

The aeroponics system, which is connected to closed soilless culture systems, is the most technologically sophisticated form of hydroponics system. According to El-Kazzaz and El-Kazzaz (2017), in this method, the roots are suspended in the air beneath the reservoir lid, and a stressful pump mists the space around the root with nutritional solution as well as, plants above the reservoir cover serve as nutritional solution reservoirs in the form of sealed root chambers (polystyrene oar other material). A system of pipes, spray nozzles, a pump, and a timer must be used to spray fertilizer solution from a storage tank. It employs a tiny internal microjet spray to provide nutrient-rich liquids to the roots in the rooting chamber through a thin, high-pressure mist. The high-pressure aeroponic pump is activated by an electronically controlled cyclic timer, and water and fertilizers are combined in a reservoir basin. The liquid is filtered before being placed in a pressurized storage tank, where it is sporadically misted over the root system. The misting is usually done around the dangling roots every few minutes. The mechanism often only has worked for a brief period every two to three minutes. Sometimes, roots can be quickly dehydrated if the sprinkling system is stopped or disrupted since they are left exposed to the air. To be functional, the aeroponic setup must have a reliable monitoring and control system for the distribution of nutrients and water in order to be effective. The nutrient pump, like in standard hydroponic systems, is controlled by a timer; however, an aeroponics system needs a brief cycle timer to run the pump briefly every few minutes. Root hair development facilitates the uptake of nutrients from moisture.

Table 17.1 Brief characteristics of hydroponic system

Methods	Components	Pros	Cons	Crops	References
Wick system	Grow tray, wick, reservoir, and aeration system	Easy to set up and use Limited water requirement and less space Reliable even during power outages	Can be less efficient Wicks can become saturated with water, leading to inadequate plant growth Require periodic monitoring and adjustment	Fast growing herbs (rosemary, peppers) or lettuce	Dubey et al. (2020)
Nutrient film technique (NFT)	Channels, net pots and growing media (perlite, coconut, rockwool)	Simple to inspect roots for a sign of disease and feed sufficiency as well as poor water and nutrient intake Environmentally friendly Easy to disinfect plant roots and can maintain a consistent pH and conductivity	If supply of nutritional solution halt, the roots will become withered and stressed Roots of vigorously growing plants can choke channels Not suitable with large tap root system like carrot Saline water is not the ideal option because salinity of the recirculating water will progressively decrease	Popular varieties of lettuce-ostinata, ruby sky, Vulcan, Rex, etc. Strawberries and herbs Roots of vigorously growing plants can clog channels	Alipio and Cruz (2019)

(continued)

Table 17.1 (continued)

Methods	Components	Pros	Cons	Crops	References
Deep water culture (DWC)	Air pump, air stone, net pots, and airline tubing	Accelerated growth due to superior uptake of nutrients and oxygen No need for much fertilizer Root aeration enhances plant uptake and cell proliferation Very little moving parts and it requires little maintenance	If the air pump does not work or broken, the plants will rapidly decline Nutrient concentration, water level, and pH may fluctuate Risk of the plant roots drowning in low oxygen nutritional solution during power outage	Shallow roots plants like radish, spinach, basil, okra, kale, cabbage, collard greens, sorrel chard, etc.	Guruchandran et al. (2022) Accelerated development as a result of improved food and oxygen intake

(continued)

Table 17.1 (continued)

Methods	Components	Pros	Cons	Crops	References
Drip irrigation system	Water delivery system, water filters, water pressure regulators, gauges, valves, chemical injectors, and system controller	Maximum use of water As water is administered locally and leaching is reduced, fertilizer loss is decreased Fertilizers can be used with high efficiency Energy cost is reduced as it is operated in lower pressure than other irrigation method	Initial cost is high and sometimes clogging occurs if water is not filtered correctly High skills are required	Orchard crops grapes, banana, pomegranate, watermelon, pineapple Vegetable plants tomato, chilly, capsicum, cabbage Cash crops sugarcane, cotton, areca nut Flower plant—rose, jasmine, marigold Plantation crops—rubber, coffee, coconut, spices, Oilseed crops—sunflower, oil palm, groundnut, forest crops teakwood, bamboo	Yildirim et al. (2016)
Ebb and flow system	Plant tray, growing container, overflow tube, nutrient solution reservoir, submersible pump	Low cost and easy to operate Offers abundance of nutrients Beneficial for plants that poorly thrive in traditional hydroponic systems, such as tomatoes, cucumbers	Fluctuation of pH levels Sensitive to power failure Take quite a bit of trial and error to identify optimum timings for fertilizers The entire system is dependent on the water pump to fill the grow tray	Mini cocktail cucumbers Lettuce-salanova green butter, green frilled tomatoes	Sharma et al. (2018b)

Furthermore, the root system makes it easier to feed the plant with different nutrients and a healthy supply of oxygen completely prevents root rot. Since the spray particles are so small, very little nutritional solution waste is generated. The chamber must be completely lightless to keep the roots functionally dark and minimize algae development, which obstructs plant growth and pollutes the system (Kumar et al. 2019). A large droplet of water may dramatically reduce oxygen levels, whereas an undersized droplet may diminish the efficacy of an aeroponic system by encouraging the formation of root hairs while inhibiting the growth of lateral roots. Since the spray particles are so small, very little nutritional solution waste is generated. Moreover, the foundation of aeroponics is the possibility of growing vegetables in containers filled with flowing plant nutrients without the need for a substratum or soil (Mangaiyarkarasi et al. 2020). The best conditions for moisture and oxygenation can be found in these containers for roots. Thus, aeroponics approach makes it easy to monitor pH and nutrient levels because of little to no contact between the plant and the support system. The farmed plants will develop more swiftly because under these conditions, nutrient absorption is higher and more balanced.

Types of Aeroponics Are Depicted Below

High-Pressure Areoponics

Typically, they are made up of a two-sided tank with an elastic separator, air, and the supplement arrangement on either side of it. An air compressor pressurizes the tank and water line from the supplement arrangement side that rushes to the clouding heads, while a solenoid uses a cycle clock to precisely open and reopen a valve in the water line. In the actual aeroponics system, high pressure is used (60–90 psi) (Savvas et al. 2013). This technique makes use of pressured nutrient delivery systems, low-mass polymers, nutrient sterilization, and air and water purification technologies.

Low-Pressure Aeroponics

The second framework, which is generally referred to as soakaponics, is the low-pressure aeroponics frameworks (El-Behairy et al. 2015). Low-pressure systems employ standard submersible water pumps, but also demand a sizable volume of water weight to prevent the sprinkler heads from just gushing water. In the majority of low-pressure aeroponic gardens, the plant roots are usually hanging above a reservoir of nutritional solution or a conduit that is within and connected to a reservoir. A low-pressure pump disperses the nutrient solution using jets or ultrasonic transducers to drain the nutrients back into the reservoir. Moreover, dry patches form in their root systems of mature plants, preventing them from properly absorbing nutrients. These machines lack qualities that would clean the nutrition solution and get rid of unwanted microorganisms, hence suitable for benchtop cultivation.

Table 17.2 Nutrient concentration used in aeroponics

Nutrient	Concentration (g/L)
N-NH_4 (ammonium)	0.54
N-NO_3 (nitrate)	0.35
P (phosphorous)	0.40
K (potassium)	0.35
Ca (calcium)	0.17
Mg (magnesium)	0.08
Na (sodium)	0.04

Ultrasonic Foggers

The third framework uses aeroponic frames to generate fog with a small quantity of water beading, but there is scarcely any actual moisture. The produced fog from the ultrasonic foggers has a tendency to sink to the holder's base, ensuring that the roots are constantly fully covered by the fog (El-Kazzaz and El-Kazzaz 2017).

Nutrients' Composition

The primary nutrients used in aeroponics (Table 17.2) are shown underneath (Sahoo 2020).

Components of Aeroponics System

Spray Misters

In spray misters, water is atomized by forcing it under high pressure via nozzles and there are distinct spray patterns and orifice sizes for nozzles. Larger nozzles and orifices are less prone to clogging even if they operate under pressure and have high flow rates. The size of a spray's droplets can vary from 100 to 1000 of microns (Chang et al. 2011).

Droplet Size

A droplet size between 20 and 100 μm is suitable for most plant species. The minuscule droplets fill the air in this region, keeping the humidity levels in the growth chamber constant. The larger droplets, measuring 30–100 μm in size, make the substantial association with the roots. Spray droplets with a diameter under 30 microns have a propensity to persist as fog (Mangaiyarkarasi et al. 2020). The majority of droplets larger than 100 μm, however, fall to the ground before they can support any roots. However, if a water droplet is determined to be excessively big, there is less oxygen accessible to the root system.

High-Pressure Water Pump

High-pressure aeroponics necessitates the use of a pump capable of pressurizing water to droplet sizes ranging from 20 to 50 μm. These pumps are frequently reverse osmosis booster pumps or diaphragm pumps and must consistently deliver 80 PSI with the necessary fertilizer flow (Buckseth and Sharma 2016).

Light and Temperature

It is fundamental to replace solar lighting with fluorescent lights with the requisite intensity of 15,000–20,000 lx as it is required for vegetative development. Fruit and flowers require between 35,000 and 40,000 lx. Along with that, 15–25 °C is the ideal temperature range for all plants (Oztazu et al. 2014).

Misting Frequency and Nutrient Reservoir

In this kind of growing system, there is no chance of water logging or an oxygen deficit because aeroponic systems can mist the root system constantly or occasionally, and both techniques are beneficial (Walters et al. 2015). Intermittent aeroponics systems have lower running expenses because the pump is only function for a brief time of period. Between mistings, the roots are nevertheless kept in a nutrient-moisture and oxygen-rich environment. Under typical conditions, a 1- to 2-min spray cycle will guarantee that the root system does not dry up.

Aeroponically Cultivated Crops

Chives, oregano, basil, sage, and rosemary are some of the herbs that can be cultivated using this approach. Other plants that can be cultivated effectively in aeroponics gardens or on farmlands include mints, skullcap, stinging nettles, ginger, Tuscan kale, red leaf lettuce, romaine lettuce, butterhead lettuce, yerba mansa, etc. (Kaldate et al. 2021). Apart from it, numerous fruits and vegetables, including cabbage, beets, broccoli, cauliflower, carrots, corn, cucumber, carrots, sweet potato, melons, onions, eggplant, grapes, tomatoes, potatoes, peas, peppers, radish, strawberry, raspberry, and watermelon, can be successfully raised by using aeroponic method.

17.3.2.4 Aqua-agriculture (Aquaponics)

A biointegrated and environmentally friendly method of growing food that blends recirculating aquaculture and hydroponic gardening is aquaponics. Unlike standard hydroponic systems, which rely on mineral fertilizers to provide plants with the nutrients they require, aquaponic systems employ readily accessible fish water that is rich in fish waste as nutrients for plant development. The system encourages the symbiotic relationship of aquatic species such as fish, bacteria, and plants, as well as the sustainable use of water and nutrients, including their recycling. In this synergistic interaction, the diverse ecological disadvantages of aquaculture and hydroponics are transformed into advantages. To develop properly, plants require macronutrients such as C, H, O, N, P, K, Ca, and Mg, as well as micronutrients such as Fe, Mn, Cl, B, Zn, Cu, and Mo. With the exception of C, H, and O, which are derived from air and water, all of these elements are present in hydroponic solutions in predetermined ratios and are added in the form of ions (Resh et al. 2022). In aquaponics systems, organic compounds from fish waste (gill discharge, urine, and faeces), which are both soluble and solid, are converted into nutrient-rich ionic form in the water and supplied to plants as a source of nutrition (Somerville et al. 2014). Additionally monitoring the concentrations of essential nutrients and trace elements is necessary to encourage proper plant growth. Various microorganism communities involved in the processing and solubilization of fish waste must be able to support in aquaponic systems.

Nitrogen Cycle

Nitrogen is a crucial component of plant development, and it is abundant in a well-maintained aquaponics system. The most essential biological process in aquaponics is nitrification, a vital stage in the nitrogen cycle. Nitrogen (N), which is a component of all amino acids and an essential inorganic nutrient for all plants, but still plants cannot use it because it only exists as molecular form in the atmosphere (N_2). As a result, nitrogen must first be transformed from its N_2 form before plants can use it for growth. This process is known as nitrogen fixation, and it is carried out by bacterial species that chemically convert nitrogen gas (N_2) by adding additional elements (such as oxygen or hydrogen) to create new chemical compounds such as ammonia (NH_3) and nitrate (NO_3^-) (Somerville and Cohen 2014). The ammonia is broken down by a type of bacterium known as nitrifying bacteria, which is crucial for aquaponics. The ammonia produced by these bacteria is first transformed into nitrite (NO_2^-) compounds, then into nitrate compounds (NO_3^-). Pantanella et al. (2014) described that nitrifying bacteria, which may be found in a variety of settings including soil, sand, water, and air, are an important element of the nitrification process, which converts plant and animal waste into readily available nutrients for plants. Hence, the same bacterial nitrification process that happens in soil occurs in water. The fish wastes in aquaponics are the excreta that the fish emits into the culture tanks. Consequently, land-based nitrifying bacteria will naturally colonize water and any damp surface, converting ammonia from fish waste into nitrate, which plants can easily absorb. Nitrification in aquaponics systems produces nutrients for plants while removing toxic ammonia and nitrite. Although nitrates are more easily absorbed by roots, plants can utilize both ammonia and nitrates to carry out their growth activities.

Several microorganism communities involved in the processing and solubilization of fish waste must be able to survive in aquaponic systems. If the ammonia (NH_4^+) from fish excrement and urine is not eliminated from the system, it can

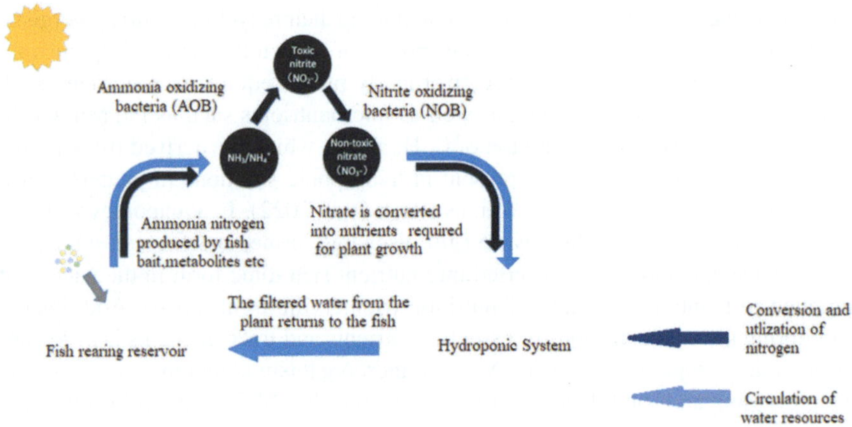

Fig. 17.2 Flowchart of nitrogen cycle in aquaponic system

accumulate to hazardous levels. This can be accomplished by a series of microbial conversions to nitrate. The nitrifying autotrophic bacterial consortium, which is predominantly composed of nitroso- and nitro-bacteria (such as *Nitrosomonas* sp.) forms a biofilm on solid surfaces. *Nitrospira* sp. and *Nitrobacter* species, are one of the most significant microbiological components in this system. Therefore, the first set of bacteria consumes the ammonia/ammonium produced by fish and converts it to nitrite, whereas the second group of bacteria assimilates the nitrites and produces nitrates, the most accessible form of nitrogen for rapid plant growth (Goddek et al. 2015).

1. Ammonia-oxidizing bacteria (AOB) bacteria, i.e. *Nitrosomonas*, convert ammonia (NH_3) into nitrite (NO_2^-).
2. Nitrite-oxidizing bacteria (NOB), i.e. *Nitrobacter*, convert nitrite (NO_2^-) into nitrate (NO_3^-).

System Description of Aquaponics

Delaide et al. (2015) showed that the aquaponics system represents the link between a typical recirculating aquaculture system (RAS) with hydroponics components (Fig. 17.2). After running in a loop from the fish tank to the filter units, water is then poured into the hydroponic beds, which serve as water reprocessing units. Mechanical filtering devices, such as drum filters or settling tanks, are used to remove solid particles, whereas moving bed or trickling biofilters are used for nitrification operations. Despite the fact, system configuration and complexity can vary significantly.

Aquaponics Techniques

Based on several types of bed grows, aquaponics systems can be built in multitude ways. Basically, the classification might be termed as media-filled, floating raft, and nutrient film method (NFT) (flood and drain) (Engle et al. 2015).

Nutrient Film Technique (NFT)

Thapa and Lamsal (2022) illustrated that plants with modest size and moderate root growth can be supported by this technique. The method provides continuous supply of water, oxygen, and nutrients ensuring the vegetables great yield in the end (Table 17.3). Large roots hinder the system recirculating system; therefore, only tiny plants are ideal. In order to avoid clogging, solid particles must be efficiently removed from a biofilter and sedimentation tank.

Floating Raft Technique/Deep Water Culture (DWC)

For large-scale commercial aquaponics, rafts or deep water culture (DWC) is commonly used. Using this method, plants are cultivated on perforated rafts that float in specialized water tanks. To allow the plant roots to collect nutrients from the water, an empty space is given to them to grow freely; therefore, the water routed in this manner does not become blocked (Liang et al. 2013). This is a highly productive technology that necessitates strong aeration and rigorous nitrification filtering to maintain the water pure and free of solid debris (Table 17.3).

Ebb and Flow

Another widely technique utilized by backyard growers is Ebb and Flow Aquaponics. A large, deep container with enough surface area serves as the growing region. This container is filled with pebbles, LECA, or another soilless growing media, and it is where the vegetables will be planted. In this process, biofilters are not fundamental because the system itself is made up of material that allows nitrification (Zou et al. 2016). A syphon is used to establish direct contact between plant roots and air, allowing the plants to get enough oxygen (Table 17.3). Despite its benefits, the main issue with an aquaponics system is blockage and an oxygen shortage.

Factors Affecting the Maintenance of Healthy Bacterial Colony of Aquaculture

A suitable surface area of biofilters and the proper water conditions are the two key factors that influence bacteria growth and should be taken into account while keeping it in good surrounding conditions (Fig. 17.3).

Table 17.3 Merits and demerits of aquaponic system

Methods	Merits	Demerits	Nutrient uptake	References
Ebb and flow	Biofiltration: the media acts as a substrate for nitrifying microorganisms Mineralization in grow bed serves as solid filtering medium and is colonized by a diverse microbiota	Heavy hydroponic infrastructure and large pump tank needed Difficulty in maintenance and cleaning thus causes water clogging, water channeling	High	Ghosh and Chowdhary (2019)
Floating raft or deep water culture (DWC)	A constant water supply and a small pump tank are required Maintenance and cleaning are simple	Separate biofilter and heavy hydroponic infrastructure to be needed Demand for an excessive volume of water and device for roots' aeration is mandatory	High	Palm et al. (2018)
Nutrient film technique (NFT)	Light hydroponic infrastructure that is ideal for roof farming Constant flow of water A small pump tank is required Ease of maintenance and cleaning	Separate biofilter is required Lower yields (demonstrated for lettuce) Expensive material; the system is less stable as there is less water	Lower (smaller root–water contact area)	Jasni et al. (2022)

Growing Media and Its Components

The growing beds are the regions where the crops will be grown. They are sometimes referred to as growing media or supporting media and can provide superior conditions than agricultural soil (in one or more ways) (Deepak et al. 2019). There are two fundamental types of aquaponic beds: media-based beds and beds that grow directly in water. Each style has pros and cons, yet they both have similar elements. Plant cultivation in soilless systems is based on the use of solid rooting material for growing plants (Gruda 2018). The first concern when choosing a material to utilize as a growing bed is to identify safe and inert materials. In addition, the growing bed material should be inert, which means that it does not leach or introduce materials into the system. This is significant because leaching material has the potential to progressively alter the chemical composition of the system. Furthermore, containers hold

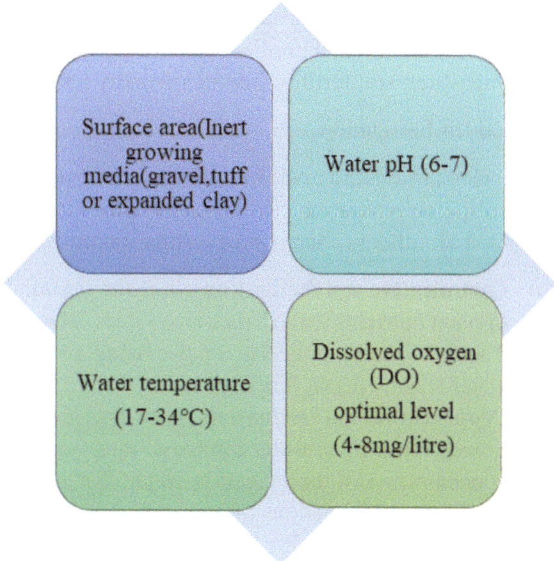

Fig. 17.3 Key factors for maintenance of aquaculture system

growing media (organic substrates, perlite, and so on). They are sometimes utilized as prepared cubes (rockwool cubes for seedling and transplant production), bags and slabs (peat-based substrates and rockwool, respectively), troughs (rockwool), and mats (polyurethane foam) (Gruda 2013). Besides this, FlexiPlugs, created by combining high-quality peat with other organics and a foam binder in Quick plug North America, is widely used in commercial aquaponic systems. On the other hand, hydrocorn, a substitute for the defunct hydroton, is clay that is moulded into little balls and then burnt at high temperatures. This media is popular because it is low in weight, drains easily, and has a constant pH (Anonymous 2014).

Soil Fertigation System

Shukla and Prasad (2018) described, fertigation is a relatively new cultural practice in which fertilizers are sprayed alongside irrigation water via drip irrigation system in order to improve usage efficiency while simultaneously increasing crop yields. Nutrient solutions are injected into irrigation water using the proper injection equipment. Fertigation provides critical components directly to the active root zone, decreasing nutrient losses and ultimately improving farm output and quality while lessening the risk of environmental pollution. Fertigation necessitates the use of specialized equipment depending on crop type and irrigation system, as well as the careful selection of an appropriate fertilizer and its mixture. Fertigation is a term used in agriculture to describe the combination of irrigation and fertilization in one step, allowing farmers to utilize both water and fertilizers more sustainably on open fields. Aquaponics merges aquatic animal culture and hence plant cultivation in recirculating systems, combining aquaculture and hydroponics in a soilless approach

(Deepak et al. 2019). Aquaponics offers an innovative approach as a solution to the potential environmental implications of aquaculture, such as climate change, lack of drinking water, soil fertility, and biodiversity loss.

Methods of Fertigation

Fertilizer solutions are made in stock tanks before being injected into irrigation water while utilizing fertigation. The crop sown, the kind of soil, and the farm management approach all influence the sort of fertigation utilized.

(a) **Quantitative soil fertigation—**it is the introduction of specific concentrations of plant nutrients to the irrigation system. This technique is often utilized in soil. The fertilizer is applied in a pulse using a fertilizer tank following a particular water sheet without fertilizer.
(b) **Proportional soil fertigation**—nutrients are applied in a consistent and proportionate ratio to the water discharge rate, resulting in a predetermined fertilizer concentration in the irrigation water. For instance, 1 l of fertilizer solution is added into 1000 gallons of irrigation water. In this case, fertilizers are administered directly using fertilizer pumps. This approach is predominantly employed in soilless media and sandy soils.

Basic Requirements for Fertigation

The amalgamation of fertilizers into the irrigation system demands the following fundamental requirements such as fertilizer injection equipment, vacuum injection (Venturi) (Chinchmalatpure 2018).

(a) **Fertilizer injection equipment**—the proper injection equipment must be chosen just as carefully as the accurate nutrient. The pressure of the injected fertilizer solution must be larger than the pressure of the surrounding air. A filter should be provided to prevent solid particles of fertilizer solution from entering and clogging the dripper. A backflow prevention valve is also necessary to hamper chemical backflow into the water supply. Fertilizer injectors are the most critical part of a fertigation equipment.
(b) **Vacuum injection (Venturi)**—this is based on the principle of the venturi tube. A pressure differential between the injector's intake and exit is needed, resulting in a reduced pressure (vacuum) that draws the fertilizer solution into the line (Table 17.4).

Advantages of Soil Fertigation System

1. High nutrient availability as a result of drip irrigation's capacity to keep soil moisture surrounding the root zone wet.
2. Nutrient loss by leaching is reduced to roughly 10%, compared to 40–55 per cent in the old system, resulting in considerable reductions in environmental contamination.
3. As minute quantity is given at regular intervals rather than in one or two big doses, fertigation results in very high nutritional uptake and absorption.
4. Significant effort and energy savings in fertilizer application.

Table 17.4 Nutrients and suitable fertilizers (Shukla and Sadhu 2018)

Nutrients	Fertilizers and nutrient concentration %
Nitrogen (N)	Urea (46% N)
	Ammonium nitrate (34% N)
	Ammonium sulphate (21% N)
Phosphorous (P)	Phosphoric acid (52% P_2O_5)
Nitrogen and phosphorous (N and P)	Urea phosphate 17% N, 44% (P_2O_5)
	Mono ammonium phosphate 12% N, 61% (P_2O_5)
Potassium (K)	Potassium chloride (60% K_2O)
	Potassium nitrate (multi K) (13% N, 46% K_2O)
	Potassium thiosulphate (25%, K_2O 17.5% S)
	Sulphate of potash (50% K2O, 17.5% S)
Nitrogen, phosphorous, and potassium (N, P, and K)	Poly feed (19-19-19% NPK)
	Urea phosphate with SOP (18–18-18% NPK)
Magnesium and calcium (Mg and Ca)	Magnesium nitrate (11% N)
	Calcium nitrate (16% N, 19% Ca)

Limitations of soil Fertigation

1. Nutrients are delivered unevenly when the irrigation system malfunctions.
2. Chemical blockage may result from fertilizer chemical interactions with Ca and Mg.
3. Phosphoric fertilizers and several micronutrients may precipitate in microirrigation systems.
4. It is necessary to have corrosion-resistant fertigation equipment.
5. Its widespread implementation is hampered by high preliminary costs and a lack of information in a variety of areas such as crop water requirements, irrigation time, and fertigation.

Filteration Methods

The circulating water treatment primarily performs the functions of water filtration, oxygen enrichment, and sterilization, as well as providing a suitable growing habitat for aquatic life. As a result, water treatment equipment is divided into three categories: physical filtration, biological filtration, and mechanical filtration.

(a) **Physical Filtration**

Due to the relatively high unit water density of farmed fish, the fish produces a substantial amount of solid waste during the aquaculture process. For every kilograms of feed consumed, the fish adds 0.25 kilos of solids to the water. In a short amount of time, the buildup of fish excreta and bait will create a poisonous environment, encouraging fish growth and eventually clogging the system and disrupting water flow, creating an unstable system (Mullins et al. 2016). Therefore, in order to remove bait, waste, and suspended solids (SS) from aquaculture water, physical filtration is fundamental. One of the essential steps in the RAS water treatment process is the removal of SS, and the efficacy of this step has a direct impacts on the quality of the water and the system's stability. The organic loading rate, which is determined by the daily feed volume and faecal formation, is a major determinant of the most efficient solids' removal mechanism in a given system (Rakocy et al. 2012). Physical filtration can be done in a variety of ways with mechanical and sedimentation processes being the most common methods for eliminating solid particles in aquaponic RAS. In aquaponic systems, clarifier tanks are usually used, and gravity is generally used in the sedimentation technique to freeze and remove solid waste particles (Bailey et al. 2017).

A straightforward device for sorting and eliminating precipitable solid fish waste and particles from water is a clarifier, commonly referred to as a solid filter. Two cylindro-conical clarifiers are the key components of UVI's commercial scale aquaponics system, and they can remove up to 50% of the system's total particle solids. According to an experiment conducted by Savidov et al. (2003), the water flowing through the clarifier will again travel through the sedimentation tank while removing particles.

By screening suspended particulate matter, Pfeiffer et al. (2008) assessed the effectiveness of removing suspended particulate matter from a cyclone separator in a tilapia RAS. Despite the fact that sedimentation technology can only remove sedimentable and large suspended particle matter from water, it has been widely used in the pretreatment stage of suspended particulate matter in recirculating aquaculture production due to its low priced and minimal water loss.

(b) **Mechanical Filtration**

Mechanical filtration is the elimination of solid waste before being broken down by biological processes. By utilizing a substance to screen the water, mechanical filtration successfully accomplishes its goal of separating particles from water. Some of its filtration methods include gravity methods like sedimentation, swirl separators/ radial flow separators, screening microscreen (drum) filter, bead filters, sand filters, etc. (Forchino et al. 2018). Drum screen filters are the most often used filters nowadays, and rotary drum microfilters are one of the most common types of equipment for removing large particulate matter. Karimanzira et al. (2016) employed spinning microscreen filters to filter water, while drum filters were used to separate water from solids. Besides this, different water treatment systems were compared and analyzed through tests and results demonstrated that the aquaponics system with mechanical filtration had the maximum plant growth rate and production, as well as adequate water quality (Wei et al. 2019).

(c) **Biological Filtration**

Biofiltration is the most integral part for maintaining adequate water quality and improving fish health. Ammonia–nitrogen, a substantial component of fish excretion, is mostly produced by fish metabolites, fish bait, and organic debris. In the water system, over 10% of the protein in fish diet is transformed to ammonia. Ammonia buildup can progress to the point that it endangers fish (0.05 mg L^{-1}). The biological filtering process, commonly referred as nitrification, is the primary method by which ammonia–nitrogen is eliminated from the circulating water system. Through an infiltration system, the nitrification process may convert 93%–96% of the ammonia–nitrogen into nitrate (Jiao and Wei 2019). The entire surface area accessible for the development of nitrifying bacteria plays a significant role in determining the ability of biofilters to remove ammonia. Nitrifying bacteria are given adherent substrates to create biofilms in fixed-membrane biofilters, which also offer sufficient surface area, an adequate temperature, degrees of acidity and alkalinity, and dissolved oxygen concentrations (Ebeling et al. 2012). In addition to being a typical technique for attaining autotrophic nitrification, it also serves as the foundation of RAS's water filtration system. As a result, biofilm filtering is commonly employed to lower water ammonia–nitrogen and nitrite concentrations. The biofiltration step of recirculating aquaculture water treatment is crucial, and the tools and technologies utilized in this stage have also been enhanced and are showing promising outcomes in real-world applications.

Ideal Conditions for Filtration Methods

According to Suhl et al. (2016), an environment and various filter media have a significant impact on how well the biofilter works. The ideal filtration conditions for the biofilter are as follows: the temperature should be between 77 and 86 °F, the pH should be between 7.0 and 9.0, the biochemical oxygen demand (BOD) should be less than 20 mg/L, the dissolved oxygen (DO) should be saturated, and the total alkalinity should be at least 100 mg/L. Aquaponic biofilters should be roughly the same size as those advised for recirculating systems. Trickling biofilters, fluidized bed filters, and moving bed biofilters are some of the biofilter components that are frequently utilized in aquaponic systems (Forchino et al. 2018; Bittsanszky et al. 2016).

The first type of biofilter that has been thoroughly investigated and used is a trickling biofilter. The benefits of this technology include a consistent high oxygen content, efficient CO_2 removal, minimal operation costs, and simplicity of operation and management. Its main objectives are nitrification and BOD eradication. The maximal rate of ammonia elimination is 0.43 g m^{-2} day^{-1}, it rises linearly as ammonia concentration rises (Pattillo et al. 2017), and its main objectives are nitrification and BOD eradication.

Pest and Disease Control

Another dimension that needs improvement is the difficulty in managing diseases and pests. Aquaponic systems have a wider range of microflora than traditional hydroponic systems because fish breeding and biofiltration happen in the same water loop. Aquaponics prohibits the use of conventional insecticides to protect the fish from damage and the desired biofilm (such as an autotrophic nitrifying biofilm). Antibiotics and fungicides are also forbidden for managing and eliminating fish infections in the aquatic environment due to the need to maintain the nitrification biofilm and other nutrient solubilizing bacteria.

Since antibiotics cannot be applied to plants, aquaponic systems must avoid using them to treat fish diseases. These limitations call for novel approaches for executing fish, plant pests, and diseases that have minimal negative effects on fish and desired microbes. Plant and fish infections are classified into four classes based on specific alternative treatment approaches. This constitutes (1) fish parasites like monogenea and cestoda, (2) plant pathogens like bacteria, fungus, and viruses that target plants, (3) fish diseases brought on by microbes and viruses that harm fish, (4) plant parasites like aphids and spider mites.

Therefore, low density of fish and/or plants, preventive sanitary methods, environmental circumstances that reduce relative humidity surrounding the plants are a few examples of crop practises that reduce the occurrence of diseases.

In addition to these techniques, there are a few cutting-edge biocontrol strategies now available for plants grown in fields or greenhouses. These strategies rely on the employment of biocontrol microorganisms with high antibacterial efficacy, such as microbe extracts or plant extracts (including essential oils). Some mutualistic microorganisms in the plant biosphere (plant growth-promoting rhizobacteria and plant growth-promoting fungus) may both decrease disease development and boost growth. Due to the existence of a wide variety of microflora in aquaponics practises, it is critical to identify the presence of pathogens and the risk they pose to human health in order to evaluate the safety of aquaponics and carry out the necessary quality control.

Monitoring Microbial Species

It is important to be aware of the general state of the bacteria in aquaponic systems because they rely so largely on microbes. Without a microscope, however, it is difficult to identify bacteria because they are minute organisms. An easy way to monitor how effectively bacteria are operating is to test for ammonia, nitrite, and nitrate as these tests provide information on the health of the bacterial colony (Pantanella et al. 2014).

While taking into this consideration, ammonia and nitrite levels in a healthy, functioning aquaponic system should always be between 0 and 1 mg/l. There are two plausible, common causes which can be occurred for improper activity of bacterial health. The foremost is that system may have been initially balanced when the fish was younger, but as they grew and are fed more with the same size biofilter, the system may have become gradually out of balance, which can lead to this issue. Second, even if the system size is balanced, the bacteria might not be operating effectively. Each of the aforementioned parameters needs to be examined because this can indicate a problem with the water quality. This frequently happens in the winter when bacterial activity slows and water temperatures start to drop.

Role of Bacterial Species in Aquaponic System

(A) ***Essential bacteria***

(a) ***Heterotrophic bacteria***

Along with other microorganisms, another significant bacterial community involved in aquaponics is heterotrophic bacteria. These bacteria mostly contribute to the breakdown of solid fish and plant waste and use organic carbon as a food source. These bacteria mostly contribute to the digestion of solid fish and plant waste and feed on organic carbon. Most fish retain only 30–40% of their food, so that 60–70% of what they consume is expelled as waste. Therefore, 50–70% of waste is disintegrated trash, which is discharged as ammonia. The leftover waste, however, is a mixture of organic materials that includes proteins, carbs, lipids, vitamins, and minerals. These solid wastes are processed by the heterotrophic bacteria through a process known as mineralization, which makes vital micronutrients accessible to plants in aquaponic systems (Lovatelli and Cohen 2014). These heterotrophic bacteria aid in the breakdown of the solid component of fish waste, along with some naturally occurring fungus and release nutrients that were trapped in the solid excrement into the water. The inability of plants to absorb solid nutrients makes this mineralization process crucial. Heterotrophic bacteria devour any organic substance, including decaying plants, dead bacteria, solid fish faeces, and uneaten fish food. Similar to nitrifying bacteria, heterotrophic bacteria need certain conditions to develop, especially in environments with high DO levels. All parts of the apparatus are colonized by heterotrophic bacteria; however, they are particularly dense where the solid waste gathers. In contrast to nitrifying bacteria, heterotrophic bacteria multiply far more quickly in hours as opposed to days.

Aquaponics requires mineralization because it releases a number of micronutrients essential to plant growth. Some plants could suffer from nutrient deficits without mineralization, in which case they would require additional fertilizer. Earthworms, amphipods, isopods, larvae, and other minute creatures can be found in aquaponic systems, particularly inside media beds, assisting heterotrophic bacteria in the decomposition of solid waste. These organisms collaborate with bacteria to break down solid waste and frequently avoid waste accumulation.

(B) ***Non-essential bacteria***

(a) ***Sulphate-reducing bacteria***

One of the unwanted bacterial species is the sulphate-reducing bacteria as these bacteria thrive in anaerobic conditions (absence of oxygen) and are prevalent in lakes, salt marshes, and estuaries, as well as an organic part of the sulphur cycle which obtain their energy from sulphate-based redox processes. The issue is that the gas created by this process, hydrogen sulphide (H_2S), is extremely hazardous to fish. The issue is that the gas created by this process, hydrogen sulphide (H_2S), is exceedingly dangerous to fish. A challenge in aquaponics occurs, when solid wastes accumulate more quickly than the accompanying population of heterotrophic bacteria. The anoxic festering conditions may then favour these sulphate-reducing bacteria. As a result, excessive amount of solid waste in systems is generated, and mechanical filters cannot be cleaned as rapidly as they should, which encourages the growth of these bacteria and the creation of their poisonous byproducts.

A degassing tank is frequently used in large aquaponic systems to safely release hydrogen sulphide back into the atmosphere (Stankus et al. 2014). Smaller systems do not need to be degassed. If a bad smell, resembling rotten eggs or raw sewage, is detected, it is crucial to take the appropriate management action, even in small-scale systems. To avoid these germs and sludge accumulation, ensure that adequate aeration is provided and that mechanical filtering is increased.

(b) **Denitrifying Bacteria**

The undesirable microorganisms that cause denitrification fall under a distinct category can also thrive in anaerobic environments. They change the nitrate back into nitrogen, which plants cannot use. Additionally widespread around the globe, these microorganisms are significant in their own way. These bacteria can, however, reduce efficiency in aquaponic systems by essentially removing the nitrogen fertilizer. Denitrification is occasionally used on purpose in large aquaponic systems. Despite the fact that it usually results in excessive nitrate levels, the feed rate ratio balances the nutrients for the plants. Water exchanges may be used to dilute this nitrate for small-scale systems. The mechanical filter could also be designed to encourage controlled denitrification (Somerville et al. 2014).

(c) **Pathogenic Bacteria**

The bacteria that infect humans, fish, and plants to cause disease make up the final category of unwanted microorganisms. Overall, it is crucial that aquaponic systems have good agricultural practises (GAPs) to reduce and minimize the risk of bacterial infections. It is crucial to avoid utilizing rainwater collected from polluted rooftops unless it has been thoroughly cleaned. The major concern from warm-blooded animals is *Escherichia coli*, and birds typically carry *Salmonella* spp., a harmful bacteria that may enter the system through animal faeces (Pantanella and Stankus 2014).

17.4 Commercial Applications of Soilless Cultivation

a. Deep water culture technologies and NFT have been widely applied for commercial cultivation of crops such as tomatoes in major agricultural nations like Spain, France, and the Netherlands. NFT techniques are widely used in countries like North America since these systems are the most effective for growing lettuce, which is in high demand there (Anonymous 2022).
b. The HVAC system is critical in the development of indoor hydroponic farms since it is in charge of cooling, dehumidifying, and maintaining the optimal temperature within the facility. The HVAC system is made to counteract the weather and offer the best conditions for year-round cultivation and growth.

The hydroponic greenhouse's equipment is crucial for delivering optimal plant development conditions (Anonymous 2022).

c. Mendoza et al. (2022) used the fertilizer drawing forward osmosis (FDFO) process to investigate the osmotic dilution of commercial nutrient solutions for hydroponics, a standard method for growing plants without soil. According to pilot-scale experiments, the FDFO technique may deliver the requisite fertilizer content and ultimate water quality (i.e. pH and conductivity) for hydroponic applications. Later on, FDFO and pressure-assisted osmosis (PAO) could be coupled to lessen operational expenses (i.e. energy and membrane replacement costs) (Kim and Kuntz 2019). Moreover, hydroponic lettuce grown using the pilot FDFO technique demonstrated similar growth patterns to the control without exhibiting any evidence of nutrient deficit after receiving a test application of nutrient solution.
d. Different dosages of humic-like substances (HLS) extracted from two digestates-sewage sludge and manure and has been employed to biostimulate hydroponic cultures of *Lactuca sativa* (lettuce), and their growth was compared to a commercial HLS and leonardite HLS (Guilayn et al. 2020). The average increment in aerial biomass ranged from 7 to 30%; however, there was significant variation across the results (about 20%).
e. Digestates have been effectively used in hydroponic systems as a nutrient supply and as a digestate treatment option (Krishnasamy et al. 2012). Additionally, it has already been shown that dissolved organic debris from digestate produces effects on plants similar to auxin (Scaglia et al. 2015). Using digestates as substrates, valuable soluble organic compounds known as HLS, soluble biopolymers, biowaste-generated soluble substances, bio-based organic substances, and other variants have already been extracted (Fascella et al. 2015).
f. HLS from fossil sources or organic materials have both been used to demonstrate the biostimulation of hydroponic systems (Conselvan et al. 2017). Additionally, with the recent implementation of a new European Union fertilizer rule (EC 2019/1009), which includes a biostimulant product category, the development of biostimulants from digestates is now a tangible short-term value-adding avenue (Massa 2018).
g. Although, recommended doses of HLS and biostimulants are often based on amounts per crop surface but are still been used to examine the dose–effect of HLS on lettuce (*Lactuca sativa* var. *Tourbillon*) roots and aerial development in a hydroponic (soilless) system (Lyons and Genc 2016). Additionally, the extraction approach is used in circumstances that are thought to be industrially scalable in the near future.
h. Hydroponic strawberry plant cultivation results in improved quality and yield because soil-borne disease problems are excluded. Recycled hydroponics is practised in farming to make it economical, ecologically effective, and long-lasting. However, hydroponic fertilizer solution is recycled, and plant root exudates build up and cause autotoxicity, i.e. an allelopathic condition that prevents growth and development (Asaduzzaman and Asao 2020). Globally,

recycled hydroponics-based commercial strawberry growing under greenhouse and plant factory conditions has gained popularity in recent years.

i. Numerous vegetable crops, including *Lactuca sativa*, *Solanum lycopersicum*, and *Cucumis sativum*, and green vegetables have been grown using aeroponics. Additionally, this method has been successful in producing yam (*Dioscoria* sp.) tubers and potato seed tubers on a big scale. By adjusting the temperature of the root zone, aeroponics is also used to grow crops from the subtropical and temperate regions in tropical environments (Sharma et al., 2018b).
j. Growing therapeutic plants with aeroponics has also been documented in recent years. Mehandru et al. (2014) described cultivation of highly valued root crops including yerba mansa, burdock (*Arctium lappa*), and medicinal rhizomes like ginger (*Zingiber officinale*). A novel chemical compound, namely 2,3-Dihydrowithaferin A-3β-O-sulphate (prodrug of withaferin A), has been discovered in *Withania somnifera* cultivated aeroponically.
k. Mehandru et al. (2014) used aeroponic culture to clonally propagate three fragile medicinal asclepiads: *Leptadenia reticulate*, *Tylophora indica*, and *Caralluma edulis*. They discovered that stem cuttings produced aeroponically have a higher rooting percentage than those cultivated in soil.
l. Salachas et al. (2015) observed the influence of accessible root zone volume on yield and nutritional quality of farmed *Ocimum basilicum* in an aeroponic system. Aeroponic-cultured medicinal plants have also been examined for their cytotoxic, antibacterial, and antiplasmodial qualities.
m. The Open Pond (OP) and Photobioreactor (PBR) systems were utilized to evaluate large-scale algal biomass production employing a unique aeroponic substrate-based growth approach. These substrate system's aeroponic characteristics have no adverse effects on the produced biomass's fatty acid composition, thus demonstrating the system great potential for producing biofuel, nutraceuticals, fish feed, and other products (Johnson et al. 2015).
n. Numerous aspects have been investigated in order to enhance the aeroponic system, including the impact of ultrasonics, nitrogen supply, root zone temperature, water stress, electrical conductivity (EC) of nutrient solutions, pH, plant density, and harvest interval (Tunio and Gao 2020). Even though aeroponics is turning into unique minitubers and high-quality potato production technology, certain conditions and specific potato cultivars still require optimal production technology (Buckseth and Muthuraj 2016). In addition, Chandra et al. (2014) found that the aeroponics technique produces plants with comparable phenolic, flavonoid, and antioxidant qualities to those grown in soil but with better yields.
o. The investigation done by Saha et al. (2016) at Georgia Southern University in Statesboro, Georgia, USA, found that using crayfish (*Procambarus* spp.) as the aquatic species in soilless farming (*Ocimum basilicum* L.) increased basil output and growth. Therefore, basil might have supplied additional nutrients that have come from crayfish waste (excreta and leftover meal).
p. Johnson (2017) evaluated the production of biomass from algae using an aeroponic system (*Parachlorella kessleri* and *Tetraselmis chuii*). They examined a novel aeroponic system and discovered that biomass output is increased without

affecting fatty acid content. This strategy has reduced energy inputs while increasing financial rewards. It also has huge potential for creating biofuel, nutraceuticals, fish feed, and a range of other products.

q. The Brussels start-up BIGH (Building Integrated Green Houses) was the first aquaponics farm, a closed-loop aquaponics system that produced fish, fruits, vegetables, and herbs. Moreover, 35 tonnes of top-quality striped bass fish can be produced by the fish farm annually (Aleksić 2020). Herbs, tomatoes, and microgreens are all produced in the greenhouse.

r. Efficient City Farm (ECF) is a food manufacturer in Germany that caters to two unique markets: supermarkets and Hotels, Restaurants, and Catering (HoReCa). In one urban environment, ECF's aquaponics farm combines fish farming (invitations) with plant cultivation (in a greenhouse). It focuses on a single species from each important domain-basil and tilapia.

s. The top and sixth floors of De Schilde, a former Philips plant in the Hague, have been taken over by urban farmers, a large European aquaponics farm. They form an ideal symbiotic system for the city's fish and vegetable production, in which dirty water from the fish tanks piped into the planter beds to feed the plants, which in turn filter the water for the fish. Irrigation systems, technical setups, and a facility for processing fish and vegetables are also located on both floors (Šušteršič 2020).

t. Tilamur, a Spanish company, was founded in 2012, and the technology in use is based on a unique design intended to facilitate the rapid and simple installation of parts that allow the manufacture of sustainably generated food for people. This aquaponics system, which is connected and divided into three sections, serves as the basis for this circular economy model (Aleksić and Šušteršič 2020). Because of its optimal design and low-power usage, this system can function autonomously utilizing solar and wind technologies. The concept may be scaled up or down indefinitely, with the simplest version having a total greenhouse area of 400 m^2 and producing around 3320 kg of tomatoes, 2000 kg of fish (tilapia), and up to 137 kg of algae (spirulina) yearly.

17.5 Conclusion and Future Prospective

Soilless farming is the rapid growing sector of agriculture and it has tremendous merits and demerits, but with the global loss in arable land, it is critical to discover creative methods for large-scale food production. Therefore, new technologies like as hydroponics, aeroponics, and aquaponics are being established to generate additional channels for agricultural production. Soilless farming, unlike conventional gardening methods, permits for year-round harvesting of fresh food. Growing plants in a nutrient solution has several advantages, including the ability to provide plant nutrients in fertilizers instantly to the roots, without facing challenges related to soil composition or fertilizer failure to reach the root system. Apart from this, there is no need to use large gear to prepare beds for seeds or planting, weeds and diseases may be handled without the use of pesticides, and fertigation requires no irrigation infrastructure. Moreover, environmental regulations in the modern era are limiting the amount of water that can be consumed or disposed. Thus, aquaponics creates a chance in the field of indoor farming by providing marginal farmers' two sources of income—fish and vegetables, so they can increase their profits while using a minimal amount of land. Even if soilless farming does not appear to be substantial at this time, it has huge promise in the future. Soilless cultivation methods and practises are referred to as next-generation agricultural science because, if properly explored and assessed, they will pave the way for the formation of a new society.

References

Aleksić N (2020) Analysis of application of aquaponic system as a model of the circular economy: a review. Recycl Sustain Dev 13(1):73–86

Alipio MI, Doria JDA, Fruto RMS (2019) On the design of Nutrient Film Technique hydroponics farm for smart agriculture. Eng Agric Environ Food 12(3):315–324

AlShrouf A (2017) Hydroponics, aeroponic and aquaponic as compared with conventional farming. Am Acad Sci Res J Eng Technol Sci 27(1):247–255

Anonymous (2014) https://na.quickplug.global/aquaponics/commercial-aquaponics-growing-media/

Asaduzzaman M, Asao T (2020) Autotoxicity in strawberry under recycled hydroponics and its mitigation methods. Hortic J 89(2):124–137

Buckseth T, Muthuraj R (2016) Methods of pre-basic seed potato production with special reference to aeroponics—a review. Sci Hortic 204:79–87

Chandra S, Khan S, Avula B, Lata H, Yang MH, ElSohly MA, Khan IA (2014) Assessment of total phenolic and flavonoid content, antioxidant properties, and yield of aeroponically and conventionally grown leafy vegetables and fruit crops: a comparative study. Evidence-Based Complement Alternative Med

Chinchmalatpure AR (2018) Fertigation-modern technique of fertilizer application. Indian Farmer 5(9):1062–1071

Conselvan GB, Pizzeghello D, Francioso O, Foggia M, Nardi S, Carletti P (2017) Biostimulant activity of humic substances extracted from leonardites. Plant Soil 420(1):119–134

Deepak P, Saminathan S, Haribabu S (2019) Experimental investigation/on fertigation in open field agriculture

Dubey N, Nain V (2020) Hydroponic—the future of farming. Int J Environ Agric Biotechnol 5(4)

El-Kazzaz KA, El-Kazzaz AA (2017) Soilless agriculture a new and advanced method for agriculture development: an introduction. Agric Res Technol 3:63–72

Fascella G, Montoneri E, Ginepro M, Francavilla M (2015) Effect of urban biowaste derived soluble substances on growth, photosynthesis and ornamental value of Euphorbia x lomi. Sci Hortic 197:90–98

Figeac A (2019) ECF FARMSYSTEMS—a circular economy business model case. http://www.r2piproject.eu/wpcontent/uploads/2019/06/ECF-Farms-CaseStudy_2.pdf. Aceesed 07 Aug 2020

Fussy A, Papenbrock J (2022) An overview of soil and soilless cultivation techniques—chances, challenges and the neglected question of sustainability. Plants 11(9):1153

Gosh K, Chowdhury S (2019) Review of aquaponics system: searching for a technically feasible and economically profitable aquaponics system. J Agric Environ Consum Sci 19:5–13

Gruda N (2013) Soilless culture. Good Agricultural Practices for greenhouse vegetable crops, p 303

Gruda N (2018) Application of soilless culture technologies in the modern greenhouse industry—a review. Eur J Hortic Sci 83(5):280–293

Guilayn F, Benbrahim M, Rouez M, Crest M, Patureau D, Jimenez J (2020) Humic-like substances extracted from different digestates: first trials of lettuce biostimulation in hydroponic culture. Waste Manag 104:239–245

Guruchandran S, Muninathan C, Ganesan ND (2022) Novel strategy for effective utilization of anaerobic digestate as a nutrient medium for crop production in a recirculating deep water culture hydroponics system. Biomass Convers Biorefinery 1–13

https://www.businesswire.com/news/home/20221025005590/en/Hydroponics-Global-Market-Report-2022-Rise-in-Demand-for-Fresh-Food-in-Urban-Areas-to-Drive-Growth---ResearchAndMarkets.com

Jasni NA, Said R, Jabridin J, Marcus FF, Hassan MH (2022) DIY nutrient film technique (NFT) aquaponic system

Johnson M (2017) Development, optimization, and economic evaluation of a novel attached growth algae cultivation system. Doctoral dissertation, Rutgers University-School of Graduate Studies

Johnson M, Villani TS, Azmat A, Simon JE (2015) Evaluation of algal biomass production on vertical aeroponic substrates. Algal Res 10:240–248

Kim JE, Kuntz J (2019) Techno-economic assessment of fertiliser drawn forward osmosis process for greenwall plants from urban wastewater. Process Saf Environ Prot 127:180–188

Lyons GH, Genc Y (2016) Uncoupling of sodium and chloride to assist breeding for salinity tolerance in crops. New Phytol 210(1):145–156

Massa N (2018) Combined bacterial and mycorrhizal inocula improve tomato quality at reduced fertilization. Sci Hortic 234:160–165

Mehandru P, Shekhawat NS, Rai MK, Kataria V, Gehlot HS (2014) Evaluation of aeroponics for clonal propagation of *Caralluma edulis*, *Leptadenia reticulata* and *Tylophora indica*-three threatened medicinal Asclepiads. Physiol Mol Biol Plants 20(3):365–373

Mendoza E, Buttiglieri G, Blandin G, Comas J (2022) Exploring the limitations of forward osmosis for direct hydroponic fertigation: impact of ion transfer and fertilizer composition on effective dilution. J Environ Manag 305:114339

Pandey K, Singh KG, Singh A (2023) Multi-sensors based smart nutrient reuse management system for closed soilless culture under protected cultivation. Comput Electron Agric 107495

Palm HW, Knaus U, Appelbaum S, Goddek S, Strauch SM, Vermeulen T, Kotzen B (2018) Towards commercial aquaponics: a review of systems, designs, scales and nomenclature. Aquacult Int 26(3):813–842

Pradhan B, Deo B (2019) Soilless farming—the next generation green revolution. Curr Sci 116(5):00113891

Saha S, Monroe A, Day MR (2016) Growth, yield, plant quality and nutrition of basil (*Ocimum basilicum* L.) under soilless agricultural systems. Ann Agric Sci 61(2):181–186

Sahoo D (2020) Aeroponics system of cultivation in horticultural crops, vol 1, pp 32–40

Sharda R, Singh A, Pandey K (2023) Paddy and maize straw-based media as an alternative for cocopeat in soilless cultivation. J Plant Nutr 1–19

Salachas G, Savvas D, Argyropoulou K, Tarantillis PA, Kapotis G (2015) Yield and nutritional quality of aeroponically cultivated basil as affected by the available root-zone volume. Emirates J Food Agric 911–918

Savvas D, Gianquinto G, Tuzel Y, Gruda N (2013) Soilless culture. FAO Plant Production and Protection. Paper (217)

Sharma N, Acharya S, Kumar K, Singh N, Chaurasia OP (2018a) Hydroponics as an advanced technique for vegetable production: an overview. J Soil Water Conserv 17(4):364–371

Sharma U, Barupal M, Shekhawat NS, Kataria V (2018b) Aeroponics for propagation of horticultural plants: an approach for vertical farming. Hortic Int J 2:443–444

Shukla M, Prasad I (2018) Fertigation-modern technique of fertilizer application. Indian Farmer 5(9):1062–1071

Shukla M, Sadhu AC (2018) Fertigation-modern technique of fertilizer application. Indian Farmer 5(9):1062–1071

Singh K, Gupta G, Ghangal R (2017) Soilless cultivation, its various types and applications in agricultural and pharmaceutical sector. World J Pharm Res 6:722–739

Sundari RS, Sulistyowati L, Noor TI, Setiawan I (2022) Soilless culture for agribusiness throughout urban farming in Indonesia

Šušteršič V (2020) Analysis of application of aquaponic system as a model of the circular economy: a review. Recycl Sustain Dev 13(1):73–86

Tunio MH, Gao J, Shaikh SA, Lakhiar IA, Qureshi WA, Solangi KA, Chandio FA (2020) Potato production in aeroponics: An emerging food growing system in sustainable agriculture for food security. Chilean J Agric Res 80(1):118–132

Underwood J, Dunn B (2016) Aquaponics. Oklahoma Cooperative Extension Service

Yildirim M, Dardeniz A, Kaya S, Ali B (2016) An automated hydroponics system used in a greenhouse. Sci Pap Ser E Land Reclam Earth Observ Surv Environ Eng 5:63–66

Chapter 18
Nutrient Management Strategies for Water and Nutrient Saving in Substrate Soilless Culture Under Protected Cultivation

Kusum Pandey

Abstract In soilless farming, nutrients solution plays a significant role for plant growth and food quality. Knowledge of accurate nutrient selection and its management is always challenging for crop growers in soilless systems. Further, there is a growing need to recirculate and reuse drained solutions in order to decrease environmental issues and economic expenses. However, one of the weakest points in soilless systems is the lack of information on managing the nutrient solution. Therefore, this chapter describes the nutrient selection and compositions, recycling of drained nutrient solution, frequency of fertigation, types of fertigation system, and their nutrient management in soilless system under protected cultivation.

18.1 Introduction

The demand for food is increasing significantly as the world's population continues to rise. According to the United Nations, at present 7.7 billion people are inhabiting the world, but it is predicted that this number would increase to 8.6 billion in 2030 and 9.8 billion in 2050 (Sagheer et al. 2020). For this, to meet the demands of the world's people by 2050, it will require 42% more cropland and 120% more water (Godfray et al. 2010). Conventional farming needs more space, high man power resources, dependency on weather and seasons, more energy, high water consumption, more economy and ecology losses, greater chances of disease and pest attack, all of which lead to poor yield and crop quality. Therefore, protected cultivation has been shown to be an effective method of food production which can increase crop yield and quality. Moreover, due to various soil-borne diseases emerged in soil, protected cultivation has not able to satisfy the grower demands (Hussain et al. 2014; Sharda et al. 2023).

K. Pandey (✉)
Scientist-C, Garhwal Regional Centre, G. B. Pant National Institute of Himalayan Environment, Srinagar, Uttarakhand, India
e-mail: khushipandey166@gmail.com

K. Pandey et al. (eds.), *Artificial Intelligence and Smart Agriculture*,
Advances in Geographical and Environmental Sciences,
https://doi.org/10.1007/978-981-97-0341-8_18

Therefore, recently, soilless cultivation arose as a solution to conventional farming method and increase our ability to tackle with the present challenges like natural resource management (Nejad and Ismaili 2014).

In soilless cultivation, irrigation is a repetitive operation that ensures all the nutrients are provided to the plants. A slight modification in the nutrient management can be considerably affected the plant growth and product quality. Nutrient management is the process of regulating the amount, source, scheduling, and method of nutrient application to enhance the food production and diminishing the nutrient losses. Therefore, the nutrient solution should fulfil the following conditions (Pandey 2021): (1) All essential macro and micronutrients are included in ion forms, (2) each and total ions concentrations are within the desired range, (3) harmful substances are not included, (4) pH and EC of the nutrient solution are around 5.5–6.5 and 2.2–2.7 dSm^{-1}, respectively, (5) the concentration of each ion should be remained constant during all the plant development stages, and (6) in the deep flow technique (DFT) and nutrient film technique (NFT), enough oxygen is dissolved for root respiration. Thus, proper knowledge of the nutrient solution is important for plant cultivation. Moreover, several researchers have designed various kinds of composition or software for the calculation of nutrient solution requirements, which depends on the crop physiology, growth stage, and quality of the product (Rouphael et al. 2016; Marques et al. 2017; Mohammed 2018; Wada 2019; Van 2020). For this context, this chapter aims to describe the nutrient selection and compositions, recycling of drained nutrient solution, frequency of fertigation, types of fertigation system, and their nutrient management in soilless system under protected cultivation.

18.2 Nutrient Selection and Its Composition in Soilless System

For vegetable production, in soilless growing media involves a steady and adequate amount of nutrient solution in appropriate proportion to assure a consistent growth rate, yield, and high postharvest quality. For soilless cultivation, every necessary plant nutrient should be provided through the nutrient solution, with the exclusion of hydrogen and carbon, which are absorbed from the water and air, respectively. For the production of cucumber, macronutrients, namely Phosphorus (P), Nitrogen (N), Potassium (K), Magnesium (Mg), Sulphur (S), and Calcium (Ca), and micronutrients, namely Manganese (Mn), Zinc (Zn), Boron (B), Molybdenum (Mo), Iron (Fe), and Copper (Cu), are required. Each time the nutrient solution (mixture of macro and micronutrients) has been made in the mixer tank. Nutrient requirement (ppm) for raising nursery and cucumber grown in soilless media under polyhouse are depicted in Table 18.1. All these nutrients were supplied in the water-soluble form to the plants. The quantity of nutrient solution applied was dependant on the plant development stage. The quantities of water-soluble fertilizers applied are presented in Table 18.2. Apart from the macro and micronutrients, plants also require oxygen, carbon, and hydrogen which were obtained by the plants from air and water. Calcium-free fertilizers and phosphate and sulphate-free fertilizers are dissolved in different fertilizer tanks, i.e. A and B. Phosphoric acid (86.0%) is used for dropping the pH value of fresh nutrient solution. The plants are fertilized according to the development stage. The prepared fresh nutrient solution is checked based on EC and pH. EC and pH of the applied nutrient solution are between from 1.5 to 2.5 dSm^{-1} and 5.5–6.5, respectively (Samarakoon et al. 2006). The preparation of fresh nutrient solution in the mixture tank is shown in Fig. 18.1.

Table 18.1 Nutrient requirement (ppm) for raising nursery and cucumber grown in soilless media under polyhouse

Crop stage	N	NH_4	P	K	Ca	Mg	Fe	Mn	Zn	B	Cu	Mo	S
Nursery raising	200	10	50	253	247	75	0.8	0.55	0.33	0.5	0.05	0.05	120
Normal feed	210	22	50	370	190	75	0.8	0.55	0.33	0.5	0.05	0.05	120

Table 18.2 Water-soluble fertilizers applied (both during autumn and spring)

Water-soluble fertilizer (straight/binary)	Slab saturation (mg $litre^{-1}$)	28–42 DAT (mg $litre^{-1}$)	Normal feed (mg $litre^{-1}$)	Heavy fruit load (mg $litre^{-1}$)
Calcium nitrate	1121.1	1010.0	1010.0	1010.0
Potassium nitrate	257.2	257.2	257.2	410.0
Monopotassium phosphate	227.0	227.0	220.0	220.0
Potassium sulphate	225.0	500.0	510.0	493.0
Magnesium sulphate	700.0	780.0	670.0	780.0
Iron chelate	6.667	6.667	6.667	6.667
Manganese sulphate	1.803	1.803	1.803	1.803
Zn EDTA	1.000	1.000	1.000	1.000
Borax (boron)	2.500	2.500	2.500	2.500
Copper sulphate	0.208	0.208	0.208	0.208
Ammonium molybdate	0.096	0.096	0.096	0.096

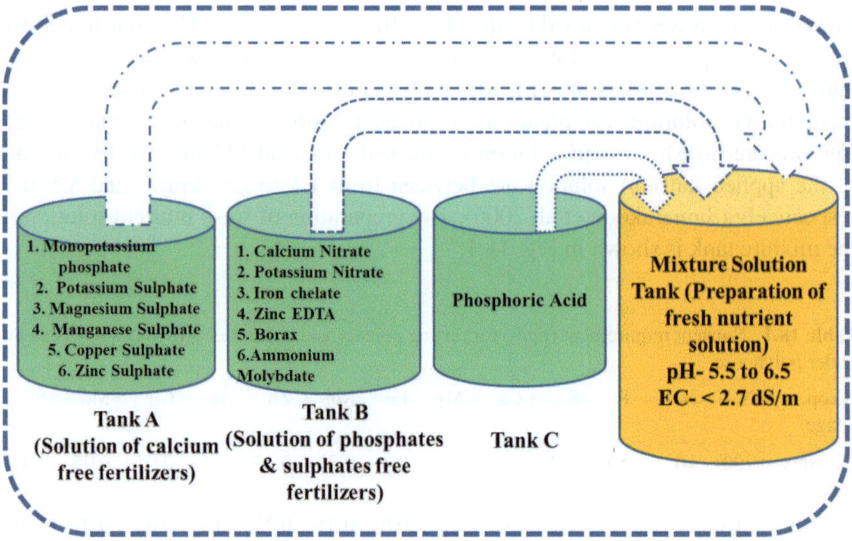

Fig. 18.1 Preparation of fresh nutrient solution

18.3 Measurement of Applied Fresh Nutrient Solution

The fresh nutrient solution was supplied on every day basis to the crop root zone through drip irrigation. The applied nutrient solution amount was measured using measuring cylinders having a capacity of 2.0 L mounted at different places in the research trial shown in Fig. 18.2.

18.4 Leachate Drainage

For safe disposal of leachate drainage water from growing media slabs, a gradient of 1.0% in the North–South direction was laid to polyhouse floor. In growing media slabs, two cut each at 45°angle on two opposite faces along length nearly at 1 inch above the bottom were given. Through an underground pipeline system, the leachate is collected and stored outside the greenhouse in a tank. The leachate reaching the end of the trough was collected through an underground pipeline system in a tank placed outside the polyhouse. The collected leachate was further analyzed for its nutrients content, purified, and used with fresh nutrients in an appropriate proportion for fertigating the crop grown in different growing media in a polyhouse. Leachate collection tank placed outside the polyhouse is shown in Fig. 18.3.

Fig. 18.2 Measurement of nutrient solution applied per plant per day on daily basis

Fig. 18.3 Leachate collection outside the polyhouse

18.5 Recycling of Drained Nutrient Solution (Leachate)

The excess water or nutrient solution draining from the substrate media is termed as leachate. The leachate contains suspended growing media particles, root matter, microorganisms, etc. In order to reuse the drained leachate solution is put through a filter so that the pollutants are filtered out and after checking the EC and pH of the filtrate leachate solution, it is supplied to the again in the polyhouse with the fresh nutrient solution. Further, the reuse of the drained nutrient solution can help in water and nutrient conservation, groundwater contamination, clogging of drippers, etc.

Fig. 18.4 Measurement of drained nutrient solution (leachate)

18.6 Measurement of Leachate Solution in the Polyhouse

At various points in experimental area, customized trays were positioned for the collection and measuring of leachate flowing out of the individual growing media slabs shown in Fig. 18.4. Then leachate was collected in a container placed outside at the outlet of the leachate collection trays. The measurement of collected leachate was taken on regular basis.

18.7 Measurement of pH and EC of Fresh Nutrient Solution and Leachate Solution

The pH and electrical conductivity (EC) of fresh nutrient solution and drained and filtered leachate solution was checked on regularly using a digital tester (Fig. 18.5).

Fig. 18.5 Digital waterproof tester for measurement of EC and pH

18.8 Frequency of Application

Fertigation application determined by quantity concept is effective in laboratories, but on field application, these tactics have hardly been accepted. In polyhouse, feedback process is used to readjust the calculation processes of plant moisture requirements. In substrate soilless system, the frequency of fertigation varies from 3 to 4 times per day; however, duration (5–10 min) of fertigation is governed by the crop growth stage, prevailing weather conditions, and physical properties of soilless media used.

18.9 Fertigation Systems Used in Soilless Cultivation

In soilless cultivation, fertigation systems are used for constant delivery of nutrient solution in the plant root zone. For this, drip fertigation system is the best available technology for efficient mannerly delivery of water and nutrients to the plants grown in soilless media. Low-volume high-frequency irrigation/fertigation concept is popularly used for different horticultural crops grown in soilless media. Generally, there are two types of soilless system: (i) open system and (ii) closed system. In the open system, once the nutrient solution is delivered to the plant roots, it is discarded; however, drained leachate is recovered and reused in the closed system (Raviv and Leith 2008). However, the drained leachate solution is a main concern in the open soilless system which had an adverse result on the surrounding ecosystems and contaminates the groundwater. The recycle measures are taken in a closed soilless system which has 20–40% less nutrient consumption compared to open system (Nederhoff and Stanghellini 2010). Further, Tuzel et al. (1999a, b) also recorded 22 and 35% less water and nutrient consumption, respectively, in closed system with perlite growing substrate. Moreover, Venezia et al. (2003) found that more than 80% of water is saving in closed system in a mixture of peat and perlite media. Similarly, Grewal et al. (2011a, b) also demonstrated a 33% decrease in water required for cucumbers grow in hydroponic whereas lowering the drainage water's environmental impact. Soilless cultivation is cost-effective and has more water and nutrient use efficiency (Van 1999; Savvas 2002a), producing higher yields and rapid harvests (Raviv and Lieth 2008; Nejad and Ismaili 2014) than soil-based cultivation (Grafiadellis et al. 2000).

Conventional drip fertigation system: Conventional drip fertigation system requires more manual attention during and after each irrigation or fertigation. The freshwater for fertigation was supplied either directly from the groundwater through a well or from water harvested in a pond constructed outside the polyhouse. The water supplied from the pond was passed through a filtration unit. Description of fertigation system components is displayed in Table 18.3. The conventional fertigation system fixed outside the greenhouse is listed in Fig. 18.6. The system comprised three fertigation tanks along with one filtrate leachate tank each having a capacity of 1000 L.

The number in the open bracket in Table 18.3 indicates the total quantity.

Semi-automated fertigation system: Semi-automatic systems involve manual consideration at each irrigation. In these systems, timer is used to control the irrigation

Table 18.3 Description of conventional fertigation system components

Component	Description
Main pipeline	Outer diameter = 32.0 mm, inner diameter = 28.0 mm
Lateral pipe	Outer diameter = 16.0 mm, inner diameter = 13.0 mm
Emitter pipe	Outer diameter = 6.0 mm, inner diameter = 4.0 mm
Tanks (4)	Capacity = 1000L
Electric motor (3)	Power = 1.0 hp
Pressure gauge (3)	Operating pressure = 1.5 kg cm^{-2}
Filter (3)	Width = 30 cm, length = 30 cm, inlet = 1 inch
Emitter (603)	Discharge = 2.0 L per hour

Fig. 18.6 Conventional fertigation system

procedures. It is less costly as compared to complete automated system. The irrigator often decides whether to start irrigation, how long it should last, and when it will reset the devices, transmit those back to their previous position, or both, for another irrigation. These systems involve interactions among the controller and system parts placed in the field.

Automated drip fertigation System: Fully automated systems run without human interface except for periodic assessment and regular maintenance. Automated fertigation system involves the use of many sensors and different kinds of controllers which require minimum manual interference in the investigation. It facilitates the actual amount of water supply in right time which improve crop production and minimize the water and nutrient consumption, labour cost, and energy. Nowadays, it is used for automatic nutrient preparation and its uniform distribution in agriculture. The developed economic version of the automated fertigation system is also known as a customized automated fertigation system.

The controller unit is an essential part of the sensor-driven system that ensures that the proper amount of water and nutrient is supplied at right time depending on the moisture level, relative humidity, ambient temperature, etc. Therefore, sensor systems have significantly part in the development of automated systems. There are two types of controller modules used to control fertigation system, viz. open control loop system and closed control loop system. The open-loop controller is relying on a predetermined control framework, with no feedback (Burman and Pochop 1994), whereas the close-loop type controller has a combination of feed-forward predetermined concept and feedback from the previously available necessary data to calculate the water amount for irrigation (Kia et al. 2009). Recently, numerous studies have demonstrated the effectiveness of automating fertigation systems for crop production in polyhouse (Domingues et al. 2012; Neto et al. 2014; Bezerra et al. 2017; Sun et al. 2015; Rahman et al. 2018; Pandey et al. 2023).

The automated fertigation system contains sensor unit, fertigation unit, and drip pipeline unit. The automated fertigation system installed outside the polyhouse is depicted in Fig. 18.7.

Sensor unit: The sensor unit consists of a water level sensor, EC sensor, and pH sensor.

(i) **Water level sensor**: A water level sensor (Fig. 18.8) is used to monitor water/nutrient levels in the tanks. Specifications of the water level sensor are shown in Table 18.4.
(ii) **EC sensor**: EC sensor is used to give information about the fertilizer amount being inserted into the fertilizer solution shown in Fig. 18.9. This sensor helps in maintaining the required range of EC of the water (2.2–2.7 dS/cm).
(iii) **pH sensor**: The pH sensor (Fig. 18.10) determines the acidity or alkalinity of a solution and ranges from 0.0 to 14.0. This sensor helps in maintaining the required range of pH of the water (5.5–6.5).

Fertigation unit: The fertigation system comprises fertilizer tanks, acid tank, water tank, mixer tank, nutrient solution tank, leachate tank, filtrate leachate tank, dosing

Fig. 18.7 Automated fertigation system

Fig. 18.8 Water level sensor

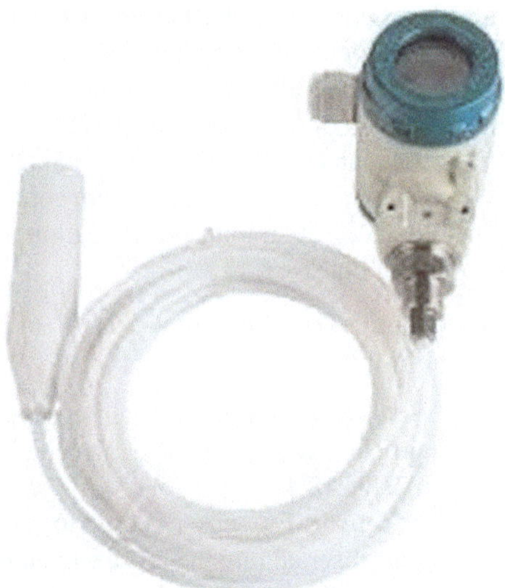

Table 18.4 Specifications of water level sensor

Type	Hydrostatic level transmitter
Measuring range	0.5–200 m
Accuracy	± 0.5%
Response time	< 5 ms
Output signal	4–20 mA,0–5 V DC
Power supply	12–36 V DC

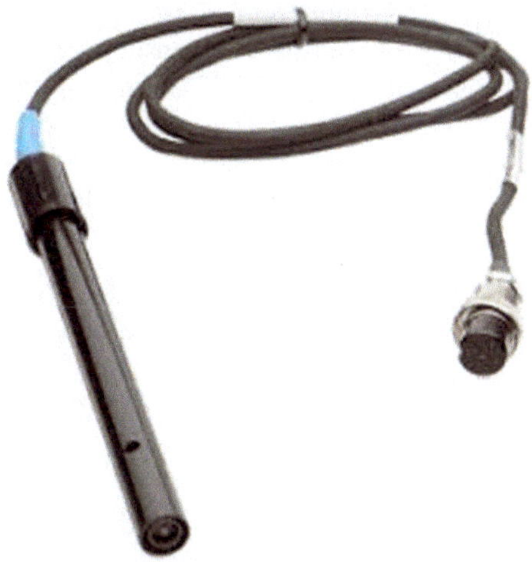

Fig. 18.9 EC sensor

Fig. 18.10 pH sensor

pumps, AC pumps, cooler pump, mixer, water level sensors, EC sensors, pH sensors, solenoid valves, and slow sand filter.

i. **Fresh water tank**: It comprises pH, EC, and water level sensors. The pump is used to lift water to the tank and the sensors measured the water level, pH, and EC of the fresh water in the tank.
ii. **Mixer tank**: It also comprises mixer, pH, EC, and water level sensors.
    (a) Water level sensor: It is used to sense level of the water transferred into the tank from the fresh water tank.
    (b) EC sensor: This sensor helps in maintaining the required range of EC of the water (2.2–2.7 ds/m).
    (c) pH sensor: This sensor helps in maintaining the required range of pH of the water (5.8–6.5).
    (d) Mixer: After dissolving the fertilizers into the tank, a mixer was used to mix them well.
iii. **Nutrient solution tank**: The mixture nutrient solution is passed on to the nutrient solution tank through pump, when the optimal dose of the nutrient solution is attained. This tank also has pH, EC, and level sensors mounted in it so that a final check of the EC and pH is done.
iv. **Fertilizer tank**: Calcium-free fertilizers and phosphate and sulphate-free fertilizers were liquefied in fertilizer Tanks A and B, respectively.
v. **Acid tank**: Phosphoric acid has contained in Tank C.
vi. **Leachate tank**: The pump of 0.5 hp is used to lift leachate to the tank.
vii. **Filtrated leachate tank**: It contains pH sensor, EC sensor, and water level sensor.
viii. **Mixer**: In mixer tank, the fertilizers are mixed with water with the help of mixer (Fig. 18.11). The mixer is working with 24 V AC from the contactor, so it is also connected through a 12 V relay.
ix. **Pump**: The pump is used to lift the water or fertilizer in the tanks. It works with 230 V AC with 400 L per hour capacity. Specifications of the pump are shown in Table 18.5 and displayed in Fig. 18.12.

Fig. 18.11 Mixer

Table 18.5 Specifications of pump

Power supply	230 V AC
Current range	280–330 mA
Operating frequency	50 Hz
Power consumption	50W
Capacity	400 lph

Fig. 18.12 Pump

x. **Dozing pump**: A dosing pump is a volumetric pump, which is designed as to deliver a fertilizer or alternative compound into the flow of water, gas, or steam and offer an absolute precisely flow rate for optimum control. The fertilizer and acid were dissolving into the mixer tank with the aid of a dosing pump so that the optimal amount of the fertilizers is formed as per the treatments required. Each pump will work sequentially with the impulse from the controller. The pumps must be calibrated before setting the operation. It works with 230 V AC with 6 L per hour capacity. Specifications of the dozing pump are shown in Table 18.6 and displayed in Fig. 18.13.

xi. **Cooler pump**: A cooler pump (Fig. 18.14) with 0.25 hp was used to flush out the excess water and the nutrient solution from the tanks.

xii. **Solenoid valve**: Solenoid valves (Fig. 18.15) are used to supply the fresh nutrient solution towards the field.

xiii. **Filter**: Filter (Fig. 18.16) is used to prevent emitters from clogging. Before every irrigation delivery to the plant, the nutrient solution is filtered via screen filter.

Table 18.6 Specifications of dozing pump

Power supply	230 V AC
Current range	280–330 mA
Operating frequency	50 Hz
Power consumption	0.25 hp
Capacity	6 lph

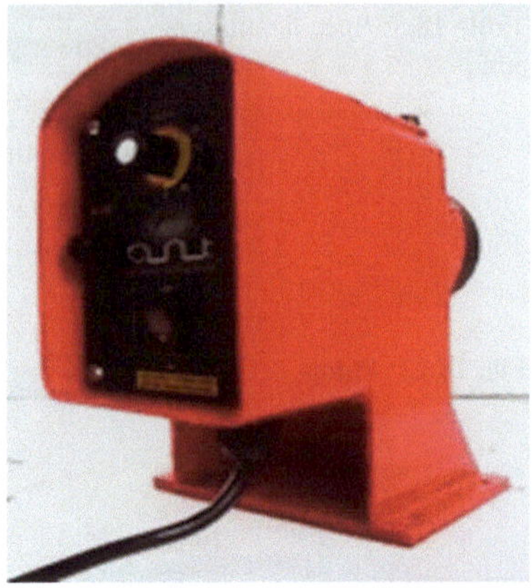

Fig. 18.13 Dozing pump

Fig. 18.14 Cooler pump

Fig. 18.15 Solenoid valve

Fig. 18.16 Filter

Pipeline Unit

i. **Main pipe**: PVC pipe of pressure rating 6 kg/cm^2 with outer diameter = 32 mm and inner diameter = 28 mm was used to convey water from source to the experimental site through sub-main.
ii. **Sub-main**: PVC pipe of pressure rating 6 kg/cm^2 with outer diameter = 32 mm and inner diameter = 28 mm was used as a sub-main pipe to convey water from main lines to laterals.
iii. **Lateral pipe**: LDPE pipe of outer diameter = 16 mm and inner diameter = 13 mm was used as laterals.
iv. **Emitters**: Emitter pipe of outer diameter = 6 mm and inner diameter = 4 mm was used. Emitters were used to supply nutrient solutions to the plant root zone with a capacity of 2 L per hour.

18.10 Effect of Nutrient Management on Growth, Productivity, and Water Consumption in Substrate Culture

Nutrients influenced plant growth and productivity positively. Tuzel et al. (1999a, b) found no significant difference in cucumber yield grown in perlite bags in both open and closed systems. In addition, 22% and 35% of water and nutrients respectively have saved in closed system, while Grewal et al. (2011a, b) demonstrated 33% of water saving in hydroponic cucumbers production. Further, Grasselly et al. (2005) found that 20–30% more nutrient solution amount was applied compared to plant requirement. Gul et al. (2007) compared the effect of organic and inorganic nutrient solution for cucumber production and recorded highest yield with organic treatment. Also, the total fruit yield was reduced in the organic solution by 31.3% as compared to the inorganic nutrient solution. Additionally, Kidoglu and Gul (2009) investigated the inorganic and organic fertilizers' effect on cucumber and found that the size and hardness of fruits were more in organic nutrition as compared to inorganic, while total soluble solids and titratable acidity were higher in the fruits with inorganic nutrient solution. Magan et al. (2008) studied the effect of salinity on the fruit yield of tomato grown in Styrofoam media under different salinities of nutrient solution from 2.5 to 8.5 dSm^{-1} and reported that as increased salinity, the total yield decreased significantly. Similar result was also found by Usanmaz and Abak (2019) in cucumber production under cocopeat. Further, yield was not significantly affected by different nutrient compositions reported by Pandey et al. (2023), Ashraf et al. (2020) in cucumber grown in cocopeat substrate.

18.11 Conclusion

Nutrient management is very essential components in substrate soilless systems for optimizing crop yield, quality, conserves water and nutrients. There are many researchers who have developed the different strategies for nutrient supply in soilless systems. But it should be different depending upon the crop, season, climatic conditions, location, availability of water, etc. The present study will help to the growers for managing and developing new nutrient systems in soilless culture.

References

Ashraf A, Singh KG, Singh A (2020) Development and evaluation of nutrient reuse system in soilless media grown cucumber under protected cultivation. J Plant Nutr 44(9):1241–1257

Bezerra RDS, Evangelista AW, Vellame LM, Alves JJ, Casaroli D (2017) Low cost automation of fertigation with programmable logic controller and gas filled sensors. Eng Agri 37:394–402

Burman R, Pochop LO (1994) Evaporation, evapotranspiration and climatic data. Developments in atmospheric science, Elsevier, Amesterdam, Netherlands, pp 22

Domingues DS, Takahashi HW, Camara CA, Nixdorf SL (2012) Automated system developed to control pH and concentration of nutrient solution evaluated in hydroponic lettuce production. Comp Electro Agri 84:53–61

Godfray HCJ, Beddington JR, Crute IR, Haddad L, Lawrence D, Muir JF, Pretty J, Robinson S, Thomas SM, Toulmin C (2010) Food security: the challenge of feeding 9 billion people. Science 327:812–818

Grafiadellis I, Mattas K, Maloupa E, Tzouramani I, Galanopoulos K (2000) An economic analysis of soilless culture in Gerbera production. Hort Sci 35(2):300–303

Grasselly D, Merlin G, Sedilot C, Vanel F, Dufour G, Rosso L (2005) Denitrification of soilless tomato crops run-off water by horizontal subsurface constructed wet lands. Acta Horti 691(1):329–332

Grewal HS, Maheshwari B, Parks SE (2011a) Water and nutrient use efficiency of a low-cost hydroponic greenhouse for a cucumber crop: an Australian case study. Agric Water Manage 98(5):841–846

Grewal HS, Maheshwari B, Parks SE (2011b) Water and nutrient use efficiency of a low-cost hydroponicgreenhouse for a cucumber crop: an Australian case study. Agric Water Manage 98(5):841–846

Gul A, Engindeniz S, Aykut N (2007) Can closed substrate culture be an alternative for small-scale farmers. Acta Horti 747:83–89
Hussain AK, Iqbal S, Aziem PM, Negi AK (2014) A review on the science of growing crop without soil (soilless culture) a novel alternative for growing crops. Int J Agric Crop Sci 7:833–842
Kia PJ, Far AT, Omid M, Alimardani R, Naderloo L (2009) Intelligent control based fuzzy logic for automation of greenhouse irrigation system and evaluation in relation to conventional systems. World Appli Sci J 6(1):16–23
Kidoglu, F, Gul A (2009) Effect of nutrient sources on fruit quality of cucumbers grown in different soilless media. In: International symposium on strategies towards sustainability of protected cultivation in mild winter climate, vol 807, pp 485–490
Magán JJ, Gallardo M, Thompson RB, Lorenzo P (2008) Effects of salinity on fruit yield and quality of tomato grown in soil-less culture in greenhouses in Mediterranean climatic conditions. Agric Water Manage 95(9):1041–1055
Marques DJ, Luz FDV, Barroso RWF, Bianchini HC (2017) Software for calculation of nutrient solution for fruits and leafy vegetables in NFT hydroponic system. In: Potassium-improvement of quality in fruits and vegetables through hydroponic nutrient management. Intech Open
Mohammed S (2018) Tomorrow's agriculture NFT hydroponics—grow within your budget. Springer
Nederhoff E, Stanghellini C (2010) Water use efficiency of tomatoes in greenhouses and hydroponics. Pract Hydroponics Greenhouses 115:52–59
Nejad AR, Ismaili A (2014) Changes in growth, essential oil yield and composition of geranium (*Pelargonium graveolens* L.) as affected by growing media. J Sci Food Agric 94(5):905–10
Neto AJS, Zolnier S, Lopes DC (2014) Development and evaluation of an automated system for fertigation control in soilless tomato production. Comp Electro Agri 103:17–25
Pandey K (2021) Development of customized automated fertigation system for soilless media in protected cultivation. Ph.D. thesis, Punjab Agricultural University, Ludhiana, India
Pandey K, Singh KG, Singh A (2023) Multi-Sensors based smart nutrient reuse management system for closed soilless culture under protected cultivation. Comput Electron Agric 204:107495
Rahman MKIA, Buyamin S, AbidinM Z, Mokji MM (2018) Development of automatic mixing process for fertigation system in rock melon cultivation. Int J Electri Comp Eng 8:1913–1919
Raviv J, Lieth H (2008) Significance of soilless cultivation in agriculture. In: Soilless culture: theory and practice. Academic Press, San Diego, pp 1–11
Rouphael Y, Raimondi G, Caputo R, De Pascale S (2016) Fertigation strategies for improving water use efficiency and limiting nutrient loss in soilless Hippeastrum production. Hort Sci 51(6):684–689
Sagheer A, Mohammed M, Riad K, Alhajhoj M (2020) A cloud-based IoT platform for precision control of soilless greenhouse cultivation. Sensors 21(1):223
Samarakoon UC, Weerasinghe PA, Weerakkody WAP (2006) Effect of electrical conductivity (EC) of the nutrient solution on nutrient uptake, growth and yield of leaf lettuce (*Lactuca sativa* L.) in stationary culture
Savvas D (2002a) Nutrient solution recycling. In: Hydroponic production of vegetables and ornamentals, pp 299–343
Sharda R, Singh A, Pandey K (2023) Paddy and maize straw-based media as an alternative for cocopeat in soilless cultivation. J Plant Nutr 1–19
Sun G, Li X, Wang X, Li Y, Chen M, Zhang Y, Yan T (2015) Design and testing of a nutrient mixing machine for greenhouse fertigation. Engg Agric Environ Food 8(2):114–121
Tuzel IH, Irget ME, Gul A, Tuncay O, Eltez RZ (1999a) Soilless culture of cucumber in Glasshouses-II. A comparison of open and closed systems on water and nutrient consumption. Acta Hortic 491:395–400
Tuzel IH, Irget ME, Gul A, Tuncay O, Eltez RZ (1999b) Soilless culture of cucumber in Glasshouses-II. A comparison of open and closed systems on water and nutrient consumption. Acta Horti 491:395–400

Usanmaz S, Abak K (2019) Plant growth and yield of cucumber plants grafted on different commercial and local rootstocks grown under salinity stress. Saudi J Biological Sci 26(6):1134–1139
Van D, Nazarideljou SH, Marcelis LF (2020) Nutrient solutions for Arabidopsis thaliana: a study on nutrient solution composition in hydroponics systems. Plant Methods 16:1–14
Van O (1999) Closed soilless growing systems: a sustainable solution for Dutch greenhouse horticulture. Water Sci Technol 39(5):105–112
Venezia A, Tonini A, Piro F, Di Cesare C, Schiavi M (2003) Water and nutrient use efficiency of tomato soilless culture as affected by irrigation method and water quality. Proc Int Symp Manag Greenhouse Crops Saline Environ 609:417–421
Wada T (2019) Theory and technology to control the nutrient solution of hydroponics. Plant Factory Using Artif Light 13:5–14

Chapter 19
Integration of AI and IoT in Soilless Cultivation to Power Sustainable Agricultural Revolution

Ajay N. Satpute, Kishor P. Gavhane, Simranpreet Kaur, Ayushi Jha, Nrusingh Charan Pradhan, and Manojit Chowdhury

Abstract The growing population and the effects of climate change have made it increasingly difficult to provide enough food, as fertile land becomes scarce. Traditional farming methods are inefficient and require large amounts of natural resources, such as land, water, and fertilizers. Maintaining the desired nutrient levels during crop growth is also a challenge. To address these issues and achieve food security, many countries are adopting smart farming concepts that utilize IoT platforms and artificial intelligence. One such technique is soilless cultivation, a cutting-edge method that allows for the growth of fruits, vegetables, and medicinal plants without soil. Soilless cultivation is highly regarded worldwide for its ability to maximize crop production in a productive and intensive manner. It utilizes closed growing systems that enable the collection and reuse of drainage, minimizing water and nutrient losses to the environment. By integrating IoT sensors, continuous monitoring of crop growth is made possible. Hydroponics, aeroponics, and aquaponics are the commonly used soilless cultivation techniques, employing nutrient solutions and inert media to support plant growth. These advancements in technology offer promising prospects for agriculture, but also present challenges. This chapter explores the current technological developments, potential opportunities, and obstacles in the field of soilless cultivation.

A. N. Satpute
ICAR-National Academy of Agricultural Research Management, Hyderabad, Telangana 500030, India

K. P. Gavhane (✉) · N. C. Pradhan · M. Chowdhury
Division of Agricultural Engineering, ICAR-IARI, New Delhi 110012, India
e-mail: kishorgavhane43@gmail.com

S. Kaur
Water Technology Centre, ICAR-IARI, New Delhi 110012, India

A. Jha
Department of Soil and Water Engineering, PAU, Ludhiana, Punjab 141004, India

K. Pandey et al. (eds.), *Artificial Intelligence and Smart Agriculture*,
Advances in Geographical and Environmental Sciences,
https://doi.org/10.1007/978-981-97-0341-8_19

Keywords Soilless cultivation · Hydroponics · Aeroponics · Aquaponics · Internet of Things · Artificial intelligence

19.1 Introduction

Traditional open-field or soil-based agriculture has faced numerous challenges throughout history, and one of the major issues is the decline in available land per person. The amount of land per capita is projected to decrease from 0.5 ha in 1960 to 0.25 ha by 2050 (Sardare and Admane 2013). This decline is due to factors such as urbanization, industrialization, and climate change, leading to a shrinking arable area for cultivation. Additionally, soil fertility has degraded to a level where it cannot be easily improved even with the use of fertilizers, resulting in reduced agricultural productivity. Various factors such as low soil fertility in certain areas, continuous cultivation hindering natural soil fertility build-up by microbes, frequent drought conditions, unpredictable climate patterns, temperature rise, river pollution, poor water management, and declining groundwater levels further threaten food production through traditional soil-based agriculture (Lakhiar et al. 2018). While soil is a vital medium for plant growth, it also poses limitations such as disease-causing organisms, unsuitable soil composition, compacted soil, poor drainage, and erosion-induced degradation (Sardare and Admane 2013). Given these circumstances, solely relying on open-field agriculture will soon make it challenging to meet the global food demand. To address this impending food crisis, it is crucial to simultaneously improve existing farming practices and develop innovative techniques to enhance agricultural productivity (Gashgari et al. 2018).

While it is not feasible to completely replace conventional soil-based cultivation, relying solely on this method in the future is not viable either. Therefore, there is a need to supplement current food production with innovative cultivation methods that can meet the increasing demand. A policy paper explored the potential of alternative approaches like Vertical Farming as a sustainable solution to meet additional food requirements for urban populations in the twenty-first century (NAAS 2019). Consequently, soilless culture, such as hydroponics or aeroponics, is becoming increasingly relevant in addressing the challenges associated with increased food production.

19.2 Soilless Cultivation

Soilless farming, also known as soilless cultivation, refers to a method of growing plants without using soil as the primary medium for root anchorage. Instead, plants are grown in nutrient-rich water solutions that provide the necessary minerals and elements for their growth. Although the principles of soilless cultivation have been employed for a considerable period, the first documented work on this subject can be traced back to Sir Francis Bacon's book "Sylva Sylvarum", published in 1627. In

1666, the Irish scientist Robert Boyle conducted experiments where he submerged plant roots in water, further exploring the concept of growing plants without soil. Woodward (1969) found that spearmint plants grew better in less-pure water than in distilled water. Later in the decade of 1860s, German botanists, Julius von Sachs and Wilhelm Knop prepared mineral nutrient solutions for growing plants in soil-lees culture. These early contributions laid the foundation for the development and advancement of soilless farming techniques.

The process through which plants produce their food is known as photosynthesis, which is represented as,

$$\begin{gathered} 6CO_2 + 6H_2O \rightarrow C_6H_{12}O_6 + 6O_2 \\ (\text{Carbon Dioxide} + \text{Water} \rightarrow \text{Glucose} + \text{Oxygen}) \end{gathered}.$$

In this reaction, there is no mention of soil, which means that plants can be grown without soil. Soil only acts as a storehouse of nutrients and a medium for support to the plant (Devvrat et al. 2017). If, by any means, we can provide these two conditions along with water outside the soil, plants will have favorable environments to grow. The same principle as discussed in the reaction is used in soilless cultivation. As the global population continues to rapidly increase, the availability of water becomes increasingly scarce, making soilless cultivation an attractive option for crop production. Soilless cultivation is a contemporary plant cultivation system that utilizes inert organic or inorganic substrates, often combined with nutrient solutions, to provide plants with the necessary nutrients (Nerlich and Dannehl 2021). This cultivation method allows for precise management of the nutrient solution's quantity and composition, as well as the selection of the growing medium, resulting in improved crop yield and quality compared to traditional soil-based cultivation (Putra and Yuliando 2015).

In soilless cultivation systems, the roots of plants do not need to penetrate to significant depths as required in soil-based agriculture. This is because the roots receive a continuous supply of all the necessary nutrients directly. Consequently, plants in soilless culture exhibit faster growth with smaller root systems and increased density. Compared to soil-based agriculture, soilless culture requires only a fraction of the water and land resources, 20% and 5%, respectively (Dholwani et al. 2018). Moreover, soilless systems eliminate the risk of disease outbreaks, weed infestations, and soil-borne pests. Additional advantages of soilless cultivation include precise control over plant nutrition, efficient water usage, improved nutrient utilization, reduced labor requirements, management of the root environment, and year-round production. Areas of land which are typically unsuitable for traditional crop production can be utilized in soilless cultivation. This approach also allows for the effective management of fertilizers, higher planting densities, increased yield per acre, and improved product quality (Hussain et al. 2014).

Soilless cultivation, in comparison to conventional plant cultivation, requires a higher level of monitoring and precise management. Continuous monitoring of various factors such as lights, temperature, and nutrient solution composition, including pH and electrical conductivity, is essential to maintain a carefully controlled growing environment. Failure to monitor these parameters accurately or human errors can result in crop failure or a decrease in both the quality and quantity of crops. To mitigate these risks and reduce the need for extensive human resources, automation emerges as the most effective solution. By employing appropriate sensors, farmers can regularly monitor environmental indicators, such as water quality and level. Automating the monitoring process enables farmers to save resources and labor, leading to cost and time reductions. The implementation of automation guarantees increased output while embracing the latest technologies that offer convenience in various aspects of life, including agriculture (Joshitha et al. 2021).

19.2.1 Factors Influencing the Soilless Cultivation

Various factors impact crop growth in soilless cultures, including electrical conductivity (EC) and pH levels. EC measures the salt concentration in the soil within the root zone, and an excessive salt concentration can negatively affect plant growth. Thus, it is essential to control EC during crop production in soilless systems. The root-zone EC in soilless culture systems needs to be maintained within specific target ranges to ensure optimal crop development and quality. Typically, root-zone EC fluctuates dynamically due to environmental changes, but it can be regulated by adjusting the nutrient solution concentration. Therefore, accurate EC prediction is crucial for managing nutrition in soilless cultures. Despite the development of various prediction algorithms (Neto et al. 2014; Kinoshita et al. 2016), most nutrient control systems rely on modern EC monitoring and are sensitive to ion balance in root-zone nutrient solutions. However, the challenge arises when attempting to transfer a control system from one location to another with different climatic conditions. To date, no studies have attempted to forecast and predict future EC under diverse climatic conditions (Moon et al. 2019).

The growth of plants can be influenced by pH, which is an important environmental factor that affects soil acidity and alkalinity. These pH levels can create various stresses in plants, leading to nutritional imbalances, toxicities, and deficiencies (Msimbira and Smith 2020). In soilless agriculture production, the nutrient solution plays a crucial role in determining crop yield and quality. Even if the nutrients are present in sufficient quantities, their availability to plants is dependent on achieving the desired pH level in the growing medium (Rajatha et al. 2019).

19.3 Soilless Cultivation Techniques

The future agriculture sector will be driven by the requirement to provide an adequate output that meets customer demands for high-quality products. In protected agriculture, soilless cultivation is utilized extensively to increase control over the growing environment and to eliminate uncertainties in the soil's water and nutrient condition (Putra and Yuliando 2015). The following section represents the main soilless cultivation techniques.

19.3.1 Hydroponics

Hydroponics, derived from the Greek words "hydro" meaning water and "ponic" meaning working, is a method of cultivating plants without soil. Instead, it utilizes mineral nutrient solutions dissolved in water to provide essential nutrients to the plants. The primary principle shared by all hydroponic systems is the delivery of an aqueous nutrient solution, either through a continuous flow or a stationary arrangement. This nutrient solution contains all the necessary elements for healthy plant growth. By ensuring a consistent and adequate supply of nutrients, hydroponics promotes accelerated plant development, resulting in higher yields compared to traditional soil-based agriculture. Hydroponic systems can be broadly categorized into two types: passive and active systems. The primary distinction lies in how the nutrient solution is delivered. In active systems, pumps are employed to supply nutrients directly to the plant roots, while passive systems utilize wicks or growing media to draw in the nutrient solution (Macwan et al. 2020). In passive systems, plant roots come into direct contact with the nutrient solution, with the plants supported by the growing media. However, supporting the plants as they grow can be challenging in this method. Nonetheless, passive systems are the simplest form of hydroponics and require minimal setup. In contrast, active systems utilize an electric pump to deliver the nutrient solution to the plant roots, while gravity facilitates the drainage of excess solution for reuse and recycling. Consequently, active systems offer greater efficiency and productivity compared to passive systems.

19.3.1.1 Types of Hydroponic Systems

Wick System

The Wick system represents a passive hydroponic approach that does not rely on electricity, pumps, or aerators. It serves as a fundamental technique where plants are cultivated in soilless mediums like vermiculite, perlite, or coconut coir. To supply the necessary nutrient solution, lamp wicks or wicks made of materials such as nylon or polyester are employed (Fig. 19.1). This method is particularly suitable for low-maintenance small plants, herbs, and spices like rosemary and lettuce, but it is not effective for plants with high water requirements. The principle behind the Wick system lies in harnessing the capillary action of water, which enables the movement of water and nutrients from a reservoir to the plants. Each plant is connected to the reservoir through at least one wick, ideally positioned directly below the growing

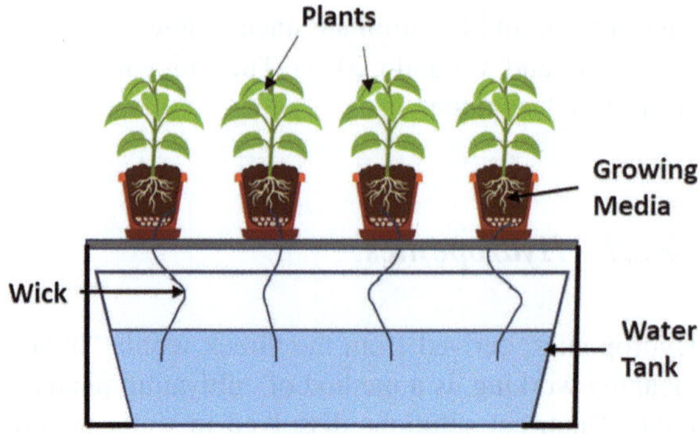

Fig. 19.1 Wick system

surface or tray to achieve optimal efficiency. The wicks facilitate the upward transfer of water and dissolved nutrients to the plants, driven by capillary action.

Nutrient Film Technique (NFT)

Cooper, in the 1960s, introduced a technique that revolutionized the global commercialization of hydroponics. This system involves plants being placed in a flowing stream of water containing all the essential dissolved nutrients for their growth.

By utilizing a water pump, the system circulates the nutrient solution through pipes and into a tray (Fig. 19.2). The shallow stream and suspended roots allow for oxygen absorption in addition to nutrient absorption. The plants are cultivated in an inverted "V"-shaped channel, which offers the advantage of maximizing productivity in a limited space (Macwan et al. 2020). The nutrient solution, consisting of the optimal nutrient blend, is stored in a primary reservoir and then distributed through the pipes. The nutrient film technique (NFT) method ensures a continuous supply of nutrients to the plants, while also enhancing water consumption efficiency through the recycling of excess nutrients and water absorbed by the plants. The NFT is most frequently used to cultivate lettuce and other smaller, quickly growing plants. Using the NFT, some commercial growers also cultivate various kinds of baby greens and herbaceous plants.

Deep-Water Culture

The deep-water culture technique involves suspending plant roots directly in a nutrient solution while utilizing air stones to provide oxygen to the roots. The plants' roots are kept afloat in nutrient-rich water and cultivated in net pots (Fig. 19.3). To prevent the growth of molds and algae in the reservoir, it is essential to carefully monitor the levels of oxygen, nutrients, pH, and salinity. This technique is particularly popular for the hydroponic production of cherry tomatoes and serves as the primary method for their cultivation.

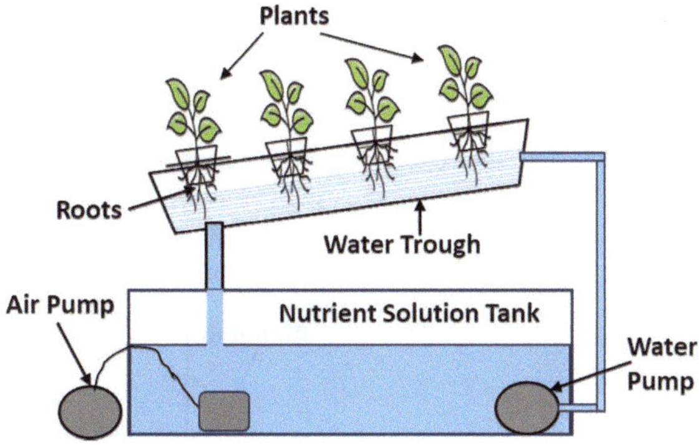

Fig. 19.2 Nutrient film technique

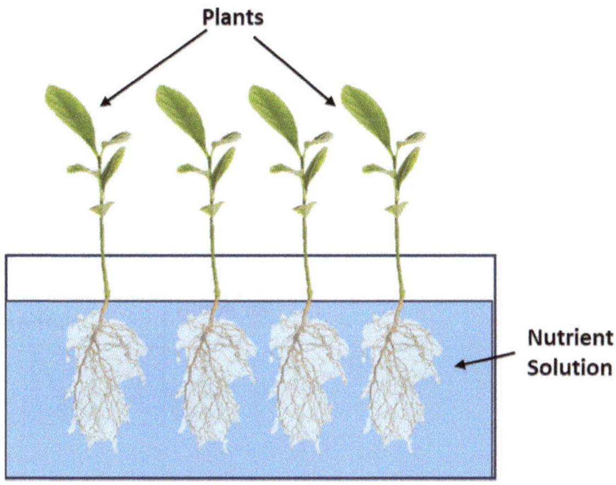

Fig. 19.3 Deep-water culture

Ebb and Flow System

The ebb and flow system, also known as the flood and drain system, was the first commercially available hydroponics system based on the flood and drain principle. It has gained popularity due to its simple maintenance and cost-effective setup. To support plant growth, a combination of water and nutrient solution is pumped from a reservoir and temporarily flooded onto the plants for a specific duration, allowing them to absorb the necessary nutrients (Fig. 19.4). However, challenges such as root rot, algae, and mold have been identified with this system, necessitating the development of a modified version with a filtering unit. Additionally, incorporating computer automation is a viable option to ensure the efficient operation of this system (Wortman 2015).

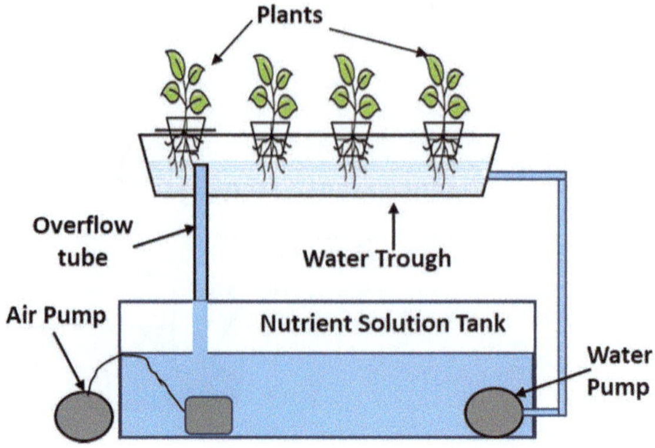

Fig. 19.4 Ebb and flow

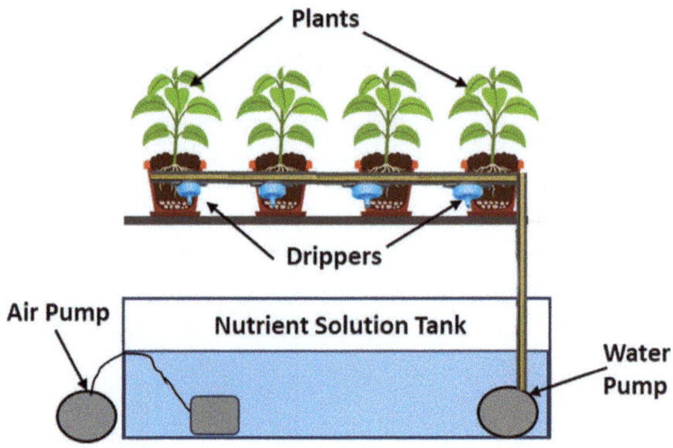

Fig. 19.5 Drip system

Drip System

The main advantages of a drip system over other types of hydroponics systems are that it conserves more water and can survive unexpected equipment or power failures. Due to the scheduled solution supply, this system can be exceedingly expensive and challenging to set up. The nutrient solution is delivered in the proper quantity to each plant's root system through a water pump (Fig. 19.5). With its widespread use, it produces a very high-quality crop of tomatoes and peppers (Sharma et al. 2018).

19.3.2 Aeroponics

The term "Aeroponics" is derived from Greek, combining "Aer" meaning "air" and "Ponos" meaning "labor". It describes a method of plant cultivation without the use of soil, instead relying on an air-based ecosystem (Fig. 19.6). This farming technique can be linked to hydroponics because it involves providing nutrients to plants through

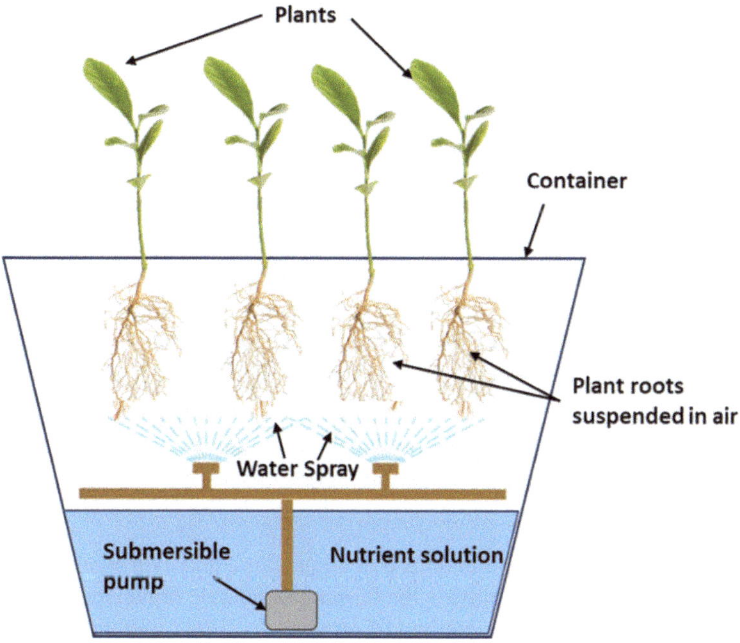

Fig. 19.6 Aeroponics system

water. In an aeroponics system, the plant roots are suspended in the air and periodically misted with a nutrient-rich solution. This results in improved oxygenation of the roots, leading to healthier plants, faster crop growth, and increased yields. Aeroponics operates on the principle that vegetables can be grown in containers using flowing plant nutrients, rather than relying on a substrate or soil as in traditional hydroponics methods.

The fundamental idea behind aeroponics gardening is to grow plants suspended in a closed or partially closed environment by misting their lower stems and roots with atomized or sprayed nutrient-rich water solutions (Kumari and Kumar 2019).

19.3.2.1 Types of Aeroponics System

Depending on the amount of pressure applied to atomize the water beads, there are two different types of aeroponics systems.

Low-Pressure Aeroponics System

The roots of the plant are suspended above a reservoir of nutritional solution or a channel that is inside and connected to a reservoir in the majority of low-pressure aeroponics gardens. A low-pressure pump disperses the nutrient solution using jets or ultrasonic transducers, which drips or drains the nutrients back into the reservoir. When plants reach maturity, their root systems develop dry patches, which prevent them from properly absorbing nutrients (Ali 2017; Kumari and Kumar 2019).

High-Pressure Aeroponics System

In high-pressure aeroponics devices, a high-pressure pump produces mist. Additionally, it is typically employed in the production of valuable crops. This technique makes use of pressured nutrient delivery systems, low-mass polymers, nutrient sterilization, and air and water purification technologies (Kumari and Kumar 2019).

19.3.3 Aquaponics

The term "aquaponics" is a portmanteau of aquaculture and hydroponics. Aquaculture, also known as aqua farming, is the farming of aquatic organisms. Aquaponics is an environmentally friendly method of growing food that uses hydroponics and aquaculture to grow plants and fish without using any soil (Fig. 19.7).

In an aquaponics system, a cost-effective symbiotic cycle exists between fish and plants. The plant bed in this system acts as a bio-filter, receiving fish waste, specifically ammonia, which is then utilized by the plants as a nutrient source. This process effectively removes nitrates from the water, creating a clean environment for the fish. The cycle continues as the purified water is returned to the fish tank, starting the process anew. Compared to conventional irrigation systems, aquaponics offers the distinct advantage of efficient water conservation. By regularly circulating water between the fish habitat and the plant bed, water usage is minimized. Another benefit of aquaponics is the organic fertilization of plants through dissolved fish excrement, providing natural nutrients for plant growth (Table 19.1).

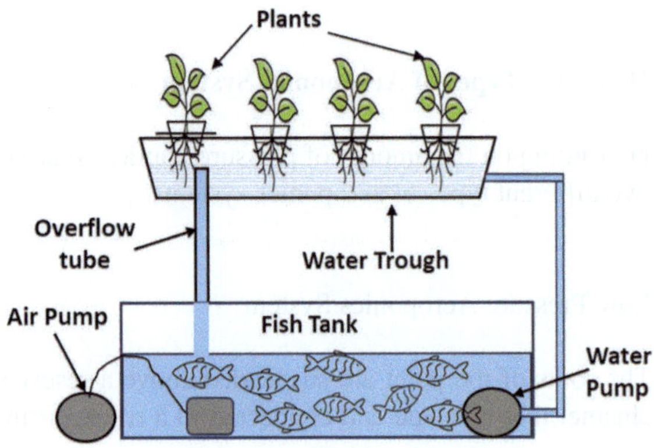

Fig. 19.7 Aquaponics system

Table 19.1 Advantages and limitations of different soilless cultivation techniques

Techniques	Advantages	Limitations
Hydroponics	1. Efficient water use	1. Requires precise control of the nutrient solution
	2. Higher yields compared to soil-based cultivation	2. Susceptible to equipment failures or power outages
	3. Faster growth and shorter crop cycles	3. Reliance on artificial nutrient solutions
Aeroponics	1. Increased plant growth	1. Initial investment and operational costs
	2. Water and nutrient efficiency	2. Technical complexity
	3. Space-efficient and indoor growing	3. Dependency on power supply
Aquaponics	1. Integration of fish farming and plant cultivation	1. Complex system setup and maintenance
	2. Reduced water usage compared to traditional agriculture	2. Sensitive to water quality and fish health
	3. Nutrient-rich waste from fish provides natural fertilizer	3. Potential for imbalances in nutrient levels

19.4 Need for IoT and AI in Agriculture

There is a growing recognition of the need for sustainable farming methods to address concerns about feeding the world's expanding population. With diminishing natural resources, shrinking arable land, and worsening weather patterns due to climate change, countries are increasingly concerned about the future of agriculture (Elijah et al. 2018). Conventional farming practices are characterized by unnecessary human intervention, leading to higher labor costs and increased exposure to severe weather conditions, primarily due to the lack of data-driven decision-making systems (Maraveas and Bartzanas 2021; Pandey et al. 2023; Sharda et al. 2023). Given the detrimental effects of global warming and climate change on agricultural productivity and safety worldwide, the role of technology in future farming is of utmost importance (Pawlak and Kołodziejczak 2020). In greenhouse environments, strict regulation and continuous monitoring are necessary to prevent extensive damage to plants caused by excessive heat or humidity. Internet of Things (IoT) technology offers solutions to these challenges by leveraging artificial intelligence (AI) and machine learning (ML) algorithms to implement water and energy conservation measures, automate farming processes, and introduce machinery for effective crop monitoring (Goel et al. 2021).

IoT technologies enable farmers to remotely monitor and control their farms, facilitating the implementation and monitoring of large-scale agricultural operations (Kassim 2020). AI algorithms can analyze weather data and identify patterns indicating increased risks of pests, diseases, or extreme weather events, enabling farmers to take preventive measures (Javaid et al. 2022). Different crops have varying growth cycles, which affect their water requirements. Traditional flood irrigation methods can be replaced with more efficient irrigation techniques using IoT technology, effectively addressing the issue of water scarcity (Xu et al. 2022). Additionally, IoT and AI can automate irrigation and weeding processes. Smart sensors and irrigation systems ensure that crops receive the right amount of water at the appropriate times, while robotic systems can be deployed to eliminate weeds and perform other tasks.

19.5 What are IoT and AI

In 1999, Kevin Ashton, a British innovator at Massachusetts Institute of Technology (MIT), coined the term "Internet of Things" (IoT). The IoT refers to a network of physical objects that are equipped with sensing devices, software, and network connectivity, enabling data collection and exchange (Boyes et al. 2018). These interconnected devices can communicate with each other, centralized servers, or cloud-based platforms, facilitating remote management and monitoring. The application of IoT in agriculture holds a great potential for farmers, as it can help to increase crop productivity, reduce costs, and contribute to resource conservation and environmental protection (Rehman et al. 2022). IoT technology can be leveraged to optimize soilless cultivation practices by providing real-time data and insights. This allows for automation of various aspects of the growing process and enables remote control of these factors.

19.5.1 Major Components of IoT

19.5.1.1 Sensors

Real-time environmental changes are captured by sensors and transmitted to a remote server or cloud platform via an IoT gateway (Subeesh and Mehta 2021). In the context of soilless cultivation, sensors play a crucial role in monitoring key environmental factors that influence plant growth, including temperature, humidity, light, and nutrient levels. These sensors can be integrated directly onto the plants or installed within the growth chamber. The collected data are then sent to a centralized system or cloud infrastructure, where it can be analyzed and utilized for decision-making purposes.

19.5.1.2 Actuators

Actuators are devices that are able to carry out a predetermined task, such as switching on an environmental control system or altering the temperature of the growing medium in response to a signal from a control system. Actuators can be used to regulate the conditions in which plants are grown in soilless cultivation.

19.5.1.3 Network Connectivity

Connectivity to a network is crucial for the IoT since it facilitates the exchange of data between devices and with the central server. To allow for remote control and monitoring of the growing environment in soilless cultivation, wireless technologies like Wi-Fi, Bluetooth, Zigbee, Long-Term Evolution (LTE), Long-Range Wide-Area Network (LoRaWAN), etc., can be used.

19.5.1.4 Central System and Server

The information gathered by sensors and actuators is processed and analyzed by central systems and remote servers, which then send out commands and control signals to other nodes. Whether on-premises or in the cloud, centralized systems use sophisticated analytics and machine learning to fine-tune operations and make important judgments. Most cloud services use Not only Structured Query Language (NoSQL) databases because they better handle unstructured data and deliver real-time insights. The use of these techniques in soilless farming helps to maintain the ideal environment for plant growth and development and guarantees that crops get the nutrients they need for optimal growth.

19.5.1.5 User Interface

User interfaces are the devices and platforms that enable users to interact with the IoT, including smartphones, tablets, and voice assistants. Controlling and monitoring the growing environment and gaining access to data and insights generated by the IoT can all be accomplished via user interfaces in soilless cultivation.

19.6 Artificial Intelligence

The term "artificial intelligence" (AI) was first used in 1956 by an American scientist, John McCarthy, who was a pioneer in computer science and mathematics. He defined AI as "the science and engineering of making intelligent machines, especially intelligent computer programs". The term AI describes the broad field devoted to studying and implementing computer systems and algorithms that can learn, solve problems, make decisions, and perceive in ways that humans can. It has the potential to greatly affect the agricultural sector by automating a variety of tasks and improving the efficiency and sustainability of farming (Kar et al. 2022).

19.6.1 Use of AI and IoT in Agriculture

Smart systems have gained significant attention in agricultural research due to their potential to enhance crop production while optimizing resource utilization. Precision farming, which involves the application of IoT technologies in urban farming, precision agronomy, greenhouses, and livestock management, showcases the most prominent applications of IoT in agriculture (Gómez-Chabla et al. 2019). By utilizing IoT-based sensors within wireless sensor networks (WSNs), farmers can gather essential data for analysis and decision-making. IoT has also been employed through the use of drones in agriculture, enabling farmers to improve nutrient efficiency, increase yields, mitigate crop damage, and implement site-specific irrigation, weed detection, and fertilizer application, among other benefits (Bagha et al. 2022).

Through IoT, administrators receive warnings whenever environmental conditions surpass predetermined thresholds, allowing them to mitigate potential threats. Real-time adjustments of environmental factors such as temperature, humidity, carbon dioxide concentration, and lighting can be made based on crop conditions (Shi et al. 2019). IoT systems equipped with cameras can detect pest infestations and crop diseases in protected cultivation, empowering farmers to identify and address issues promptly through targeted measures. The emission of excessive greenhouse gases contributes to rising temperatures, negatively impacting agricultural output. To monitor greenhouse gases and CH_4, unmanned aerial vehicle (UAV) systems powered by solar energy and WSNs have been introduced (Malaver Rojas et al. 2015).

Agriculture incurs substantial financial losses due to both wildlife predation and crop theft by humans. To combat theft and damage, various sensors and tracking devices have been deployed in the fields. For instance, Roy et al. (2015) presented a technique for detecting intrusions, which sends text message alerts to farmers whenever suspicious activity is detected in crop fields. Several IoT-based approaches have also been proposed for tracking animal locations and behaviors, while biosensors and computer programs aid in monitoring contagious diseases and general cattle health. Stress has a negative impact on commercial chickens and imposes significant costs on the poultry industry. Monitoring the noise levels of chickens in commercial facilities can serve as an alert for potentially stressful situations (Lee et al. 2015). Wireless monitoring of body temperature and behavioral shifts enables the determination of cows' pregnancy status (Sharma and Koundal 2018).

Artificial intelligence (AI) and machine learning algorithms have been employed in various agricultural activities, including seed classification, soil moisture estimation, insect and weed detection, nutrient analysis, fruit and vegetable grading, and crop yield prediction and price estimation. Techniques such as Support Vector Machine (SVM), Deep Learning (DL), Artificial Neural Network (ANN), Convolutional Neural Network (CNN), Naïve Bayes, and Tree-based models have been utilized for these tasks (Osorio et al. 2020; Loddo et al. 2021). The application of AI techniques enables farmers to reduce water and fertilizer overuse, mitigate crop diseases and pests, increase crop yields, and maximize profits from agricultural produce.

19.7 Application of IoT and AI in Protected Cultivation

Protected farming using greenhouse technology offers a more controlled and stable environment for crop growth compared to open-field cultivation. This method minimizes the impact of environmental events, resulting in increased efficiency and productivity of agricultural resources (Shi et al. 2019). In smart agriculture and greenhouses, the IoT is applied in several key areas: optimizing plant growth conditions, improving irrigation and fertigation efficiency, monitoring crop health, maximizing yields, preventing disease spread, and enhancing greenhouse security (Krishnaveni et al. 2021). IoT architectures rely heavily on sensor systems that provide data on both crops and environmental conditions within the greenhouse. Smart greenhouse sensors encompass a range of measurements, including plant growth, temperature, humidity, insect detection, soil temperature, pH, soil moisture, solar radiation, atmospheric pressure, wind speed, and CO_2 (Maraveas and Bartzanas 2021). These sensors can be categorized into biosensors, which directly record plant-related data, and sensors based on physical parameters that measure environmental factors such as humidity, temperature, and gas levels.

Observing and tracking plant development through IoT systems can lead to improvements in crop yield and production. Khan and Hussain (2019) developed a sensor platform weighing 0.44 g, which was placed on plant leaves rather than near the roots. Additionally, they created a flexible strain sensor capable of measuring plant growth at a micrometer level. The integration with IoT enables real-time visualization and monitoring of sensor data. Despite the controlled environment in greenhouses, crops still face challenges from pests and diseases. To maintain high yields, farmers need to promptly detect diseases and make informed prevention decisions. Early disease detection can be achieved using algorithms like random forest, IoT-enabled environmental sensors, camera modules for image capture, and machine learning techniques such as KNN, Random Forest, Logistic Regression, and Linear Regression (Patil and Kale 2016).

Robotic technology in greenhouse cultivation offers improved performance, production efficiency, labor quality, and safety. Robots can replace human labor in tasks like spraying chemicals and pesticides, performing repetitive operations safely, and operating in hazardous environments. In protected cultivation, robots are utilized for various farming tasks, including planting, fertilizing, watering, grafting, weed control, harvesting, pollination, and multitasking.

19.7.1 *Smart Agriculture*

The emergence of smart automation technologies has revolutionized various aspects of life, including agriculture, offering the potential for advancement and improvement in different agricultural systems. Smart agriculture, facilitated by Information and Communication Technology (ICT), smart devices, and AI implementation, brings numerous benefits such as enhanced process control, data-driven decision-making, and reduced reliance on manual labor (Martinez et al. 2019). Within the context of soilless cultivation techniques, there is significant potential for the application of smart systems. In this context, a smart system refers to the utilization of intelligent devices that perform various functions such as monitoring, operation, and control either individually or simultaneously. These devices encompass sensors, actuators, network connectivity devices, storage units, and computing units (Taha et al. 2022).

Artificial intelligence (AI) has found diverse applications in soilless cultivation techniques which are discussed in the following sections.

19.7.2 *Current AI Research Trends in Soilless Cultivation*

AI entails the creation of computer-based systems that are capable of carrying out operations that need human intellect like decision-making and sensory perception. AI is aided in this task by the IoT, big data analytics, and cloud computing. The digitalization of agriculture which is commonly called smart agriculture is mainly backed by AI. More specifically, the machine learning (ML) algorithms are the main drivers behind it. AI technologies are capable of improving every stage of crop production from sowing, harvesting, and processing to marketing through real-time monitoring of the crops. There are now several intelligent farming systems that are developed for artificial lightning, smart irrigation, weed detection, disease identification, or yield prediction. These systems mainly use the machine learning (ML) algorithms of AI. These algorithms proved better in offering high-accuracy results (Wills et al. 2013) (Table 19.2).

The following sections deal with the AI-based tools, techniques, and technologies that are being used in soilless cultivation.

19.7.2.1 AI Approaches to Predict Growth Parameters

The interactions between plants and their environment, as well as the physiological processes within plants such as photosynthesis, transpiration, and respiration, are highly complex. To effectively manage the cultivation of crops, it is necessary to have techniques that can accurately represent these intricate growth mechanisms. Crop growth models have been developed to describe these processes in a simplified manner, providing quantitative information on plant growth and development, as

Table 19.2 Advantages and limitations of AI in soilless cultivation

Advantages	Limitations
Increased efficiency: AI can optimize resource usage, including water, nutrients, and energy, resulting in higher yields and reduced waste	**Initial investment**: Implementing AI systems in soilless cultivation requires significant upfront investment in infrastructure and technology
Real-time monitoring: AI enables continuous monitoring of environmental factors such as temperature, humidity, pH levels, and nutrient concentrations	**Technical complexity**: AI systems in soilless cultivation can be complex, requiring specialized knowledge and expertise for setup and maintenance
Predictive analytics: AI algorithms can analyze vast amounts of data to predict crop growth patterns, disease outbreaks, and optimal harvesting times	**Reliance on data quality**: AI's effectiveness relies on accurate and reliable data. Inaccurate or incomplete data can lead to erroneous predictions
Autonomous Decision-making: AI can make automated decisions regarding irrigation, nutrient delivery, and climate control, minimizing human error	**Lack of Human Intuition**: AI may lack the intuition and adaptability of human farmers, making it challenging to respond to unexpected situations
Crop personalization: AI systems can tailor environmental conditions to specific crop requirements enabling customization and optimized growth	**Ethical considerations**: The use of AI in agriculture raises ethical concerns related to data privacy, ownership of intellectual property, and potential displacement of human labor

well as predicting crop yield. However, the empirical nature of these models limits their accuracy in predicting outcomes. Fortunately, the emergence of the Internet of Things (IoT) and artificial intelligence (AI) technologies has opened up new possibilities for assessing complex plant processes using continuous data. By leveraging these technologies, large amounts of data on various environmental parameters like temperature and humidity can be collected and utilized for predicting the status of crops, particularly in soilless cultivation (Muangprathub et al. 2019).

AI techniques such as fuzzy logic (FL), neural networks (NN), and deep learning (DL) have gained prominence in predicting plant growth parameters such as quality, yield, and harvest time. In a study by Chang et al. (2021), an AI-based approach was proposed to predict the quality of lettuce grown in a hydroponic system controlled by the IoT. The study utilized three AI techniques: Neural Network (NN), Neural Fuzzy (NF), and Fuzzy Logic (FL). Data for the analysis were collected through an environment module consisting of sensors (e.g., temperature, humidity, CO_2, and light sensors), a microcontroller, a wireless transmission device, and a power supply. The sensor data were stored in a cloud database and analyzed using the AI techniques mentioned above. Six intelligent models (three FL, one NN, and two NF) were adopted and implemented using MATLAB software. These models were utilized to predict various parameters such as the number of leaves, contour area of leaves, dry mass, net transpiration rate, and photosynthesis rate of lettuce grown in hydroponics. By utilizing data from the sensors, these models estimated the quality of the lettuce and predicted its harvest time. Statistical analysis demonstrated the accuracy of these models, highlighting their importance in developing descriptive plant growth models for specific environmental conditions.

The application of IoT and AI technologies provides a promising avenue for assessing and predicting complex plant processes. These technologies enable the collection and analysis of large volumes of data, leading to more accurate predictions and the development of plant growth models tailored to specific environmental conditions.

19.7.2.2 AI for Optimizing the Long-Term Plant Growth

Different AI approaches or techniques can be used for optimizing plant growth in hydroponics. Morimoto et al. (1996) developed an intelligent control system using a combination of neural network and genetic algorithm techniques in order to optimize plant growth in hydroponics. This control system incorporates two additional systems: an expert system and a hybrid system, each serving specific purposes based on the plant's growth requirements. The expert system is responsible for determining appropriate nutrient concentration values at different stages of plant growth. On the other hand, the hybrid system is utilized to establish the optimal set points or levels of nutrient concentration that will maximize stem diameter (SD) and total leaf length (LL) during the seedling stage.

The hybrid system operates by initially identifying the nutrient concentration values that have an impact on SD and total LL, utilizing neural networks for this purpose. Subsequently, the genetic algorithms are employed to simulate the selected model and determine the optimal nutrient concentration value. By maintaining the optimal nutrient concentration levels, this AI-based intelligent control system can effectively optimize long-term plant growth in hydroponics.

19.7.2.3 AI for Disease Detection

In soilless cultivation methods, such as hydroponics, the rapid spread of waterborne diseases poses a significant risk and can lead to substantial losses. The common nutrient solution used in these setups can easily transmit diseases among all the plants. Therefore, early detection of diseases is crucial in soilless cultivation techniques. To address this challenge, researchers have developed AI-based systems for disease detection. One such intelligent system, called JAPISET, has been specifically designed by López-Morales et al. (2008) to assist tomato growers in making informed management decisions. JAPISET diagnoses diseases in hydroponically grown tomatoes by analyzing the characteristics of harmful pest species, as well as the biological and morphological properties of the plants. It calculates the vapor pressure deficit, which indicates the plant's susceptibility to diseases and pests. The system also provides visual representations of the detected pests and diseases, allowing farmers to validate the findings. Moreover, JAPISET offers treatment recommendations from a database tailored to the specific detected disease.

By leveraging AI technology, this intelligent system enables the early detection of diseases in tomatoes cultivated in hydroponics, thereby significantly reducing potential losses.

19.7.2.4 AI for Fault Detection

Artificial intelligence methods are also used for fault detection in soilless cultivation. Neural network-based systems are capable to detect biological as well as mechanical faults. These systems utilize two different fault detection systems for this purpose. The first fault detection system needs input from a neural network from the sensors for detecting the faults. The sensors involved here are the EC sensor, pH sensor, air temperature sensor, humidity sensor, nutrient solution temperature, and light intensity sensor. Different biological faults such as transpiration rate are detected by the second fault detection system. The neural network-based model can detect mechanical problems through the first system within 20–40 min in most cases. This time is sufficient for both aeroponics and hydroponics setups. This time interval is increased to 2–3 h for the detection of biological faults. The time interval taken by the system proves the efficacy of the AI-based fault detection system for soilless cultivation (Ferentinos and Albright 2003).

19.7.2.5 AI for Intelligent Farming

Intelligent farming is a modern technology that makes use of AI for continuous monitoring of the different processes, takes decisions accordingly, and executes those actions on its own. Considering the factors involved in the soilless cultivation of a crop and their complexity, AI-enabled intelligent farming has a great scope for its implementation (Pala et al. 2014). The typical architecture of intelligent farming involves a communication layer that connects all the sensors and any other devices. The sensors used in the greenhouse collect information regarding the status of temperature, humidity, radiation, and CO_2 level. Also, the sensors which monitor the plant growth status are employed. The camera also aids the information by having continuous visual monitoring of the plants. The next layer in the architecture is the acting layer. This layer is responsible for collecting the data. Based on these data, the feedback is provided by the system by performing a certain action. The last layer is the processing layer which mainly handles and stores the data. The decision is made by this layer. The decision is executed by the actuators. Intelligent soilless farming will reduce the dependency on human labor and will improve product quality (Liu et al. 2007).

19.7.2.6 AI for Remote Monitoring and Control of Operations

The concept of remote control refers to the ability of a system operator to modify the state of specific parameters using signals. Remote control encompasses not only monitoring but also extends to the control of systems and actuators, as described by Taha et al. (2022). In soilless cultivation, maintaining various parameters is essential for enhancing the quality and quantity of production. While specific parameters may vary depending on the technique employed (hydroponics, aquaponics, or aeroponics), factors such as electrical conductivity (EC), pH, and nutrient solution temperature are common across all soilless cultivation methods.

Soilless cultivation techniques are susceptible to potential failures due to the numerous parameters that must be maintained and the various factors that can influence them. Issues such as nutrient solution preparation and distribution, pump malfunctions, pipe blockages, and power failures are frequently encountered. Consequently, continuous attention is necessary to ensure uninterrupted operations. (AI) can be effectively utilized to monitor and manage these parameters with great efficiency.

By utilizing various sensors, real-time data are collected and transmitted to growers through cloud-based services. This information empowers growers to take necessary actions using actuators. The remote-control capabilities of AI enable operators to manage artificial lights, pumps, fans, and other relevant actuators, thereby providing comprehensive control over the cultivation environment in soilless systems.

19.7.2.7 AI for Nutrient Optimization

The emergence of AI and the IoT has helped farmers in automating the hydroponic culture and aeroponic culture. The IoT and AI facilitate the machine-to-machine (M2M) interaction. This interaction is beneficial in monitoring pH, water level, EC, temperature, and light intensity and intelligent as well as autonomous controlling of the system (Mehra et al. 2018). This will help in the optimization of resources in soilless cultivation. Dhal et al. (2022) used AI to optimize aquaponic nutrient concentration. The system was designed to alert the users of an aquaponics setup about the nutrient concentration levels. The system also recommends the values of the nutrient concentration that should be maintained to achieve the maximum yield in aquaponics. This optimization helped in selecting the best algorithm based on different error estimation techniques for taking decisions regarding nutrient regulation. Finally, the decision support system was developed which incorporated the support vector machine (SVM) algorithm of machine learning to maintain the nutrient concentration.

19.8 Future Prospects

AI in soilless cultivation, also known as "Smart Hydroponics", holds great promise and offers numerous potential advantages for the future. The following key points outline the potential benefits of AI in soilless cultivation:

1. Optimal Growth and Plant Monitoring: AI-driven systems can analyze data collected from sensors to determine the best growth conditions for plants. By continuously monitoring various parameters such as pH levels, electrical conductivity, nutrient supply, light density, temperature, and humidity, AI can provide real-time insights into plant health and growth.
2. Precision Automation: AI software systems can automate various processes in soilless cultivation. They can control LED lighting, adjust light intensity, and manage irrigation systems based on the specific requirements of plants. This precision automation reduces manual tasks and ensures optimal resource utilization.
3. Predictive Analytics and Decision Support: AI algorithms, including deep neural networks, can analyze complex data patterns and make predictions regarding plant growth and control actions. By leveraging historical data and real-time sensor inputs, AI can assist farmers in making informed decisions, optimizing crop yields, and improving resource efficiency.
4. Remote Monitoring and Control: AI-powered systems enable remote monitoring and control of hydroponics farms. Farmers can access real-time data and manage cultivation operations from anywhere, enhancing convenience and efficiency.
5. Enhanced Resource Efficiency: The application of AI in soilless cultivation can lead to improved resource efficiency. By precisely monitoring and controlling factors such as water, nutrients, and light, AI systems can minimize waste and optimize resource utilization, resulting in higher yields and reduced environmental impact.
6. Advanced Disease and Pest Detection: AI algorithms can analyze visual data from cameras placed within hydroponic setups to detect early signs of diseases or pest infestations. This early detection allows for timely intervention, preventing the spread of diseases and minimizing crop damage.
7. Increased Adoption and Market Growth: Although the adoption of AI in hydroponics is still in its early stages, there is a growing interest in smart farming technologies worldwide. Startups and agripreneurs are exploring AI-driven solutions, and the hydroponics market is expected to expand further.

With ongoing research and development, AI has the potential to revolutionize crop cultivation in controlled environments by offering increased efficiency, improved yields, and sustainable practices.

19.9 Conclusion

The field of smart agriculture is rapidly progressing, primarily driven by advancements in data management. This modern form of agriculture relies on relevant data for informed decision-making and is known by various names such as Smart Farming, Agriculture 4.0, and Digital Farming. Artificial intelligence (AI) and Internet of Things (IoT) technology serve as the foundation of this concept and have already made significant contributions to various aspects of agriculture. With the changing landscape of climate change and a growing global population, conventional agriculture is exploring new avenues. Soilless cultivation methods such as hydroponics, aeroponics, and aquaponics are providing alternative perspectives on food production. However, they also present challenges related to precise management of operations, production factors, disease and pest control, overall management, and profitability.

AI offers solutions to these management challenges in soilless cultivation. Implementing AI in soilless cultivation techniques facilitates early detection of diseases and pests, swift fault identification, precise monitoring and control of nutrient levels, and optimization of plant growth. Intelligent farming approaches aim to grow plants in soilless environments without human intervention, thereby enabling more efficient and profitable soilless cultivation. The application of AI and IoT in soilless cultivation is currently limited by high initial costs associated with technology development and hardware installation. The required software packages for data storage, processing, and management of smart devices and equipment further contribute to these constraints. However, as the demand for safe and sustainable food continues to rise, AI-enabled technologies in soilless cultivation will play a crucial role in achieving this goal.

References

Ali A (2017) Hydroponics, aeroponic and aquaponic as compared with conventional farming. Am Sci Res J Eng Technol Sci 27(1):247–255

Bagha H, Yavari A, Georgakopoulos D (2022) Hybrid sensing platform for IoT-based precision agriculture. Future Internet 14(8):233. https://doi.org/10.3390/fi14080233

Boyes H, Hallaq B, Cunningham J, Watson T (2018) The industrial internet of things (IIoT): an analysis framework. Comput Ind 101:1–12. https://doi.org/10.1016/j.compind.2018.04.015

Chang CL, Chung SC, Fu WL, Huang CC (2021) Artificial intelligence approaches to predict growth, harvest day, and quality of lettuce (Lactuca sativa L.) in an IoT-enabled greenhouse system. Biosys Eng 212:77–105. https://doi.org/10.1016/j.biosystemseng.2021.09.015

Devvrat RR, Ajay K (2017) Automated hydroponics system to deliver nutrients directly to the root of plants. IJERT (Special Issue):327–330. ISSN No. 2278-0181

Dhal SB, Bagavathiannan M, Braga-Neto U, Kalafatis S (2022) Nutrient optimization for plant growth in Aquaponic irrigation using machine learning for small training datasets. Artif Intell Agric 6:68–76. https://doi.org/10.1016/j.aiia.2022.05.001

Dholwani SJ, Marwadi SG, Patel VP, Desai VP (2018) Introduction of hydroponic system and it's methods. Int J Recent Technol Eng 3(3):69–73

Elijah O, Rahman TA, Orikumhi I, Leow CY, Hindia MN (2018) An overview of Internet of Things (IoT) and data analytics in agriculture: benefits and challenges. IEEE Internet Things J 5(5):3758–3773. https://doi.org/10.1109/JIOT.2018.2844296

Ferentinos KP, Albright LD (2003) Fault Detection and diagnosis in deep-trough hydroponics using intelligent computational tools. Biosys Eng 84(1):13–30. https://doi.org/10.1016/S1537-5110(02)00232-5

Gashgari R, Alharbi K, Mughrbil K, Jan A, Glolam A (2018) Comparison between growing plants in a hydroponic system and soil-based system. In: Proceedings of the 4th world congress on mechanical, chemical, and material engineering. ICMIE, Madrid, pp 1–7

Goel RK, Yadav CS, Vishnoi S, Rastogi R (2021) Smart agriculture–urgent need of the day in developing countries. Sustain Comput Inf Syst 30:100512. https://doi.org/10.1016/j.suscom.2021.100512

Gómez-Chabla R, Real-Avilés K, Morán C, Grijalva P, Recalde T (2019) IoT applications in agriculture: a systematic literature review. In: 2nd international conference on ICTs in agronomy and environment. Springer, Cham, pp 68–76. https://doi.org/10.1007/978-3-030-10728-4_8

Hussain A, Iqbal K, Aziem S, Mahato P, Negi AK (2014) A review on the science of growing crops without soil (soilless culture)-a novel alternative for growing crops. Int J Agric Crop Sci 7(11):833

Javaid M, Haleem A, Khan IH, Suman R (2022) Understanding the potential applications of artificial intelligence in agriculture sector. Adv Agrochem. https://doi.org/10.1016/j.aac.2022.10.001

Joshitha C, Kanakaraja P, Kumar KS, Polavarpu A, Gurduru S (2021) An eye on hydroponics: the IoT initiative. In: 7th International conference on electrical energy systems. https://doi.org/10.1109/ICEES51510.2021.9383694

Kar AK, Choudhary SK, Singh VK (2022) How can artificial intelligence impact sustainability: a systematic literature review? J Clean Prod 134120. https://doi.org/10.1016/j.jclepro.2022.134120

Kassim MRM (2020) IoT applications in smart agriculture: issues and challenges. In: 2020 IEEE conference on open systems (ICOS). IEEE, pp 19–24. https://doi.org/10.1109/ICOS50156.2020.9293672

Khan S, Hussain MM (2019) IoT-enabled plant sensing systems for small and large-scale automated horticultural monitoring. In: 2019 IEEE 5th world forum on Internet of Things (WF-IoT). IEEE, pp 303–308. https://doi.org/10.1109/WF-IoT.2019.8767309

Kinoshita T, Yamazaki H, Inamoto K, Yamazaki H (2016) Analysis of yield components and dry matter production in a simplified soilless tomato culture system by using controlled-release fertilizers during summer–winter greenhouse production. Sci Hortic 202:17–24. https://doi.org/10.1016/j.scienta.2016.02.019

Krishnaveni K, Radhamani E, Preethi K (2021) 7 Internet of Things platform for smart farming. In: Internet of Things and machine learning in agriculture. De Gruyter, pp 131–158. https://doi.org/10.1002/9781119752165.ch13

Kumari R, Kumar R (2019) Aeroponics: a review on modern agriculture technology. Indian Farmer 6(4):286–292

Lakhiar IA, Gao J, Syed TN, Chandio FA, Buttar NA (2018) Modern plant cultivation technologies in agriculture under controlled environment: a review on aeroponics. J Plant Interact 13(1):338–352. https://doi.org/10.1080/17429145.2018.1472308

Lee J, Noh B, Jang S, Park D, Chung Y, Chang HH (2015) Stress detection and classification of laying hens by sound analysis. Asian Australas J Anim Sci 28(4):592–598. https://doi.org/10.5713/ajas.14.0654

Liu B, Wang FY, Geng J, Yao Q, Gao H, Zhang B (2007) Intelligent spaces: an overview. In: IEEE international conference on vehicular electronics and safety, ICVES, pp 1–6

Loddo A, Loddo M, Di Ruberto C (2021) A novel deep learning-based approach for seed image classification and retrieval. Comput Electron Agric 187:106269. https://doi.org/10.1016/j.compag.2021.106269

López-Morales V, López-Ortega O, Ramos-Fernández J, Muñoz L (2008) JAPIEST: an integral intelligent system for the diagnosis and control of tomatoes diseases and pests in hydroponic greenhouses. Expert Syst Appl 35(4):1506–1512. https://doi.org/10.1016/j.eswa.2007.08.098

Macwan J, Pandya D, Pandya H, Mankad A (2020) Review on the soilless method of cultivation: hydroponics. Int J Recent Sci Res 11(1):37122–37127. https://doi.org/10.24327/ijrsr.2020.1101.505

Malaver Rojas J, Gonzalez F, Motta N, Villa TF, Etse V, Puig Garcia E (2015) Design and flight testing of an integrated solar-powered UAV and WSN for greenhouse gas monitoring emissions in agricultural farms. In: Proceedings of the IROS workshop on agri-food robotics 2015. Institute of Electrical and Electronics Engineers Inc, pp 1–6

Maraveas C, Bartzanas T (2021) Application of Internet of Things (IoT) for optimized greenhouse environments. AgriEngineering 3(4):954–970. https://doi.org/10.3390/agriengineering3040060

Martinez P, Ahmad R, Al-Hussein M (2019) A vision-based system for pre-inspection of steel frame manufacturing. Autom Constr 97:151–163. https://doi.org/10.1016/j.autcon.2018.10.021

Mehra M, Saxena S, Sankaranarayanan S, Tom RJ, Veeramanikandan M (2018) IoT-based hydroponics system using deep neural networks. Comput Electron Agric 155:473–486. https://doi.org/10.1016/j.compag.2018.10.015

Moon T, Ahn TI, Son JE (2019) Long short-term memory for a model-free estimation of macronutrient ion concentrations of root-zone in closed-loop soilless cultures. Plant Methods 15(1):1–12. https://doi.org/10.1186/s13007-019-0443-7

Morimoto T, Hatou K, Hashimoto Y (1996) Intelligent control for a plant production system. Control Eng Pract 4(6):773–784. https://doi.org/10.1016/0967-0661(96)00068-8

Msimbira LA, Smith DL (2020) The roles of plant growth promoting microbes in enhancing plant tolerance to acidity and alkalinity stresses. Front Sustain Food Syst 4:106. https://doi.org/10.3389/fsufs.2020.00106

Muangprathub J, Boonnam N, Kajornkasirat S, Lekbangpong N, Wanichsombat A, Nillaor P (2019) IoT and agriculture data analysis for smart farm. Comput Electron Agric 156:467e474. https://doi.org/10.1016/j.compag.2018.12.011

NAAS (2019) Vertical farming. In: Policy paper no. 89, National Academy of Agricultural Sciences, New Delhi, 20pp

Nerlich A, Dannehl D (2021) Soilless cultivation: dynamically changing chemical properties and physical conditions of organic substrates influence the plant phenotype of lettuce. Front Plant Sci 11:601455. https://doi.org/10.3389/fpls.2020.601455

Neto AJS, Zolnier S, de Carvalho Lopes D (2014) Development and evaluation of an automated system for fertigation control in soilless tomato production. Comput Electron Agric 103:17–25. https://doi.org/10.1016/j.compag.2014.02.001

Osorio K, Puerto A, Pedraza C, Jamaica D, Rodríguez L (2020) A deep learning approach for weed detection in lettuce crops using multispectral images. AgriEngineering 2(3):471–488. https://doi.org/10.3390/agriengineering2030032

Pandey K, Singh KG, Singh A (2023) Multi-Sensors based smart nutrient reuse management system for closed soilless culture under protected cultivation. Comput Electron Agric 107495

Pala M, Mizenko L, Mach M, Reed T (2014) Aeroponic greenhouse as an autonomous system using intelligent space for agriculture robotics. In: Kim J-H, Matson ET, Myung H, Xu P, Karray F (eds) Robot intelligence technology and applications 2, vol 274. Springer, pp 83–93. https://doi.org/10.1007/978-3-319-05582-4_7

Patil KA, Kale NR (2016) A model for smart agriculture using IoT. In: 2016 international conference on global trends in signal processing, information computing and communication (ICGTSPICC). IEEE, pp 543–545. https://doi.org/10.1109/ICGTSPICC.2016.7955360

Pawlak K, Kołodziejczak M (2020) The role of agriculture in ensuring food security in developing countries: considerations in the context of the problem of sustainable food production. Sustainability 12(13):5488. https://doi.org/10.3390/su12135488

Putra PA, Yuliando H (2015) Soilless culture system to support water use efficiency and product quality: a review. Agric Agric Sci Procedia 3:283–288. https://doi.org/10.1016/j.aaspro.2015.01.054

Rajatha KD, Prasad RS, Kumar UM, Nethra NMN (2019) Standardization of pH of the Solution under soil-less agriculture system for Tomato (Solanum lycopersicum L.). Int J Curr Microbiol App Sci 8(12):1111–1118. https://doi.org/10.20546/ijcmas.2019.812.140

Rehman A, Saba T, Kashif M, Fati SM, Bahaj SA, Chaudhry H (2022) A revisit of internet of things technologies for monitoring and control strategies in smart agriculture. Agronomy 12(1):127. https://doi.org/10.3390/agronomy12010127

Roy SK, Roy A, Misra S, Raghuwanshi NS, Obaidat MS (2015) AID: a prototype for agricultural intrusion detection using wireless sensor network. In: 2015 IEEE international conference on communications (ICC). IEEE, pp 7059–7064. https://doi.org/10.1109/ICC.2015.7249452

Sardare DM, Admane SV (2013) A review on plant without soil—hydroponics. Int J Res Eng Technol 2(3):299–304. https://doi.org/10.15623/ijret.2013.0203013

Sharda R, Singh A, Pandey K (2023) Paddy and maize straw-based media as an alternative for cocopeat in soilless cultivation. J Plant Nutrition 1–19

Sharma B, Koundal D (2018) Cattle health monitoring system using wireless sensor network: a survey from innovation perspective. IET Wirel Sens Syst 8(4):143–151. https://doi.org/10.1049/iet-wss.2017.0060

Sharma N, Acharya S, Kumar K, Singh N, Chaurasia OP (2018) Hydroponics as an advanced technique for vegetable production: an overview. J Soil Water Conserv 17(4):364–371. https://doi.org/10.5958/2455-7145.2018.00056.5

Shi X, An X, Zhao Q, Liu H, Xia L, Sun X, Guo Y (2019) State-of-the-art internet of things in protected agriculture. Sensors 19(8):1833. https://doi.org/10.3390/s19081833

Subeesh A, Mehta CR (2021) Automation and digitization of agriculture using artificial intelligence and internet of things. Artif Intell Agric 5:278–291. https://doi.org/10.1016/j.aiia.2021.11.004

Taha MF, ElMasry G, Gouda M, Zhou L, Liang N, Abdalla A, Rousseau D, Qiu Z (2022) Recent advances of smart systems and Internet of Things (IoT) for aquaponics automation: a comprehensive overview. Chemosensors 10(8):303. https://doi.org/10.3390/chemosensors10080303

Wills A, Schön TB, Ljung L, Ninness B (2013) Identification of hammerstein-wiener models. Automatica 49:70–81. https://doi.org/10.1016/j.automatica.2012.09.018

Woodward J (1969) II Some thoughts and experiments concerning vegetation. Philos Trans Roy Soc Lond 21(253):193–227

Wortman SE (2015) Crop physiological response to nutrient solution electrical conductivity and pH in an ebb-and-flow hydroponic system. Sci Hortic 194:34–42. https://doi.org/10.1016/j.scienta.2015.07.045

Xu J, Gu B, Tian G (2022) Review of agricultural IoT technology. Artif Intell Agric. https://doi.org/10.1016/j.aiia.2022.01.001

Chapter 20
Plant Disease Detection, Diagnosis, and Management: Recent Advances and Future Perspectives

Priyanka Negi and Shubham Anand

Abstract A vast range of plant diseases and pests are known to affect agricultural crop and cause significant crop losses. For over many years the researchers are working on enabling a better approach for disease detection, diagnosis and control. The development of modern technology such as implementing sensors, machine learning has made it possible to identify plant pathogens quickly, precisely, and sensitively. However, the application of such advanced technologies has always been a challenge as the green house and field applications are complex and require different design studies. Selection of the type of sensor to be used, the sensor platform, and other accessory techniques require great technical knowhow since all the pathosystems are unique and differ in host–pathogen–environment interactions. Therefore, large database needs to be created to enable deeper insight into these interactions. Hence, the use of a quartz crystal microbalance (QCM), surface-enhanced Raman spectroscopy (SERS), radiofrequency identification (RFID) microchips, microfluidic systems, and smartphone-based fingerprinting of leaf volatiles are some of the examples of use of sensors and other modern technologies. Wearable sensors can be used to detect and manage biotic stress. Modern machine learning techniques are created to achieve quick data analysis of such complicated datasets, which not only saves time but also permits an objective examination of the data. The potential to find likely coherent parameters during plant-pathogen-environment interactions is particularly great for deep learning systems. This aids farm management in the context of smart agriculture and assists pathologists in disease diagnosis. The powerful computers, sharp displays, and comprehensive built-in accessory sets, including high-definition cameras and smartphones are a very new way to help diagnose diseases. For the purpose of diagnosing diseases, this method makes use of image processing technology, which is typically integrated with artificial intelligence such as neural networks. So, this chapter will present the work of several researchers regarding application of sensors and other technologies in plant disease detection, diagnosis, and control.

P. Negi (✉) · S. Anand
Punjab Agricultural University, Ludhiana, Punjab 141004, India
e-mail: shubham-1904001@pau.edu

K. Pandey et al. (eds.), *Artificial Intelligence and Smart Agriculture*,
Advances in Geographical and Environmental Sciences,
https://doi.org/10.1007/978-981-97-0341-8_20

Keywords Artificial intelligence · Conventional techniques · Crop diseases · Smart agriculture

20.1 Introduction

Diseases are referred to the manifestation of conditions that lead to abnormal function of normal physiological processes of plants (i.e. uptake of water and nutrients, translocation of photosynthates and metabolism of synthesized substances, reproduction, etc.) caused by biotic and/or abiotic factors in a manner that plant either loses the accepted appearance, productivity, or both. Different signs and symptoms can be associated with the disease development in the plants. Therefore, the branch of science that deals with the study of plant diseases with respect to their causation, development, and management is referred to as phytopathology. For understanding the host plant interaction, it is important to study the various factors affecting disease development along with the various disease detection and diagnosis tools. Once the cause of the disease is successfully established, we can further study the epidemiology as well as management of the disease. Low production due to plant diseases jeopardises the safety of the food supply. The world is concerned about both nutritional and food security. A variety of biotic as well as abiotic factors are known to adversely affect the crop productivity of which diseases as well insect pests cause major losses to the agricultural production and reduce the farmers' profit. It is reported that the diseases, pests, and weeds together contribute to around 20–40% loss of productivity annually, with the pathogens alone contributing up to 10–15% of crop losses (Savary et al. 2012; Mitra 2021). More specifically, plant diseases cause annual economic losses to the agricultural sector of over $220 billion, which affects more than 800 million people's access to food (Mitra 2021). According to Godfray et al. (2010), by 2050, the world's population is estimated to increase at an alarming rate of 70%, necessitating a rise in agricultural production in order to comply with the food security. Plant diseases are among the biotic issues that cause the most worry. In the case of major cereals, biotic stress causes a loss of about 23% of the anticipated yield that can be achieved (FAO 2009), in addition to postharvest losses of 30–40% related to illnesses and subpar output quality. Accurate diagnosis and identification of plant diseases are essential for guaranteeing food security and preventing the spread of invasive pathogens and pests in the face of environmental change and globalisation. Additionally, accurate, sensitive, and specific diagnosis is required for the efficient and cost-effective control of plant diseases.

Before understanding the various techniques and methods of identifying the plant diseases one must understand what is meant by disease detection and diagnosis. Plant disease detection is the identification of the microorganism/pathogen, whereas plant disease diagnosis refers to conclusively draw the relation between the microorganism associated and the symptoms produced or in other words to establish the cause of the disease in relation to the symptoms produced by the disease. Visual inspection for the identification of diseases of plants have been one of the most acceptable methods for disease assessment using traditional scales. Recognising the visual symptoms have been indispensable for the diagnosis of plant diseases, however, this method is rather more subjective to be used. New technologies have given way to light photography, digital image analysis, high-throughput serological techniques like enzyme-linked immunosorbent assays (ELISA) and molecular techniques like polymerase chain reactions in the diagnosis of plant diseases (PCR) which are more objective offering reliability, precision, and efficiency. The application of bioinformatics in plant pathology has made it feasible to identify certain motifs and DNA sequences, increasing the accuracy of existing techniques for the detection of plant diseases. Though improvements in every aspect of plant disease diagnosis have been made, improving sensitivity and specificity has remained the main topic for advancement (Balodi et al. 2017).

However, the search for effective diagnostic tools to identify and control plant infections must overcome additional challenges in an era characterised by climate change and increased international trade. In addition to trade of goods, globalisation is also responsible for introduction of a large number of harmful "alien" species, such as viruses, phytoplasmas, bacteria, fungi, insects, nematodes, and weeds into the new areas and poses serious challenges to agriculture. Early detection of plant pathogens is thus becoming increasingly crucial in plant health monitoring in order to manage disease infections at various phases of development, decrease the risk of disease propagation, and avoid the introduction of new ones (Buja et al. 2021).

20.2 Importance of Advanced Techniques

The fundamental procedures in farming, which have been carried out for centuries by farmers to prevent crop loss and maintain crop quality, are disease identification and selecting the proper control measures for managing crop diseases. However, the draconian increase in several agricultural diseases has been observed as a result of global climatic changes, altered human lifestyles, pollution, and numerous other reasons. It becomes crucial to understand the diseases that affect the crops in order to manage them successfully. Many crop diseases and infections typically go undiagnosed due to lack of expertise in disease identification, which eventually impacts the crops' overall output and quality. Therefore, automatic crop disease identification and categorisation is necessary and urgently needed in order to address such concerns. Technology advancements have made detection and diagnosis procedures more sensitive and accurate, which is particularly necessary presently in order to prevent the spread of invasive, harmful infections to previously uninvaded places. The detection procedures that are utilised for diagnostic or quarantine measures are more consistent, reproducible, and have a low rate of false positives. With advanced techniques, pathogen's genetic variability, its concentration, and resemblance to other organisms are taken into account (Candresse et al. 2014).

Machine learning and computer vision are already being used in agriculture by researchers and farmers from all over the world for a number of purposes. Using machine learning-based techniques, agricultural diseases can be promptly and reliably identified and categorised. This will increase crop quality and productivity while lowering labour costs and improving farmer accuracy in crop cultivation. Early detection of the disease in the crop helps to reduce potential productivity losses and enables targeted and specialised management strategy.

20.3 Methods of Plant Pathogen Detection

For many plant pathogen systems, there has been a long-standing dearth of understanding of the physiological mechanism of host–pathogen interaction. There have recently been created quantitative high-throughput image-based methods for phenotyping plant growth and disease development. These techniques aid in understanding how pathogen infection alters plant physiology, which in turn aids in understanding how disease symptoms occur. Proximate detection, immunological and DNA-based methods, approaches based on analysis of volatile compounds and genes as biomarkers of disease, remote sensing (RS) technologies along with spectroscopy-based methods, sensors-based techniques, and biophotonics are all current and future methods for the detection of plant diseases.

20.4 Traditional Methods

These are the traditional techniques that have been used for plant disease detection, diagnosis, and management. These procedures have assisted the scientists in determining the origin of a disease and in suggesting an appropriate countermeasure. Depending on their manner of detection, they have evolved with time and with the help of technology. Some of the traditional methods used are:

i. Based on symptoms
ii. Based on biological properties
iii. Based on morphological and/or cultural properties
iv. Based on proteins or nucleic acids
v. Based on sequencing.

20.5 Advanced Methods

Modern approaches and high-throughput technology are used in advanced methods, which are faster and more accurate than earlier techniques. The field of plant pathology has advanced in the modern era of deep learning, machine learning, and other artificial intelligence technologies. Here are a few cutting-edge techniques:

20.6 Visual Inspection

Visual inspection of plant germplasm and selection of healthy material is one of the simplest methods for detecting plant diseases. For the visual inspection, a variety of techniques can be utilised, such as visible light imaging, chlorophyll florescence imaging, hyperspectral imaging, and thermal imaging (Mutka and Bart 2015). The biological and environmental elements have a significant influence on these procedures. Furthermore, due to a lack of homogeneity and the need for technical proficiency, traditional techniques of grading disease severity are unreliable for estimating disease in the community (Mutka and Bart 2015). When dealing with asymptomatic plants or less virulent diseases, such as *Xylella fastidiosa*, visual inspection may not be beneficial. Visual inspection could have trouble further separating strains that are closely linked to one another. This approach of diagnosis is greatly impacted by the ability to recognise symptoms and distinguish them from similar symptoms brought on by other variables (Riley et al. 2002).

20.7 Serodiagnostics

Generally, the visual symptoms of any disease are helpful in deciding the further diagnostic techniques to be applied for developing the cause or infection and suggesting the management strategies. However, the chances of successfully getting a good diagnosis are extremely low when the pathogen is found in asymptomatic items (such as potato tubers, wood, etc.) or seed materials and diseases with similar symptoms. Therefore, the development of serodiagnosis strategies have provided a better diagnostic approach. The identification of disease based on antibodies backed by a colour shift in the assay is known as serological detection of plant pathogens. Immunoglobulins (Ig) are typically foreign proteins, complex carbohydrates, polynucleotides, or lipopolysaccharides, which form the building blocks of antibodies that are obtained from animal's body in response to the presence of antigens. Each antibody has a unique binding site for a particular antigen. The first and most popular serological method for the identification of plant viruses was called enzyme linked immunosorbent assay (ELISA) (Clark 1981). Thereafter, several variants of ELISA were used. The two ELISA kinds that were modified to meet the particular needs of the detection test were indirect ELISA (antigen does not directly bind to antibody-enzyme conjugate) and direct ELISA (conjugated antibody-enzyme complex directly binds to antigen) (Paul Khurana 2006). Currently, monoclonal and polyclonal antisera are widely employed in ELISA to detect bacteria and fungi that cause plant diseases.

With the use of these commercially available recombinant antibodies, ELISA's accuracy has greatly risen, practically improving the reliability of the diagnosis (López et al. 2001). For instance, the tomato mosaic virus, the potyvirus family, *Ralstonia solanacearum*, and many other pathogens are frequently detected using the lateral flow-ELISA. Some of them can even be used in the field or at any stage of the production and distribution process (Danks and Baker 2000). Similar site-specific detection methods, including tissue print ELISA and other lateral flow devices are commonly used.

Other supplemental serological techniques have had a big impact on the revolution in plant disease detection and at the same time stimulated the identification of many phytopatho systems, allowing for precise quantification of pathogen participation in a diseased sample. Additionally, serological approaches can detect isolates with higher virulence or the ability to detect isolates with pesticide resistance. The number or extent of the pathogen, however, is not always correlated with the severity of the disease since pathogen biomass is not always directly proportionate to the outward manifestation of disease symptoms. It is essential to precisely and quantitatively detect the pathogen in the injured plant organ in order to properly diagnose and quantify the pesticide dose prior to application.

20.8 Flow Cytometry (FCM)

The optical method of flow cytometry makes use of laser technology. Using this method, a sizable population of fluorescently marked cells in a fluid stream is analysed. It is possible to monitor the amount of light dispersed and the emission of a fluorescence label as the particles move past a focused light source. This method is widely used for protein engineering, cell sorting, biomarker detection, and cell counting. Multiple samples can be processed simultaneously by electronic detection equipment, and the method measures many parameters at once (Fang and Ramasamy 2015). Flow cytometry is widely used for the assessment of microorganisms in drinking and marine water as well as food processing. Although it may be helpful for multiplexed pathogen detection, estimating the genome size of bacterial, oomycete, and fungal illnesses, and establishing the pathogens' viability, FCM is still not commonly employed in plant pathology as a study and detection approach (D'Hondt et al. 2011). Although pathogens like bacteria can be found using serological techniques like ELISA, these techniques cannot determine whether the pathogens are still alive. A key criterion for determining the risk of disease development is to check viability in addition to the presence of the pathogen. FCM may therefore be used in plant pathology to detect diseases and determine their viability, particularly bacterial pathogens.

20.9 Sensors in Plant Disease Detection

Detecting events or changes in the physical environment such as light, temperature, moisture, pressure, heat, etc., is the function of a sensor, a device, or detector. It can save data as electromagnetic signals or as raw data, and these data are then analysed using a variety of image classifiers, including ANN, SVM, SVAM, and others (Petersson et al. 2016). Depending on the interactions between the signal and the object, several noninvasive sensor types can evaluate various parameters of the targeted objects.

Optics-based sensors, often known as electro-optical sensors, detect light. Wide-ranging light waves (VIS, NIR, and SWIR) are detected by these sensors, which then transform the light waves into electromagnetic signals that produce spectral signatures or spectral profiles. According to their unique characteristics and applications, optical sensors are classified into many types, such as RGB sensors, hyperspectral and multispectral sensors, thermal sensors, etc. (Mahlein et al. 2018).

RGB image sensors: Plant pathology uses digital photographs as crucial instruments for determining the health of plants. Digital cameras offer an accessible supply of RGB (red, green, and blue) digital images for detection, identification, and quantification. Plant biotic stress has been identified using the red, green, and blue channels of RGB-colour pictures (Bock et al. 2010). Understanding the characteristics of plant diseases requires spatial data in addition to colour data in RGB, LAB (L for lightness, A and B for colour-opponent dimensions, based on nonlinearly compressed coordinates), YCBCR (colour compression scheme Y is luma component and CB and CR are the blue-difference and red-difference chroma components, respectively), or HSV (hue, saturation, and value). Additionally, colour, grey level, texture, dispersion, connectivity, and form parameters can be used to identify criteria for the detection and identification of disease signals in plants (Camargo and Smith 2009; Neumann et al. 2014). Digital infrared thermography can be used to track variations in the aggressiveness of *Venturia inaequalis* isolates, host tissue resistance, and degree of apple scab infection (Oerke et al. 2011).

Hyperspectral or multispectral sensors: Multispectral sensors, which typically evaluate an object's spectral information across a number of rather large wavebands, were the first spectrum sensors to be developed. Hyperspectral optical sensors collect pictures in a collection without harming anything (Bergstrasser et al. 2015). These photos are combined to form a three-dimensional (x, y, λ) hyperspectral data cube for processing and analysis, where x and y stand for the scene's two spatial dimensions and λ for the spectral dimension (comprising a spectrum of wavelengths). Hyperspectral cubes are produced using a variety of platforms, including UAV and SAR (Wendel and Underwood 2017). To locate things, identify materials, or detect processes, hyperspectral imaging (HSI) seeks to obtain the spectrum for each pixel in an image of a scene. Spectral imagers can be divided into four categories, including whisk broom, push broom, filter wheel, and snapshot sensors (Thomas et al. 2017). Human eye perceives visible light (400–700 nm) as having colour in primarily three bands (red, green, and blue), whereas the spectrum is divided into several more bands by HSI from the VIS to SWIR range (400–2500 nm). Images can be divided into bands using this technique outside of the visible spectrum. Fine wavelengths are included in the HSI recorded spectra (more bands). Continuous spectral bands measured by HSI are known as multiband imaging (MSI). It will also produce jumbled pixels if it identifies a significant number of relatively narrow frequency bands (false positive result).

HSI and multispectral imaging are connected. Sometimes it is inappropriate to establish the distinction between hyper and multi band on the number of bands or the kind of measurement. Multiple pictures are handled by MSI at distinct and fairly constrained bands. Multispectral imaging at the visible wavelength is distinct and somewhat narrower than colour photography. Due to these exceptional qualities, HSI technology has begun to be used in phytopathology for a variety of research, including disease detection, phenotyping, and plant pathogen interaction (Leucker et al. 2016).

It has been demonstrated that both imaging and non-imaging sensors may be used to detect changes in plant vigour. Modifications to the biophysical and metabolic properties of plant tissue cause changes in reflectance. According to West et al. (2010), diseases impact the optical properties of leaf tissue through altering crop canopy morphology and density, transpiration rate, tissue colour and form, and how plants interact with solar radiation. Reflection of leaves is influenced by plant stress due to colour alterations, hypersensitive reactions, and cell wall breakdown. A progression of chlorotic and necrotic tissue with various optical and chemical characteristics, as well as typical fungal structures such powdery mildews, rusts, and downy mildews, may also be able to be recognised as disease-specific symptoms. The makeup and concentration of the leaf pigments tend to change when pathogens that cause chlorosis and necrosis symptoms attack plants. The pattern of reactions and the extent of up- and down-regulation of physiological systems are influenced by interactions between the host and pathogen. Necrotrophs quickly kill plant cells and then take the nutrients released from the dead tissue, in contrast to biotroph infections, which create haustoria to take nutrients from living cells. Considering that each disease has its own set of symptoms, several wavebands may be effective in the diagnosis of each. Utilising sensing tools, monitoring of vegetation still has trouble recognising diseases and separating them from other diseases and abiotic pressures. Spectral reflectance data cannot currently be interpreted without an understanding of the spectral characteristics of leaves and related symptoms. According to Mahlein et al. (2012), the VIS and NIR spectrum ranges produce excellent results for disease identification. For instance, sugar beet leaves affected by the diseases Cercospora leaf spot, powdery mildew, and sugar beet all showed distinctive and disease-specific spectral signatures. Their dependence on disease severity highlighted the causal connection between diseases and specific spectral changes. Spectral data could be used to identify *Phytophthora infestans* on tomatoes (Zhang et al. 2003), *Magnaporthe grisea* on rice (Kobayashi et al. 2001), and *V. inaequalis* on apple trees (Zhang et al. 2003). It becomes feasible to recognise agricultural damage caused by viruses, illnesses, or insects using spectral sensors.

Thermal sensors: IRT measures plant temperature and has links to microclimate in crop stands, variations in transpiration brought on by early infections by plant diseases, and plant water stress. The thermographic and infrared camera can identify IR emissions in the thermal range of 8–12 m, and these emissions are shown in false-color images with each image pixel containing the temperature of the object being examined. In plant science, IRT can be used in a variety of temporal and spatial contexts, including aerial and small-scale applications. However, it is regularly impacted by external factors like temperature, sunlight, rainfall, or wind speed. According to Jones et al. (2002), there is a significant relationship between leaf temperature and plant transpiration, which is impacted variably by a number of illnesses. Root diseases (like *Pythium* spp. or *Rhizoctonia solani*) or systemic infections (like *Fusarium* spp.) typically alter the transpiration rate and water flow throughout the entire plant or its organs, in contrast to many foliar pathogens, including leaf spots and rusts, which cause particular, localised changes. Local temperature changes brought on by pathogen infection or defensive mechanisms have been connected to plant-virus interactions in tobacco and *Cercospora beticola* in sugar beetroot (Chaerle et al. 2004). IRT was effective in keeping an eye on cucumber downy mildew and apple scab. Gomez (2014) observed *Peronospora sparsa* infection and its dissemination on various *Rosa* cultivars.

Fluorescence imaging: Different metrics of chlorophyll fluorescence are used to estimate changes in photosynthetic activity in plants. Chlorophyll fluorescence imaging equipment often uses active sensors with an LED or laser light source to assess photosynthetic electron transport (Murchie and Lawson 2013). The differences in photosynthetic activity caused by biotic and abiotic stressors over the leaf area have been investigated using this method (Burling et al. 2011). Fluorescence imaging coupled with an image analysis method can aid in classifying and quantifying various fungi infections (Konanz et al. 2014). Fluorescence imaging of plant samples, however, necessitates adherence to a specific methodology, making it challenging to employ this approach in agricultural greenhouses or field settings.

20.10 Scope of Sensors in Plant Disease Detection

Scientists have been working on developing sensor technologies and their advancements that could be used in agricultural applications over the past four to five decades. The availability of a wide range of sensor systems has given plant pathologists fantastic opportunities to gather high-resolution data, early detection, and identification of plant diseases. Advances in plant phenotyping and precision agriculture have produced complex methods, of which sensors are the most efficient. It is one of the noninvasive field-based evaluations of crop nutrient status. Two examples of experimental prototypes include the creation of a robot with multispectral cameras and an online machine vision pipeline for the detection of the tulip breaking virus (TBV). Bravo et al. (2003) created a hyperspectral imaging platform prototype for the purpose of detecting wheat yellow rust caused by *Puccinia striiformis*.

Resistance screening has also been done using sensors. The advancements began with the controlled investigation of a single plant (Chaerle et al. 2007), and more recently, sophisticated field platforms have been developed to characterise plant performance across many plots or the entire canopy (Walter et al. 2015).

20.11 Biosensor Technology for Plant Pathogen Detection

In several research disciplines, such as environmental monitoring, aerobiology, medical sciences, and the detection of pesticide residues in foods and beverages, biosensors have emerged as cutting-edge detection techniques (Liu et al. 2018). A biosensor is a type of diagnostic device that frequently combines a physicochemical transducer with a biological sensing component. The biosensor generates an electrical signal in response to a specific analyte of interest or pathogen in solution. The result of a biomolecular interaction is subsequently converted into a digital output via a transducer (Elmer and White 2018). Selective character of bioreceptors such as antibodies, DNA, enzymes, tissue types, whole cells, etc., provides recognition specificity to the biosensors. A biosensor can be classified as either an electrochemical, optical, thermal, or piezoelectric biosensor, depending on the kind of transducer it employs (Sawant 2017). For the purpose of detecting plant pathogens in the field, these biosensors are low cost, simple to use, and quick to identify target diseases. The biosensors are also very sensitive and precise. Different biosensors are discussed below:

Electrochemical biosensor: The analyte and bio-recognition component actively interact in the biosensors, producing a signal that is translated into an electrical signal for quantitative analysis. This specific subset of biosensors may detect specific diseases in the air, water, on seeds, etc. The fundamental premise of antibody-based immunosensors is the biomolecular interaction between the analyte and antibody that occurs on the transducer surface. The essential concept of DNA-based biosensors is hybridization or H-bonding between a target DNA sequence and a DNA probe sequence installed on a sensing platform. In point-of-care diagnostics, plant antibodies and DNA are employed to detect plant infections because they are more reliable than other biological sensing components that can be coupled to a transducer.

For the detection of plant diseases, various electrochemical biosensors, including voltammetry, quartz crystal microbalance-based, and electrochemical impedance spectroscopy are utilised. The study of plant diseases that have severely affected agricultural production all over the world has made substantial use of EIS and QCM. When voltage is applied to the cell membrane, redox reactions and chemical interactions at the electrode surface oppose the flow of current, operating the EIS (Shoute et al. 2018). The premise of QCM immunosensors is that chemical adsorption on a quartz crystal's surface alters the resonance oscillation frequency of the crystal. At the surface of a biosensor's recognition element, it is an acoustic label-free sensing device with remarkable real-time monitoring capabilities (Chen et al. 2018). It is crucial to remember that QCM biosensors are incredibly sensitive since they can detect even the smallest mass changes in real time (Table 20.1).

1. **Optical biosensors**: Optical biosensors use a light source, an optical transmission medium, an immobilised bio-recognition component, and a signal detection system for the quantification of the interaction between target analyte and ligand. Bio-recognition process creates physicochemical conversion (change)

Table 20.1 Different electrochemical biosensors developed for the plant pathogens detection (Dyussembayev et al. 2021)

Bio-recognition element	Technique(s)	Crop(s)	Pathogen(s)	References
Antibody	Electrochemical impedance spectroscopy	Plum and tobacco	*Plum pox virus*	Jarocka et al. (2011)
			Candidatus, Phytoplasma aurantifolia	Ebrahimi et al. (2019)
	QCM-based detection	Maize	*Maize chlorotic mottle virus*	Huang et al. (2014)
	Microfluidic immunosensor	Walnut	*Xanthomonas arboricola*	Regiart et al. (2017)
		Cacao	*Phytophthora palmivora*	Franco et al. (2019)
	Label-free impedimetric method employing gold nanoparticles-modified SPCE	Citrus	*Citrus tristeza virus*	Khater et al. (2019)

which is measured as an amplitude, phase, and frequency change of the provided light (Ray et al. 2017). Most widely used optical biosensors used for detection of plant pathogens are colorimetric biosensors, fluorescence-based assays, and surface plasmon resonance-based biosensors. In Table 20.2, a few illustrations of optical biosensors are shown.

20.12 Volatile Organic Compounds

Numerous volatile organic compounds (VOCs) that are vital for a plant's development, communication, defence, and survival are released by plants into their immediate environment (Baldwin et al. 2006). The host plant's terminal metabolites, or VOCs, which are released from leaf surfaces, can reveal the physiological state of the host plant. VOCs often exist at extremely low concentrations that are below the threshold for human olfaction because they have a low boiling point and a high vapour pressure, which makes it easy for them to exist in the gaseous phase at standard pressure and temperature. The ability to quickly and non-intrusively assess the health condition of high-value commodity crops using the emerging science of VOC profiling in plants would be welcomed by growers as it would eliminate the need for chemical reagents.

Historically, technologies based on gas chromatography-mass spectrometry (GC-MS) were used to identify VOCs, but GC-MS is difficult, expensive, time-consuming, and requires substantial training to operate correctly (Sharma et al. 2019). Recently,

Table 20.2 Different optical biosensors developed for the detection of plant pathogens

Bio-recognition element	Technique	Crop	Pathogen(s)	References
Antibody	Lateral flow immunoassay	Potato	*Potato virus X*	Drygin et al. (2012)
		Maize	*Pantoea stewartia* subsp. *stewartii*	Feng et al. (2015)
	Surface plasmon resonance immunoassay	Orchid	*Cymbidium mosaic virus*	Lin et al. (2014)
	Microsphere immunoassay	Watermelon	*Acidovorax avenae* subsp. *citrulli* *Watermelon silver mottle virus*	Charlemroj et al. (2013)
DNA	Bridging flocculation	–	*Pseudomonas syringae*	Wee et al. (2015)
	Gold nanoparticles-based colorimetric detection	–	*Pseudomonas syringae*	Vaseghi et al. (2013)
	Gold nanoparticles-based lateral flow assay	Banana	*Banana bunchy top virus*	Wei et al. (2014)
		Potato	*Phytophthora infestans*	Zhan et al. (2018)
	Electrochemiluminescence-based DNA analysis	Banana	*Banana bunchy top virus* *Banana streak virus*	Tang et al. (2007)
	Microfluidic microarray assembly method	Cucumber	*Botrytis cinerea*	Wang and Li (2007)

electronic nose (EN) has been developed for the detection of VOCs. EN is a noninvasive, quick, and affordable alternative to GC-MS procedures for a variety of applications (Wilson et al. 2018) having components such as a multisensory array, an information processing unit (in the form of an artificial neural network), software with digital pattern-recognition algorithms, and reference-library databases, that mimics the human olfaction (Persaud and Dodd 1982). Each of the specific metal oxide sensors used by these systems is selective for a particular class of VOCs. Individual sensors within the e-nose produce impedance responses in reaction to the addition of volatiles, which are concurrently recorded and presented (Gardner and Bartlett 1994; Doleman et al. 1997; Fend et al. 2006). E-nose can quickly provide a VOC mixture's overlay spectral response in close to real time by using the complete VOC profile without separating out individual components. Plant health status can be correlated with specific physiological changes using e-noses (Laothawornkitkul et al. 2008).

As a developing technology, e-nose still has repeatability, resolution, and robustness problems that need to be resolved. Because of how complex the data produced by e-nose are, multivariate statistical approaches must be used to analyse and interpret them effectively. If these issues are resolved, e-nose technology can be promising device for field analysis.

Classification of VOC species and contents was done with the help of several mathematical models. Due to lack of specificity, this technology is still challenging to use in the field. The scientists suggested concentrating on the mixes of VOCs released by sick citrus trees in order to make the system possibly acceptable for real-time detection. Li et al. (2019) designed methodology in order to facilitate noninvasive diagnosis of late blight caused by *P. infestans*, developed a smartphone-based fingerprinting of leaf volatiles. Plasmonic nanoparticles were particularly sensitive to the main late blight marker, green leafy aldehyde, (E)-2-hexenal, down to sub-ppm levels when used as chemical sensors transducers. The method differentiated between 10 different plant volatiles, allowing an earlier diagnosis before the onset of symptoms, both in laboratory-inoculated and field-collected tomato leaves, thanks to an algorithm with a disease detection accuracy above 95%. The presence of a disease can be indirectly detected by VOCs, but not by the specific virus that is causing it. To satisfy the requirements for online monitoring of plant diseases, the next sections introduce wearable sensors and Internet of Things (IoT) technologies.

20.13 Wearable Sensors

Plants are exposed to a variety of environments in the field, with some stresses combining both biotic and abiotic factors. The plants' responses to these combinations of stresses, however, are different from how they react to each of the stresses administered separately (Suzuki et al. 2014), making procedures that are not based on diagnostic techniques less reliable. In order to gain a thorough understanding of plant health, wearable sensors can be used to research the host–pathogen interaction in the outdoor setting. Such technologies have a number of benefits, like aiding farm management within the context of smart agriculture or assisting botanists in comprehending growth requirements.

Advanced technologies like nanotechnology is now possible as a non-intrusive tool for plants due to the development of wearable sensors. Plant-implanted flexible wearable nanoelectronic circuit-based sensor networks offer real-time wireless communication using small amounts of volatile organic chemicals. With the use of integrated arrays of SWCNT channels and graphite electrodes placed to the plant surface, chemicals in the air may be monitored. These sensors are useful tools for the production of agricultural products since they may collect information in real time regarding the physiological health of plants. A wearable plant sensor may be readily affixed to the surface of a plant due to its excellent mechanical compatibility (Zhao et al. 2020). Nanomaterial-based wearable sensors can bend around objects with a small radius of curvature, functioning as a flexible skin (Giraldo et al. 2019). Wearable radio frequency (RF)-based SWCNT-graphite sensors can be combined with electronic devices for wireless monitoring of gàs molecules up to 5 ppm without consuming any power (Lee et al. 2014). Wearable carbon nanotube sensors are commercially available for sensing plant VOCs, such as polyethylene, but they have not yet been used for electronic monitoring. Lan et al. (2020) developed a wearable humidity sensor based on laser-induced graphene to monitor plant leaf transpiration. Due to its high flexibility, this wearable sensor can measure transpiration directly on the plant leaf surface without causing long-lasting or unstable physical damage. In near future wearable sensors can be used in combination with the laser technology along with the nanoparticles being used. There is scant information about the use of wearable sensors for diagnosing and managing plant diseases, a few examples have been provided in Table 20.3.

Using graphene nanomaterial, Oren et al. (2017) developed a flexible microscale plant sensor that was patterned onto various kinds of tapes. This sensor uses variations in the electrical resistance of graphene in different humidity conditions to determine the hydraulic conductivity by monitoring the time for water to flow from the root to the leaf and the rate at which plants transpire. Hydraulic conductivity and plant transpiration rate can be used as markers of ambient temperature and drought stress in plants. Kim et al. (2019) in their study carried out direct vapour printing of polymer electrodes onto living plant tissue and demonstrated that the sensor can identify a variety of biological and abiotic stressors in plants. Additionally, during a 130-day period, this novel electrode demonstrated great adherence to plant tissues without impairing the plant's biocompatibility or normal development pattern.

Another kind of wearable sensors are microneedles, which have demonstrated the capacity to penetrate plant vasculature with the least amount of invasion. Both of these sensors have the ability to extract sap and analyse its physiochemical characteristics, such as pH and electric conductivity, in addition to sensing sap flow rate. The modified Granier approach was used by Baek et al. (2018) to create a microneedle sap flow sensor for monitoring water transport in stem of tomato plant. Changes in

Table 20.3 Application of wearable-based for plant monitoring

Technology used	Crop	Mode of signal	Nano material	Application(s)	References
Microneedle patch-based sensor	Tomato leaves	UV absorption spectra	–	Pathogen/ *Phytophthora infestans*	Paul et al. (2019)
Smartphone-based	Tomato leaves	GC-MS spectra and chemometric analysis	GNP	*P. infestans*	Li et al. (2019)
Lab-on chip device	Olive trees	EIS	–	*Xylella fastidiosa* detection	Chiriacò et al. (2018)
Chemi-resistive sensor array	Tomato leaves	–	Graphene	Monitoring of late blight	Li et al. (2021)

environmental parameters including soil water content, wetness, and sun irradiance intensity may be shown with this wearable sensor (SWC). Thin-film wearable sensors have various drawbacks despite enabling quick prefabrication and mass production. When attaching to irregular sections and complicated plant tissue, they may result in low accuracy and compromised normal plant function. Due to the large increase in plant tissue size during growth, this constraint will be significantly made worse. This issue greatly hinders long-term monitoring, especially for plants that grow quickly, like vegetables. One way to get over this limitation is the development of flexible wearable sensors with improved adhesion to uneven surfaces and increased deformation capabilities. Because the device components can be positioned along uneven surfaces, this compatibility challenge may be overcome by directly constructing such wearables on plant living tissues (Wong et al. 2017). Among its desired characteristics are wearables' nontoxicity and controlled degradability, particularly when they are used with fruits, vegetables, and agricultural products. Portable devices that can instantly print, write, or deposit on plants must be developed in order to be used in large-scale in-field operations (Huang et al. 2020). Wearables for plants that can only detect local signals risk missing distant signals or signals originating from various channels. Microneedle-based wearable sensors do not experience mismatching on the plant's uneven surface unless they have used an array of microneedles. Nevertheless, depending on the size of the microneedle and the size of the plant, the invasive detection of microneedles may cause plant harm. Callus formation caused by wounding may be lessened by utilising smaller microneedles, biocompatible materials (such as pectin, cellulose, hemicellulose, and lignin), and other techniques (such as degradable microneedles) (Wong et al. 2017).

20.14 IoT and Remote Sensing Technologies

The goal of sustainability in contemporary agriculture can be considerably aided by the application of nanotechnologies and the Internet of Things (IoT) (Maksimovic and Omanovic-Miklicanin 2017). Smart farming is the use of modern information and communication technologies (ICTs) like remote sensing (Bastiaanssen et al. 2000), cloud computing (Hashem et al. 2015), and the Internet of Things (IoT) (Weber and Romana 2010) to help farmers monitor field conditions from anywhere or with in-field high. In particular, "real-time communication" and "wireless sensing" are considered to be key components of smart farming. It is also important to emphasise the assistance provided by robotics, particularly in the areas of fruit picking, plant protection, and weed control (Ampatzidis et al. 2017). The term "remote sensing" describes the process of acquiring quantitative and qualitative information about an environment or an object that is located far from a sensor (such as a satellite, an aeroplane, a UAV/UGV, or a probe). Understanding the regional and temporal variability of agricultural systems and how this kind of non-destructive sensing might lessen negative environmental effects by controlling resource depletion are essential for achieving our aims. In addition to being able to detect (even at early stages) changes induced in plants under stress conditions (leaf area index, chlorophyll content, or surface temperature), analysis of molecular interactions between crop stress and biophysical or biochemical characteristics can also be done (Mulla 2013). This results in different fingerprints compared with the healthy plant (Meroni and Colombo 2010). The 1980s saw the beginning of the use of remote sensing in precision farming, which at first concentrated on just a few visible or near-infrared bands but eventually evolved into hyperspectral remote sensing. Depending on the pathogen, various spectral regions can be used to monitor plant-related events, such as pathogen propagules in the VIS, chlorophyll degradation (necrotic or chlorotic lesions) in the VIS and red-edge (550 nm; 650–720 nm), photosynthesis disruption as fluorescence (450–550 nm; 690–740 nm), and senescence in the TIR (8000–14,000 nm). They can certainly be used to detect disease in field crops. Recently, utilising a hyperspectral radiometer and the ratio Fv/Fm between two leaf ChlF-derived parameters, the intrinsic efficiency of photosystem II photochemistry was determined from leaf reflectance (Peng et al. 2017). Stressed leaves demonstrated a rapid reduction in Fv/Fm while maintaining a constant level of chlorophyll. It was found that the slope of reflectance in the spectral range 700–900 nm rose concurrently with this drop, and there was a good link between the first derivative reflectance in the NIR regions and Fv/Fm. Unmanned aerial vehicles (UAVs), also referred to as agricultural drones, can be useful. They can support surveillance operations and assist with human endeavours like crop planting, pest control, and crop monitoring. For instance, the eBee SQ farm drone from "Sense Fly" (Sensefly 2020) employs multispectral image analysis in conjunction with eMotion Ag software. Direct multispectral image uploading from a drone covering hundreds of acres to cloud services for accurate crop monitoring and analysis is one of its capabilities. Aerial hyperspatial data is used for crop growth estimation or farmland analysis, while aerial lidar data is utilised for 3-D remote sensing for topography and forests. These are only a few of the agricultural applications that involve aircraft or satellite technologies. For multispectral and thermal photography, the HiSystems MK-Okto mini-UAV was created at a low cost (Bendig et al. 2012). The georeferencing technique works well for collecting multispectral and thermal images while guaranteeing data comparability. Applications on a modest scale are possible with the 15-min time trip. Agricultural land use monitoring, crop yield forecasts, yield optimization using crop monitoring, and ecosystem services are other remote sensing applications (Weiss et al. 2020). Remote sensing for environmental monitoring, which also gives information on ecosystem sustainability, drought mitigation, human health, and other environmental studies, explains Earth's surface monitoring and characterisation. IoT, a new technology based on connectivity, is becoming more and more well-known due to the many industries it is used in. The word was initially coined in the context of supply chain management by Kevin Ashton in 1999. It is now widely used in technologies including radio frequency identification (RFID) (Luvisi et al. 2016), wireless sensor networks, and cloud computing. It provides a wide range of ground-breaking solutions in industries like agricultural, retail, transportation, healthcare, and security. Three fundamental building elements serve as the foundation for IoT technologies: a sensor component, a network-like data transit and storage system, and a manipulation component (Tzounis et al. 2017). Due to ongoing technology advancements, even in the sphere of agriculture, the IoT scenario is being developed in order to assist the researchers as well as farmers in taking preventive steps to fight phytopathogens. In particular, Liqiang et al. (2011) developed a system based on wireless sensors as a crop monitoring network, allowing the gathering of data regarding temperature, humidity, and also crop growth images, through which it is possible to analyse crops in an intuitive and understandable manner. Even blueberry planting regions have utilised wireless sensor-based systems (Jing et al. 2017). Instead, it was demonstrated by Song et al. (2012) that the sensor nodes' capacity to gather variables including temperature, greenhouse humidity, carbon dioxide content, and lighting intensity could be accomplished with the use of a solar energy supply system.

Obtaining industrial data to enhance traceability or enabling sanitary certification are only two examples of the advantages of IoT. Radiofrequency identification (RFID) microchips were inserted, according to Luvisi et al. (2011). In the latter, all plants have been successfully identified using RFID technology during ampelographic, genetic, and sanitary inspections. RFID has the potential to build mother plant vineyards, retrieve propagated material, and tag derived certified material, according to Luvisi et al. (2012). Robotics, cloud computing, and video processing can be used to detect pests such as tomato borers in order to handle phytosanitary treatments carefully (Rupanagudi et al. 2015). A cloud-hosted programme receives a real-time video of tomato crops for processing in which the robot is given instructions to spray pesticides using completely automated system that can undertake continuous surveillance of the farm based on the findings of picture analysis. A wireless sensor network with autonomous, self-powered nodes dispersed around a vineyard was also suggested (Pérez-Expósito et al. 2017). The hardware and software platform VineSens makes use of epidemiological models to prevent diseases like downy mildew, aiding farmers in management and offering significant savings, such as reducing the frequency of phytosanitary treatments. Additionally, the software enables the collection of weather data from several vineyard locations, and access to it is guaranteed via a web-based interface on desktop or mobile devices. Another monitoring approach for crop management and insect pest identification is the use of remote sensing data and geospatial image processing via unmanned aerial vehicles (UAVs) (Vanegas and Bratanov 2018). The UAVs incorporate cutting-edge sensors (such multispectral, hyperspectral, or digital RGB sensors) along with information about the terrain. For agricultural monitoring, systems like "Arable" and "Semios" are also employed. The former sends meteorological and plant information to the cloud, allowing users to access the data remotely and in real time. It also continuously monitors stress, insect pest, and disease markers. It is possible to monitor pathogens using airborne platforms equipped with multispectral and thermal cameras, as was done for *X. fastidiosa* (*Xf*) (Poblete et al. 2020), where the sensitivity of spectral bands indicated the severity of symptoms. In a different study, Hornero et al. (2020) used a 3D radiative transfer modelling approach (3D-RTM) to investigate the spatiotemporal evolution of *Xf* infections in olive orchards using airborne hyperspectral imagery and Sentinel-2 satellite data. The results showed that Sentinel-2 time-series photos provided useful spatiotemporal indicators to monitor the losses brought on by *Xf* infections over large areas.

20.15 Merits of Advanced Techniques

- Automated techniques with higher sensitivity, precision for disease detection
- Some techniques like VOCs offer noninvasive methods for monitoring plant health, remote sensing and cloud computing can analyse the disease situation with the help of the large database and provide a quick decision.
- Real-time data about plant physiology can be gathered with the help of wearable plant sensors that could be quickly mounted to the plant surface due to their mechanical compatibility.
- On-site detection of the plant health condition as well as the disease can be done using the advanced technologies.

20.16 Drawbacks

- High-cost technologies are still not affordable by many farmers and also certain methods require technical know-how which lacks among many farmers.
- Certain techniques like wearable sensors might become a constraint resulting in low accuracy and compromised normal plant functioning which might aggravate with growth of the plant tissue or organ. Long-term monitoring might be hampered for quick growing plants such as vegetables.
- Large-scale production of portable on-site operation techniques is still a challenge.

20.17 Discussion and Future Prospects

The potential of these technologies to identify diseases has not yet been thoroughly studied. It may be possible to improve our understanding of the biophysical and biochemical processes involved in disease development by using multisensory distinctive signatures of plant diseases. The relationship between molecular methods and non-intrusive optical methods might be a fresh tactic that has to be looked at in more depth. The interpretation of sensor data, particularly spectral data, places existing limitations on the identification, diagnosis, and quantification of illnesses in crop production. Precision crop protection will become more widely adopted and used as a result of cross-disciplinary collaboration on its incorporation into trustworthy decision support systems. The rapid development of nanotechnology and nano-based production methods over the past decade has led to substantial progress in a range of biosensors, wearable sensors, bioelectronic noses, and nanostructured substrates for the analysis of plant diseases.

20.18 Conclusion

An overview of modern techniques for identifying plant diseases as well as cutting-edge strategies for recognising symptoms and stopping the spread of pathogens have been provided. Taking into consideration more recent advancements in wearable sensing and Internet of Things (IoT) technologies, we have collated breakthroughs in sensor and microfluidics technologies. Instead, the merging of sensing and robotics technologies could result in the development of new "lab-on-a-drone" analysis platforms. These technologies would eliminate the requirement for sample collection and processing, allow quick in-flight assays with smartphone connectivity, and offer scenarios for emergency response, agricultural bio-surveillance, and veterinary field care. Smartphones can help create more accurate, sophisticated, and portable detection systems that might support institutions and farmers around the world in the fight against plant diseases, thanks to connectivity, high-quality images, and processing capacity. These developments can all talk to one another, opening up fresh possibilities for sensible and efficient plant pathogen control.

References

Ampatzidis Y, De Bellis L, Luvisi A (2017) IPathology: robotic applications and management of plants and plant diseases. Sustainability 9:1010

Baek S, Jeon E, Park KS, Yeo KH, Lee J (2018) Monitoring of water transportation in plant stem with microneedle sap flow sensor. J Microelectromech Syst 27(3):440–447

Baldwin IT, Halitschke R, Paschold A, von Dahl CC, Preston CA (2006) Volatile signalling in plant-plant interactions: 'talking trees' in the genomics era. Science 311:812–814

Balodi R, Bisht S, Ghatak A, Rao KH (2017) Plant disease diagnosis: technological advancements and challenges. Indian Phytopathol 70(3):275–281

Bastiaanssen WGM, Molden DJ, Makin IW (2000) Remote sensing for irrigated agriculture: examples from research and possible applications. Agric Water Manag 46:137–155

Bendig J, Bolten A, Bareth G (2012) Introducing a low-cost mini-UAV for thermal-and multispectral-imaging. Int Arch Photogramm Remote Sens Spat Inf Sci XXXIX-B1:345–349

Bergstrasser S, Fanourakis D, Schmittgen S, Cendrero-Mateo MP, Jansen M (2015) HyperArt: noninvasive quantification of leaf traits using hyperspectral absorption-reflectance-transmittance imaging. Plant Methods 11:1

Bock CH, Poole GH, Parker PE, Gottwald TR (2010) Plant disease severity estimated visually by digital photography and image analysis and by hyperspectral imaging. Crit Rev Plant Sci 29:59–107

Bravo C, Moshou D, West J, McCartney A, Ramon H (2003) Early disease detection in wheat fields using spectral reflectance. Biosyst Eng 84:137–145

Buja I, Sabella E, Monteduro AG, Chiriacò MS, De Bellis L, Luvisi A, Maruccio G (2021) Advances in plant disease detection and monitoring: from traditional assays to in-field diagnostics. Sensors 21:2129. https://doi.org/10.3390/s21062129

Burling K, Hunsche M, Noga G (2011) Use of blue-green and chlorophyll fluorescence measurements for differentiation between nitrogen deficiency and pathogen infection in wheat. J Plant Physiol 168:1641–1648

Camargo A, Smith JS (2009) Image pattern classification for the identification of disease causing agents in plants. Comput Electron Agric 66:121–125

Candresse T, Filloux D, Muhire B, Julian C, Galzi S (2014) Appearances can be deceptive: revealing a hidden viral infection with deep sequencing in a plant quarantine context. PLoS ONE 9:e102945

Chaerle L, Hagenbeek D De Bruyne E, Valcke R, Van Der Straeten D (2004) Thermal and chlorophyll-fluorescence imaging distinguish plant-pathogen interactions at an early stage. Plant Cell Physiol 45:887–896

Chaerle L, Hagenbeek D, De Bruyne E, van der Straeten D (2007) Chlorophyll fluorescence imaging for disease-resistance screening of sugar beet. Plant Cell Tissue Organ Cult 9:97–106

Charlermroj R, Himananto O, Seepiban C, Kumpoosiri M, Warin N, Oplatowska M et al (2013) Multiplex detection of plant pathogens using a microsphere immunoassay technology. PLoS ONE 8:e62344. https://doi.org/10.1371/journal.pone.0062344

Chen JY, Penn LS, Xi J (2018) Quartz crystal microbalance: sensing cell substrate adhesion and beyond. Biosens Bioelectron 99:593–602. https://doi.org/10.1016/j.bios.2017.08.032

Chiriacò MS, Luvisi A, Primiceri E, Sabella E, De Bellis L, Maruccio G (2018) Development of a lab-on-a-chip method for rapid assay of *Xylella fastidiosa* subsp. *pauca* strain CoDiRO. Sci Rep 8(1):1–8

Clark MF (1981) Immunosorbent assays in plant pathology. Annu Rev Phytopathol 19:83–106

D'hondt L, Hofte M, Van Bockstaele E, Leus L (2011) Applications of flow cytometry in plant pathology for genome size determination, detection and physiological status. Mol Plant Pathol 12:815–828

Danks C, Baker I (2000) On-site detection of plant pathogens using lateral-flow devices. EPPO Bull 30:421–426

Doleman BJ, Severin EJ, Lewis NS (1997) Trends in odour intensity for human and electronic noses: relative roles of odorant vapor pressure vs. molecularly specific odorant binding. PNAS 95:5442–5447. https://doi.org/10.1073/pnas.95.10.5442

Drygin YF, Blintsov AN, Grigorenko VG, Andreeva IP, Osipov AP, Varitzev YA et al (2012) Highly sensitive field test lateral flow immunodiagnostics of PVX infection. Appl Microbiol Biotechnol 93:179–189. https://doi.org/10.1007/s00253-011-3522-x

Dyussembayev K, Sambasivam P, Bar I, Brownlie JC, Shiddiky MJA, Ford R (2021) Biosensor technologies for early detection and quantification of plant pathogens. Front Chem 9:636245. https://doi.org/10.3389/fchem.2021.636245

Ebrahimi M, Norouzi P, Safarnejad MR, Tabaei O, Haji-Hashemi H (2019) Fabrication of a label-free electrochemical immunesensor for direct detection of Candidatus Phytoplasma Aurantifolia. J Electroanal Chem 851:113451. https://doi.org/10.1016/j.jelechem.2019.113451

Elmer W, White JC (2018) The future of nanotechnology in plant pathology. Annu Rev Phytopathol 56:111–133. https://doi.org/10.1146/annurevphyto-080417-050108

Fang Y, Ramasamy RP (2015) Current and prospective methods for plant disease detection. Biosensors 4:537–561

FAO (2009) Global agriculture towards 2050–high level expert forum. FAO, UN, Rome, Italy. http://www.fao.org/fileadmin/templates/wsfs/docs/Issues_papers/HLEF2050_Global_Agriculture.pdf

Fend R, Kolk A, Bessant C, Buijtels P, Klatser PR, Woodman AC (2006) Prospects for clinical application of electronic-nose technology to early detection of *Mycobacterium tuberculosis* in culture and sputum. J Clin Microb 44:2039–2045. https://doi.org/10.1128/JCM.01591-05

Feng M, Kong D, Wang W, Liu L, Song S, Xu C (2015) Development of an immunochromatographic strip for rapid detection of *Pantoea stewartii* subsp. *stewartii*. Sensors 15:4291–4301. https://doi.org/10.3390/s150204291

Franco AJD, Merca FE, Rodriguez MS, Balidion JF, Migo VP, Amalin DM et al (2019) DNA-based electrochemical nanobiosensor for the detection of *Phytophthora palmivora* (Butler) Butler, causing black pod rot in cacao (*Theobroma cacao* L.) pods. Physiol Mol Plant Pathol 107:14–20

Gardner JW, Bartlett PN (1994) A brief history of electronic noses. Sens Actuators B Chem 18:210–211. https://doi.org/10.1016/0925-4005(94)87085-3

Giraldo JP, Wu H, Newkirk GM, Kruss S (2019) Nanobiotechnology approaches for engineering smart plant sensors. Nat Nanotechnol 14(6):541–553

Godfray HCJ, Beddington JR, Crute IR, Haddad L, Lawrence D, Muir JF, Pretty J, Robinson S, Thomas SM, Toulmin C (2010) Food security: the challenge of feeding 9 billion people. Science 327:812–818

Gomez S (2014) Infection and spread of *Peronospora sparsa* on *Rosa* sp. (Berk.)—a microscopic and a thermographic approach. Dissertation, University of Bonn, Germany

Hashem IAT, Yaqoob I, Anuar NB, Mokhtar S, Gani A, Ullah Khan S (2015) The rise of "big data" on cloud computing: review and open research issues. Inf Syst 47:98–115

Hornero A, Hernandez Clemente R, North PRJ, Beck PSA, Boscia D, Navas Cortés J, Zarco-Tejada P (2020) Monitoring the incidence of *Xylella fastidiosa* infection in olive orchards using ground-based evaluations, airborne imaging spectroscopy and Sentinel-2 time series through 3-D radiative transfer modelling. Remote Sens Environ 236:111480

Huang CH, Singh GP, Park SH, Chua NH, Ram RJ, Park BS (2020) Early diagnosis and management of nitrogen deficiency in plants utilizing Raman spectroscopy. Front Plant Sci 11:663

Huang X, Xu J, Ji HF, Li G, Chen H (2014) Quartz crystal microbalance-based biosensor for rapid and sensitive detection of maize chlorotic mottle virus. Anal Methods 6:4530–4536. https://doi.org/10.1039/c4ay00292

Jarocka U, Wa sowicz M, Radecka H, Malinowski T, Michalczuk L, Radecki J (2011) Impedimetric immunosensor for detection of plum pox virus in plant extracts. Electroanalysis 23:2197–2204. https://doi.org/10.1002/elan.201100152

Jing Y, Yuzhi Z, Dan D, Xiao W, Ping Y, Ling-fang C, Yue-fang S, Ze-tao L (2017) An early warning system of diseases and pests for blueberry based on WSN. In: Proceedings 36th Chinese control conference (CCC), Dalian, China, 26–28 July 2017, pp 8885–8889

Jones HG, Stoll M, Santoa T, de Sousa C, Chaves MM, Grant OM (2002) Use of infrared thermography for monitoring stomatal closure in field: application to grapevine. J Exp Bot 53:2249–2260

Khater M, de la Escosura-Muñiz A, Quesada-González D, Merkoçi A (2019) Electrochemical detection of plant virus using gold nanoparticle-modified electrodes. Anal Chim Acta 1046:123–131. https://doi.org/10.1016/j.aca.2018.09.031

Kim JJ, Allison LK, Andrew TL (2019) Vapor-printed polymer electrodes for long-term, on-demand health monitoring. Sci Adv 5(3):eaaw0463

Kobayashi T, Kanda E, Kitada K, Ishiguro K, Torigoe Y (2001) Detection of rice panicle blast with multispectral radiometer and the potential of using airborne multispectral scanners. Phytopathology 91:316–323

Konanz S, Kocsanyi L, Buschmann C (2014) Advanced multicolour fluorescence imaging system for detection of biotic and abiotic stresses in leaves. Agriculture 4:79–95

Lan L, Le X, Dong H, Xie J, Ying Y, Ping J (2020) One-step and large-scale fabrication of flexible and wearable humidity sensor based on laser-induced graphene for real-time tracking of plant transpiration at bio-interface. Biosens Bioelectron 165:112360

Laothawornkitkul J, Moore JP, Taylor JE, Malcom G, Tim D, Hewitt CN, Paul ND (2008) Discrimination of plant volatile signatures by an electronic nose: a potential technology for plant pest and disease monitoring. Environ Sci Technol 42:8433–8439. https://doi.org/10.1021/es801738s

Lee K, Park J, Lee MS, Kim J, Hyun BG, Kang DJ, Na K, Lee CY, Bien F, Park JU (2014) In-situ synthesis of carbon nanotube–graphite electronic devices and their integrations onto surfaces of live plants and insects. Nano Lett 14(5):2647–2654

Leucker M, Mahlein AK, Steiner U, Oerke EC (2016) Improvement of lesion phenotyping in *Cercospora beticola*—sugar beet interaction by hyperspectral imaging. Phytopathology 106:177–184

Li Z, Paul R, Ba Tis T, Saville AC, Hansel JC, Yu T, Ristaino JB, Wei Q (2019) Non-invasive plant disease diagnostics enabled by smartphone-based fingerprinting of leaf volatiles. Nat Plants 5:856–866

Li L, Zhang S, Wang B (2021) Plant disease detection and classification by deep learning—a review. IEEE Access 9:56683–56698

Lin HY, Huang CH, Lu SH, Kuo IT, Chau LK (2014) Direct detection of orchid viruses using nanorod-based fiber optic particle plasmon resonance immunosensor. Biosens Bioelectron 51:371–378. https://doi.org/10.1016/j.bios.2013.08.009

Liqiang Z, Shouyi Y, Leibo L, Zhen Z, Shaojun WA (2011) Crop monitoring system based on wireless sensor network. Procedia Environ Sci 11:558–565

Liu C, Xu C, Xue N, Sun JH, Cai H, Li T et al (2018) Enzyme biosensors for point-of-care testing. MEMS Sens Des Appl 49:49–70. https://doi.org/10.5772/INTECHOPEN.73249

López MM, Llop P, Cubero J, Penyalver R, Caruso P, Bertolini E, Penalver J, Gorris MT, Cambra M (2001) Strategies for improving serological and molecular detection of plant pathogenic bacteria. In: Plant pathogenic bacteria. Springer, Berlin, pp 83–86

Luvisi A, Panattoni A, Bandinelli R, Rinaldelli E, Pagano M, Triolo E (2011) Implanting RFIDs into Prunus to facilitate electronic identification in support of sanitary certification. Biosyst Eng 109:167–173

Luvisi A, Panattoni A, Bandinelli E, Rinaldelli MP, Triolo E (2012) Propagative material of grapevine: RFID technology for supporting traceability of "basic" and "certified" material along the wine production chain. Adv Hortic Sci 26:39–43

Luvisi A, Ampatzidis Y, De Bellis L (2016) Plant pathology and information technology: opportunity for management of disease outbreak and applications in regulation frameworks. Sustainability 8:831

Mahlein AK, Oerke EC, Steiner U, Dehne HW (2012) Recent advances in sensing plant diseases for precision crop protection. Eur J Plant Pathol 133:197–209

Mahlein AK, Kuska MT, Behmann J, Polder G, Walter A (2018) Hyperspectral sensors and imaging technologies in phytopathology: state of the art. Annu Rev Phytopathol 56:535–558

Maksimovic M, Omanovic-Miklicanin E (2017) Green internet of things and green nanotechnology role in realizing smart and sustainable agriculture. In: Kovacevic D (ed) Proceedings of the VIII international scientific agriculture symposium "AGROSYM 2017", Jahorina, Bosnia, 5–8 Oct 2017, University of East Sarajevo, Sarajevo, Bosnia, pp 2290–2295

Meroni MRM, Colombo R (2010) Characterization of leaf physiology using reflectance and fluorescence hyperspectral measurements. In: Maselli FMM, Brivio PA (eds) Optical observation of vegetation properties and characteristics. Research Signpost, Trivandrum, India

Mitra D (2021) Emerging plant diseases: research status and challenges. Emerging Trends Plant Pathol 1–17

Mulla DJ (2013) Twenty five years of remote sensing in precision agriculture: key advances and remaining knowledge gaps. Biosyst Eng 114:358–371

Murchie EH, Lawson T (2013) Chlorophyll fluorescence analysis: a guide to good practice and understanding some new applications. J Exp Bot 64:3983–3998

Mutka AM, Bart RS (2015) Image-based phenotyping of plant disease symptoms. Front Plant Sci 5:734

Neumann M, Hallau L, Klatt B, Kersting K, Bauckhage C (2014) Erosion band feature for cell phone image based plant disease classification. In: Proceeding of the 22nd international conference on pattern recognition (ICPR), Stockholm, Sweden, 24–28 Aug 2014, pp 3315–3320

Oerke EC, Fröhling P, Steiner U (2011) Thermographic assessment of scab disease on apple leaves. Precis Agric. https://doi.org/10.1007/s11119-010-9212-3

Oren S, Ceylan H, Schnable PS, Dong L (2017) High-resolution patterning and transferring of graphene-based nanomaterials onto tape toward roll-to-roll production of tape-based wearable sensors. Adv Mater Technol 2(12):1700223

Paul R, Saville AC, Hansel JC, Ye Y, Ball C, Williams A, Chang X, Chen G, Gu Z, Ristaino JB Wei Q (2019) Extraction of plant DNA by microneedle patch for rapid detection of plant diseases. ACS Nano 13(6):6540–6549

Paul Khurana SM (2006) Detection of plant pathogens: development and applications. Indian Phytopathol 59:1–15

Peng Y, Zeng A, Zhu T, Fang S, Gong Y, Tao Y, Zhou Y, Liu K (2017) Using remotely sensed spectral reflectance to indicate leaf photosynthetic efficiency derived from active fluorescence measurements. J Appl Remote Sens 11:026034

Pérez-Expósito JP, Fernández-Caramés TM, Fraga-Lamas P, Castedo L (2017) VineSens: an eco-smart decision-support viticulture system. Sensors 17:465

Persaud K, Dodd G (1982) Analysis of discrimination mechanisms in the mammalian olfactory system using a model nose. Nature 299:352–355

Petersson H, Gustafsson D, Bergstrom D (2016) Hyperspectral image analysis using deep learning-a review. Paper presented at the 6th international conference on image processing theory, tools and applications (IPTA 2016), Oulu, Finland, 12–15 Dec. http://ieeexplore.ieee.org/document/7820963/

Poblete T, Camino C, Beck PSA, Hornero A, Kattenborn T, Saponari M, Boscia D, Navas-Cortes JA, Zarco-Tejada PJ (2020) Detection of *Xylella fastidiosa* infection symptoms with airborne multispectral and thermal imagery: assessing bandset reduction performance from hyperspectral analysis. ISPRS J Photogramm Remote Sens 162:27–40

Ray M, Ray A, Dash S, Mishra A, Achary KG, Nayak S et al (2017) Fungal disease detection in plants: traditional assays, novel diagnostic techniques and biosensors. Biosens Bioelectron 87:708–723. https://doi.org/10.1016/j.bios.2016.09.032

Regiart M, Rinaldi-Tosi M, Aranda PR, Bertolino FA, Villarroel-Rocha J, Sapag K et al (2017) Development of a nanostructured immunosensor for early and in situ detection of *Xanthomonas arboricola* in agricultural food production. Talanta 175:535–541. https://doi.org/10.1016/j.talanta.2017.07.086

Riley MB, Williamson MR, Maloy O (2002) Plant disease diagnosis. Plant Health Instructor. https://doi.org/10.1094/PHII-2002-1021-01

Rupanagudi SR, Ranjani BS, Nagaraj P, Bhat VG, Thippeswamy G (2015) A novel cloud computing based smart farming system for early detection of borer insects in tomatoes. In: Proceedings international conference on communication, information & computing technology (ICCICT), Mumbai, India, 15–17 Jan 2015, pp 1–6

Savary S, Ficke A, Aubertot JN, Hollier C (2012) Crop losses due to diseases and their implications for global food production losses and food security. Food Secur 4(4):519–537

Sawant S (2017) Development of biosensors from biopolymer composites. In: Biopolymer composites in electronics. Elsevier, Amsterdam, The Netherlands

SenseFly (2020) The professional's mapping drone. Available online: https://www.sensefly.com/. Accessed on 7 Oct 2020

Sharma R, Zhou M, Hunter MD, Fan X (2019) Rapid in situ analysis of plant emission for disease diagnosis using a portable gas chromatography device. J Agric Food Chem 67:7530–7537

Shoute LC, Anwar A, Mackay S, Abdelrasoul GN, Lin D, Yan Z et al (2018) Immuno-impedimetric biosensor for onsite monitoring of ascospores and forecasting of sclerotinia stem rot of canola. Sci Rep 8:1–9. https://doi.org/10.1038/s41598-018-30167-5

Song Y, Ma J, Zhang X, Feng Y (2012) Design of wireless sensor network-based greenhouse environment monitoring and automatic control system. J Netw 7:838–844

Suzuki N, Rivero RM, Shulaev V, Blumwald E, Mittler R (2014) Abiotic and biotic stress combinations. New Phytol 203:32–43

Tang Y, Xing D, Zhu D, Liu J (2007) An improved electrochemiluminescence polymerase chain reaction method for highly sensitive detection of plant viruses. Anal Chim Acta 582:275–280. https://doi.org/10.1016/j.aca.2006.09.021

Thomas S, Wahabzada M, Kuska MT, Rascher U, Mahlein AK (2017) Observation of plant-pathogen interaction by simultaneous hyperspectral imaging reflection and transmission measurements. Funct Plant Biol 44:23–34

Tzounis A, Katsoulas N, Bartzanas T, Kittas C (2017) Internet of Things in agriculture, recent advances and future challenges. Biosyst Eng 164:31–48

Vanegas F, Bratanov DA (2018) Novel methodology for improving plant pest surveillance in vineyards and crops using UAV-based hyperspectral and spatial data. Sensors 18:260

Vaseghi A, Safaie N, Bakhshinejad B, Mohsenifar A, Sadeghizadeh M (2013) Detection of *Pseudomonas syringae* pathovars by thiol-linked DNA-Gold nanoparticle probes. Sens Actuators B Chem 181:644–651. https://doi.org/10.1016/j.snb.2013.02.018

Walter A, Liebisch F, Hund A (2015) Plant phenotyping: from bean weighing to image analysis. Plant Methods 11:14

Wang L, Li PCH (2007) Flexible microarray construction and fast DNA hybridization conducted on a microfluidic chip for greenhouse plant fungal pathogen detection. J Agric Food Chem 55:10509–10516. https://doi.org/10.1021/jf0721242

Weber R, Romana W (2010) Internet of Things. Legal perspectives. Springer, Berlin

Wee EJH, Lau HY, Botella JR, Trau M (2015) Re-purposing bridging flocculation for on-site, rapid, qualitative DNA detection in resource-poor settings. Chem Commun 51:5828–5831. https://doi.org/10.1039/c4cc10068a

Wei J, Liu H, Liu F, Zhu M, Zhou X, Xing D (2014) Miniaturized paper-based gene sensor for rapid and sensitive identification of contagious plant virus. ACS Appl Mater Int 6:22577–22584. https://doi.org/10.1021/am506695g

Weiss M, Jacob F, Duveiller G (2020) Remote sensing for agricultural applications: a meta-review. Remote Sens Environ 236:111402

Wendel A, Underwood J (2017) Illumination compensation in ground based hyperspectral imaging. ISPRS J Photogramm Remote Sens 129:162–178

West SJ, Bravo C, Oberti R, Moshou D, Ramon H, McCartney HA (2010) Detection of fungal diseases optically and pathogen inoculum by air sampling. In: Oerke EC, Gerhards R, Menz G, Sikora RA (eds) Precision crop protection—the challenge and use of heterogeneity. Springer, Dordrecht, pp 135–150

Wilson A (2018) Applications of electronic-nose technologies for noninvasive early detection of plant, animal and human diseases. Chemosensors 6:45

Wong MH, Giraldo JP, Kwak SY, Koman VB, Sinclair R, Lew TTS, Bisker G, Liu P Strano MS (2017) Nitroaromatic detection and infrared communication from wild-type plants using plant nanobionics. Nat Mater 16(2):264–272

Zhan F, Wang T, Iradukunda L, Zhan J (2018) A gold nanoparticle-based lateral flow biosensor for sensitive visual detection of the potato late blight pathogen, *Phytophthora infestans*. Anal Chim Acta 1036:153–161. https://doi.org/10.1016/j.aca.2018.06.083

Zhang M, Qin Z, Liu X, Ustin S (2003) Detection of stress in tomatoes induced by late blight disease in California, USA, using hyperspectral remote sensing. Appl Earth Obs Geoinfo 4:295–310

Zhao F, He J, Li X, Bai Y, Ying Y, Ping J (2020) Smart plant-wearable biosensor for in-situ pesticide analysis. Biosens Bioelectron 170:112636

Chapter 21
Disease and Pest Control Through Advance Technology

Shalaka Ahale, Gaurav Rakhonde, Sheenam Bhateja, J. Kuppuraj, and Soumya Mishra

Abstract The dynamic change in global environment along with natural hindrances leads to several economic losses by causing several outbreaks and epidemics all over the world resulting in major threat to plants and human health quantitatively and qualitatively. To mitigate all these challenges, it is necessary to break the chain of these co-epidemic conditions which can be successfully accomplished by prior detection of the upcoming risk. Several advanced techniques have been emerged to face these extremities. Smart agriculture concept mainly deals with the modernisation of conventional detection and management tactics which would increase chances of early detection, precision monitoring, and occurrence of pest and diseases. Combination of two or more techniques such as remote sensing with artificial intelligence helps to diagnose characterises and monitors diseases and pest across different spatial scales. Besides remote sensing, there are other myriad techniques such as hyperspectral imaging, UAVs, information and communication, Internet of Things (IoTs), and imaging are available to cope with several disease and pests. Genetic engineering methodologies including CRISPR-Cas, RNAi, and nanotechnology-based proper application of fertilisers and agrochemicals control the plant disease and pest emergence before crossing their respective thresholds. This chapter provides a new perspective for sustainable agriculture from the current researches, new technology for observation of several crops, and their habitat since it not only helps to manage the pests and diseases but also aids in biophysical control on their emergence.

S. Ahale (✉) · J. Kuppuraj
Department of Plant Pathology, Punjab Agricultural University, Ludhiana, Punjab 141004, India
e-mail: shalakaahale135@gmail.com

G. Rakhonde
Department of Plant Pathology, University of Agricultural Sciences, GKVK, Bangalore, Karnataka 560065, India

S. Bhateja
Department of Entomology, Maharana Pratap University of Agriculture and Technology, Udaipur, Rajasthan 313001, India

S. Mishra
Department of Extension Education, Punjab Agricultural University, Ludhiana, Punjab 141004, India

K. Pandey et al. (eds.), *Artificial Intelligence and Smart Agriculture*,
Advances in Geographical and Environmental Sciences,
https://doi.org/10.1007/978-981-97-0341-8_21

Keywords Artificial techniques · Internet of Things · Remote sensing · Sustainable agriculture

21.1 Introduction

Global agricultural techniques were greatly impacted by the swift rise in the human population and quick, corresponding fall in the area under cultivation. As a result, smart agriculture has gradually replaced traditional farming methods in recent years. One of the most significant issues in modern agriculture is the spread of various plant diseases and pests, especially invasive ones. Due to the occurrence of disease and pest, majority of world is facing reduction in food production. Eventually, they cause yield losses that account for up to 30% of the global economy each year (Savary et al. 2017; Rizzo et al. 2021). Therefore, early monitoring is required to address these pertinent issues. Traditionally farmers were unable to identify the exact cause of the yield losses but after the green revolution a great emphasise were given towards grain production at larger extent which was result of proper management practices against the prevailing pathogen. In case of today's scenario, the time has changed, and with this advent of time, several pathogens have been emerged with new races and biotypes (pathogen and pest, respectively) with respect to their preferential host. For the purpose of making a decision for a later management approach, the precise detection of the primary infection and disease evolution is crucial. Pests and pathogens interact with hosts to cause diseases and pest damage to crops, which serves as the physical foundation for their remote monitoring. It should be noted that not all plant diseases and pests may be detected remotely because some of them lack distinctively identifying traits. Plant responses towards the infection can be easily recognised by using sensor-based advanced technology. Therefore, detection of precise responses by a particular sensor or sensor system is a crucial prerequisite for identifying, monitoring, and managing plant diseases and pests via remote sensing. Prior detection, new-generation fungicides, use of LiDAR, RADAR, genetic engineering approaches, finding resistance gene, genome editing, nanotechnology in pesticide and fungicide combinations, computer-based surveillance, and IoT all are the viable choices for raising smart as well as sustainable agriculture. Therefore, it is very important in the actual world to seek for quick, automatic, affordable, and efficient methods to control disease and pest outbreaks.

21.2 Sensor-Based Technologies: A Multifaceted Tool in Agriculture

Remote sensing has revolutionised research in myriad sectors of ecology and environment science. Application of remote sensing in agriculture has been around for a long time and is trending. Remote sensing is defined as the detection of emitted or reflected electromagnetic radiations from afar using ground-based satellites and aerial platforms. Sensors can serve as early forecasting and warning systems to counteract the climatic or biological changes before having negative consequences on crop yield. Remote sensing applications have been playing a pivotal role in agriculture for evaluating crop health, yield loss (%) estimation, irrigation management, nutrient requirement, water stress management, weed as well as pest detection and weather forecasting, etc. (Shanmugapriya et al. 2019).

Producers around the globe hunting for novel techniques like remote sensing, global position system (GPS), and geographic information system (GIS) for utilizing them to minimise the production losses caused due to several pest and pathogens. Sensors used in remote sensing technology help to record different observations by using electromagnetic radiations emitted from ground-based targets (El-Ghany et al. 2020). Major components of remote sensing system compose of (1) energy source to illuminate the target with electromagnetic energy, (2) sensors for tracking electromagnetic radiation from the target, (3) platforms that includes different ground-based, airborne, or spaceborne sensor carriers, (4) process of transmission and reception, (5) data interpretation, analysis, and its further applications (Natural Resources Canada 2019).

Optical sensors used in remote sensing help to identify the pathogen or pest accurately depending on the nature of damage they caused. This approach thus eventually helps in process of plant phenotyping (Kuska and Mahlein 2018). A plant's ability to reflect, absorb, or transmit electromagnetic energy from the sun depends on the energy's wavelength and the traits of that plant specifically. Bawden (1933) and Colwell (1956) were the first to use visible aerial photography to detect viral infections in potato and tobacco crops. Although interior leaf structural changes in cereal crops caused by diseases may be distinguished by using infrared imaging.

21.3 What are Sensors?

Sensors are mechanical devices that are capable of detecting and responding to a variety of environmental inputs. The precise input might include any one of the natural phenomena like light, heat, motion, moisture, or pressure. The output in general is a signal that is converted to a human-decipherable view at the sensor location or that is electronically transmitted over a network for reading or additional processing utilising image processing technology. It can store the information as electromagnetic signals or raw data and these data are analysed with various image

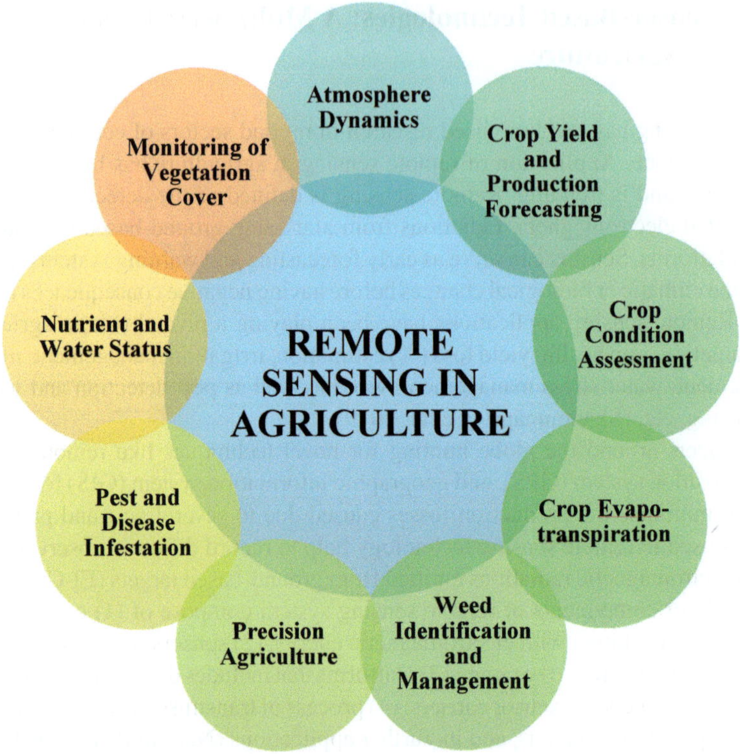

Fig. 21.1 Agricultural applications of remote sensing

classifiers, for example, *Support Vector Machine* (SVM), Artificial Neural Networks (ANN), etc. (Petersson et al. 2016).

21.3.1 RGB Image Sensors

RGB-based imaging is substituting the human vision system towards identification of any diseases and pests, to evaluate the quality of various food items (processed or fresh), weed and crop mapping, determination of secondary metabolite, non-invasive estimation of nutrient status such as nitrogen content and calculating the amount of chlorophyll in micro-propagated seedlings (Aldea et al. 2006; Mercado-Luna et al. 2010; Yadav et al. 2010; Dutta Gupta et al. 2013). Red, green as well as blue colours from a visible spectrum of light reflected from a scene are specifically captured by RGB picture sensors. A focal scene is captured using the RGB sensor array as the mechanism. The sensor array is made up of clustered rows and columns of photodiodes covered in red, green, blue, and no (clear) filters. There are 16 red, green, and blue filters with photodiodes in every ordinary 8×8 RGB array. Only red incident light can fall on the photodiode because of red filter layer; all other light spectrums are blocked. As a result, a photodiode with a red filter generates a photocurrent that corresponds to red light. Other photodiodes generate similar photocurrents. The composite current values are transformed by supporting electronics into an equivalent frequency square wave, which is then processed and transmitted by a computer or an Internet of Things (IoT) device. The RGB imaging technique uses sensors that can generate picture data in the red (550–750 nm), green (500–549 nm), and blue (400–499 nm) wavelength ranges of the spectrum (Lowe et al. 2017). Red, green, and blue are the three fundamental colour values used by these sensors to quantify pictures, which produce an integrated response across spectral bands and generate a range of colours. The plant optical characteristics are influenced by light reflection, light absorption by pigments, sugars, water, and light transmission by tissues. The digital images generated by sensors either examined in field condition or on the farm where they can be utilised to spot various anomalies along with some changes in tissues of plants (Mahlein 2016). RGB indices are used by researchers for the past ten years to determine the severity of losses brought on by fungi and the degree of resistance to particular diseases (ex. wheat yellow rust) and to estimate grain production (Maloney et al. 2014; Zhou et al. 2015; Vergara-Diaz et al. 2015). The RGB sensors are similarly efficient in the acquisition of high-resolution aerial photos from satellites or aircraft which are used to monitor plant diseases. To assess the effects of the *Phymatotrichops* causing root rot disease in alfalfa, a system was emerged for the supervised categorization of RGB aerial data (Mattupalli et al. 2018). RGB pictures can also be used to measure the concentration of plant pathogen-produced secondary metabolites like mycotoxins which can directly affect food quality control and consequently human health. Similarly, RGB imaging analysis was used to identify a toxin (deoxynivalenol) in *Fusarium graminearum*-affected plants. It was practically demonstrated that RGB image analysis is an effective substitute for traditional methods for analysing the impact of pathogens on plant growth and microbial toxin contamination since it is very inexpensive, simple to use, and effective (Cambaza et al. 2019).

21.3.2 Thermal Sensors

Temperature is a key physical factor in many biological and commercial processes and the thermography technique has completely changed the field of plant science. Infrared (IR) thermal imaging was originally created for military use, but it is now widely used in a variety of industries including engineering, aerospace, medical, agriculture, and veterinary sciences (Danno et al. 1977; Oerke et al. 2006). Materials like InGaAs (an alloy of indium arsenide and gallium arsenide) are used which are more sensitive to shortwave infrared (wavelength 0.78–1.7 m) in thermal imaging sensors rather than visible spectrum of light. Every object we encounter in daily life including ice emits heat (IR waves). A measurement of an object's emitted radiations (IR) is also known as thermography. The electromagnetic spectrum's infrared (IR) region is the emission range for all objects above − 273.15 °C. Thermal imaging transforms an object's unseen radiation patterns into observable, comprehensible images. A thermal camera with an IR detector, a signal processor, and an image-acquisition computer make up an IR thermal imaging system. Thermography has been used to successfully identify diseases caused by several kinds of pathogens (including bacteria, fungi, viruses, phanerogamic parasites, etc.) and pests too at the early stages of infection. A similar method was used to identify powdery mildew disease early in wheat using a thermal imaging sensor (Bayoumi and Abdullah 2016). The early stages of the peanut leaf spot disease were identified using thermal imaging techniques and image spectroscopy based on differences between spectral reflectance components in healthy and diseased leaves (Omran 2017). The detection of wheat stripe rust's pre-symptomatic monitoring was also done using thermal imaging (Yao et al. 2018). Further, a fast infrared-based thermal image processing method was created by Yang and colleagues to monitor tea diseases in tea orchards (Yang et al. 2019). An infrared camera has been used to capture photos of both healthy and diseased tea plants, and based on these thermal images, the tea canopy was categorised. Furthermore, utilising a thermal image processing technique, a computer-based vision system was created to identify tea diseases. Additionally, a severe viral disease is caused by the co-infection of the sweet potato (*Ipomoea batatas* L.) with both the sweet potato chlorotic stunt virus and the sweet potato feathery mottle virus which was detected (Wang et al. 2019). There are numerous uses for IR thermography in plant sciences particularly in plant disease at various spatial and temporal scales. When a plant is infected with a pathogen, its physiological processes change including transpiration, photosynthesis, stomatal conductance, resistance reactions, and cell death which are determined using different technologies, which thus helps in finding solution over ongoing losses of crop plant system (Xu et al. 2006; Hashim et al. 2020).

21.3.3 Thermal Imaging Using Infrared Rays

Visibility of reference items in complete darkness is enhanced by recognising them with infrared radiation used in imaging technique, and subsequently, a picture is generated based on the data. Thermal imaging, near-infrared lighting, and low-light imaging are the three most extensively used night vision technologies. The distribution and size of the droplets in drone-based pesticide sprays can be evaluated using infrared thermal imaging.

Lv et al. (2019) accurately measured droplet deposition using infrared thermal imaging as the temperature of the leaves will vary after the drones spray. The infrared thermal imaging technique uses optoelectronic technology to detect the specific signals of an object's thermal radiation as an additional method for measuring droplet accumulation. Following the spray test, enclosed environmental chambers can be used in conjunction with thermal pictures acquisition to eliminate outside interference. Temperature fluctuations can be avoided by sampling in the middle of the fields using infrared thermal imaging before and after spraying.

21.3.4 Fluorescence Imaging

Plant photosynthetic activity fluctuations are computed using several chlorophyll fluorescence metrics. Active sensors with an LED or lasers are frequently used in chlorophyll fluorescence imaging equipment to assess photosynthesis electron transport (Bauriegel et al. 2014). The changes in photosynthetic activity brought on by living and non-living stressors over the leaf area have been studied using this technique (Scholes and Rolfe 2009). Both fluorescence imaging and image analysis technologies together allow for accurate detection and measurement of pathogen infections (Konanz et al. 2014). The plant preparation process must adhere to precise guidelines, making it challenging to use current chlorophyll fluorescence imaging techniques in typical agricultural field or greenhouses environments.

21.3.5 Electro-optical Sensors

Light-based sensors are called optical sensors or electro-optical sensors. This sensor converts light energy into electronic signal. They can detect electromagnetic radiations from broad range such as *visible region*, near-infrared (NIR), short-wave IR. It uses a mechanical device to transfer the physical characteristics of light into readable forms. In most cases, an optical sensor is a component of a larger system that includes a light source, a gauging device, and the optical sensor and is frequently attached to a specific electrical trigger that responds to an alteration in the illumination signal of the sensors. A sensor that uses light may detect differences in one or more light beams. The light sensor functions as a photoelectric trigger when a change takes place, raising or lowering the electrical output in turn to form a spectral signature or spectral profile (Mahlein et al. 2013). Based on their specific properties and utility, optical sensors are separated into different types. They are three main groups—through bream sensors, retro reflective senors, and diffusion reflection sensors. In plant monitoring point of view RGB sensors, multi and hyperspectral sensors, thermography or chlorophyll sensors and fluorescence based senors. Among all these sensors, hyperspectral sensors are quite advanced and it is working with retroreflective senors' principle.

21.3.6 Hyperspectral Sensors

Optical sensors which detect more spectral band in higher spatial resolution (more than 300 bands) with wide spectral light range (VIS to SWIR range: 400–2500 nm) are called hyperspectral sensors (HS). Hyperspectral sensing is a passive remote sensing method that measures reflected sun radiation which has a lot of potential for tracking abiotic as well as biotic plant stress. This method collects and stores information from an object's spectroscopy in a three-dimensional spectral cube that includes both spatial information and a large number of contiguous wavelengths. Hyperspectral imaging offers many alternatives for the early plant disease identification since it provides early signs in the form of minute variations in spectrum reflectance brought on by absorption or reflection. Comparatively speaking, multispectral imagers record the spectral information of objects or sceneries in a number of broad wave bands. By using standard multispectral camera, image data is obtained in the form of R (red), G (green), B (blue), near-infrared (NIR), and UV wavebands. Multispectral camera systems use a hybrid composite set of spectral filters and image sensors to take several pictures spanning various electromagnetic spectrum bands. Optical sensors can measure plant pathogen-induced changes in the physiological process, by both non-invasively and objectively, which is not able to do with most common disease rating and detection methods. Several studies exposed that particularly hyperspectral sensors are the most valuable tool for detecting, identifying, and quantifying plant disease at various scales from the tissue towards the plant canopy level. Bohnenkamp et al. (2017) reported wheat yellow rust in winter wheat cultivars using HSI, UAV, and phytobike Bussard and Asano. They stated that the field deployment of hyperspectral imaging technologies for plant disease identification remains difficult. Currently existing instrumentation and database techniques are designed for highly controlled laboratory environments. But only real-time field data can direct plant protection strategies and improve resource efficiency. Field-level variables including the proper measurement of geographical distance, relevant data sets, and well-designed algorithms still need to be assessed. Using a ground-based vehicle and an unmanned aerial vehicle, the identification and quantification of yellow rust in wheat using hyperspectral imaging were investigated. The identification and quantification of yellow rust in wheat using hyperspectral imaging were examined using two platforms: a ground-based vehicle and UAV. These sensors use spectral sensors, particularly in the electromagnetic range from 400 to 2500 nm, to measure light reflected throughout pathogenic infection and disease progression in severity levels of yellow rust in wheat fields and detection of crop canopy. Crop plant diseases have been seen using high spatial resolution satellite technology and airborne imaging, and the impact on productivity has been assessed (Zheng et al. 2018). In a plot-based field trial, appropriate elements for online disease assessments at the field level were identified and repeated on various disease developing phases and severity levels. These assessments would incorporate sensor technologies and simple analysis models with rapid evaluation.

Mahlein et al. (2012a, b) offered a non-invasive diagnostic technique with excellent potential for diagnosing powdery mildew, cercospora leaf spot, and leaf rust on sugar beet at various stages of growth. Continuous monitoring of disease progress during the pathogenesis process was carried out under controlled settings using an HSI line scanning spectrometer with a 2.8 nm spectral resolution from 400 to 1000 nm and a 0.19 mm spatial resolution. The identification and comprehensive reporting of sick tissue across leaf level were made possible by a mapping spectral reflectance of pixel images in visible as well as near-infrared region. Spectral reflectance patterns of leaves, also known as leaf spectral signatures, were connected to leaf structure. Plant-to-plant, disease-to-disease, and other environmental factors all have different spectral signatures. Variation in spectral signatures is associated with functions of disease development phases and leaf area occupied by pathogen which affects reflectance pattern for image. The process of extracting HSI deep pixelwise spectral fingerprints aids in a better understanding that how plant diseases affect variations in leaf reflectance patterns. Thus, the accuracy and precision of plant disease proximate sensing are considerably enhanced by HSI technology. Similar to this, the apple scab disease detection data set shows two sets of hyperspectral images of healthy and diseased apple tree leaves. These images were obtained each day from two days after the first infection and correspond to an advanced stage of infections (11 days following inoculation). By using reflection correction to calibrate them, the hyperspectral pictures have been assigned to fit the reference image's geometry; thus on final day of the experiment, scab positions were provided (Nouri et al. 2018).

Reports given by many scientists help us to understand and need of HSI in plant pathology which are as follows. Mahlein et al. (2018) worked with HSI on *Cercospora* leaf spot during pathogenesis and mentioned the space resolution of 0.19 nm. The sensitivity and accuracy percentage (96.25%) were also observed with early identification of *Tomato spotted wilt virus*—TSWV (Ashourloo et al. 2014)—with 83 bands per pixel along with false positivity percentage (1.47%) (Wang et al. 2019). After acquiring the image, it will process with various algorithms (OR-AC-GAN) for classifying the image. Even though HSI has lot of importance in phytopathology research, still it has their own drawback like mixed pixel spectrum (Zhu et al. 2017). Possible applications of HSI on different levels for plant protection are discussed and evaluated. Both the advantages and disadvantages on each scale, in addition to the external impact such as light, wind, viewing angle of measuring object, greenhouses, and field conditions, are critically evaluated in order to support pathologist and agriculture technicians. As digital technology advances, a vision that incorporates smart and user-friendly methods for phenotyping plant traits and making judgements about plant protection strategies against a background of phytopathology is increasingly being addressed in society and industry. Furthermore, a comprehensive review about optical sensors can be used on different levels of disease detection which emulates the possibilities of non-invasive detection. This focuses advantages of hyperspectral sensors over other optical sensors when investigating plant pathogen interactions through different examples and by some approaches' detection before visible symptoms appearance is feasible.

21.4 Principles in Monitoring Plant Diseases and Pests

Different agricultural pests and disease organisms produce different symptoms and damage to crop when seen in pathogen and host interactions, which serves as the physical foundation for their remote surveillance. But the fact that some plant diseases and pests lack readily recognised features means that not all of them can be found offsite. Alternatively, many root and soil-borne diseases that have a systemic impact on a plant's physiology can be found easily. As a result, specific responses are acquired by a particular sensor or sensor system is a crucial prerequisite for remote sensing-based plant disease and pest detection and monitoring. It has been discovered that there are four harms associated with remote sensing (Fig. 21.2) which includes (1) LAI decline and biomass reduction among the signs and physiological changes brought on by diseases or pests in plants. Usually, attacks by pests result in this kind of devastation. Pests like armyworms in maize can devour plant materials including leaves and stalks, which results in a large loss of biomass and leaf area (Zhang et al. 2016). But because this destruction lacks spectrum specificity, monitoring it inevitably runs into problems with significant ambiguity. (2) Necrotic lesions and pustules which are specific with respect to particular causal agent or disease. In terms of colour and shape, lesions and pustules might vary between diseases and pests. So, conclusively distribution of lesions on canopy of crop plants such as evenly scattered throughout the plant or only at bottom plays an important role in their ability to detect the cause (Cao et al. 2013; Moshou et al., 2004). (3) Destroyed pigment system. Several instances, pest infestation and disease infection can result in the death of chloroplasts or other organelles, causing fluctuations in the amounts of pigments like chlorophyll, carotenoid, and anthocyanin. To detect this kind of response, hyperspectral RS measurements are typically necessary (Grisham et al. 2010; Zhang et al. 2013). (4) Wilting. Although it is not a particularly mere presence of biotic agent, the loss of plant rigidity caused upon with dehydration can be interpreted for drought stress. Plants will wilt because of some pests' piercing and sucking behaviours (such as those of beetles or aphids) (Cheng et al. 2010). Additionally, severe infections damages plant vascular systems like formation of tyloses and aggregation of fungal mass inside vascular bundles results in obstruction of water flow and dehydration of entire plant (Calderón et al. 2013). It is considered that plant diseases and pests can result in a single or numerous types of destruction, as shown above. In reality, the accumulation of symptoms can be recorded by remotely sensed monitoring. Infections or attack of pests often takes place over time, and several symptoms may overlap, interact, and manifest at various stages with varying degrees of severity (Huang et al. 2007). For instance, in winter yellow rust first results in lesions on the leaves; later on, it results in loss of chlorophyll. Damage of photosynthetic system, if not controlled, will reduce biomass and impede a plant's ability to use water for its metabolism. As a result, the plants that have been extensively affected will eventually display the symptom of withering. Identifying and keeping an eye on certain diseases and pests will be made more difficult by the temporal pattern of symptom changes. As a result, it is important to recognise particular remote sensing traits at various stages of a disease or insect invasion. Positively, distinct variations in the symptom manifestation may also make it possible to extract similar temporal characteristics, which when put together to create a unique signature for the disease or pest can be used to reduce monitoring error (Eklundh et al. 2009).

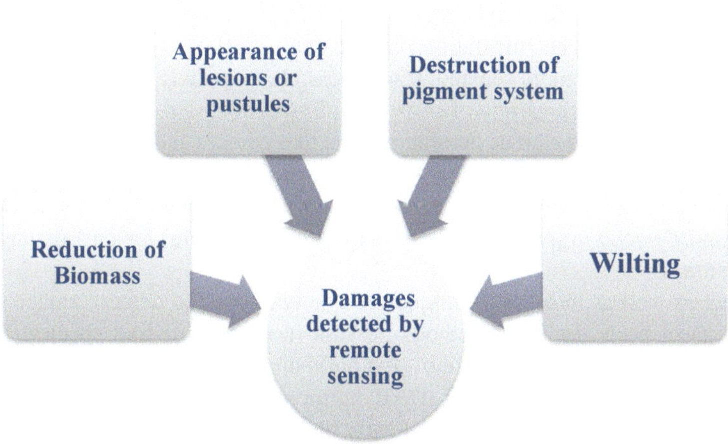

Fig. 21.2 Major types of damages detected by using remote sensing

21.5 Remote Sensing in Disease and Pest Management

The exact diagnosis of primary sites of infection and dynamics of disease is critical for making a management decisions. Optical sensors in this situation can offer a precise and accurate identification of plant diseases. Maize infected with the Maize Streak Gemini virus (MSV) has just been identified and mapped using RapidEye multispectral sensors combined with vegetation indices. RapidEye's use of strange clusters led to an increased accuracy of 82.75% in the detection and mapping of MSV. Finally, the satellite-based plants' signals increased the accuracy of the categorization. Remote sensing is significant in agricultural pests control and has become a popular use in insect taxonomic identification, physiology, and biology (Nansen and Elliott 2016). Depending on the ecological damage done by targeted insect pest, such as the presence of sooty mould deposits produced by honeydew secretion, defoliation, change in colour, and geometric disruption of the plant shape, various techniques such as videography, radio-telemetry, X-ray radiography, radio frequency identification, computed tomography, thermal imaging, radar, and satellites have emerged in the form of new devices for pest surveillance, detection of infestation level, and pre-treatment. Remote sensing also includes spatial statistics techniques for finding and modelling geographical linkages and correlations between climatic and environmental variables (Fernando et al. 2018).

Many successful remote sensing applications have been documented. For example, an ASD Field spec 3 radiometer is used to identify and distinguish between rice canopies and those without BPH infestation, as well as to investigate the sensitive bands and the most useful and effective eigenvalues for doing so under greenhouse conditions. Remote sensing can be utilised to identify the stress brought on by BPH in rice, according to reflectance data and estimated eigenvalues from the spectral graphs from the radiometer. At canopy measuring level, 1813–1836 nm may be the most vulnerable to BPHs infestation (Zhiyan et al. 2010).

With the help of remote sensing, it was possible to quickly identify the presence and spread of pests on a variety of crops including cotton, rice, wheat, citrus, grapes, date palms, walnuts as well as on ornamentals like gerbera and vegetables like potatoes, mustard, and chillies. Satellite remote sensing data can be used to create crop maps for vast agricultural systems infested by a polyphagous pests like *Bemisia tabaci* and *Bemisia argentifolii*. These maps are very helpful for pest management in large areas.

Direct as well as indirect techniques can be employed to detect damage caused by the insect pests. Indirect methods possess feeding signs like spectral feeding signs, structural also nest structure of pest build up by them in plants whereas direct detection includes entomological LiDAR, harmonic radar, VLR and passive sensing of insects. Although the advantages of early identification of crop damage are clear, there are not many examples of remote sensing being used practically to evaluate crop condition. Utilising spectral data as an indicator of biotic stress necessitates a change in the plant's reflectance properties in visible region from 380 and 700 nm in the visible spectrum, along with 700 and 2500 nm in the infrared spectrum (near IR). An example of interruption in physiological system of plants is physical invasion of maize by *Ostrinia nubilalis* larval stage affects nutrition, and water transport inside the maize plants increases the frequency of anthracnose and stem rot diseases and can cause premature ageing and hence reflectance pattern of light. Interruption in nutrient translocation influences plant pigment dynamics and is interrelated with a plants physiological condition. The amount of water in the leaf influences how light interacts with wavelength-independent components such as the kind, number, and organisation of the cells as well as their shape and ultrastructure (Matthew et al. 2008).

21.6 Innovative Technologies for Management of Disease and Pest

Global disturbances and climatic changes have already paved the way for the severe co-epidemics of crop pest and diseases which resulted in serious deterioration of crop yield and quality. The most crucial aspect for successful pest control programmes is basic information of damage level caused by insects, weed population, and plant pathogens causing diseases as it gives a clear picture for management programme and decision-making. Conventionally, crop pest and diseases are monitored by manual scouting and visual inspection of individual plant, which is ineffective and time consuming (Wenjiang et al. 2019). Around the world, remote sensing methods are being tested to see if they could support this system by monitoring, mapping, and forecasting outbreak patterns. By collecting and analysing multispectral visuals, remote sensing technology can assume command (Sivanpillai and Latchininsky 2013). The different applications of digital technology in IPM include use of remote sensing, GIS, GPS, E-pest surveillance, drones, and robotics along with artificial intelligence (AI) that can aid in success of an IPM programme.

21.6.1 UAVs

In remote sensing technologies, satellites and drones (UAV) are used to monitor farms for the presence of pests and diseases. Drones have the advantages of great flexibility and good image quality. This can fulfil the needs for crops' pest and disease monitoring. The information regarding pests and diseases affecting crops should be standardised and digitalized before drones are employed to detect their presence. Recent developments in remote sensing using a drone and processing of geographical images have accelerated the expansion of technologies for detection/surveillance of disease and insect pest for crop management (Vanegas et al. 2018). High-definition cameras mounted on UAVs or mounting racks are used to create a public real-time video monitoring system for diseases and pest (Yuan et al. 2012) which forces farmers to visit the farmland to inspect the crops regularly and increase overall working efficiency (Rademacher 2015). An UAV is equipped with a Light Detection and Ranging (LiDAR) sensor to map the overflown environment of farms in point clouds (UAV) in order to avoid external interference and the effects of drone flying speed on droplet size, shape, dispersion, and uniformity. An accurate and high-throughput solution for crop disease identification in practical settings can be achieved by combining aerial picture data with AI-based techniques (Selvaraj et al. 2020).

Aphids and mites are major crop-damaging pests that feed on cotton plants and are of great concern. Aerial sprays employing drones were tried to combat these pests and it was discovered that they were effective against spider mites and cotton aphids. Insecticides are sprayed aerially to manage pests including Brown plant hopper (BPH) in rice, aphids, and spiders in cotton and fall armyworms in sugarcane using drone technology which works on the artificial intelligence (AI) principle. This technique is more effective than conventional spraying at controlling pests in various agricultural and horticultural crops.

21.6.2 RADAR

The habitat and environmental circumstances affect the effectiveness of luminous marking and portable harmonic radar systems (Kho et al. 2018). Insect radar echoes have been used in migration research. Recent advancements in polar metric capabilities have boosted the precision observation regarding pests (Leskinen et al. 2011). High-power electromagnetic pulses are transmitted by the Radar Meteorologico Argentino-1 (RMA1) that are produced from radar's antenna and sends the pulses to the targets which is being observed. The pulses are returned to the radar in modest amounts. The radar receiver digitises, amplifies, and processes the received echo so that it can be stored and shown. The target is revealed in great detail by the dual-polarisation technique. Poffo et al. (2018) described during the day, as the swarm of the White butterfly, *Asciamonuste* L., approached from east to west, horizontal reflectivity factor echo occurred in a series. The migratory locust pest *Chortoicetes terminifera* Walker and the height at which the swarms of the locust *S. cancellata* were recorded by radar in Argentina were both present in Australia. Line RMA1 is a reliable method for swarm detection. The radar results showed that the swarming of *A. monuste* was divided into multiple bands, while that of locust was uniform with concentric Doppler radar. Radar made early detection possible. The RMA1 was very good at estimating the direction of swarm movement and assessing migration factors like height and swarm form. Since 40 years ago, researchers have used radars to track moth migration routes. Radars that study insects are better suited than metre radars for monitoring pest populations.

21.6.3 LiDAR System

LiDAR uses light for detecting and ranging in a remote sensing device in the form of a pulsed laser to measure ranges and distances to the earth. It measures the amount of time it takes for a light beam to return to its source after being focused on a surface. When one shines a torch on a surface, they observe the light reflecting off the surface and travelling back to their retinas. Mei et al. (2012) used a LiDAR system to remotely and in situ detect distinct species of pest moths and rice plant hoppers in China. Three species of moths, namely *Helicoverpa armigera, Spodoptera litura, and Spodoptera exigua*, as well as three species of plant hoppers, i.e. *Nilaparvata lugens,* white-backed plant hopper, *Sogatella furcifera*, and *Laodelphax striatellus*, were studied. Using LiDAR, a few mm-long coloured rice plant hopper could be located at a distance of 50 m. Researchers through their research advised employing LiDAR to study insect migration and movement around pheromone traps. LiDARs may be used near to the ground for observing insects, unlike radars.

21.6.4 Genetic Engineering

Scientists have created alternative pest management strategies that focus on the genetics of the organisms as agricultural technology has evolved under the burden of increasing population and increased standard of living. Genetic control is a pest management strategy in which crops are genetically modified to become resistant to pests and diseases caused by pests. Crops can be genetically modified to develop chemical or physical barriers that protect them from pests. Pest damage can now be reduced thanks to genetic control. Crops can be genetically modified to create chemical or physical defences against pests. Owing to genetic modifications, pest damage can be decreased. Genetic engineering in most cases depends on the conventional approaches or on recent genome editing techniques. It mainly emphasises on addition or deletion of particular genomic regions or gene to control the pertaining disease or pest hazard. Genetically modified crops are available in market to avoid the occurrence of pest on particular host plant by modifying either host or pest genomic sequences like insertion of gene from other non-host, toxin-producing gene, etc. One of the examples is of *Bt* gene introduced in cotton plant from *Bacillus thuringiensis* against bollworm. With the identification of microRNAs (miRNAs), a subclass of endogenous noncoding RNAs, genetic engineering for viral resistance has become even more sophisticated (Ambros 2001). Modifications of susceptibility gene *SWEET11* with the help of CRISPR-Cas has brought resistance to *Xanthomonas oryzae* pv. *oryzae* in rice lines (Zhou et al. 2014).

21.6.5 Artificial Intelligence

Machines that simulate cognitive processes like learning and problem-solving that people connect with the human mind are referred to as having artificial intelligence. Advanced technical solutions are needed in agriculture to identify plant pests early and prevent unintended herbicide consumption. Automated pesticide spraying across a field is now possible because of UAVs with computer vision AI. UAV sprayers can work with excellent precision in terms of area and amount sprayed thanks to real-time detection of target spraying areas. The computer vision model can count them with 92.5% accuracy and can recognise bees, flies, mosquitoes, moths, chafers, and fruit flies with an accuracy of 90.18%. Agricultural issues have been solved utilising computer systems and image processing employing AI approaches, such as machine learning and Artificial Neural Networks (ANNs).

21.6.6 Nanotechnology

Formulations with nanoscale particles and their distinctive features are referred to as nanopesticides. Both organic and inorganic substances can be found in nanopesticides. Now a days nanopesticides used to protect plants must meet three requirements: (1) they must be in the nanosize range, (2) contain the "nano" prefix, and (3) have innovative nanomaterial qualities. It is possible to create nanocapsules out of the nanoparticles. Nanocapsules allow pesticides and crop protection chemicals to penetrate effectively into plant cuticular cells and tissues. Broad-spectrum efficacy against pests is exhibited by nano-based pesticides including Cu, Ag, SiO_2, and ZnO as well as their formulations which also significantly reduce chemical, air, water, and soil contamination. In comparison to synthetic nanopesticides, the encapsulated plant-based nanopesticides are more active and easily absorbed by the pests they are intended to control. Since they degrade over time, plant-based nanopesticides are environmentally friendly. It is possible that polyphagous pests and vectors would not build resistance quickly or easily. These formulations can be used to make slow-release medications (Singh et al. 2018). In addition to having antifungal characteristics, nanoparticles according to Ponmurugan et al. (2016) also assist in preserving the balance of soil nutrients. On various crops, nano-biosensors are used to find pests and pathogens (Rai and Ingle 2012). Nanoparticles were used by Rai et al. (2018) to review the literature on ginger disease. The use of nanoformulation, in particularly *Pythium* and *Fusarium* spp. (fungal pathogens), in treating soft rot of ginger is now being assessed. *Fusarium graminearum* is resistant to synthetic characteristics of ZnO nanoparticles, which have been demonstrated by Dimkpa et al. (2013). The formulation of the water-soluble pesticide Novaluron (30–100 nm nanoparticles) which targets *Spodoptera littoral is* larvae of the cotton leafworm, which was found to be effective against these pests. Nanoparticles can more effectively penetrate into plants, pests, and in the delivery system for insecticides (Margulis-Goshen and Magdassi 2012). The rice crop affected by the blast fungus was pushed to respond favourably to a chemical compound like pyra-clostrobin. Porous SiO_2 nanospheres enhance the pathogen-fighting, stability, and activity of pesticides (Huang et al. 2018). An emerging field of nanotechnology called nanoscale carriers allows for the loading and coating of microorganisms like *Trichoderma* sp. or *Pseudomonas* sp. with materials like nanoclay polymer nanocomposites (Ammar 2018).

21.6.7 Internet of Things (IoTs)

The Internet of Things (IoT) is becoming the primary way of data gathering as well as transmission and it will be an important technology across a variety of sensors to collect, analyse, transmit, and manage all data linked to plant disease and insect pests. The whole process starts from acquiring information about an object or phenomenon using various satellites, aircraft, etc., and lastly storing, analysing, managing, and presenting data connected to places using GIS software. Wireless sensor networks (WSN) and general packet radio service are the most often utilised network technologies in agriculture for information transfer. In order to monitor physical or environmental parameters such as pressure, temperature, sound, motion, and pollution among others, WSNs are generally made up of spatially distributed autonomous sensors. The automatic data collection module primarily collects wind speed and its direction, temperature, humidity, CO_2 concentration, surface temperature and humidity, conductivity, and video data. The wireless transmission network module is responsible for quickly, proficiently, and reliably delivering collected data to the total node. The remote data transmission module is responsible for delivering a summary of monitoring data points via the GPRS network via Internet. Data are at the heart of IoT monitoring systems.

The general concept of the monitoring information system is separated into four areas based on the architecture of agricultural disease and insect pests: network sensor

Table 21.1 Advantages and limitations of advanced technologies

Advantages	Limitation
1. Prior detection of prevailing pest and disease epidemics 2. Precise and accurate information is produced 3. Less time is required as compared to labourial work in field conditions 4. Eco-friendly and promote organic farming 5. Safe, self-limited, and targeted approach 6. High quality and better income generation	1. Ambiguity in detection 2. Lack of farmers' knowledge about modern techniques 3. Need of much easier programme to be run by farmers 4. Changing climate affects the life cycle and dispersal of invading pathogen; thus, flexibility of techniques is much needed 5. Adoption and acceptance by farming community of developing country 6. Sensor calibration for appropriate results and understanding the analysis, interpretation, and retrieval of data is a difficult process

node, monitoring information and its acquisition, sensor and wireless communication technology, and intelligent data processing. In order to access the real-time data, historical data, and remote parameter settings collected by the monitoring sensing element, they can log in to the remote server using PCs, mobile phones, tablet computers and other devices. IoT sensors and drones are used in an early warning and response system to combat cotton pests (Azfar et al. 2023). Agricultural specialists can then analyse the data and create an expert database. The IoT-based disease and insect pest monitoring information platform would be a unified system in farming culture (Table 21.1).

21.7 Future Prospective

Farmers and researchers have a serious difficulty in the strategic project of preventing and controlling agricultural pests and diseases, which calls for extensive data collection and analysis. Over past few years, developing countries have adopted several techniques and instruments which prove that they are far better than traditional methods for management and help in improving crop production. A long-term pest and disease prediction model should be created through more data collection and extensive data analysis. Information communication technologies such as smartphones and the IoT can change agriculture by facilitating access to markets and providing problem-solving information. These technologies can also bridge the rural–urban gap, creating awareness among the next generation of growers. Further study is required to assess how the latest tools and technologies can be practically used to support agriculture. Development of high-resolution, reasonably priced, and portable spectral sensors is necessary because hyperspectral data have mainly demonstrated their ability to help with early stress detection, such as a slight decline in chlorophyll and leaf water content, giving forest managers an opportunity to perform effective disease control. Since it has been demonstrated that leaf-level spectroscopy can be used successfully as a ground reference or training input for airborne-based platforms, more research should focus on connecting spectral data acquired at various scales. Thus, integration with artificial intelligence and IoT techniques will help to raise agriculture individually as a progressive sector. The agricultural industry could begin a new era of super-convergence with the help of modern equipment including agricultural robots, cloud computing, artificial intelligence, and big data.

21.8 Conclusion

Diseases and pest are always being a hurdle for development of agriculture to flourish at its maximum potential. Growing world population, urbanisation of rural areas, and mechanisation of traditional approaches dealing towards the disease and pest control are somehow responsible for food losses direct or indirectly, and hence, they became an important challenge to meet the ever-increasing hunger of the growing population. Farmers' acceptance of any new technology is mainly depending upon its effectiveness to combat the disease and durability of the solution on the existing problem. Conventional approach of using chemical pesticide, fungicides, and fertilisers needs to be change into advanced approach. Diagnostic techniques are always been a non-destructive method mainly helps in decision-making for the management process or type of treatment further to be used under on field conditions on the prior basis. Artificial intelligence and computational automated techniques are much more advanced than the farming community and will take time to reach towards every farmer in person. Reasons behind this can be a lack of knowledge to the poor farmers and communication barriers with the expertise community. In recent decades, novel technologies like remote sensing, genome editing, and molecular approaches have advanced the agriculture particularly from the prospective of monitoring, detection, and control as compared to conventional farming. Current chapter mainly focuses on such different techniques and tools and provides a better eyesight to wrap up the epidemics by the pest and pathogens.

References

Aldea M, Frank TD, Delucia EH (2006) A method for quantitative analysis for spatially variable physiological processes across leaf surfaces. Photosynth Res 90:161–172. https://doi.org/10.1007/s11120-006-9119-z

Ambros V (2001) MicroRNAs: tiny regulators with great potential. Cell 107:823–826. https://doi.org/10.1016/S0092-8674(01)00616-X

Ammar AS (2018) Nanotechnologies associated to floral resources in agri-food sector. Acta Agronóm 67(1):146–159. https://doi.org/10.15446/acag.v67n1.62011

Ashourloo D, Mobasheri MR, Huete A (2014) Evaluating the effect of different wheat rust disease symptoms on vegetation indices using hyperspectral measurements. Remote Sens 6(6):5107–5123. https://doi.org/10.3390/rs6065107

Azfar S, Nadeem A, Ahsan K, Mehmood A, Almoamari H, Alqahtany SS (2023) IoT-based cotton plant pest detection and smart-response system. Appl Sci 13(3):1851. https://doi.org/10.3390/app13031851

Bauriegel E, Brabandt H, Gärber U, Herppich WB (2014) Chlorophyll fluorescence imaging to facilitate breeding of *Bremia lactucae*-resistant lettuce cultivars. Comput Electron Agric 105:74–82. https://doi.org/10.1016/j.compag.2014.04.010

Bawden FC (1933) Infrared photography and plant virus detection. Nature 132:168. https://doi.org/10.1038/132168a0

Bayoumi TY, Abdullah AA (2016) Application of thermal imaging sensor to early detect powdery mildew disease in wheat. J Middle East North Afr Sci 10(3907):1–8. https://doi.org/10.12816/0032657

Bohnenkamp D, Thomas S, Kuska MT, Brugger A, Alisaac E (2017) Benefits of hyperspectral imaging for plant disease detection and plant protection: a technical perspective. J Plant Dis Prot 125:5–20. https://doi.org/10.1007/s41348-017-01246

Calderón R, Navas-Cortés JA, Lucena C, Zarco-Tejada PJ (2013) High-resolution airborne hyperspectral and thermal imagery for early detection of Verticillium wilt of olive using fluorescence, temperature and narrow-band spectral indices. Remote Sens Environ 139:231–245. https://doi.org/10.1016/j.rse.2013.07.031

Cambaza E, Koseki S, Kawamura S (2019) Why RGB imaging should be used to analyze *Fusarium graminearum* growth and estimate deoxynivalenol contamination. Methods Protoc 2(1):25. https://doi.org/10.3390/mps2010025

Cao X, Luo Y, Zhou Y, Duan X, Cheng D (2013) Detection of powdery mildew in two winter wheat cultivars using canopy hyperspectral reflectance. Crop Prot 45:124–131. https://doi.org/10.1016/j.cropro.2012.12.002

Cheng T, Rivard B, Sánchez-Azofeifa GA, Feng J, Calvo-Polanco M (2010) Continuous wavelet analysis for the detection of green attack damage due to mountain pine beetle infestation. Remote Sens Environ 114(4):899–910. https://doi.org/10.1016/j.rse.2009.12.005

Colwell RN (1956) Determining the prevalence of certain cereal crop diseases by means of aerial photography. Hilgardia 26:223–286. https://doi.org/10.3733/hilg.v26n05p223

Danno A, Miyazato M, Ishiguro E (1977) Quality evaluation of agricultural products by infrared imaging method: grading of food bioprocess technology fruits for bruise and other surface defects. Memoirs Fac Agric Kagoshima Univ 14:123–138

Dimkpa CO, McLean JE, Britt DW, Anderson AJ (2013) Antifungal activity of ZnO nanoparticles and their interactive effect with a biocontrol bacterium on growth antagonism of the plant pathogen *Fusarium graminearum*. Biometals 26(6):913–924. https://doi.org/10.1007/s10534-013-9667-6

Dutta Gupta S, Ibaraki Y, Pattanayak A (2013) Development of a digital image analysis method for real-time estimation of chlorophyll content in micro propagated potato plants. Plant Biotechnol Rep 7:91–97. https://doi.org/10.1007/s11816-012-0240-5

Eklundh L, Johansson T, Solberg S (2009) Mapping insect defoliation in Scots pine with MODIS time-series data. Remote Sens Environ 113(7):1566–1573. https://doi.org/10.1016/j.rse.2009.03.008

El-Ghany A, Nesreen M, El-Aziz A, Shadia E, Marei SS (2020) A review: application of remote sensing as a promising strategy for insect pests and diseases management. Environ Sci Pollut Res 27(27):33503–33515. https://doi.org/10.1007/s11356-020-09517-2

Fernando V, Dmitry B, Kevin P, John W, Felipe G (2018) A novel methodology for improving plant pest surveillance in vineyards and crops using UAV-based hyperspectral and spatial data. Sensors 8(1):260–281. https://doi.org/10.3390/s18010260

Grisham MP, Johnson RM, Zimba PV (2010) Detecting Sugarcane yellow leaf virus infection in asymptomatic leaves with hyperspectral remote sensing and associated leaf pigment changes. J Virol Methods 167(2):140–145. https://doi.org/10.1016/j.jviromet.2010.03.024

Hashim IC, Shariff ARM, Bejo SK, Muharam FM, Ahmad K, Hashim H (2020) Application of thermal imaging for plant disease detection. IOP Conf Ser: Earth Environ Sci 540(1):012052. https://doi.org/10.1088/1755-1315/540/1/012052

Huang B, Chen F, Shen Y, Qian K, Wang Y, Sun C, Cui H (2018) Advances in targeted pesticides with environmentally responsive controlled release by nanotechnology. Nanomaterials 8(2):102. https://doi.org/10.3390/nano8020102

Huang W, Lamb DW, Niu Z, Zhang Y, Liu L, Wang J (2007) Identification of yellow rust in wheat using in-situ spectral reflectance measurements and airborne hyperspectral imaging. Precision Agric 8(4):187–197. https://doi.org/10.1007/s11119-007-9038-9

Kho J-W, Jung M, Lee D (2018) Evaluating the efficacy of two insect detection methods with *Riptortus pedestris*: portable harmonic radar system and fluorescent marking system. Pest Manag Sci 75:224–233. https://doi.org/10.1002/p25106

Konanz S, Kocsányi L, Buschmann C (2014) Advanced multi-color fluorescence imaging system for detection of biotic and abiotic stresses in leaves. Agriculture 4(2):79–95. https://doi.org/10.3390/agriculture4020079

Kuska MT, Mahlein AK (2018) Aiming at decision making in plant disease protection and phenotyping by the use of optical sensors. Eur J Plant Pathol 152(4):987–992. https://doi.org/10.1007/s10658-018-1464-1

Leskinen M, Markkula I, Koistinen J, Pylkkö P, Ooperi S, Siljamo P (2011) Pest insect immigration warning by an atmospheric dispersion model, weather radars and traps. J Appl Entomol 135(1–2):55–67. https://doi.org/10.1111/j.1439-0418.2009.01480.x

Lowe A, Harrison N, French AP (2017) Hyperspectral image analysis techniques for the detection and classification of the early onset of plant disease and stress. Plant Methods 13:80. https://doi.org/10.1186/s13007-017-0233-z

Lv B, Xu H, Wu J, Tian Y, Zhang Y, Zheng Y, Tian S (2019) LiDAR-enhanced connected infrastructures sensing and broadcasting high-resolution traffic information serving smart cities. IEEE Access 7:79895–79907. https://doi.org/10.1109/ACCESS.2019.2923421

Mahlein AK (2016) Plant disease detection by imaging sensors–parallels and specific demands for precision agriculture and plant phenotyping. Plant Dis 100(2):241–251. https://doi.org/10.1094/PDIS-03-15-0340-FE

Mahlein AK, Kuska MT, Behmann J, Polder G, Walter A (2018) Hyperspectral sensors and imaging technologies in phytopathology: state of the art. Annu Rev Phytopathol 56:535–558. https://doi.org/10.1146/annurev-phyto-080417-050100

Mahlein AK, Rumpf T, Welke B, Dehne P, Plümer L, Steiner U, Oerke EC (2013) Development of spectral indices for detecting and identifying plant diseases. Remote Sens Environ 128:21–30. https://doi.org/10.1016/j.rse.2012.09.019

Mahlein AK, Steiner U, Hillnhütter C, Dehne HW, Oerke EC (2012a) Hyperspectral imaging for small-scale analysis of symptoms caused by different sugar beet diseases. Plant Methods 8(1):1–13. https://doi.org/10.1186/1746-4811-83

Mahlein AK, Oerke EC, Steiner U, Dehne HW (2012b) Recent advances in sensing plant diseases for precision crop protection. Eur J Plant Pathol 133:197–209. https://doi.org/10.1007/s10658-011-9878-z

Maloney PV, Petersen S, Navarro RA, Marshall D, McKendry AL, Costa JM, Murphy JP (2014) Digital image analysis method for estimation of Fusarium-damaged kernels in wheat. Crop Sci 54(5):2077–2083. https://doi.org/10.2135/cropsci2013.07.0432

Margulis-Goshen K, Magdassi S (2012) Organic nanoparticles from microemulsions: formation and applications. Curr Opin Colloid Interface Sci 17(5):290–296. https://doi.org/10.1016/j.cocis.2012.06.005

Matthew C, Glaser JA, Hellmich RL, Hunt TE, Sappington TW, Calvin DD, Copenhaver K, Fridgen J (2008) Use of spectral vegetation indices derived from airborne hyperspectral imagery for detection of European corn borer infestation in Iowa corn plots. J Econ Entomol 101:1614–1623

Mattupalli C, Moffet CA, Shah KN, Young CA (2018) Supervised classification of RGB aerial imagery to evaluate the impact of a root rot disease. Remote Sens 10(6):917. https://doi.org/10.3390/rs10060917

Mei L, Guan ZG, Zhou HJ, Lv J, Zhu ZR, Cheng JA, Somesfalean G (2012) Agricultural pest monitoring using fluorescence LiDAR techniques. Appl Phys B 106(3):733–740. https://doi.org/10.1007/s00340-011-4785-8

Mercado-Luna A, Rico-García E, Lara-Herrera A, Soto-Zarazúa G, Ocampo-Velázquez R, Guevara-González R, Herrera-Ruiz G, Torres-Pacheco I (2010) Nitrogen determination on tomato (*Lycopersicon esculentum* Mill.) seedlings by color image analysis (RGB). Afr J Biotechnol 9:5326–5332

Moshou D, Bravo C, West J, Wahlen S, McCartney A, Ramon H (2004) Automatic detection of 'yellow rust' in wheat using reflectance measurements and neural networks. Comput Electron Agric 44:173–188. https://doi.org/10.1016/j.compag.2004.04.003

Nansen C, Elliott N (2016) Remote sensing and reflectance profiling in entomology. Annu Rev Entomol 61:139–158. https://doi.org/10.1146/annurev-ento-010715-023834

Natural Resources Canada (2019). https://www.nrcan.gc.ca/earthsciences/geomatics/satellite-imagery-air-photos/satellite-imageryproducts/educational-resources/9363

Nouri M, Gorretta N, Vaysse P, Giraud M, Germain C, Keresztes B, Roger JM (2018) Near infrared hyperspectral dataset of healthy and infected apple tree leaves images for the early detection of apple scab disease. Data Brief 16:967–971. https://doi.org/10.1016/j.dib.2017.12.043

Oerke EC, Steiner U, Dehne HW, Lindenthal M (2006) Thermal imaging of cucumber leaves affected by downy mildew and environmental conditions. J Exp Bot 57(9):2121–2132. https://doi.org/10.1093/jxb/erj170

Omran ESE (2017) Early sensing of peanut leaf spot using spectroscopy and thermal imaging. Arch Agron Soil Sci 63(7):883–896. https://doi.org/10.1080/03650340.2016.1247952

Petersson H, Gustafsson D, Bergstrom D (2016) Hyperspectral image analysis using deep learning—a review. Paper presented at the 6th International Conference on Image Processing Theory, Tools and Applications (IPTA 2016), Oulu, Finland, 12–15 Dec 2016. https://doi.org/10.1109/IPTA.2016.7820963

Poffo DA, Beccaece HM, Caranti GM, Comer RA (2018) Migration monitoring of *Ascia monuste* (Lepidoptera) and *Schistocerca cancellata* in Argentina using RMAI weather radar. ISPRS J Photogramm Remote Sens. https://doi.org/10.1016/j.isprsjprs.2018.05011

Ponmurugan P, Manjukarunambika K, Elango V, Gnanamangai BM (2016) Antifungal activity of biosynthesised copper nanoparticles evaluated against red root-rot disease in tea plants. J Exp Nanosci 11(13):1019–1031. https://doi.org/10.1080/17458080.2016.1184766

Rademacher W (2015) Plant growth regulators: backgrounds and uses in plant production. J Plant Growth Regul 34:845–872. https://doi.org/10.1007/s00344-015-9541-6

Rai M, Ingle A (2012) Role of nanotechnology in agriculture with special reference to management of insect pests. Appl Microbiol Biotechnol 94(2):287–293. https://doi.org/10.1007/s00253-012-3969-4

Rai M, Ingle AP, Paralikar P, Anasane N, Gade R, Ingle P (2018) Effective management of soft rot of ginger caused by *Pythium* spp. and *Fusarium* spp.: emerging role of nanotechnology. Appl Microbiol Biotechnol 102(16):6827–6839. https://doi.org/10.1007/s00253-018-9145-8

Rizzo DM, Lichtveld M, Mazet JAK (2021) Plant health and its effects on food safety and security in a One Health framework: four case studies. One Health Outlook 3:6. https://doi.org/10.1186/s42522-021-00038-7

Savary S, Mcroberts N, Esker PD, Willocquet L, Teng PS (2017) Production situations as drivers of crop health: evidence and implications. Plant Pathol 66(6):867–876

Scholes JD, Rolfe SA (2009) Chlorophyll fluorescence imaging as tool for understanding the impact of fungal diseases on plant performance: a phenomics perspective. Funct Plant Biol 36(11):880–892. https://doi.org/10.1071/FP09145

Selvaraj MG, Vergara A, Montenegro F, Ruiz HA, Safari N, Raymaekers D, Blomme G (2020) Detection of banana plants and their major diseases through aerial images and machine learning methods: a case study in DR Congo and Republic of Benin. ISPRS J Photogramm Remote Sens 169:110–124. https://doi.org/10.1016/j.isprsjprs.2020.08.025

Shanmugapriya P, Rathika S, Ramesh T, Janaki P (2019) Applications of remote sensing in agriculture—a review. Int J Curr Microbiol Appl Sci 8(01):2270–2283. https://doi.org/10.20546/ijcmas.2019.801.238

Singh P, Kumari K, Vishvakarma VK, Aggarwal S, Chandra R, Yadav A (2018) Nanotechnology and its impact on insects in agriculture. In: Trends in insect molecular biology and biotechnology. Springer, Cham, pp 353–378. https://doi.org/10.1007/978-3-319-61343-7_17

Sivanpillai R, Latchininsky AV (2013) Special section on advances in remote sensing applications for locust habitat monitoring and management. J Appl Remote Sens 7(1):075001–075033. https://doi.org/10.1117/1.JRS.8.084801

Vanegas F, Bratanov D, Powell K, Weiss J, Gonzalez F (2018) A novel methodology for improving plant pest surveillance in vineyards and crops using UAV-based hyperspectral and spatial data. Sensors 18:260. https://doi.org/10.3390/s18010260

Vergara-Diaz O, Kefauver SC, Elazab A, Nieto-Taladriz MT, Araus JL (2015) Grain yield losses in yellow-rusted durum wheat estimated using digital and conventional parameters under field conditions. Crop J 3:200–210. https://doi.org/10.1016/j.cj.2015.03.003

Wang L, Poque S, Valkonen JP (2019) Phenotyping viral infection in sweet potato using a high-throughput chlorophyll fluorescence and thermal imaging platform. Plant Methods 15(1):1–14. https://doi.org/10.1186/s13007-019-0501-1

Wenjiang H, Yue S, Yingying D, Huichun Y, Mingquan W, Bei C, Linyi L (2019) Progress and prospects of crop diseases and pests monitoring by remote sensing. Smart Agric 1(4):1–11. https://doi.org/10.12133/j.smartag.2019.1.4.201905-SA005

Xu H, Zhu S, Ying Y, Jiang H (2006) Early detection of plant disease using infrared thermal imaging. In: Optics for natural resources, agriculture and foods, vol 6381, pp 638110. International Society for Optics and Photonics. https://doi.org/10.1117/12.685534

Yadav SP, Ibaraki Y, Gupta SD (2010) Estimation of the chlorophyll content of micropropagated potato plants using RGB based image analysis. Plant Cell Tissue Organ Culture (PCTOC) 100(2):183–188. https://doi.org/10.1007/s11240-009-9635-6

Yang N, Yuan M, Wang P, Zhang R, Sun J, Mao H (2019) Tea diseases detection based on fast infrared thermal image processing technology. J Sci Food Agric 99(7):3459–3466. https://doi.org/10.1002/jsfa.9564

Yao Z, He D, Lei Y (2018) Thermal imaging for early nondestructive detection of wheat stripe rust. In: 2018 ASABE annual international meeting. American Society of Agricultural and Biological Engineers, p 1. https://doi.org/10.3964/j.issn.1000-0593(2018)10-3303-07

Yuan L, Zhang J, Wang J (2012) Research progress of crop diseases and pests monitoring-based on remote sensing. Trans Chin Soc Agric Eng 28:1–11. https://doi.org/10.3969/j.issn.1002-6819.2012.20.001

Zhang J, Huang Y, Yuan L, Yang G, Chen L, Zhao C (2016) Using satellite multispectral imagery for damage mapping of armyworm (*Spodoptera frugiperda*) in maize at a regional scale. Pest Manag Sci 72(2):335–348. https://doi.org/10.1002/ps.4003

Zhang L, Zhang L, Tao D, Huang X, Du B (2013) Hyperspectral remote sensing image subpixel target detection based on supervised metric learning. IEEE Trans Geosci Remote Sens 52(8):4955–4965. https://doi.org/10.1111/tgis.12164

Zheng Q, Huang W, Cui X, Shi Y, Liu L (2018) New spectral index for detecting wheat yellow rust using Sentinel-2 multispectral imagery. Sensors 18:868. https://doi.org/10.3390/s18030868

Zhiyan Z, Zang Y, Zuoxi Z, Luo X, Zhou X (2010) Canopy hyperspectral reflectance feature of rice caused by Brown Plant-hopper (*Nilaparvata lugens*) infestation. In: American Society of

Agricultural and Biological Engineers annual international meeting 2010, ASABE 2010. https://doi.org/10.13031/2013.29933

Zhou B, Elazab A, Bort J, Vergara O, Serret MD, Araus JL (2015) Low-cost assessment of wheat resistance to yellow rust through conventional RGB images. Comput Electron Agric 116:20–29. https://doi.org/10.1016/j.compag.2015.05.017

Zhou H, Liu B, Weeks DP, Spalding MH, Yang B (2014) Large chromosomal deletions and heritable small genetic changes induced by CRISPR/Cas9 in rice. Nucleic Acids Res 42:10903–10914. https://doi.org/10.1093/nar/gku806

Zhu H, Chu B, Zhang C, Liu F, Jiang L, He Y (2017) Hyperspectral imaging for presymptomatic detection of tobacco disease with successive projections algorithm and machine-learning classifiers. Sci Rep 7:4125. https://doi.org/10.1038/s41598-017-04501-2

Chapter 22
A Critical Analysis of Convolutional Neural Networks for Leaf Disease Detection in Plants

Gurdit Singh

Abstract Deep learning techniques have just recently been used to picture identification. Convolutional neural networks, a sort of deep learning technology, have shown outstanding breakthroughs in this discipline. Convolutional neural networks have shown effective in applications such as object identification, face recognition, handwritten digit recognition, and traffic sign recognition. The success of CNN in image recognition has prompted researchers to investigate their application in agriculture for a broader range of purposes, including plant identification, crop monitoring, weed detection, water and soil management, fruit counting, disease diagnosis, and crop health assessment. The availability of different research works on the use of deep learning models in agriculture makes it difficult to pick an appropriate model based on the kind of dataset and experimental setting. This paper focuses on the plant disease prediction using leaf pictures collected with a deep convolutional neural network. This paper also examines and discusses several preprocessing approaches, convolutional neural network models, frameworks, and optimization strategies for recognising and categorising plant illnesses in their leaves. We also discusses the datasets and performance indicators that are used to assess the efficacy of a model. The strengths and downsides of various techniques and models discovered in the existing research are highlighted. This survey will be useful to researchers interested in applying deep learning approaches to the problem of leaf disease diagnoses.

Keywords Convolutional neural networks · Deep learning · Agriculture · Leaf · Disease · Survey

G. Singh (✉)
Department of CSE, Punjab Agricultural University, Ludhiana, India
e-mail: gurditsinghpec@gmail.com

K. Pandey et al. (eds.), *Artificial Intelligence and Smart Agriculture*,
Advances in Geographical and Environmental Sciences,
https://doi.org/10.1007/978-981-97-0341-8_22

22.1 Introduction

The world's population is growing at an alarming rate. According to (Zhang 2008), the world's population is expected to reach 8.5 billion by the end of 2030. For this reason, increasing agricultural output is crucial to meet the needs of a growing population. Temperature, humidity, and precipitation all contribute to the optimal development circumstances for bacteria, viruses, fungus, nematodes, and other microorganisms. Because of the abundance of pathogens in their environment, plants become more susceptible to illness. The decline in agricultural productivity is largely attributable to insect and disease attacks. In order to take the necessary precautions to avoid and protect plants against illness, accurate predictions of the onset of such diseases are essential. As a result, it contributes to better crop yields and higher production overall.

There are a number of symptoms that can be used to diagnose plant diseases, including lesions, colour changes, leaf damage, stem damage, aberrant growth of the stem, leaf, bud, flower, and/or root. Spots, dryness, premature leaf drop, etc., are also visible on leaves and serve as an early warning sign of disease (Chouhan et al. 2020). Understanding the causes of these outward signs is a reliable method for identifying plant diseases. The conventional method of plant disease identification relies on a skilled or experienced eye(s). However, thorough understanding and experience in disease identification are necessary for success with this strategy. In addition, it may cause inaccurate forecasts as a consequence of cognitive biases and optical illusions (Bock et al. 2010). As a result, the method is impractical for extensive farmland.

The researchers have recognised the shortcomings of the conventional method and have proposed technological methods for plant disease prediction. Artificial neural networks (ANNs) (Huang 2007; Huang et al. 2018), support vector machines (SVMs) (Bauer et al. 2011; Li et al. 2016), Bayes classifiers (Bayes nets) (Tan et al. 2016; Pound et al. 2017), k-nearest neighbours (KNNs) (Singh et al. 2016; Ampatzidis et al. 2017), and other machine learning techniques are used to develop automatic disease detection models. Acquiring photos of plant leaves, performing image preprocessing, segmenting those images, extracting features from those segments, and finally classifying diseases are the five stages of a disease detection model's workflow. Disease detection using current models is quite good. However, accurate disease classification requires human experience in acquiring leaf images (Pound et al. 2017). So, depending on the nature of the datasets and the experimental settings, there is room for development in these models. Therefore, in (Singh et al. 2016), the authors aim to improve the effectiveness and precision of the aforementioned models.

It is difficult for researchers to select an acceptable model in the light of their dataset, parameters, hardware setup, and experimental settings due to the abundance of literature on the issue of applying machine learning and deep learning models to the prediction of plant diseases. Therefore, scientists require easy access to a comprehensive literature review that can point them in the direction of a feasible model for preprocessing, predicting, and categorising plant disease data. To address this, the authors give an in-depth analysis of the various preprocessing techniques, deep convolutional neural network (DCNN) designs, DCNN frameworks, and optimization techniques currently available. In addition to recognising and classifying plant leaf diseases, the paper discusses the merits, limits, and prospective applications of deep learning models developed.

22.2 Materials and Methods

Nutritional deficiencies, microbial and rodent attacks, adverse environmental circumstances, etc., are identified as major root causes of plant illnesses in the literature (Ampatzidis et al. 2017). Damage to a plant's structure or functionality results from several causes of stress. Stress in the plants are of two types:

(1) **Biotic**—it is caused by various organisms like viruses, bacteria, fungus, etc.
(2) **Abiotic**—it is caused from non-living factors like unpleasant environment, poor soil, intense sunlight, fluctuating temperatures, an inadequate amount of precipitation, or an abundance of an important mineral(s). Stress caused by living things is more perilous than that caused by non-living things (Kaur et al. 2019). Insects' voracious eating can significantly reduce a leaf's surface area, slowing photosynthesis. The nitrogen deficiency stress in plants was identified using a deep learning-based method published by (Azimi et al. 2021). On the other hand, (Noon et al. 2020) reported the results of a deep learning-based survey on the identification of plant leaf stress.

Symptoms of a diseased plant include a deviation from the norm in terms of appearance (colour, form, size, growth rate, etc.). When an illness is in its early, middle, or late stages, its symptoms will change accordingly. In the transition phase, a healthy plant is compromised by the disease-causing factor(s), and the first signs of the disease begin to show. There is little to no way to tell a healthy plant from a diseased one at this point. There is also a good probability that the plant's immune system will be weakened by one illness, leading to an attack by multiple diseases. As a result, symptoms for multiple diseases may appear similar (Barbedo 2016).

Disease symptoms can also be affected by weather conditions such as temperature, wind speed and direction, humidity, and sunlight exposure. These factors cause the diseased area to vary in size, colour, and shape. When this happens, it is difficult to diagnose a disease by looking at a plant or plant part for the first time (Chouhan et al. 2018). However, cutting-edge AI methods like convolutional neural networks (CNNs) and deep convolutional neural networks (DCNNs) can reduce the need for humans to get involved while still providing reliable results in disease diagnosis. However, diagnosing disease is complicated by the presence of extraneous details in an image, such as dead leaves, soil, branches, other plant leaves, weeds, insects, and so on.

In addition, factors like lighting, cloud cover, the angle of reflection, the direction of the sun, and so on have a significant impact on the final quality of a photograph of a plant. Poor-quality, low-contrast images do not provide enough information to accurately diagnose an illness. The construction of an appropriate model for the dataset, the collection of a large dataset for training a model, the identification of the optimal number of layers in a model, and the identification of the optimal number of neurons in each layer are all additional difficulties in the application of AI and DCNNs. Similarly, determining how many parameters should be fed into a CNN system is difficult.

During the training phase, a common issue for deep learning models is disappearing gradients. Adjusting the network's parameters can be challenging since the gradient of the loss function could approach zero, making training more challenging. The first several layers are crucial for understanding the input data. If the gradient is low, then the parameters of the first few layers would not be updated very well, and the whole network will be off. This means that going beyond a certain network depth will result in a decline in the model's accuracy. According to (Wang et al. 2017), increasing the depth of a network past the initial eight nodes reduces the accuracy of a shallow network. When deep neural networks are trained, the output of each layer is used as the input for the subsequent layer. So, it is a contributing factor to the issue of internal covariate shift. The difficulty makes it take longer to learn something new. It calls for slower learning rates and the careful setting of parameters. To reduce the impact of the internal covariate shift issue, the authors used the batch normalisation method (Ioffe and Szegedy 2015). With batch normalisation, it is possible to use far faster learning rates with little to no regard for initialisation. The inputs to each hidden layer are attempted to be standardised. As a result, the allocation of resources is quite stable. This boosts deep networks' accuracy and learning speed generally (Too et al. 2019).

22.3 Comparative Analysis

Here, we take a look at the many methods used to automate disease prediction, such as preprocessing techniques, CNN trained on different datasets, network architectures, strategies of optimization, and underlying frameworks.

22.3.1 Preprocessing Techniques

Noise, unwanted backdrop, low illumination, etc., can be found in captured photos of plant leaves. No reliable findings are obtained (Chouhan et al. 2020) when classification techniques are used to these recorded images. Preprocessing procedures must be applied to raw datasets before input to a model. It is crucial to speed up the training process and boost the classification precision.

Images may be resized, converted to grayscale, normalised, enhanced, cropped, and a region of interest (ROI) are extracted as part of the preprocessing phase. Most commonly employed preprocessing methods are grouped together in the categories:

- Augmentation.
- Normalisation.
- Standardisation.
- Annotation.
- Resizing.
- Outlier rejection.
- Denoising.

Research into the relevant literature reveals a variety of preprocessing methods fall under each of following subsections. It is most crucial step in order for a CNN model to accurately and efficiently classify a given dataset. The training process can be simplified and speed up by converting RGB images to grayscale (Amara et al. 2017; Oppenheim and Shani 2017; Mohanty et al. 2016) since a mono colour channel has fewer processing requirements than three colour channels. Like principal component analysis (PCA), zero-phase component analysis (ZCA) is a whitening technique. It is used to bring attention to characteristics and structures in order to facilitate learning (Lu et al. 2017b). Images are cropped to focus on the region of interest (ROI) (Ferentinos 2018; Sladojevic et al. 2016; Ha et al. 2017). Before applying the correlation coefficient method of segmentation to a region, contrast stretching is used (Khan et al. 2018). Segmenting photographs of leaves is done using Otsu's technique (Cruz et al. 2017). Following are an in-depth look at the various preprocessing methods currently in use.

22.3.1.1 Resizing

The sizes of images collected from various sources vary. Therefore, prior to using a CNN model, it is necessary to resize the captured and/or gathered images. An effective method for achieving and maintaining image size consistency is the padding of zeros in small images and the cropping of large photos. In the Inception-v3 model (Szegedy et al. 2016) and the Xception model (Chollet 2017), for instance, images are downsized to 299×299 in accordance with an input layer of the CNN model. Images are scaled to 224×224 in the VGGNet architecture, as described in (Simonyan and Zisserman 2014), the ResNet architecture, as provided in (He et al. 2016), and the DenseNet architecture, as described in (Huang et al. 2017). It is recommended to choose a resolution of 227×227 when working with AlexNet (Boulent et al. 2019). Image resizing was automated by researchers in (Mohanty et al. 2016; Ferentinos 2018; Sladojevic et al. 2016; Cruz et al. 2017; Brahimi et al. 2017) using Python and JavaScript. Any image larger than 256 pixels on either side will have its dimensions automatically halved by the code. The RGB image sizes were lowered from 5760×3840 to 512×512 while the authors of (Lu et al. 2017b) retained the model with the features of the input images. Images were resized by the authors of (Oppenheim and Shani 2017; Kawasaki et al. 2015), from original dimensions of 2048×1536 and 2736×1824 pixels to an even more manageable 224×224 pixels. When writing (Amara et al. 2017), the authors relied on 60×60 pixel photographs of plant leaves.

The authors in (Ha et al. 2017) utilised 200 × 200 pixel images, (Ma et al. 2018) used 800 × 600 pixel images, and (Lu et al. 2017a) used 832 × 832 pixel images to maximise the effectiveness of image processing methods and reduce computational cost. As we have seen, shrinking photos is crucial to speeding up the model training and classification processes.

22.3.1.2 Data Augmentation

Deep neural networks' effectiveness is very sensitive to the volume of training data. If only a limited dataset is at hand, then additional information will have to be gleaned from elsewhere. Brightness modification, horizontal and vertical shift, cropping, PCA jittering (Liu et al. 2017), shearing, flipping, and zooming of training images (Wang et al. 2017), rotation (Oppenheim and Shani 2017; Durmuş et al. 2017), and affine transformations (Sladojevic et al. 2016) are all examples of data augmentation approaches. Along with the aforementioned methods, deep learning is also complemented by more cutting-edge approaches, such as Generative Adversarial Networks (GANs) and Neural Style Transfer (NST). These methods intentionally generate changed versions of images to increase the size of a training dataset (Ghosal et al. 2018). That is why they are so effective at improving model performance and keeping networks from overfitting (Pawara et al. 2017; Barbedo 2018).

22.3.1.3 Normalisation and Standardisation

The term "normalisation" refers to the procedure of scaling values or dimensions within a certain range. Pixel values in an 8-bit RGB image, for example, are integers between 0 and 255. When big input values are compounded by low weights, it throws off a CNN's learning process. Therefore, it is necessary to standardise pixel values. To accomplish this, we divide every pixel value by the highest value, which is 255. It provides all channel values between 0 and 1 without altering the appearance of images (Too et al. 2019). Light and dark areas in an image might generate distortions that can be corrected by using normalisation. Normalising can be accomplished in numerous ways, including the Decimal scaling, Min-Max normalisation, and Z-score normalisation.

22.3.1.4 Annotation

Image labelling, or "annotation," is a method used in model training. Annotation for disease detection in plant leaves is performed by professionals with in-depth knowledge of the respective illnesses (Barbedo 2018). Annotation strategies range from bounding box annotations to annotations made pixel by pixel in an image. Bounding box annotation is a popular method of annotation. In this method, the object of interest is enclosed within a compact rectangle or cube. The method has the problem of increasing the overall amount of background noise within the contained area. Images can be annotated pixel by pixel, allowing for precise object selection. Moreover, it is more time-consuming than annotating using bounding boxes (Sa et al. 2016) due to the need for edge-by-edge selection of items. This method is labour-intensive, time-consuming, and fraught with the possibility of inaccuracy.

22.3.1.5 Outlier Rejection

To reject outliers, a dataset must be checked for and removed of any erroneous or superfluous pictures. Examples of image rejection criteria (Sladojevic et al. 2016) include low resolution, low intensity values, blurriness, noise, irrelevance, and duplicates. The authors of (Nguyen et al. 2019) created a convolutional neural network (CNN) model called organNet to filter out erroneous or irrelevant pictures from a dataset.

22.3.1.6 Denoising

Denoising is a technique for cleaning up images by removing noise while preserving the integrity of the original image. It improves an image classification method's performance in a noisy environment. Researchers employ a wide range of denoising algorithms, such as:

- Mean filter.
- Median filter.
- Gaussian filter.
- Small-window median filter, etc.

Commonly used in denoising processes, the Gaussian filter blurs an image to remove noise and related information. The median filter is an example of a nonlinear filter. It filters out irrelevant information. The median filter iteratively scans the pixel matrix of an image with a filter size of 7×7, 5×5, or 3×3. The median of all pixel values is used to replace the centre window pixel. It does a fantastic job of preserving blade edge. Removes "salt and pepper" noise in image processing. The term "salt-and-pepper noise" is commonly used to describe the appearance of pixels that are equally spaced between black and white. Salt-and-pepper noise causes an image's bright areas to have dark pixels and its dark areas to have light pixels. Both the mean filter and the Wiener filter are quite effective in removing unwanted Gaussian noise from audio. Using a median filter with a small window, (Cruz et al. 2017) were able to remove distracting noise from their collection of leaf images.

22.3.2 Convolutional Neural Networks (CNNs)

Hubel and Wiesel (Hubel and Wiesel 1962) invented convolutional neural networks in the year 1962. According to a review of the relevant literature, convolutional neural networks are the most widely used deep learning models for the classification of visual data (Traore et al. 2018). Convolutional neural networks (CNNs) take their design cues from the human visual cortex's receptive field. CNNs are a type of feed-forward neural network. A hierarchical network has several levels, including a convolution layer, a pooling layer, and a fully linked layer. As can be seen in Fig. 22.1, a CNN model has a certain structure.

Backpropagation techniques are used to fine-tune the weights and biases of a network in a CNN architecture during training. Because of this, the cost function becomes less useful. It was in the late 1980s that CNN models were first used to decipher the US Postal Service's handwritten zip codes (LeCun et al. 1989). When CNN is used to detect diseases in plant leaves, it first examines a dataset made up of photos of leaves, then extracts the most important features from those images, and uses those features to categorise the diseases. CNN uses an approach to decision-making that is analogous to that of humans. It records the colour changes and textured areas of plant leaves that are caused by diseases (Toda and Okura 2019). Using a wide variety of nonlinear filters, CNN's feature learning power can automatically discover a leaf image's main properties. As a result, a model's total performance improves, surpassing that of similarly constructed models (Singh et al. 2018).

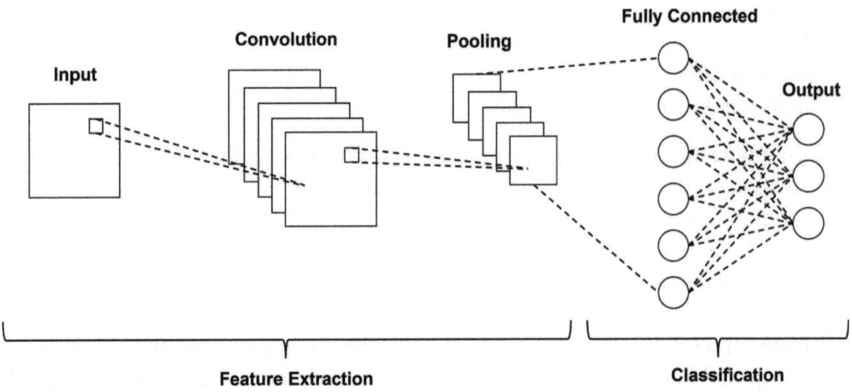

Fig. 22.1 CNN architecture

22.3.3 *Common CNN Architectures*

CNN has progressed since its inception in 1998, when it was first introduced with the help of LeNet (Mohanty et al. 2018). AlexNet's 2012 victory in the ImageNet Large-Scale Visual Recognition Challenge (Ngugi et al. 2021) catapulted it into the public consciousness. As architectures continue to evolve, their accuracy and error rates will improve. There are distinct datasets and experimental settings that are optimal for various architectural approaches (Kundu et al. 2020). The following are some of the most popular structures used in image categorisation.

22.3.3.1 LeNet-5

As can be seen in Fig. 22.2, LeNet-5 is an innovative, seven-tier architecture that is rather easy to understand. A fully connected layer (F6), two pooling layers (S2 and S4), an output layer, and three convolutional layers (C1, C3, and C5) are included. The specifics are laid down in (Kundu et al. 2020). LeNet-5 architecture was initially used for 32 $\times$ 32 pixel images of handwritten numerals. The convolution was carried out with six 5 $\times$ 5 filters and a stride of 1. The resulting feature map has the dimensions of 28 $\times$ 28 $\times$ 6. When using the same filter width and stride 2, it performs an average pooling. It shrank the measurements down to 14 $\times$ 14 $\times$ 6. There was also a second convolutional layer utilised, this time with 16 filters of size 5 $\times$ 5 and a stride of 1. In contrast, a pooling layer with a filter size of 2 $\times$ 2 and a stride of 2 produced a feature map with dimensions of 5 $\times$ 5 $\times$ 16. C5, and the following convolutional layer includes 120 feature maps. Connected to 120 neurons are the 400 nodes of the preceding layer, which had dimensions of (5, 5, 16). There are 84 units in the F6th layer. Convolutional layer 5 (C5) is fully linked and has 120 units. There are 84 classes represented in the output layer, and each class is represented by a unit of Euclidean radial basis function (RBF). The result of the ith RBF unit is denoted by y_i, the jth input vector unit by x_j, and the ijth parameter vector value by w_{ij}. According to the article (Liu and Wang 2021), LeNet is advantageous because of how quickly it can be deployed, how well it can work with parallel hardware, and how well it can handle low-volume image recognition tasks. For the purpose of categorising Black Sigatoka and Banana Speckle in banana leaves, LeNet's fundamental architecture is utilised. It captured photos with a 92–99% accuracy rate despite complex backdrops, varying sizes, dim lighting, and wrong orientation (Amara et al. 2017). According to (Cruz et al. 2017), a change was made to LeNet for spotting olive rapid decline syndrome, and the resulting model had a 99% detection rate. Its superiority is demonstrated by a comparison with three other feature vector representations: videlicet, SIFT, and Local Binary Patterns and the Gabor filter. Using photos of diseased cucumber leaves, a network architecture similar to LeNet-5 is trained to detect the presence of the pathogen. This network consists of three convolutional layers, two pooling layers, and one fully connected layer. It produced superior outcomes compared to other classifiers such as support vector machines (SVMs), Random Forest, and AlexNet

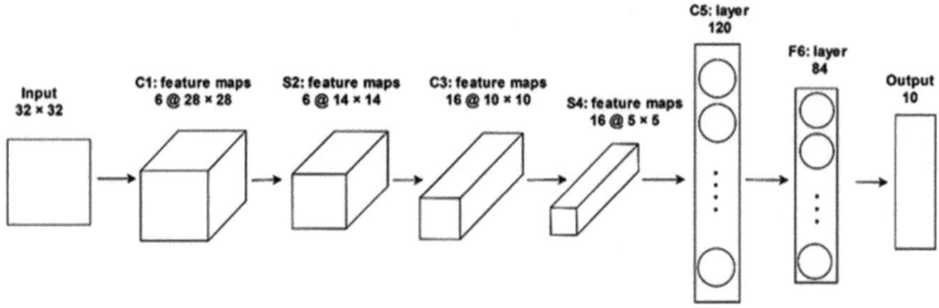

Fig. 22.2 LeNet-5 architecture

(Ma et al. 2018). When looking for signs of Esca disease in vineyards, (Ma et al. 2018) also employed LeNet-5 on photos captured by an unmanned aerial vehicle (UAV). Ground, healthy, potentially diseased, and diseased were the four classes provided. It was able to attain a 95.80% success rate in its classifications (Yadav et al. 2021). LeNet models need less data storage and less time to train; however, its training is subpar. By using the ReLU activation function, the vanishing gradient issue is introduced. As a result, it has a limited rate of success in predicting plant diseases and should not be used for this purpose.

22.3.3.2 AlexNet

With its spectacular performance in the ImageNet Large-Scale Visual Recognition Challenge-2012, AlexNet architecture was born. This structure is analogous to LeNet; however, it is more in-depth and employs almost 60 million parameters. According to the analysis in (Krishnamoorthy et al. 2021), it achieved a top-five mistake of 15.3%, which is better than the 26.2% error rate of the second-best entry. AlexNet is built on top of five convolutional and three fully linked layers. The ReLU activation function is used. It trains networks using several GPUs, the MaxPooling layer, and local response normalisation. To begin, a 224 × 224 pixel input picture is passed through 96 kernels of size 11 × 11 × 4 in the first convolutional layer. The first convolutional layer's response and pooling output are normalised before being sent to the second layer. It implemented filters with 256 kernels, each with size of 5 × 5 × 48. The following three layers are linked together without a standardisation or pooling layer in between. The third convolutional layer has 384 × 3 × 3 256-dimensional kernels. The fourth convolutional layer makes use of 384 × 3 × 3 192-dimensional kernels. Fifth-layer convolutions use 256 × 3 × 3 192-dimensional kernels. In a completely linked neuronal layer, 4096 neurons are found. A softmax receives the output from the final completely linked layer. The fifth and final convolutional layer was the ReLU layer. After the nonlinearity of ReLU, the local response is normalised. In order to facilitate generalisation, these normalising layers are layered on top.

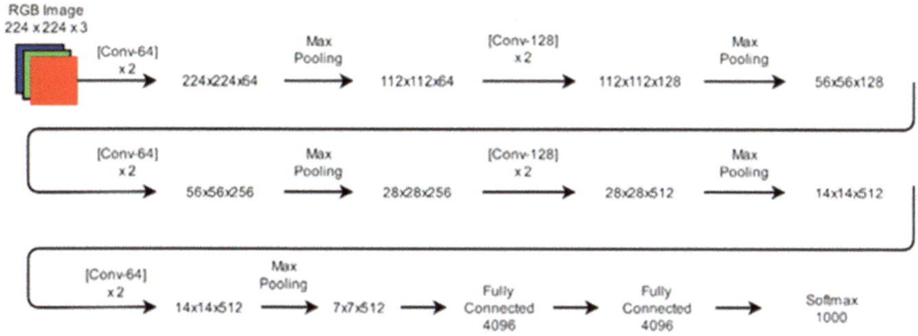

Fig. 22.3 VGG-16 architecture

The fully connected layers in AlexNet help prevent overfitting, so some people decide not to use regularisation techniques as much. At this point, the regularisation approach starts to soak up all the extraneous details, such outliers and noise, that are present in the dataset. Overfitting occurs when a model is trained with a huge dataset. It is possible that a model will perform well on the data contained in its training set, yet struggle when presented with novel data. There are still obstacles to be solved, despite AlexNet's superior accuracy over LeNet, such as correcting for imbalances in datasets and fine-tuning models for certain data types.

22.3.3.3 VGGNet

The Visual Geometry Group at Oxford University developed VGGNet (Simonyan and Zisserman 2014). It finished in second place in the ILSVRC-2014 competition. To a greater extent, this model's localization and classification accuracy improves with model depth. When compared to other networks, VGGNet is less complex. It employs 33 convolutional filters with a one-step stride across all layers. A max-pooling layer with stride 2 is included. Figure 22.3 depicts VGGNet's overall structure. The programme takes as input a 224 $\times$ 224 pixel RGB image. Each pixel in the training dataset has its RGB mean removed to do preprocessing. An image that has already been processed goes through a series of convolutional layers and then five max-pooling layers. It makes use of the first two fully linked levels, which have a total of 4096 channels between them, as well as the third fully connected layer, which has a total of 1000 channels. Finally, a 1000-way classification is carried out in the final layer. Finally, a softmax layer is used to calculate the probability of many classes.

22.3.3.4 GoogLeNet

GoogLeNet (Chen et al. 2021) was conceived as a result of the success of Inception module (Joshi et al. 2021), the winner of ILSVRC-2014. There are more depth and

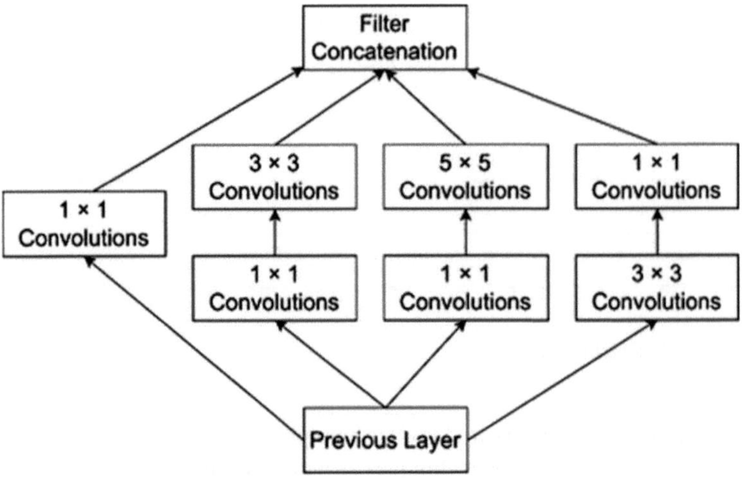

Fig. 22.4 Module of an Inception

breadth to the structure. It has 22 layers, each with 224 × 224 receptive fields, with tiny convolutions of 1 × 1, 3 × 3, and 5 × 5. To extract features, GoogLeNet combines 1 × 1 convolutions, 3 × 3 convolutions, 5 × 5 convolutions, and 3 × 3 max-pooling layers. There are nine inception modules in a vertical stack. The Inception module combines the results of three separate convolutional layers (1 × 1, 3 × 3, and 5 × 5) into a single output vector. As depicted in Fig. 22.4, this results in an input for the succeeding layer. Compared to the original inception module, GoogLeNet introduces two major changes. A 1 × 1 convolutional layer is used as the initial layer, followed by additional layers. Second, it employs a max-pooling layer that runs in parallel. GoogLeNet's dimension reduction significantly lessens the need for expensive processing. Finally, instead of using fully connected layers, it employs global average pooling at the end of inception. Overfitting is a problem that arises when networks have a lot of layers and plenty of neurons in each layer. In the convolutional layers of GoogLeNet, sparsely connected network architectures are used rather than fully connected ones.

22.3.3.5 ResNet

ResNet, developed by Xie and colleagues in 2017, outperformed other models by achieving a low 3.57% error rate, securing the first position in the ILSRVC-2015 competition. The deterioration issue brought on by adding more and more layers to a network inspired the development of this deep residual learning architecture. The vanishing gradient problem makes training a deep network problematic. Loss gradients for the weights are determined during the backpropagation process. As one travels backwards via a network, the gradients typically diminish. As a result,

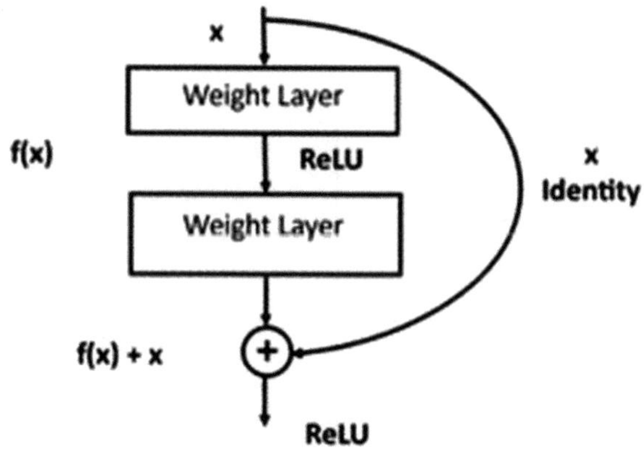

Fig. 22.5 Basic working block of residual learning

the network's performance levels out or even drop. Thus, it may be concluded that a network's lower levels are less adept at picking up new information quickly than its higher ones. An additional challenge of deeper networks is the need to optimise over a wide range of parameters. As a result, the number of training errors increases as new layers are added. As can be seen in Fig. 22.5, ResNet is a network-building method for deep learning purposed that uses residual modules.

ResNet-50 is a 50-layer convolutional network consisting of residual blocks. About 25.6 million parameters are included in it. ResNet-101 has an unprecedented 44.5 million parameters. There are around 60.2 million parameters in ResNet-152 (Bhatt et al. 2017).

22.3.3.6 ResNeXt

This new approach called "split-transform-merge" is an upgrade from ResNet. Instead of using ResNet blocks, it uses ResNeXt blocks. Unlike ResNet, it doesn't generate filters for the entire channel depth of the input. Instead, it divides the input into separate categories. In this case, each section stands in for a different channel. According to studies conducted in (Zhang et al. 2018), boosting a network's cardinality yields better results in terms of improved model performance than boosting its depth. For accurate disease diagnosis and pest recognition in tomato leaves (Sibiya and Sumbwanyambe 2019), try using ResNeXt architecture alongside a quicker R-CNN detector.

22.3.3.7 DenseNet

DenseNet, proposed by (Huang et al. 2017), is a highly interconnected system. As can be seen in Fig. 22.6, it is made up of two different modules: dense blocks and transition layers. All nodes in this network have feed-forward connections to all other

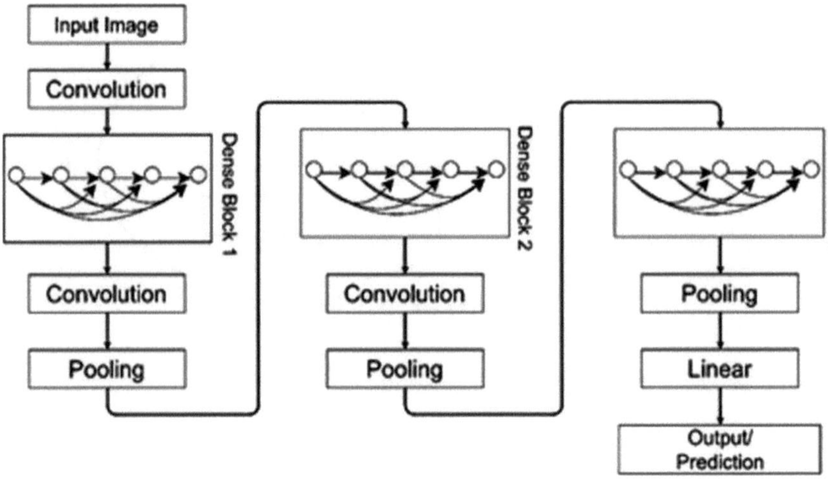

Fig. 22.6 Working model of DenseNet architecture

nodes. An N-layer DenseNet has $N(N + 1)/2$ edge-to-edge links (Huang et al. 2017). All previous layers' feature maps are fed into the current layer as inputs. Together, Batch Normalisation, ReLU activation, and 3 × 3 convolution form a dense block. In between two solid masses are transitional layers the whole architecture is beak down into three parts: Batch Normalisation, convolution, and an average pooling. The feature maps of each layer are the same size so that they can be combined within each thick block.

22.3.3.8 SqueezeNet

According to Fig. 22.7, SqueezeNet (Joly et al. 2017) is made up of eight fire modules sandwiched between two convolutional layers. The "fire" module is made up of a 1 × 1 dimensional filter squeeze convolutional layer. It is being pumped into a bloated layer. The convolutional filters in this layer range in size from 1 × 1 to 3 × 3. From the start to the finish of a network, the number of filters for each fire module grows. Max-pooling is done two steps after the convolution1, fire4, and fire8 layers. A ReLU activation function is used. In the wake of the Fire9 module, it switches to dropout. Some scholars, drawing inspiration from the aforementioned structures, created their own.

22.3.3.9 LeafNet

In order to detect illnesses in tea leaves, (Chen et al. 2019) created a CNN model called LeafNet (Ghazi et al. 2017). When compared to AlexNet, this model performs

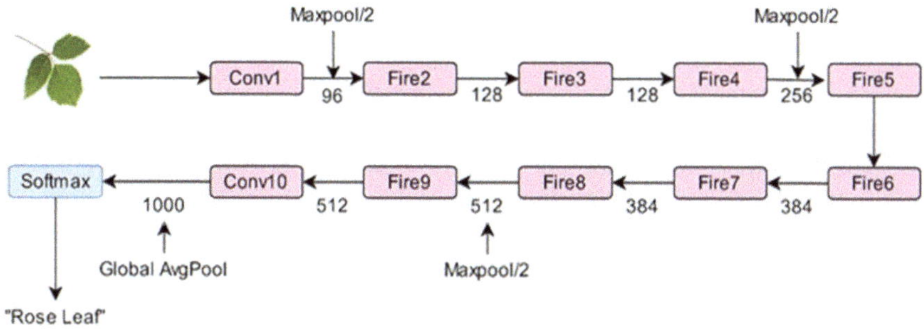

Fig. 22.7 Architecture of SqueezeNet

better. It is constructed by cutting down on the number of convolutional filters and fully connected layer nodes. With fewer network parameters to work with, overfitting can be mitigated.

22.3.3.10 M-bCNN

To detect illnesses in wheat leaves by image classification, (Lin et al. 2019) presented a unified CNN model, matrix-based CNN (M-bCNN). Compared to VGG-16 and AlexNet, the model performs better thanks to a significant rise in data streams, connection channels, and neurons at a reasonable growth in CPU requirements (DeChant et al. 2017). Two different CNN architectures were proposed by (Barbedo 2019) for the detection of rice blast illnesses. That is because they are really like LeNet-5. The first convolutional neural network (CNN) has four convolutional layers, three fully connected layers, and four max-pooling layers. Whenever a new layer is added, ReLU is also incorporated into this network. Convolutional and max-pooling layers are shared between both CNNs, although the second CNN has just two fully connected layers. Each network incorporates dropout layers to prevent overfitting.

22.3.4 Datasets and CNN Models

Researchers applied deep learning models to many datasets in order to identify and categorise plant diseases. Using their own dataset called "PlantVillage," (Deng et al. 2009) trained a CNN model. The scientists collected leaves from plants and stored them on a sheet with a dark grey or black background to create this dataset. Using a digital camera (Sony DSC—Rx100/13, 20.2 megapixels), they took pictures of leaves in a variety of lighting settings. They also determine the best time of day, background, distance, and camera angle for taking photographs. In order to train the CNN model, the authors used 54,306 pictures featuring 14 crop diseases. (Deng et al. 2009) ran a series of tests on three unique versions of the PlantVillage dataset. The first group represents the full-colour version of the original "PlantVillage" dataset. The second group is the "PlantVillage" dataset in grayscale. The third group is comprised segmented leaf pictures from the PlantVillage collection, in which unnecessary context has been eliminated. The majority of the dataset was used for training the CNN model, whereas just a small subset was used for testing. Using GoogLeNet architecture and transfer learning, they improved accuracy to 99.35% on coloured leaf images (Mohanty et al. 2016). Experiments on three different PlantVillage datasets are described in full in (Deng et al. 2009).

Everingham et al. (2010) also provided a comprehensive review of image processing methods for plant leaf disease recognition. Ten distinct DL models were evaluated on the Plant Village dataset. After analysing the models' results, the scientists came to the conclusion that deep networks, such as DenseNet-201, are the most effective for computationally demanding plant disease detection tasks. On the other hand, time-sensitive mobile programmes benefit greatly from shallow networks like Shuffle Net and SqueezeNet.

Several deep learning models were tested by (Picon et al. 2019) using a publicly available dataset of bell pepper plants. These included videlicet, VGG-16, VGG-19, ResNet-50, ResNet-101, ResNet-152, InceptionResNetV2, and DenseNet121. Through their analysis, the authors claim that the "DenseNet" model is more effective than the aforementioned methods at predicting diseases in bell pepper. They further claimed that the model may be utilised for real-time forecasting and that it is easier on computers to operate (Picon et al. 2019).

PlantVillage is a dataset that includes 3700 photos of banana leaves that were used in research by (Amara et al. 2017; Mohanty et al. 2016; Khan et al. 2018), who used the LeNet architecture to obtain a maximum classification accuracy of 96%, captured their images under diverse lighting, size, backdrop, stance, and orientation settings. Disease hotspots, automatic segmentation, and bacterial identification in peach leaves were all reported in a deep learning model by (Li 2017). We found a total classification accuracy of 98.75% based on their experimental assessments on both synthetic datasets and real-world farming. The authors of the paper (Mishkin and Matas 2015) used transfer learning with InceptionResNetV2 to forecast diseases in rice leaf samples. Transfer learning was also used by (Yu et al. 2016) to successfully identify illnesses in rice plants. In order to learn the connections between different channels, MobileNet-V2 was selected as the backend and an attention mechanism was implemented. It showed an overall accuracy of 99.67% when compared to comparable cutting-edge public datasets. Identification of leaf diseases in plants using deep transfer learning is becoming increasingly commonplace. With the use of three distinct CNN architectures (videlicet, VirLeafNet-1, and VirLeafNet-3), the authors of (LeCun et al. 1998) developed an automatic disease diagnostic technique for the Vigna Mungo plant. In experimental evaluations on a custom dataset, VirLeafNet-3 achieved 97.40% accuracy.

Wang et al. (2017) conducted a series of experiments on a subset of the PlantVillage dataset consisting of 552 photos of apple leaves. Images of the four stages of apple black rot disease are included in the data collection. There are 110 photographs of healthy leaves, 137 images of early disease, 180 images of intermediate disease, and 125 images of late disease from black rot in apple leaves. To the aforementioned dataset, the authors trained the VGG-16 model. Through transfer learning, they were able to fine-tune the model. Maximum accuracy of 90.4% was attained by the model. Using a modest training dataset, the deep learning model "VGG-16" was able to accurately predict the severity of plant diseases (Wang et al. 2017).

To analyse 14,828 tomato leaves from the PlantVillage dataset, (Brahimi et al. 2017) used GoogLeNet architecture. They were able to get as close to 99.18% as possible. AlexNet was used on the same dataset by (Durmuş et al. 2017). As a group, they were 95.65% accurate. Applying the ResNet model to the aforementioned dataset, (Canziani et al. 2016) found a 97.28% accuracy rate. All three studies performed by the researchers on the same dataset show that the GoogLeNet architecture is superior to both the AlexNet and ResNet models. Experiments were carried out on maize leaves by researchers in (Ding and Taylor 2016). They created their own CNN architecture, and it performed exceptionally well on a dataset of 400 photos of maize leaves, with an accuracy of 92.85%. The LifeCLEF data collection was collected by (Kerkech et al. 2018). Both the training and testing datasets total 91,758 and 21,446 photos, respectively.

The studies conducted by Ferentinos (2018) employed a publicly available database including 87,848 photos of leaves. Pictures were taken of both healthy and sick plants. Peach, pepper, pumpkin, raspberry, soybean, and many more are among the 58 classes and 25 species represented in the dataset. It has 70,300 images in its training dataset and 17,548 in its testing dataset. The dataset was subjected to five different models that were applied by the author. After applying AlexNet, they got an accuracy of 99.06%; AlexNetOWTBn, 99.49%; GoogLeNet, 92.27%; over feat, 98.96%; and the VGG model, 99.556%. The authors conducted a series of tests on both raw photos and photographs that had already been processed. As testing on raw data, all models improved upon performance when compared to preprocessed photos. It has been found, however, that using the aforementioned models on preprocessed, down-scaled, and squared images significantly reduces the required computational time (Ferentinos 2018).

In order to compile their database, (Sladojevic et al. 2016) retrieved 4483 pictures from the Internet. There are fifteen distinct groupings for these pictures. Of the 14 total classes, 13 are for afflicted vegetation, 1 for healthy foliage, and 1 for a blank backdrop. Experimental findings demonstrate an overall accuracy of 96.3%. Watermelon, orange, corn, grape, cherry, and blueberry leaf photos from the PlantVillage were used as a dataset for studies by the authors of (Too et al. 2019). With the VGG-16 architecture, they got an accuracy of 82%; with Inception-V4, they got 98%; with ResNet-50, they got 99.6%; with ResNet-101, they got 99.7%; and with the DenseNet121 CNN design, they got 99.75%.

Photographs of real farmland were taken by the authors of (Lu et al. 2017a; Lin et al. 2013). These photos were used to create a machine that can spot plant diseases on its own. To solve their problems, (Lin et al. 2013) tapped into the processing power of CNNs. The primary convolutional neural network (CNN) model is educated to identify disease lesions on leaves and label small regions inside images. Another CNN model in the pipeline, this one trained for overall picture classification, was fed the predictions concerning the presence of lesions. Laboratory-generated photos with carefully manipulated lighting and intensity levels are used by the research team in (Mohanty et al. 2016). The leaves were segmented, and the backgrounds were removed. Authors of (Zagoruyko and Komodakis 2016) used the GoogLeNet model on the Expanded dataset (XDB). The photographs in this dataset are authentic. Individual lesion and spot detection led to the discovery of 79 illnesses across 14 plant types.

In order to train deep learning models, enormous datasets are necessary. The shortage of training datasets is mitigated via transfer learning. Transfer learning beats models learned from scratch when only few datasets are available for analysis. As a result, scientists often rely on models that have already undergone extensive training. (Xie et al. 2017) used the ImageNet model, and (Fuentes et al. 2017) used the Pascal VOC datasets to pre-train deep learning architectures. These scientists used pre-trained models for categorization and the transfer learning principle. In the event that the dataset is tiny and there is a high degree of data similarity with a pre-trained model, the pre-trained model is used as a feature extractor. The characteristics are obtained by the models' learning weights. When the dataset is huge in size, it is necessary to retrain the network with updated weights. In addition, the number of nodes in the final layer of the network must be adjusted to account for the varying numbers of classes present in a given problem (Szegedy et al. 2015).

Diseases on apple and banana leaves were found using contrast stretching and segmentation by (Khan et al. 2018). The next step is to evaluate how bad the black rot illness is on the apple leaves. Finally, they categorised the leaves into healthy, early, middle, and late phases of the disease. To measure the performance of deep learning models, researchers split datasets into training and testing portions with varying percentages in: 60:40, 80:20, 70:30, etc. To reduce bias, the dataset is split into thirds, with 80% dedicated to training, 10% to validation, and 10% to testing, as shown in (Iandola et al. 2016).

Deep learning models are far superior to those that rely on feature learning. These models automatically extract features from given images. Classification accuracy can be enhanced, however, with large datasets (Chen et al. 2019). When researchers are working with insufficient datasets, they frequently turn to employing pre-trained models. The outcomes of training deeper networks are better, as stated in the paper (Huang et al. 2017). Methods such as skip connections (He et al. 2016), optimization procedures (Lin et al. 2019), transfer learning (Chen et al. 2019), initialisation strategies (Liang et al. 2019), layer-wise training (Krishnamoorthy et al. 2021), and Batch Normalisation (Ioffe and Szegedy 2015) have been developed to address problems in deep neural networks such as vanishing gradients, degradation, and internal covariate shift.

22.3.5 Optimization Techniques

Improving a CNN model's performance requires implementing an appropriate optimization method. Most commonly used optimization methods are briefly covered in following sections.

22.3.5.1 Batch Gradient Descent (BGD) Optimization

Using gradient calculation to sift through millions or billions of samples in a training dataset is labour intensive. Also, it is difficult to feed all samples to a model simultaneously because of restricted computer power. Batch gradient descent is not recommended for updating the parameters of a deep learning model due to its limitations, as noted in (Chen et al. 2019). The most recent model parameter is denoted by x_k + 1, the parameter at the kth iteration by x_k, the step size at the kth iteration by t_k, and the loss function for a given instance of the training data by $f\ (x_k)$. This is done infinitum, or until all n parameters have been identified.

22.3.5.2 Stochastic Gradient Descent (SGD) Algorithm

Using a downwards gradient, the Stochastic Gradient Descent technique revises a model's parameters. For each sample used in training, it determines the model's gradient and modifies its parameters accordingly. It introduces uncertainty by computing the cost of just one data point and its gradient, as opposed to the cost of computing all of the data points. In a short amount of time, the steps are updated, and the threshold is reached. As a result, it is a popular choice among optimization methods. However, because a small step size aids SGD in converging to a favourable location, training time is extended. In addition, the frequent commutation of data between the GPU memory and the local memory reduces the efficiency when GPUs are utilised to do the calculations. The updated model parameters are denoted by x_k + 1, the initial parameters are denoted by x_k, the step size for iteration k is denoted by t_k, the loss function for the current iteration is denoted by $f(x_k)$, and the iteration index is denoted by i.

22.3.5.3 AdaGrad

AdaGrad is a method of optimization that allows for customizable learning rates via input parameters. The frequency with which a certain training parameter is updated influences the tempo at which these lessons are learned. For the parameters that represent the most often occurring features, it performs fewer updates or learns at slower rates. When it comes to factors with rare characteristics, it learns quickly or makes more substantial adjustments. Because of this, AdaGrad can handle sparse data effectively.

22.3.5.4 Root Mean Square Propagation (RMSprop)

By default, RMSprop does not alter the learning rate, but instead aims to stabilise the system by decreasing the frequency of oscillations. Whenever there is a lot of variation in the gradients of multiple mini-batches, RMSprop is employed. To make fine-tuned adjustments to the weights, the learning rate is divided by the typical exponential decay of squared gradients. The learning rates are modified automatically using RMSprop. All of the parameters are taught at different rates. The RMSprop optimizer can be compared to a "gradient descent with momentum" technique. In contrast, the technique through which gradients are computed is different. It dampens vertical oscillations to a minimum. Due to this, the algorithm makes more horizontally oriented, enormous steps towards convergence.

22.3.5.5 Adaptive Moment Estimation (Adam) Optimizer

Each parameter's average gradient and squared gradient are calculated using the formulas described in (Liu et al. 2020). Adam is a hybrid optimization technique that combines gradient descent with momentum and the RMSprop algorithm. The squared gradients are used to adjust the learning rate, just like RMSprop. However, it uses the moving average of the gradient rather than the gradient itself to reap the benefits of SGD with momentum. This optimizer's key benefit is that it can usually function properly with very minor modification of hyperparameters and consumes very little memory. The SGD optimization approach was utilised by (Wang et al. 2017) to train and fine-tune VGG-16, VGG-19, Inception-V3, and ResNet-50 models. When utilising the ResNet architecture, the SGD optimizer causes subpar generalisation and instead promotes a specialised form of optimization. Furthermore, it prematurely sets the residual mapping of ResNet's building blocks to zero. The VGG-16 model using transfer learning achieves 90.4% accuracy (Wang et al. 2017).

For this research, (Zhang et al. 2018) used SGD and Adam optimization on the AlexNet, GoogLeNet, and ResNet designs independently. In all tested networks, SGD optimization outperformed Adam optimization. Using the SGD optimization approach, ResNet is able to achieve a claimed 96.51% accuracy, as seen in (Canziani et al. 2016). In order to update the parameters' weights and reduce the cross-entropy loss, (Krizhevsky et al. 2017) used the SGD, Adam optimizer, and RMSprop optimizers on the VGG-19, Inception-v3, Xception, and ResNet-50 architectures. Since the optimum local problem affects SGD, (Liu et al. 2017) favoured Nesterov's Accelerated Gradient.

22.3.6 *Frameworks*

Thanks to AI advancements, deep learning may be applied to a wide range of domains, using a wide variety of tools and platforms. From Sects. 3.6.1, 3.6.2, 3.6.3, 3.6.4, 3.6.5, 3.6.6, 3.6.7, and 3.6.8, a brief overview of the most popular frameworks is provided.

22.3.6.1 TensorFlow

The article (Li et al. 2015) claims that TensorFlow is the most popular deep learning framework which is widely used library for many different open-source projects. It is built using C++ and Python. Google creates the library for artificial neural networks which provides support for performing numerical computations on a CPU, GPU, server, desktop, or mobile device by means of data flow diagrams. Adjabi et al. (2021) used TensorFlow to develop an image recognition code for the diagnosis of cassava disease. Iandola et al. (2016) used TensorFlow in a smartphone app to categorise agricultural diseases. In (Krizhevsky et al. 2017), the authors use the Keras deep learning framework as a frontend to TensorFlow in order to assess crop vitality.

22.3.6.2 Theano

Theano (Bastien et al. 2012; Bergstra et al. 2010) is an impressive Python library. It provides a more effective means of defining, evaluating, and optimising numerical operations, especially those involving multidimensional arrays (Al-Rfou et al. 2016). This software can generate dynamic C code, optimise performance, employ symbolic differentiation effectively, and use a GPU transparently. Comprehending Theano's syntaxes is challenging. Its highly optimised performance, however, means that it is still widely used. With Keras or other deep learning packages, researchers employ Theano as a backend. Researchers in (Lu et al. 2017a) used the Theano framework to automate wheat disease diagnostics by executing their code on a Geforce GTX 1080 GPU. In order to automatically evaluate the severity of diseases from the image collection, Wang et al. conducted experiments using an Ubuntu workstation equipped with an Intel Core i5 6500 CPU and a GeForce GTX TITAN X GPU. It has been shown that this method has a citation accuracy of 90.4% (Wang et al. 2017).

22.3.6.3 Keras

When it comes to putting deep learning into practise, an open-source, straightforward, and user-friendly library like Keras is invaluable. This Python-based API is aimed for application developers. It is useful for trying out new ideas quickly, since the same code runs on both the CPU and the GPU. It is compatible with the Microsoft Cognitive

Toolkit and the TensorFlow (Chollet 2017) frameworks. Popular architectures like VGGNet, AlexNet, and GoogLeNet can be found as classes or libraries in popular tools like Caffe and Theano.

22.3.6.4 Caffe

Caffe (Shustanov and Yakimov 2017) is an open-source deep learning framework developed by UC Berkeley's Yangqing Jia. The intuitive design is useful for Matlab, C++, and Python. Because of its efficiency, it can be used in the development of convolutional neural networks for the classification of images. This useful tool makes it feasible to use neural networks without resorting to specialised code. When it comes to measuring the efficiency of deep structures, this method is by far the simplest to implement. Caffe was used by (Ha et al. 2017) with the NVIDIA DIGITS 5 toolset to train a CNN model for detecting radish Fusarium wilt. System Specifications: Linux Ubuntu 14.04, Intel Core i7-5930K Central Processing Unit, three NVIDIA Titan X 12 GB Graphics Processing Units, and four 3072 Cuda Cores. As described in (Brahimi et al. 2017; Kawasaki et al. 2015; Liu et al. 2017), CNN-based models for plant disease identification are developed using Caffe. It also simplifies the evaluation of common deep architectures in comparison to other frameworks like Theano and Torch. Caffe models could potentially benefit from increased speed thanks to the addition of the cuDNN library (Xu et al. 2018).

22.3.6.5 Torch

Torch7 is an open-source toolkit for scientific and numerical computations that is based on the Lua programming language. This programme has useful functions like efficient GPU support, array and linear algebra procedures, indexing, transposing, and slicing operations, and more. The Torch7 architecture was utilised by Ferentinos (2018) to evaluate various CNN models for their ability to identify and diagnose plant diseases. This was implemented on an NVIDIA GTX1080 graphics processing unit. PyTorch is a framework created by Facebook that is quite similar to Torch. If you are looking for a framework, this is among the best options. Among its many appealing features are a simple application programming interface (API), Python language support, and interactive calculation graphs.

22.3.6.6 Neuroph

This structure serves as a unified setting. Included are a Java Neural Network library and an integrated development environment (IDE) based on the NetBeans Platform for training and deploying neural networks. Without much experience with languages like C, Python, or MATLAB, this framework can be used to construct a CNN model (Ding and Taylor 2016).

Table 22.1 CNN frameworks' comparison

Framework	Compatible operating system	Programming language used for development
TensorFlow	Linux, macOS Windows, Android	C++, Python, CUDA
Theano	Cross-platform	Python
Keras	Linux, macOS, Windows	Python
Caffe	Linux, macOS, Windows	C++
Torch	Linux, macOS, Windows, Android, iOS	C, Lua
Deeplearning4j	Linux, macOS, Windows, Android	Java, C++
DL Matlab Toolbox	Linux, macOS, Windows	MATLAB, Java, C, C++

22.3.6.7 Deeplearning4j

Java's Deeplearning4j is a free and open-source deep learning library that uses a distributed architecture. Scala, Python, Kotlin, and Clojure are just some of the API languages it is compatible with. When using multiple graphics processing units (GPUs), Deeplearning4j achieves the same speed as Caffe when used for image identification. In addition, users can bring in models from TensorFlow and other Python libraries if Deeplearning4j is built with Keras. In order to measure how well their model for categorising banana leaf illnesses performed, (Amara et al. 2017) leveraged the deeplearning4j platform.

22.3.6.8 DL MATLAB Toolbox

With the help of the MATLAB toolbox, you can easily build a system to use deep neural networks in your projects. This system works on different types of computers like Linux, macOS, and Windows. It's been created using programming languages like MATLAB, Java, C, and C++. The toolbox also features 612 a library of pre-trained models, such as SqueezeNet, Inception-v3, and ResNet-101, that can be used for 613 transfer learning.

Table 22.1 provides the authors' comparison of the various frameworks.

22.3.7 Analysis of DCNNs for Plant Leaf Disease Identification

In the past five years, the scientist has provided numerous methods for identifying plant leaf diseases 617 using DCNN as a foundation. There is solid evidence that deep learning can accurately and efficiently identify plant leaves. Residual Teacher/Student (ResTS) is a unique architecture for categorisation and visualisation that was studied by (Dyrmann et al. 2016). The ResTeacher classifier helps break down pictures into hidden patterns before putting them back together with a decoder. The ResStudent then takes duplicated images and reconstructs them. PlantVillage is a dataset of 54,306 photos of 14 distinct plants, and when compared to other state-of-the-art methods, the findings are encouraging. Disease detection in peach plant leaves was addressed by (Han et al. 2015), who presented a hybrid model combining convolutional autoencoder (CAE) and convolutional neural networks (CNNs). The suggested model has been tested on the PlantVillage dataset, and with only a few parameters, it has provided testing accuracy of 98.38%. In contrast, (Chen et al. 2017) demonstrated a CNN ensemble that was 99.04% accurate on 1774 citrus leaves. Researches are increasingly turning to transfer learning instead of constructing a whole new CNN model from start, and this practise is becoming extremely widespread. In the field of transfer learning, researchers will employ a pre-existing CNN model that has already been trained to do a task. It was first proposed that transfer learning may be used to detect diseases in plants by analysing photographs of their leaves (Pethybridge and Nelson 2015; Tran et al. 2019; Shakoor et al. 2017). On the PlantVillage dataset consisting of 20,639 photos, (Pethybridge and Nelson 2015) showed that by fine-tuning the hyperparameters of the pre-existing ResNet-50 for illness classification, they were able to increase the accuracy to 99.26%. (Tran et al. 2019) on the other hand trained VGG-16 to identify illnesses in rice and wheat. Overall accuracy on rice and wheat plants was 97.22% and 98.75%, respectively, in experimental assessments and compared with other state-of-the-art methods.

In contrast, (Shakoor et al. 2017) used Inception-V3 to study apple plant disease identification, and they found that it had an overall accuracy of 97% on the PlantVillage dataset. Six distinct CNN models were compared by (Minh Dang et al. 2017) to determine which one was most effective in detecting powdery mildew on strawberry leaves. After analysing the data, we were able to recommend the models that satisfied our criteria with regards to the videlicet, accuracy, speed, and minimum system requirements.

While AlexNet completed tasks in the shortest amount of time and SqueezeNet-MOD2 used the least amount of memory, ResNet-50 achieved the best classification accuracy (98.11%). Table 22.2 displays various DCNN implementations based on the ideas of novel architecture and transfer learning.

Disease severity analysis using deep learning for plant leaves is becoming increasingly popular as an alternative to simple disease detection. When it comes to diagnosing the severity of cucumber plant disease, (Nachtigall et al. 2016) suggested a novel Conv2D model. Improvements in disease identification were seen when compared to other CNN models and to machine learning methods.

Table 22.2 Plant leaf disease detection using DCNN comparison

S. No.	Refs. #	Plant	Disease	Results
1	Amara et al. (2017)	Banana	Black sigatoka and black speckle	Accuracy: 99%
2	Wang et al. (2017)	Apple	Black rot on apple leaves	VGG-16: 90.4% VGG-19: 90.0% Inception-v3: 83.0% ResNet-50: 80.0%
3	Mohanty et al. (2016)	Fourteen different crop species	Twenty-six different diseases	AlexNet: accuracy: 99.28% GoogLeNet: accuracy: 99.35%
4	Sladojevic et al. (2016)	Six different fruit plant species	Thirteen different diseases	Accuracy: 96.3%
5	Brahimi et al. (2017)	Tomato	Nine different diseases in tomato	GoogleNet: accuracy: 99.18% AlexNet: accuracy: 98.66%
6	Kawasaki et al. (2015)	Cucumber	Melon Yellow Spot Virus (MYSV), Zucchini Yellow Mosaic Virus (ZYMV)	Average accuracy: 94.9% MYSV sensitivity: 96.3% ZYMV sensitivity: 89.5%
7	Ioffe and Szegedy (2015)	Rice	Ten different diseases	Accuracy: 95.48%
8	Fuentes et al. 2017	Tomato	Nine different types of diseases and pests	VGG-16: 83.06% ResNet-50: 75.37% ResNet-101: 59.0% ResNet-152: 66.83% ResNetXt-50: 71.1%
9	Ferentinos (2018)	Twenty-five different plant species	Nineteen different plant diseases	AlexNet: 99.06% AlexNetOWTBn: 99.49% GoogLeNet: 92.27% Overfeat: 98.96% VGGNet: 99.53%
10	Liu et al. (2017)	Apple	Mosaic, rust, brown spot, and alternaria leaf spot	Proposed Model: 97.62% AlexNet: 91.19% GoogLeNet: 95.69% ResNet-20: 92.76% VGGNet-16: 96.32%
11	Cruz et al. (2017)	Olive	Olive Quick Decline Syndrome (OQDS)	Accuracy of 99%
12	Durmuş et al. (2017)	Tomato	Nine different types of diseases of tomato plant	AlexNet: 95.65% SqueezeNet: 94.3%

(continued)

Table 22.2 (continued)

S. No.	Refs. #	Plant	Disease	Results
13	Lu et al. (2017a)	Wheat	Six different diseases of wheat	VGG-FCN-VD16: 97.95% VGG-FCN-S: 95.12% VGG-CNN-VD16: 93.27% VGG-CNN-S: 73.00%
14	Ma et al. (2018)	Cucumber	Anthracnose, Downy mildew, powdery mildew, and Target leaf spots	Proposed model: 93.4% SVM: 81.9% RF: 84.8% AlexNet: 94.0%
15	Ha et al. (2017)	Radish	Fusarium wilt	Accuracy: 93.3%
16	Barbedo (2019)	Fourteen different plant species	Seventy-nine different diseases	Average accuracy: 94%
17	Oppenheim and Shani (2017)	Potato	Black scurf disease, silver scurf, common scab, and black dot disease	Accuracy: 96.00%
18	Bhatt et al. (2017)	Tomato	Early blight, late blight, yellow leaf curl virus, spider mite damage, and bacterial spot	ResNet-50: 99.7% Xception: 98.6% Inception-v3: 98.4% VGG-19: 98.2%
19	Picon et al. (2019)	Wheat	Septoria, Tan spot and rust	Accuracy: 96.00%
20	Ramcharan et al. (2017)	Cassava	Three diseases: brownleaf spot, brown streak, and cassava mosaic 2	Accuracy: 93.00%
21	Too et al. (2019)	Fourteen different plant species	Not mentioned	VGG-16: 82% Inception-V4: 98% ResNet-50: 99.6% ResNet-101: 99.6% ResNet-152: 99.7% DenseNet121: 99.75%
22	Singh et al. (2016)	Maize	Eight different diseases	GoogLeNet: 98.9% Cifar10: 98.8%
23	Lin et al. (2019)	Wheat	Six different diseases	Accuracy: 90.1%
24	DeChant et al. (2017)	Maize	Northern Leaf Blight	Accuracy: 96.7%
25	Nachtigall et al. (2016)	Apple	Six different diseases	Accuracy: 97.3%
26	Minh Dang et al. (2017)	Radish	Fusarium wilt of radish	Accuracy: 90%

(continued)

Table 22.2 (continued)

S. No.	Refs. #	Plant	Disease	Results
27	Zhang et al. (2018)	Tomato	Eight different diseases	ResNet: 97.28%
28	Liang et al. (2019)	Rice	Rice blast disease	First CNN: 95.37% Second CNN: 95.83%
29	Selvaraj et al. (2019)	Banana	Five major diseases along with a pest class	Accuracy between 70 and 90%
30	Khan et al. (2018)	Apple, banana	Apple scab, apple rot, banana sigotka, banana cordial leaf spot, banana diamond leaf spot, and Deightoniella leaf and fruit spot	Accuracy: 98.6%
31	Kerkech et al. (2018)	Grapevine	Esca disease	The best results were obtained with the combination of ExR, ExG, and ExGR vegetation indices using (16 × 16) patch size reaching 95.80%
32	Sibiya and Sumbwanyambe (2019)	Maize	The northern corn leaf blight, common rust, and grey leaf spot	Accuracy: 92.85%
33	Chen et al. (2019)	Tea	Seven diseases: red leaf spot, algal leaf spot, bird's eye spot, grey blight, white spot, anthracnose, brown blight	Accuracy: 90.16%

22.4 Discussion and Conclusions

This paper presents a review of the research on deep convolutional neural networks for disease detection and classification in plant leaves (DCNN). The authors chose to give a comparative analysis of several DCNN designs by selecting relevant research articles. Datasets, dataset sizes, picture preprocessing methods, convolutional neural network (CNN) architectures, performance measures, and experimental findings of several models deployed to identify and categorise plant leaf diseases are all included in this survey.

Traditional methods of enhancement were used by the researchers for the classification of plant leaf diseases. These methods included rotation (Chollet 2017; Durmuş et al. 2017), flipping, scaling, cropping, translation (Wang et al. 2017), and the addition of Gaussian noise (Sladojevic et al. 2016). Using the methods described above, they were able to produce work that met with their approval. This explains why methods for enhancing deep learning, such as Generative Adversarial Networks and Neural Style Transfer, have been seldom used in the published literature.

The authors draw the conclusion from the available literature that DCNN is a superior feature extractor to traditional feature engineering. With DCNN, researchers do not have to manually extract features because it does it automatically. In addition, these models perform well even when presented with difficult data, like photographs with a complicated background, images of varying sizes and orientations, images acquired in low light, etc. There are, however, a number of problems and difficulties with these methods, including the need for a very large dataset and a longer training time to achieve the desired results. While current methods have reduced the need for human intervention in disease prediction for plants, they may still have limited success because to the availability of just a tiny dataset from which to learn. Also, learning how to identify plant diseases and then collecting and labelling your own samples are time-consuming and laborious. And there is no standard procedure for determining the best number of layers and parameters to use for a given dataset's format and size. Current studies suggest that consideration can be given to the number and composition of samples while developing the model's architecture. However, the contaminated area may be difficult to see during image capture, making region of interest extraction a difficult task. DCNN models' efficacy is proportional to the quality and quantity of training data labels provided. Results from fine-tuned pre-trained models can sometimes be superior to those from models generated from scratch when data for training the model is scarce. Common deep convolutional neural network (DCNN) designs are discussed, demonstrating that choosing a good CNN or DCNN model is contingent on the dataset's nature, size, and experimental settings. SqueezeNet is a fantastic option if you need a tiny network size and high accuracy. Adding more layers causes it to become less effective. ResNet, which makes use of skip connections, is the superior choice when building a network with a greater depth. When all the layers are directly coupled with each other in a feed-forward way, the DenseNet architecture excels. It needs fewer numerical parameters and has a smaller impact on the vanishing-gradient problem. To construct convolutional neural networks (CNNs) for image classification, Caffe provides a fast and effective framework. Also, neural networks can be used with this tool without the need for custom coding. As a result, Caffe is the easiest framework to implement. The authors found that approximately 33% of the researchers in the reviewed literature used the Caffe framework to assess DCNN performance. Looking at the current literature, we see that researchers used CNN models on images taken from above the leaves, either cameras or drones. While there has been a lot of work done to identify and categorise diseases, there has been less effort put into providing information regarding where exactly these diseases are concentrated. It points the way towards creating workable solutions for boosting crop output by illuminating the benefits and difficulties of using computational technologies in agriculture. It accomplishes this by directing researchers to the best illness prediction framework, architecture, model, optimization method, and data collection. In order to better implement preventative and diagnostic strategies, early-stage illness prediction is invaluable. Automated assessments of the severity of plant diseases have the potential to aid farmers in determining the best course of action to prevent the spread of a disease.

References

Adjabi I, Ouahabi A, Benzaoui A, Jacques S (2021) Multi-block color-binarized statistical images for single-sample face recognition. Sensors 21(3):728

Al-Rfou R, Alain G, Almahairi A, Angermueller C, Bahdanau D, Ballas N, Bastien F, Bayer J, Belikov A, Belopolsky A et al (2016) Theano: a python framework for fast computation of mathematical expressions. arXiv e-prints, p arXiv-1605

Amara J, Bouaziz B, Algergawy A (2017) A deep learning-based approach for banana leaf diseases classification. Datenbanksysteme fu¨r Business, Technologie und Web (BTW 2017)-Workshopband

Ampatzidis Y, De Bellis L, Luvisi A (2017) Ipathology: robotic applications and management of plants and plant diseases. Sustainability 9(6):1010

Azimi S, Kaur T, Gandhi TK (2021) A deep learning approach to measure stress level in plants due to nitrogen deficiency. Measurement 173:108650

Barbedo JGA (2018) Factors influencing the use of deep learning for plant disease recognition. Biosys Eng 172:84–91

Barbedo JGA (2016) A review on the main challenges in automatic plant disease identification based on visible range images. Biosys Eng 144:52–60

Barbedo JGA (2019) Plant disease identification from individual lesions and spots using deep learning. Biosys Eng 180:96–107

Bastien F, Lamblin P, Pascanu R, Bergstra J, Goodfellow I, Bergeron A, Bouchard N, Warde-Farley D, Bengio Y (2012) Theano: new features and speed improvements. arXiv preprint arXiv:1211.5590

Bauer SD, Korcˇ F, Fo¨rstner W (2011) The potential of automatic methods of classification to identify leaf diseases from multispectral images. Precision Agric 12(3):361–377

Bergstra J, Breuleux O, Bastien F, Lamblin P, Pascanu R, Desjardins G, Turian J, Warde-Farley D, Bengio Y (2010) Theano: a cpu and gpu math compilerin python. In: Proceedings of 9th python in science conference, vol 1, pp 3–10

Bhatt P, Sarangi S, Pappula S (2017) Comparison of cnn models for application in crop health assessment with participatory sensing. In: 2017 IEEE Global Humanitarian Technology Conference (GHTC). IEEE, pp 1–7

Bock CH, Poole GH, Parker PE, Gottwald TR (2010) Plant disease severity estimated visually, by digital photography and image analysis, and by hyperspectral imaging. Crit Rev Plant Sci 29(2):59–107

Boulent J, Foucher S, The´au J, St-Charles P-L (2019) Convolutional neural networks for the automatic identification of plant diseases. Front Plant Sci 10:941

Brahimi M, Boukhalfa K, Moussaoui A (2017) Deep learning for tomato diseases: classification and symptoms visualization. Appl Artif Intell 31(4):299–315

Canziani A, Paszke A, Culurciello E (2016) An analysis of deep neural network models for practical applications. arXiv preprint arXiv:1605.07678

Chen J, Liu Q, Gao L (2019) Visual tea leaf disease recognition using a convolutional neural network model. Symmetry 11(3):343

Chen J, Zhang D, Zeb A, Nanehkaran YA (2021) Identification of rice plant diseases using lightweight attention networks. Expert Syst Appl 169:114514

Chen SW, Shivakumar SS, Dcunha S, Das J, Okon E, Qu C, Taylor CJ, Kumar V (2017) Counting apples and oranges with deep learning: a data-driven approach. IEEE Robot Autom Lett 2(2):781–788

Chollet F (2017) Xception: deep learning with depthwise separable convolutions. In: Proceedings of the IEEE conference on computer vision and pattern recognition, pp 1251–1258

Chouhan SS, Kaul A, Singh UP, Jain S (2018) Bacterial foraging optimization based radial basis function neural network (BRBFNN) for identification and classification of plant leaf diseases: an automatic approach towards plant pathology. IEEE Access 6:8852–8863

Chouhan SS, Singh UP, Jain S (2020) Applications of computer vision in plant pathology: a survey. Arch Comput Methods Eng 27(2):611–632

Cruz AC, Luvisi A, De Bellis L, Ampatzidis Y (2017) X-fido: an effective application for detecting olive quick decline syndrome with deep learning and data fusion. Front Plant Sci 8:1741

DeChant C, Wiesner-Hanks T, Chen S, Stewart EL, Yosinski J, Gore MA, Nelson RJ, Lipson H (2017) Automated identification of northern leaf blight-infected maize plants from field imagery using deep learning. Phytopathology 107(11):1426–1432

Deng J, Dong W, Socher R, Li L-J, Li K, Fei-Fei L (2009) Imagenet: a large-scale hierarchical image database. In: 2009 IEEE conference on computer vision and pattern recognition. IEEE, pp 248–255

Ding W, Taylor G (2016) Automatic moth detection from trap images for pest management. Comput Electron Agric 123:17–28

Durmuş H,Gu¨nes¸ EO, Kırcı M (2017) Disease detection on the leaves of the tomato plants by using deep learning. In: 2017 6th international conference on agro-geoinformatics. IEEE, pp 1–5

Dyrmann M, Karstoft H, Midtiby HS (2016) Plant species classification using deep convolutional neural network. Biosyst Eng 151:72–80

Everingham M, Van Gool L, Williams CKI, Winn J, Zisserman A (2010) The pascal visual object classes (VOC) challenge. Int J Comput Vis 88(2):303–338

Ferentinos KP (2018) Deep learning models for plant disease detection and diagnosis. Comput Electron Agric 145:311–318

Fuentes A, Yoon S, Kim SC, Park DS (2017) A robust deep-learning-based detector for real-time tomato plant diseases and pests recognition. Sensors 17(9):2022

Ghazi MM, Yanikoglu B, Aptoula E (2017) Plant identification using deep neural networks via optimization of transfer learning parameters. Neurocomputing 235:228–235

Ghosal S, Blystone D, Singh AK, Ganapathysubramanian B, Singh A, Sarkar S (2018) An explainable deep machine vision framework for plant stress phenotyping. Proc Nat Acad Sci 115(18):4613–4618

Ha JG, Moon H, Kwak JT, Hassan SI, Dang M, New Lee O, Park HY (2017) Deep convolutional neural network for classifying fusarium wilt of radish from unmanned aerial vehicles. J Appl Remote Sens 11(4):042621

Han S, Mao H, Dally WJ (2015) Deep compression: compressing deep neural networks with pruning, trained quantization and huffman coding. arXiv preprint arXiv:1510.00149

He K, Zhang X, Ren S, Sun J (2016) Deep residual learning for image recognition. In: Proceedings of the IEEE conference on computer vision and pattern recognition, pp 770–778

Huang G, Liu Z, Van Der Maaten L, Weinberger KQ (2017) Densely connected convolutional networks. In: Proceedings of the IEEE conference on computer vision and pattern recognition, pp 4700–4708

Huang K-Y (2007) Application of artificial neural network for detecting phalaenopsis seedling diseases using color and texture features. Comput Electron Agric 57(1):3–11

Huang T, Yang R, Huang W, Huang Y, Qiao X (2018) Detecting sugarcane borer diseases using support vector machine. Inf Process Agric 5(1):74–82

Hubel DH, Wiesel TN (1962) Receptive fields, binocular interaction and functional architecture in the cat's visual cortex. J Physiol 160(1):106

Iandola FN, Han S, Moskewicz MW, Ashraf K, Dally WJ, Keutzer K (2016) Squeezenet: Alexnet-level accuracy with 50x fewer parameters and¡ 0.5 mb model size. arXiv preprint arXiv:1602.07360

Ioffe S, Szegedy C (2015) Batch normalization: accelerating deep network training by reducing internal covariate shift. In: International conference on machine learning. PMLR, pp 448–456

Joly A, Goe¨au H, Glotin H, Spampinato C, Bonnet P, Vellinga W-P, Lombardo J-C, Planque´ R, Palazzo S, Mu¨ller H (2017) Lifeclef 2017 lab overview: multimedia species identification challenges. In: International conference of the cross- language evaluation forum for European languages. Springer, pp 255–274

Joshi RC, Kaushik M, Dutta MK, Srivastava A, Choudhary N (2021) Virleafnet: automatic analysis and viral disease diagnosis using deep-learning in vigna mungo plant. Ecol Inform 61:101197

Kaur S, Pandey S, Goel S (2019) Plants disease identification and classification through leaf images: a survey. Arch Comput Methods Eng 26(2):507–530

Kawasaki Y, Uga H, Kagiwada S, Iyatomi H (2015) Basic study of automated diagnosis of viral plant diseases using convolutional neural networks. In: International symposium on visual computing. Springer, pp 638–645

Kerkech M, Hafiane A, Canals R (2018) Deep leaning approach with colorimetric spaces and vegetation indices for vine diseases detection in uav images. Comput Electron Agric 155:237–243

Khan MA, Akram T, Sharif M, Awais M, Javed K, Ali H, Saba T (2018) Ccdf: Automatic system for segmentation and recognition of fruit crops diseases based on correlation coefficient and deep cnn features. Comput Electron Agric 155:220–236

Krishnamoorthy N, Narasimha Prasad LV, Pavan Kumar CS, Subedi B, Abraha HB, Sathishkumar VE (2021) Rice leaf diseases prediction using deep neural networks with transfer learning. Environ Res 198:111275

Krizhevsky A, Sutskever I, Hinton GE (2017) Imagenet classification with deep convolutional neural networks. Commun ACM 60(6):84–90

Kundu N, Rani G, Dhaka VS (2020) A comparative analysis of deep learning models applied for disease classification in bell pepper. In: 2020 Sixth international conference on parallel, distributed and grid computing (PDGC). IEEE, pp 243–247

LeCun Y, Boser B, Denker JS, Henderson D, Howard RE, Hubbard W, Jackel LD (1989) Backpropagation applied to handwritten zip code recognition. Neural Comput 1(4):541–551

LeCun Y, Bottou L, Bengio Y, Haffner P (1998) Gradient-based learning applied to document recognition. Proc IEEE 86(11):2278–2324

Li H, Lin Z, Shen X, Brandt J, Hua G (2015) A convolutional neural network cascade for face detection. In: Proceedings of the IEEE conference on computer vision and pattern recognition, pp 5325–5334

Li P (2017) Optimization algorithms for deep learning. Department of Systems Engineering and Engineering Management, The Chinese University of Hong Kong

Li Y, Cao Z, Lu H, Xiao Y, Zhu Y, Cremers AB (2016) In-field cotton detection via region-based semantic image segmentation. Comput Electron Agric 127:475–486

Liang W-J, Zhang H, Zhang G-F, Cao H-X (2019) Rice blast disease recognition using a deep convolutional neural network. Sci Rep 9(1):1–10

Lin M, Chen Q, Yan S (2013) Network in network. arXiv preprint arXiv:1312.4400

Lin Z, Mu S, Huang F, Mateen KA, Wang M, Gao W, Jia J (2019) A unified matrix-based convolutional neural network for fine-grained image classification of wheat leaf diseases. IEEE Access 7:11570–11590

Liu B, Zhang Y, He DongJian, Li Y (2017) Identification of apple leaf diseases based on deep convolutional neural networks. Symmetry 10(1):11

Liu J, Wang X (2021) Plant diseases and pests detection based on deep learning: a review. Plant Methods 17(1):1–18

Liu L, Ouyang W, Wang X, Fieguth P, Chen J, Liu X, Pietika¨inen M (2020) Deep learning for generic object detection: a survey. Int J Comput Vis 128(2):261–318

Lu J, Hu J, Zhao G, Mei F, Zhang C (2017a) An in-field automatic wheat disease diagnosis system. Comput Electron Agric 142:369–379

Lu Y, Yi S, Zeng N, Liu Y, Zhang Y (2017b) Identification of rice diseases using deep convolutional neural networks. Neurocomputing 267:378–384

Ma J, Keming D, Zheng F, Zhang L, Gong Z, Sun Z (2018) A recognition method for cucumber diseases using leaf symptom images based on deep convolutional neural network. Comput Electron Agric 154:18–24

Minh Dang L, Syed IH, Suhyeon I (2017) Drone agriculture imagery system for radish wilt. J Appl Remote Sens 11:16006

Mishkin D, Matas J (2015) All you need is a good init. arXiv preprint arXiv:1511.06422

Mohanty A (2018) Algorithm and hardware design for efficient deep learningl inference, PhD diss., Arizona State University

Mohanty SP, Hughes DP, Salathe ´ M (2016) Using deep learning for image-based plant disease detection. Front Plant Sci 7:1419

Nachtigall LG, Araujo RM, Nachtigall GR (2016) Classification of apple tree disorders using convolutional neural networks. In: 2016 IEEE 28th international conference on tools with artificial intelligence (ICTAI). IEEE, pp 472–476

Ngugi LC, Abelwahab M, Abo-Zahhad M (2021) Recent advances in image processing techniques for automated leaf pest and disease recognition–a review. Inf Processing Agric 8(1):27–51

Nguyen TT-N, Le T-L, Vu H, Hoang V-S (2019) Towards an automatic plant identification system without dedicated dataset. Int J Mach Learn Comput 9(1):26–34

Noon SK, Amjad M, Qureshi MA, Mannan A (2020) Use of deep learning techniques for identification of plant leaf stresses: a review. Sustain Comput Inform Syst 28:100443

Oppenheim D, Shani G (2017) Potato disease classification using convolution neural networks. Adv Anim Biosci 8(2):244–249

Pan SJ, Yang Q (2009) A survey on transfer learning. IEEE Transac Knowl Data Eng 22(10):1345–1359

Pawara P, Okafor E, Schomaker L, Wiering M (2017) Data augmentation for plant classification. In: International conference on advanced concepts for intelligent vision systems. Springer, pp 615–626

Pethybridge SJ, Nelson SC (2015) Leaf doctor: a new portable application for quantifying plant disease severity. Plant Dis 99(10):1310–1316

Picon A, Alvarez-Gila A, Seitz M, Ortiz-Barredo A, Echazarra J, Johannes A (2019) Deep convolutional neural networks for mobile capture device-based crop disease classifica- tion in the wild. Comput Electron Agric 161:280–290

Pound MP, Atkinson JA, Townsend AJ, Wilson MH, Griffiths M, Jackson AS, Bulat A, Tzimiropoulos G, Wells DM, Murchie EH et al (2017) Deep machine learning provides state-of-the-art performance in image-based plant phenotyping. Gigascience 6(10):gix083

Ramcharan A, Baranowski K, McCloskey P, Ahmed B, Legg J, Hughes DP (2017) Deep learning for image-based cassava disease detection. Front Plant Sci 8:1852

Sa I, Ge Z, Dayoub F, Upcroft B, Perez T, McCool C (2016) Deepfruits: a fruit detection system using deep neural networks. Sensors 16(8):1222

Selvaraj MG, Vergara A, Ruiz H, Safari N, Elayabalan S, Ocimati W, Blomme G (2019) AI-powered banana diseases and pest detection. Plant Methods 15(1):1–11

Shakoor N, Lee S, Mockler TC (2017) High throughput phenotyping to accelerate crop breeding and monitoring of diseases in the field. Curr Opin Plant Biol 38:184–192

Shustanov A, Yakimov P (2017) CNN design for real-time traffic sign recognition. Procedia Eng 201:718–725

Sibiya M, Sumbwanyambe M (2019) A computational procedure for the recognition and classification of maize leaf diseases out of healthy leaves using convolutional neural networks. AgriEngineering 1(1):119–131

Simonyan K, Zisserman A (2014) Very deep convolutional networks for large-scale image recognition. arXiv preprint arXiv:1409.1556

Singh A, Ganapathysubramanian B, Singh AK, Sarkar S (2016) Machine learning for high-throughput stress phenotyping in plants. Trends Plant Sci 21(2):110–124

Singh AK, Ganapathysubramanian B, Sarkar S, Singh A (2018) Deep learning for plant stress phenotyping: trends and future perspectives. Trends Plant Sci 23(10):883–898

Sladojevic S, Arsenovic M, Anderla A, Culibrk D, Stefanovic D (2016) Deep neural networks based recognition of plant diseases by leaf image classification. Comput Intell Neurosci 2016

Szegedy C, Liu W, Jia Y, Sermanet P, Reed S, Anguelov D, Erhan D, Vanhoucke V, Rabinovich A (2015) Going deeper with convolutions. In: Proceedings of the IEEE conference on computer vision and pattern recognition, pp 1–9

Szegedy C, Vanhoucke V, Ioffe S, Shlens J, Wojna Z (2016) Rethinking inception architecture for computer vision. In: Proceedings of the IEEE conference on computer vision and pattern recognition, pp 2818–2826

Tan W, Zhao C, Huarui W (2016) Intelligent alerting for fruit-melon lesion image based on momentum deep learning. Multimedia Tools Appl 75(24):16741–16761

Toda Y, Okura F (2019) How convolutional neural networks diagnose plant disease. Plant Phenomics 2019

Too EC, Yujian L, Njuki S, Yingchun L (2019) A comparative study of fine-tuning deep learning models for plant disease identification. Comput Electron Agric 161:272–279

Tran T-T, Choi J-W, Le T-TH, Kim J-W (2019) A comparative study of deep cnn in forecasting and classifying the macronutrient deficiencies on development of tomato plant. Appl Sci 9(8):1601

Traore BB, Kamsu-Foguem B, Tangara F (2018) Deep convolution neural network for image recognition. Ecol Inf 48:257–268

Wang G, Sun Y, Wang J (2017) Automatic image-based plant disease severity estimation using deep learning. Comput Intell Neurosci 2017

Xie S, Girshick R, Dolla´r P, Tu Z, He K (2017) Aggregated residual transformations for deep neural networks. In: Proceedings of the IEEE conference on computer vision and pattern recognition, pp 1492–1500

Xu R, Li C, Paterson AH, Jiang Y, Sun S, Robertson JS (2018) Aerial images and convolutional neural network for cotton bloom detection. Front Plant Sci 8:2235

Yadav S, Sengar N, Singh A, Singh A, Dutta MK (2021) Identification of disease using deep learning and evaluation of bacteriosis in peach leaf. Ecol Inform 61:101247

Yu D, Xiong W, Droppo J, Stolcke A, Ye G, Li J, Zweig G (2016) Deep convolutional neural networks with layer-wise context expansion and attention. In: Interspeech, pp 17–21

Zagoruyko S, Komodakis N (2016) Wide residual networks. arXiv preprint arXiv:1605.07146

Zhang K, Wu Q, Liu A, Meng X (2018) Can deep learning identify tomato leaf disease? Adva Multimedia 2018

Zhang WJ (2008) A forecast analysis on world population and urbanization process. Environ Dev Sustain 10(6):717–730

Chapter 23
Robotic Seeding or Sowing System in Smart Agriculture

Nrusingh Charan Pradhan, Mude Arjun Naik, Manojit Chowdhury, Ajay Kushwah, K. R. Asha, Tushar Dhar, Kishor P. Gavhane, Sumit B. Urhe, and Ajay N. Satpute

Abstract One of the most promising approaches to smart farming is the use of autonomous agricultural robots. Farms are increasingly being automated in order to save time and effort on repetitive operations and to boost yield by treating each crop separately in accordance with the precision farming approach. Seed sowing is a critical step in the farming process. It is labour intensive and takes a long time to complete. Different advance technologies like electronic seed metering system, pneumatic seed metering unit, seed chocking detection and alarm system, down force measurement, and uniform depth of placement system have been developed in recent years. However, planters and seed drills are hitched on to the back of the tractor to do the actual sowing and planting. This results in increased compaction of the farm field and increase in cost of fuel for prime mover. A seed sowing robot is a machine that helps farmers to save both time and money by accurately planting seeds in the ground. The main aim of this chapter is to give an idea about role of robotic seeding in smart agriculture, the design aspects of different components of seed sowing robot, different navigation and localization methods, electronic components and sensors used in precision seeding robot, and economic perspective of robotic seeding.

Keywords Automation · Smart agriculture · Robotic seeding · Sensor

---

N. C. Pradhan · M. A. Naik · M. Chowdhury (✉) · A. Kushwah · K. R. Asha · T. Dhar · K. P. Gavhane
Division of Agricultural Engineering, ICAR-Indian Agricultural Research Institute, New Delhi 110012, India
e-mail: manojitchowdhury13@gmail.com

S. B. Urhe
ICAR-Central Institute of Post Harvest Engineering and Technology, Ludhiana 141004, Punjab, India

A. N. Satpute
ICAR- Indian Grassland and Fodder Research Institute, Jhansi 284003, UP, India

K. Pandey et al. (eds.), *Artificial Intelligence and Smart Agriculture*, Advances in Geographical and Environmental Sciences,
https://doi.org/10.1007/978-981-97-0341-8_23

23.1 Introduction

The increase in population is accompanied the increase of demand for food production. As per FAO statistical report, it is expected to reach the world population to 9.73 billion by 2050, this population further grows up to 11.2 billion by 2100 (FAO 2017). In order to feed this growing population, agricultural production, productivity and efficient utilization of resources particularly the field crops cultivation must increase 70% by 2050, when the world population is expected to exceed 9 billion people (Crist 2017; Sylvester 2018). To achieve this, there is a need of new and upgraded technologies as the agriculture was a one of the major energy consuming sectors (in terms of land, water, and other energy resources) in the world which will reduces the cost of production and increase the productivity.

In any crop production system several unit operations need to be performed like seed bed preparation, sowing/transplanting, spraying, intercultural operations, harvesting, threshing, and post-harvest operations. Sowing or planting is one of the most important unit operations among all the unit operations performed in any crop production as it consumes more energy and labour requires precise placement of seeds to facilitate other successive operations like intercultural operations. In the traditional method of sowing/transplanting, seeds/seedlings are placed in soil by manual labour, keeping in view of row-to-row and plant-to-plant distances in mind. This is more prone to error as there is no metering mechanism involved (Parish 2005; Kumar and Raheman 2008, 2011) and requires more labour force. To overcome these challenges associated with the traditional method, semi-automatic and automatic seedling transplanters were introduced. It consists of variety of mechanically driven components, such as metering mechanism, furrow openers, ground wheel, and hopper, to place seed. In these drive to the metering devices is through the common shaft, so the varying flow rates can be obtained by varying ground speed, but the common shaft can drive the set of row units, so either all units are turned on or off (Iacomi and Popescu 2015). Even in this has a possibility of to rotate single row unit on or off, however it requires a lot of complicated mechanism.

A further drawback of mechanically driven metering mechanisms is that due to difference in single seed in the metering unit there is chance of some cells may fail to charge seed metering or may contain several seeds. Also, due to drive wheel driven metering mechanism, wheel slippage during operation can affect bulkiness of the system and seeding rate and also the amount of material present in the seed box can also affect the seed uniformity. So, in present mechanical metering mechanisms control of only the planting population is possible but not the right timing and placement of the seed. Now, in the present era of smart farming, robotics, and artificial intelligence the automatic controlled seed simulation devices can address many of the ineffectiveness experienced by the traditional and mechanically driven semi-automatic and automatic seed metering units have potential to increase yield rates and productivity dramatically. Hence in this book chapter different seed/planting sowing robots, their navigation and cost economics are presented.

23.2 Smart Farming Systems

The smart agriculture is a modern farming system that utilizes the integration of communication technologies (ICT) and modern information with agriculture, such as drones, IOT sensors, GPS, machine learning, and artificial intelligence for efficient utilization of resources to optimize crop production, reduce waste, and increase efficiency. The major advantages of smart farming agriculture include ability to monitor crop and soil parameters in real time, and make decision based on the real data and instant planning about planting, irrigation, harvesting, and application of fertilizers and pesticides only where needed in the field. But the limitation of the technology includes it has high initial investment cost and requires technically skilled person to monitor operate the technology involved in it.

23.2.1 Application of Smart Farming

Due to recent food crises, the public is mostly concerned about food safety. So, these challenges can be overcome with the help of a new concept called "smart farming," in which the user can get the right input resources (like information about market needs, irrigation, and machinery) at the right time for farming operations to get a better yield. All the unit operations in crop production, like seed bed preparation, sowing, and intercultural operations pre-harvesting, harvesting, and post-harvesting are the major categories of important agricultural activity. The advanced technologies like machine learning technologies, IOT have contributed to boosting agricultural gains. These are the recent technologies that are assisting farmers in reducing farm losses by offering detailed crop suggestions and insights. These technologies are becoming more efficient and accurate by using deep learning algorithms. It is shown in Fig. 23.1.

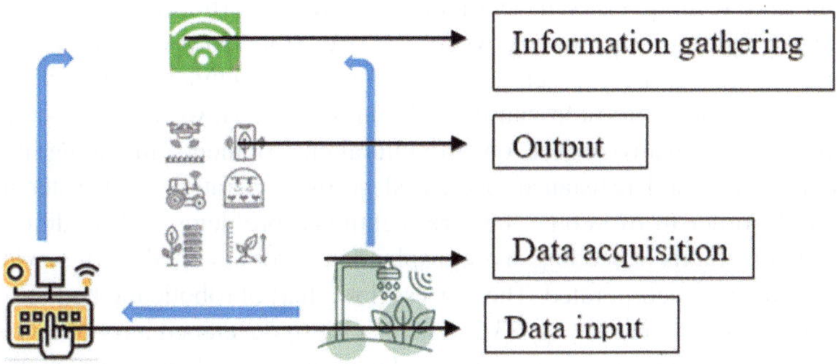

Fig. 23.1 Smart farming system

23.2.2 Precision Agriculture

Precision agriculture (PA) is a modern farming approach that uses advanced technologies to increase agricultural yields and profitability while minimizing the use of resources like land, water, and fertilizers. PA involves applying the right farming methods and techniques, in the right location, at the right time, in the right manner using data and technology to make informed decisions. The strategy involves collecting and analysing data to manage farm resources and plan crop cultivation. This approach has the potential to increase yields, improve environmental quality, and automate critical farming processes, helping farmers make smarter decisions. It is a long-term and comprehensive approach where farmers adjust their farming methods like seedbed preparation, sowing, irrigation, harvesting, and post-harvest operations to match commute crop and soil conditions requirements within field. Precision agriculture differs from conventional farming practises by precisely identifying variations in crop and linking spatial and temporal data to field actions. This allows farmers to effective management of their farms, crops, and practises from a new approach, resulting in reduced input cost, decreased environmental hazards, better management of input resources, and optimized yields and quality concerning the productive capacity of each site.

Precession agriculture has the potential to revolutionize agriculture and bring numerous benefits such as increased crop quality, yield, profitability, productivity, environmental protection, sustainability, crop quality, food safety, and rural economic development. On a regional level, it helps to identify new patterns in factors that affect sustainability in physically neighbouring regions or particular agro-ecological zones, and opens possible solutions.

23.2.3 Robotics Application in Smart Agriculture

Planting and seeding tasks have traditionally been performed manually or by specialized planting machinery powered by a tractor. However, the use of tractors and machinery are characterized by being big machines, and their constant mobility throughout the field causes soil compaction. However, soil compaction has a number of negative effects on agricultural environments, including increased apparent density, soil resistance, decreased porosity, aeration, and water infiltration rate. In order to overcome the fore mentioned problems with traditional and mechanical methods of seeding and transplanting, a robotic method of sowing, and transplanting was incorporated. The process flow chart of robotic method of sowing/transplanting is depicted in Fig. 23.2. Sakaue (1996) developed a robotic system to fully automate sowing/planting process in Japan, which included four variations of a wheeled robot. One of the prototypes, number four, was capable of planting up to 2200 plants per hour, including lettuce, broccoli, cabbage, cauliflower, or celery. However, the system was not autonomous, and it required human intervention. To

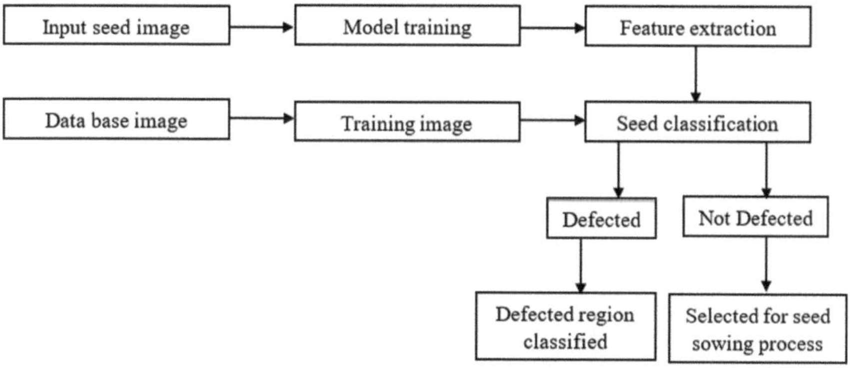

Fig. 23.2 Process flow chart of image processing-based seed sowing method

address this issue, Haibo et al. (2015) developed the Lumai-5 robot, to ensure that the sowing process remained constant, even when the detachment speed varied. The Lumai-5 robot is equipped with a four-wheel steering (4WS) system, a closed-loop control system, speed, angle, and pressure sensors to perform wheat sowing with high precision. Quality of the seeding was affected by factors such as the size of the planting tray, the vacuum chamber pressure, and the planting speed.

In 2015, the Australian Centre for Field Robotics (ACFR) embarked on the development of a robotic system based on readily available digital and physical components. The result was the creation of the Di-Wheel robot, a two-wheel-drive (2WD) robot that moves and operates solely on two wheels. The use of only two wheels offers a number of advantages, including reduced weight, size, and mechanical complexity of robot, as well as simplified transport and assembly, which takes only about 15 min. Additionally, the distance between the wheels can be adjusted to accommodate different crop types.

The Di-Wheel robot was developed to perform automatic precision seeding, weeding, spraying and tasks, with all electronic devices located in its central part. To address the financial barrier that prevents the use of robotic systems by small land holders, the Di-Wheel robot has support to fix smartphones at a height and angle which enables the use of the smartphone's internal sensors. These sensors include temperature, humidity and light sensors, RGB cameras, gyroscopes, accelerometers, and GNSS devices. The author of the article concludes that the constant advances in computing technology can contribute to better crop control and management through use of open-source ML algorithms such as Open AI, Tensor Flow, and Paddle. With the help of these tools, the Di-Wheel robot can provide more efficient and effective crop management solutions, making it a valuable asset for the agricultural industry.

Small and low-cost robots have the potential to revolutionize farming practises on small farms, where heavy vehicles like tractors may not be feasible. In Pakistan, a 4WD seeding robot prototype has been developed for sowing corn using an individual seed selector which able of disperse a suitable number of seeds for planting. This prototype covers sowing at a rate five times faster than traditional methods, sowing at 90 seeds min^{-1}, with a total cycle of 0.66 acres h^{-1}. In order to overcome and to maximize the weight versus soil compaction ratio of robots, Indian researchers developed a prototype of a small-sized sowing robot in 2016. This robot has the capacity of 17 kg of payload using a locomotion system with tracked drive to work non-uniform soils. However, as this is still a prototype in the initial research phase, and also proposed to use of a photovoltaic panel to recharge the robot's electrical system, as well as the implementation of a Kalman filter to improve the estimation of the robot's position and location along with the crop. The use of small and low-cost robots in agriculture has the potential to greatly improve farming efficiency and productivity on small-scale farms. These robots can perform tasks such as planting and sowing at a much faster rate than conventional methods, reducing labour, and time costs. Moreover, the use of innovative technologies like photovoltaic panels and Kalman filters can further enhance the performance of these robots, making them more adaptable and efficient in various farming conditions.

23.3 Different Navigation and Localization Methods of Robotic Seeder

Autonomous mobile robots face a challenging task in navigating through outdoor environments that are both anonymous and heterogeneous. This requires finding safe and efficient paths to reach the goal while avoiding obstacles along the way. While standalone Global Navigation Satellite System (GNSS) has been a popular method for navigation, it may not be reliable due to signal blocking and multi-reflection, particularly in agricultural and forestry areas (Perez-Ruiz et al. 2012; Guo et al. 2018). To overcome this, advanced autonomous system designs rely on algorithms that can fuse sensor outputs to derive the desired navigation output. Precision is a crucial element in the success of autonomous vehicles, and the achievement of precision heavily relies on the pivotal contribution of algorithms (Mousazadeh 2013). Table 23.1 represents a variety of navigation algorithms and methods available for robotic vehicles, along with their respective advantages and limitations.

23.3.1 Dead Reckoning Algorithm-Based Navigation

Autonomous agricultural machinery commonly uses two positioning systems: relative positioning via dead reckoning and absolute location via GPS. Dead reckoning involves the accumulation of distance travelled in the direction of vehicle movement using wheel sensors like encoders to determine the vehicle's speed. This speed is then multiplied by time to calculate the distance covered, resulting in the calculation of the machinery's position within a given field. This approach is generally durable and highly precise within brief intervals, but the positional accuracy may

Table 23.1 Advantages and limitations of different navigations algorithms/methods

Navigation algorithms/ method	Advantages	Limitations	References
Dead reckoning algorithm-based navigation	• Straightforward and reliable • Quite durable and accurate over short periods of time • Runs on low-cost instruments	• Encounters positional error resulting from sensor inaccuracies and slip issues, which requires periodic recalibration or alignment • Applicable to only exceptional instances • Slip is frequently unquantifiable and produces significant inaccuracy in judgement	Mattos (1994), Tavakoli et al. (2015)
Navigation using image processing	• Able to extract useful features or parameters from images, which can aid in decision-making • Efficiently can recognize and navigate rows, avoid obstacles, and perform autonomous operations • Mobile cameras can be used in agricultural applications to capture a variety of images, including variable Field-Of-View (FOV) images, which can aid in navigation	• Complicated and require costly algorithms • Traditional signal-processing techniques used in image processing are limited by their two-dimensional nature • Machine vision may not be as accurate as dead reckoning in some situations	Arif and Butt (2014), Sirigool and Kesvarakul (2020)

(continued)

Table 23.1 (continued)

Navigation algorithms/ method	Advantages	Limitations	References
Statistical navigation technique	• Uses low-cost sensors and implementations such as DGPS • The navigation software used in statistical navigation is not overly complicated or clunky • Can make decisions based on the vehicle's past and current status, allowing for more accurate navigation	• May not be as accurate as other methods, such as RTK-GPS and IMU systems • The sensor fusion vehicle dynamics model used in statistical navigation may omit important factors such as lateral slip • Require the use of equations for error extraction, which may be complicated or difficult to develop	Mousazadeh (2013), Trautman et al. (2015)
Fuzzy system-guided navigation	• Fuzzy logic provides a way to deal with uncertainty and ambiguity in data, allowing for more flexible decision-making • Fuzzification can derive membership functions from raw data, providing a solid foundation for the fuzzy system • FLC controllers using fuzzy logic can be designed with basic, low-cost sensors and can be trained with a set of training data • Fuzzy-based controllers can be strong enough to avoid obstacles	• Fuzzy logic can be complex and computationally expensive, depending on the size and complexity of the system • Fuzzy logic requires training data for the controller, which may not always be available in practical situations • Fuzzy-based navigation is rated based on the vehicle's previous and current state, with no estimation conducted, which may not be accurate enough for some applications	Omrane et al. (2016), Nattharith and Güzel (2016)

(continued)

Table 23.1 (continued)

Navigation algorithms/ method	Advantages	Limitations	References
NN and GA-based navigation	• Neural networks can detect and match complex, vague, incomplete patterns and are useful for prediction, classification, data association, data conceptualization, filtering, and planning • The Genetic Algorithm (GA) is a problem-solving technique that mimics biological evolution and can be used to solve automation challenges	• Neural networks and GA require a significant amount of training data to perform well, which may not always be available • The performance of NN and GA heavily depends on the quality of input data • NN-based algorithms may require significant computing power and may not be suitable for low-power devices	Kala et al. (2009), Savage et al. (2013)
Kalman navigation algorithm	• Can be used to regulate noisy systems • The optimal estimation for linear systems and Gaussian errors • Capable of controlling the vehicle accurately using the vehicle movement model • The implemented Kalman filter can model the steering actuation using EKF in order to estimate the tractor's position, direction, attitude, and speed	• The base Kalman filter assumes Gaussian (white) noise • The measurement and the state space may not be aligned • The algorithm is limited to template-matching function that represented the motion of a detected obstacle in sensor coordinates and reconstructed the obstacle's silhouette	Reina et al. (2007), Chen (2011), Xing et al. (2017)

(continued)

Table 23.1 (continued)

Navigation algorithms/ method	Advantages	Limitations	References
SLAM	• Simultaneously calculates a robot's condition using input sensor data while creating a map of the surrounding environment • Provides a solution for estimating unknown robot motion while mapping the unknown surroundings	• Mapping the environment may not be desirable • More complex than alternatives, such as visual odometry	Thrun (2008), Khairuddin et al. (2015)
Visual odometry	• Can calculate the motion of the on-board camera(s) using only image data as input • Can be done in a variety of ways, including 2D to 2D, where the characteristics for both images are provided in 2D coordinates and 3D to 3D motion estimate can be done using depth information from stereo cameras	• It is sensitive to the quality of input image data • It can be subject to large cumulative errors over time, leading to the need for additional corrections or re-initialization	Yousif et al. (2015), Agel et al. (2016)

be compromised by the drifting of sensors and slippage of the vehicle, requiring periodic recalibration or alignment. While dead reckoning is a straightforward and reliable method that can run on low-cost instruments, it has some limitations that restrict its use to exceptional cases. The primary limitation is slip, which is often difficult to quantify and can result in significant inaccuracies. Dead reckoning can be justified for short journeys, when the experimental speed is well determined, or when a location updating mechanism is used. However, its fundamental limitations make it less applicable in many scenarios (Guo et al. 2002).

23.3.2 Navigation Using Image Processing

Image processing encompasses the manipulation and analysis of input images, which can include photographs or frames from a video, and producing an output image or image-related features or parameters. Image processing techniques typically consider the image as a 2D signal and apply conventional signal-processing methods. Digital image processing is widely utilized, although optical and analog image processing methods are also viable alternatives. Machine vision is used in two-dimensional (2D) or three-dimensional (3D) analysis. 2D vision systems employ area or line scan cameras along with appropriate lighting to evaluate the characteristics of an object, such as surface quality and feature position. In agriculture, 2D vision is useful for sorting by colour, shape, and size. 3D analysis uses stereo vision or LED/laser triangulation to precisely measure the geometric features of an object's exterior.

Researchers have used mobile cameras instead of stationary CCD cameras to capture images. As an example, Xue et al. (2012) developed a row guidance system for an agricultural robot, which employed a machine vision approach. The system utilized adjustable Field-Of-View (FOV) along with pitch and yaw motion control mechanisms. The robot switched between a far FOV, a near FOV, and a lateral FOV based on the number of pixels in the images. Fuzzy Logic Control (FLC) was used to control the robot's speed. Machine vision is used in autonomous agricultural robots for avoiding obstacle, row recognition, and navigation. It operates in exploratory mode and is a complex technology that requires sophisticated algorithms. Machine vision-based guidance demonstrated satisfactory performance across various speeds and paths, exhibiting average errors below 3 cm. Combining machine vision with laser radar was proposed as a means to enhance guidance robustness and enable obstacle detection capabilities (Subramanian and Burks, 2005; Subramanian et al. 2006).

23.3.3 Statistical Navigation Technique

Different statistically based navigation techniques are being evaluated for autonomous agricultural robots. Some of these techniques use known statistical equations, while others develop new algorithms based on statistical calculations. Mizushima et al. (2004) devised a self-guided agricultural navigation system employing cost-effective sensors. They used two inclinometers and three vibratory gyroscopes to compute inclination, heading, and roll/pitch angles to correct position from DGPS. They employed the Least Square Method (LSM) to calculate gyroscope drift inaccuracy and heading estimation. However, the sensor fusion vehicle dynamics model did not account for the vehicle's lateral slip. In several field tests, the sensor unit estimated the heading angle (1.59°), roll angle (0.41°), and pitch angle (0.65°). When using a high-resolution analog to digital converter and a laptop, the accuracy was comparable to that of the RTK-GPS and IMU systems. Statistical navigation technology uses the vehicle's past and current status to make decisions. It employs linear or nonlinear equations to extract errors and takes advantage of low-cost sensors and implementations such as DGPS and IMU. The navigation software is not excessively complex or cumbersome.

23.3.4 Fuzzy System-Guided Navigation

Fuzzy systems use truth values within the range of 0 and 1 to represent uncertainty, with 0 denoting complete falseness and 1 indicating complete truth. On the other hand, crisp logic deals only with absolute true or false values. Fuzzification builds the foundation of the fuzzy logic framework by defining IF THEN rules and creating a membership function from raw data. Rule assessment appraises all rules and calculates their truth values using Mamdani and Sugeno, two popular fuzzy inferences. In situations where an input fails to satisfy an IF–THEN rule precisely, partial matching is used to interpolate a response. Fuzzy values obtained from rule outputs are converted to crisp values during defuzzification. The operational principle of fuzzy systems can be understood by referring to the diagram depicted in Fig. 23.3.

Anderson et al (2005) developed a Fuzzy Logic Control (FLC) controller that utilizes basic, low-cost sensors such as sonar rangefinders and optical encoders. By training the controller with a set of training data, which involved guiding the robot down a course, the dead reckoning module was designed to closely follow a trained track. The objective of the controller was to navigate through an unstructured forest environment while avoiding obstacles. Distinct fuzzy sets were employed for obstacle avoidance, corresponding to each pair of sonar rangefinders. The proposed fuzzy-based controller was reportedly effective in avoiding obstacles.

FLC has various applications, including row navigation, greenhouse aisle operation, and edge detection. It can also be used for obstacle avoidance, utilizing simple and low-cost sensors, with the data evaluated using fuzzy rules. In outdoor farms, fuzzy GA, which incorporates online learning, is sometimes used to employ this powerful technique. Analogous to statistical techniques, navigation based on FLC relies on the vehicle's past and present state without estimation.

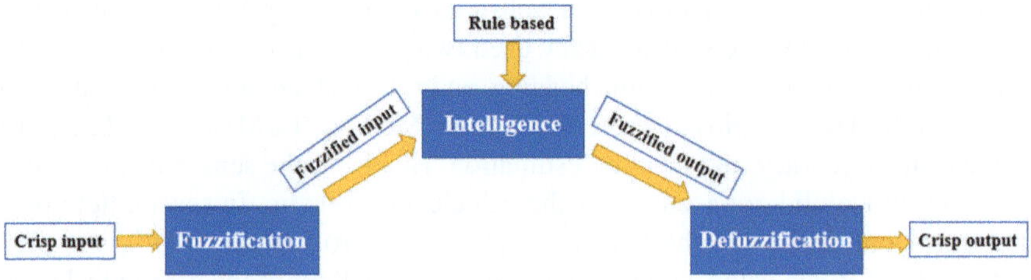

Fig. 23.3 Operational principle of FLC controller

23.3.5 *Neural Network (NN) and Genetic Algorithm (GA)-Based Navigation*

The NN is inspired by the human brain and processes information similar to nervous system. The neural network processes n inputs by performing a weighted sum of each input, followed by the application of an activation function to the resulting sum. NN is composed of a vast network of densely interconnected processing units, known as neurons that collaborate and learn, like human, to modify weights through experience. The framework of artificial neural network is given in Fig. 23.4. Learning involves changes in synaptic connections between neurons, wherein unsupervised learning methods like Kohonen's learning, as well as reinforcement learning or backpropagation (BP), serve as notable examples. NN is useful for detecting and matching complex, vague, incomplete patterns, unreliable data, data noise issues, forecasting, categorization, data linkage, data abstraction, data filtering, and strategizing. Besides neural networks, the Genetic Algorithm (GA) is also utilized to solve automation challenges as a subset of Artificial Intelligence (AI). In brief, a GA is a problem-solving programming approach that mimics biological evolution. When presented with a particular problem that requires a solution, the Genetic Algorithm (GA) receives input consisting of a group of potential solutions encoded in a specific manner, along with a fitness function metric. This metric facilitates the quantitative evaluation of each candidate.

Luo et al. (2007) fabricated a platform for rice transplanter to investigate automatic guiding technology, which used RTK-DGPS, an electronic compass, and a gyroscope to determine location and orientation. Additionally, the rice transplanter's navigation in a geographic information system was established. To calculate the required path, the input, hidden, and output layers employed four, six, nine, and one ANN nodes,

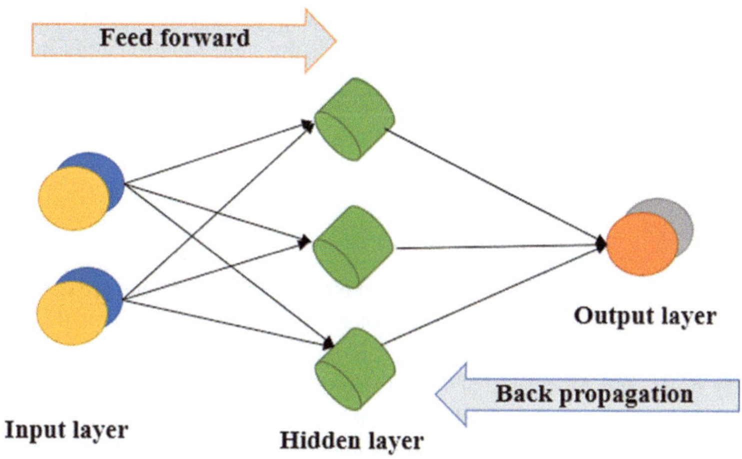

Fig. 23.4 Architecture of artificial NN

respectively. A skilled operator trained the NN and computed the weights using the supervised learning approach (BP algorithm) by moving the transplanter along a specified path. The flexible proportional-derivative (PD) turning mechanism is designed to enable efficient turning manoeuvres at headlands. The experimental results confirmed that at a speed of 0.6 m/s, the average offset error for path tracking was less than 0.15 m.

A neural network-based algorithm takes inputs and produces outputs, such as vehicle moving direction and time consumed, based on the current situation, with training based on previous data. The NN algorithm requires relatively simple computations.

23.3.6 Kalman Navigation Algorithm

The Kalman filter is a versatile and efficient iterative data processing method that can effectively control noisy systems. Its fundamental principle involves feeding noisy data into the filter and obtaining relatively less noisy data as output. The Kalman filter provides optimal estimation for linear systems and errors following a Gaussian distribution based on all previous measurements, while the optimality of nonlinear systems is subject to qualification. It eliminates the requirement of storing all prior measurements and reprocessing data at each step, making it ideal for many real-world applications viz. self-driving vehicles, which operate as linear systems.

To apply the Kalman filter, one needs to follow a set of steps: Firstly, understanding the problem and simplifying it mathematically. Secondly, modelling the state progression. Thirdly, modelling the measuring process, even if the measurement space differs from the state space. Fourthly, modelling the noise, where Gaussian noise is assumed by the Kalman filter. Fifthly, testing the filter, and finally, refining the filter by adjusting the noise parameters. The following two equations can describe a linear system, where Eq. (23.1) is the state equation and Eq. (23.2) is the output equation:

$$x_{k+1} = Ax_k + Bu_k + w_k \tag{23.1}$$

$$y_k = Cx_k + z_k \tag{23.2}$$

The equations mentioned above utilize matrices A, B, and C that are derived from the system model. The time index is denoted by k, while x represents the system's state, u denotes a known system input, y is the output to be measured, and w and z signify noise and processes of the measurement, respectively. These quantities are vectors that consist of multiple elements. The vector x contains all the relevant information regarding the system's current state, such as position, velocity, and acceleration in the context of automobile navigation. However, direct measurement of x is not feasible. Instead, y is measured, which is a function of x and is subject to the influence of noise z. Additionally, it is assumed that there is no relationship between w and z, and k, w_k, and z_k are independent random variables at all times.

Inoue (2008) employed an RTK-GPS and an IMU to equip an autonomous semi-crawler tractor for precise soybean sowing. The primary objective was to use an EKF to model steering actuation and estimate the tractor's position, direction, attitude, and speed. Based on field testing, the researchers concluded that the implemented Kalman filter effectively controlled the vehicle's movement using the vehicle movement model. The algorithm's primary function involved a template-matching approach that illustrated the motion of an identified obstacle in sensor coordinates and reconstructed the obstacle's silhouette.

23.3.7 Simultaneous Localization and Mapping (SLAM)

SLAM is a state-of-the-art approach for mapping and localization (Durrant-Whyte et al. 2006). This approach entails utilizing sensor data to ascertain the state of a robot while concurrently generating a map of its operating environment (Bailey et al. 2006). The robot model generally includes its position, velocity, calibration parameters, and sensor offsets, among other factors. The map can be described as a multi-dimensional representation of objects detected by the sensors mounted on the robot, serving as references throughout the localization process. Generating maps is often necessary to provide information about the surroundings.

Smith and Cheeseman (1986) were the initial proponents of SLAM as robotics was being approached in creative ways with a probabilistic perspective. This new approach to robotics sparked a discussion on how to handle geometric uncertainty (Durrant-Whyte 1988), which played a pivotal role in advancing mapping techniques. For instance, Smith et al. (1990) gave the idea of a stochastic map, which is essentially the same as the modern-day map, but considers the interactions between objects and their uncertainties based on available sensor data sets. All of these ideas contributed to the development of the well-known SLAM problem, which is to estimate the location of map items and the robot's position in a single joint state (Leonard and Durrant-Whyte 1991). According to Thrun (2008), the SLAM problem can be stated as follows: A self-guided mobile robot begins its expedition from a predetermined location and explores an unfamiliar environment. SLAM provides a way for estimating the unknown motion of robot while mapping the unknown surroundings.

23.3.8 Visual Odometry (VO)

When there is a preference to avoid mapping the environment, other options apart from SLAM are considered. One such alternative is visual odometry (VO), which computes the movement of the camera(s) on-board solely based on input from image data. This method involves processing a sequence of photos, calculating the relative transformation between the two entities and subsequently integrating each transformation, the absolute camera pose can be determined. VO is an algorithmic approach that addresses the challenge of reconstructing the three-dimensional (3D) environment and determining camera positions using a collection of unordered image frames. The VO algorithm is composed of a series of processes, which includes feature extraction, feature matching, and motion estimation between images. The motion determination can be done in a variety of ways, such as 2D to 2D or 3D to 3D using depth information from stereo cameras. Bundle adjustment approaches are utilized in certain scenarios to refine the local trajectory estimation by concurrently optimizing the camera parameters and the 3D landmark parameters (Yousif et al. 2015).

The success of localization in robotic platforms depends on the quality of the mapping method, which is determined by the precision of the on-board sensors and the quality of the data post-processing algorithms. Mapping methodologies used to solve the mapping procedures inherent in the SLAM context are separated into three major sets: metric maps, topological maps, and semantic maps. To solve the problem of dividing the geometric space into local maps, topological maps are a logical solution. These techniques display the world map as a collection of linked local maps kept in nodes. The robot navigation path is represented by the pathways between nodes in the SLAM context.

Adding semantic data to maps can improve the effectiveness of localization technologies. This presents a new challenge for perception systems, which must be able to identify and categorize objects and environments. The specifics of semantic notions are crucial and mostly depend on the robot's task, and the semantic categorization should be able to identify objects at the appropriate level, depending on the robot's intended purpose.

23.4 Electronic Components and Sensors Used in Precise Seeding Robots

The manual seed sowing method has different types of problems as those based on human power and conventional old techniques. Humans cannot work in an uncertain environment and they need more time and effort. Seeding robots reduce time and human efforts, increasing agriculture's productivity and efficiency. To accomplish these goals, modern seeding robots comprise several systems. Various electronic components and sensors are integrated into these systems to make these robots more precise and intelligent. In this section, we will discuss these systems as well as their electronic components and sensors (Fig. 23.5).

Fig. 23.5 Composition of hardware of a seeding robot (*source* https://farmdroid.dk/en/product/)

23.4.1 Guidance System

Like drones, the autonomous seeding robot is also designed to follow a pre-defined path with the help of a navigation system. This path is optimized in such a manner that the robot can cover most of the field with minimum power requirement and minimum seed fillings. And an adaptive seed rate can be achieved according to a soil fertility map. The coordinates of the trajectory are fed with the help of a graphical user interface by the users. Therefore, the robots are equipped with a positioning system that is used to locate robots in the field precisely and accurately. In modern robots, a differential global positioning system (DGPS) or real-time kinematic GPS is mostly used.

DGPS: It is a modification to the global positioning system (GPS) that increases position accurateness from the nominal 15 m to roughly 1–2 m. To obtain the advantages of differential GPS, two receivers are required. This involves comparing two distinct GPS signals to determine the robot's location more precisely. The DGPS receiver estimates its pseudo ranges after receiving signals from satellites. It also gets correction signals from a base station or reference source that is aware of its precise location. These adjustments are made by the robots' receivers to their computed pseudo ranges, producing a location that is far more precise.

RTK-GPS: 'Real-time' GPS is the most prevalent form of differential GPS. By comparing a code generated by the satellite with a code created internally in the receiver, RTK-GPS receivers calculate distance. The distance is calculated by multiplying the two codes' timing differences by the speed of light. The RTK uses a base station and a rover as its two receivers. The ground station is a GPS receiver that has been deployed at a location whose precise location has been determined through a different independent accurate method of survey. It also uses GPS to determine

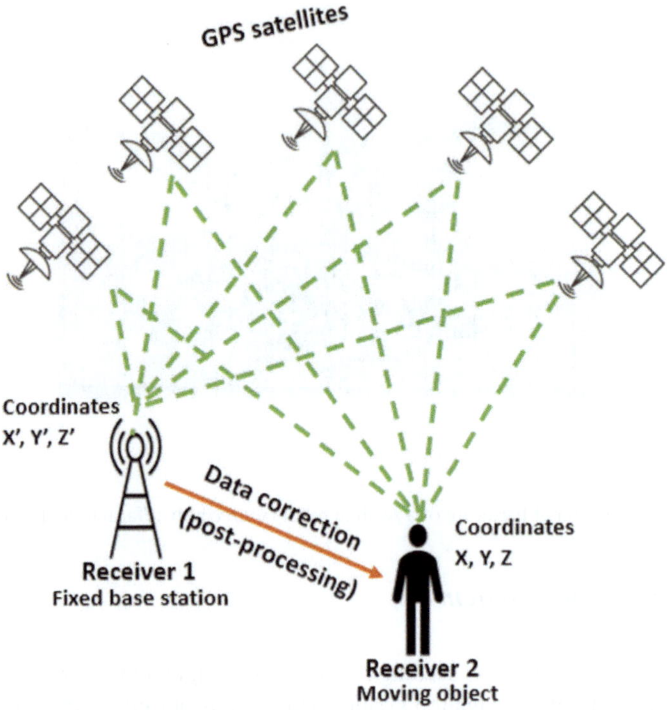

Fig. 23.6 DGPS

its location, and by comparing the result to its precise location, it determines the inaccuracy in the measurement. The rover receives this inaccuracy and applies the corrective data to enhance its own estimated location that it derives from the GPS in real time. Thus, able to achieve centimetre (up to 1 cm) accuracy (Figs. 23.6 and 23.7).

In semiautonomous robots, robots are operated with the help of a transmitter and receiver. The transmitter is a remote-control device, which transmitted the signal, and the receiver is equipped on the robot, which received the signal. They both work on the same frequency. This shows the information regarding seed rate, seed quantity, battery power level, etc. Many communication protocols are used for this purpose. Bluetooth, radiofrequency, GPS, and Wi-Fi are the major platform for telecommunication.

23.4.2 Self-Propelling System

In recent times, most agribots' propelling systems are either wheel-type or crawler-type. In order to produce different robot motions, electric drives use a range of electric

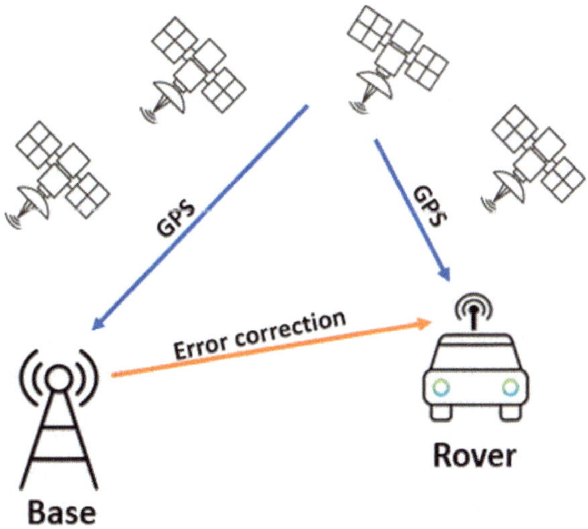

Fig. 23.7 RTK-GPS

motors to directly or indirectly transmit mechanical power to wheels, creating forces and moments (Tu et al. 2019). Because of their smooth speed regulation range, large overload capacity, and low sensitivity to electromagnetic interference, direct current (DC) motors are frequently employed in driving systems. They also have good starting and speed regulation features. The pulse width modulation (PWM) control method and the control method both limit how a DC motor can be controlled. Hybrid stepper motors and brushless DC motors are predominantly used nowadays.

Hybrid stepper motor: A hybrid stepper motor combines the advantages of stepper motors with permanent magnets and variable reluctance. In comparison to variable reluctance stepper motors, hybrid stepper motors have a shorter step length and more torque per unit volume. Sometimes these motors are used with a worm gear mechanism. In such robots, steering can be achieved by either breaking the movement of one wheel or using stepper motors. Castor wheels are also useful for making steering easy.

Brushless DC motor: Since brushless DC motors (BLDC) are increasingly the preferred option in many applications, particularly in the realm of drones and robots, many motor manufacturers have concentrated on this technology. In many aspects, BLDC motors are superior to brushed DC motors, including their capacity for high speeds, high levels of efficiency, and greater heat dissipation. These motors are very effective at generating a lot of torque across a wide speed range. They are renowned for their quiet operation and ability to maintain torque at rest. Permanent magnets move around a fixed armature in brushless motors, which solves the issue of connecting current to the armature. Crawler robots typically employ BLDC motors.

23.4.3 Obstacle-Detecting System

The safety of seeding robots is one of the main concerns on the field so that the robots can perform their specific task without any hindrance. Presently, many technologies are used for this purpose such as SONAR, LiDAR, and machine vision techniques. In this section, we will discuss the above-mentioned technologies.

SONAR: Sonar is a navigation and ranging technique that uses sound to communicate with or identify objects, measure distances, and navigate. Sonar systems use a range of acoustic frequencies, from incredibly low (infrasonic) to incredibly high (ultrasonic). This technology computes time from the transmitting acoustic signal to receiving signal and based on this time it can predict the obstacle distance. Transmitters, receivers, and transceivers are the three basic categories into which ultrasonic sensors, a form of sound sensor, can be divided. Receivers reverse the process, transforming electrical signals into ultrasonic, while transmitters do the inverse with electrical signals. On the other hand, transceivers are versatile enough to both transmit and receive ultrasound signals.

LiDAR: Due to its many benefits, such as high measurement precision and the capacity to deliver a significant amount of information about the distance at a high frequency, light detection and ranging (LiDAR) has grown in popularity in mobile robot navigation. LiDAR has consequently become a well-liked option for mobile robot navigation in recent years. LiDAR is primarily used by agricultural robots to avoid obstacles during operation. This is done by laser-targeting an object or surface and timing how long it takes for the reflected light to come back to the receiver. LiDAR scanning can be utilized as navigational data for mobile robot navigation (Nehme et al. 2021).

Machine vision technique: In recent times, agricultural robots mainly use machine vision, spectral imaging, and other related technologies for the localization, classification, and distances of targets, obstacles, etc. Stereo cameras and depth cameras play an important role in this.

23.4.4 Metering System

With the advent of the Intelligent Planting Solution, the seeding process has become considerably more streamlined and efficient. Through the use of sensor data, the system is able to accurately calculate and apply the optimal number of seeds for each individual row, resulting in more consistent and reliable planting across the entire field. Pick and place and continuous metering are the major advances in robotic seeding.

Continuous metering: The Electronic Control Unit (ECU) collects data regarding implement speed and using this digital information, it determines the rotational speed of the electric motor that powers the feed shaft in the seed hopper. For improving the seeding rate, a metering plate is attached to the feed shaft, which in turn sets the number of seeds delivered every rotation for specific rows.

A. Speed sensor: The hall effect sensor and proximity sensor are used to compute the implement speed in a non-destructive way. The rotary encoder is also used for calculating speed by connecting to the driving wheels. The rotary encoder is frequently employed as an error sensing feedback control for monitoring seed quantity and implements speed, which is used for optimizing a system's performance.
B. Stepper motor: Electric motor mostly used to power the feed shaft is stepper type because its rotation of small equal steps results in precise and desired metering of seeds.
C. Seed monitoring: The quantity of seed available in the seed box is accurately measured by an ultrasonic sensor and infrared sensor. An infrared (IR) sensor functions as a proximity sensor or a 'nearness' sensor by detecting the presence of objects in close proximity. The IR sensor utilizes a white LED as the transmitter, emitting infrared light, and a photodiode placed adjacent to the IR LED as the receiver.

Pick and place type: Recently, robotic arm type pick and place metering is the most common and promising technique. The robotic arm consists of arms and an end effector. The arms are used to navigate the end effector near the seeding pod so that the end effector grips the pod and places it directly into the soil or in the seed tube funnel. The movement of robotic arms and the end effector is achieved by servo motors, electric linear actuators, hydraulic actuators, or pneumatic actuators.

A. Servo motor: A most common way to control the movement of arms and end effectors is by using servo motors. It has quite quick and easy to instal and control. A DC motor, a gearbox, a position sensor (potentiometer), and control electronics combine to create a servo motor. The potentiometer feeds information about the motor's present position to the sensor from the gearbox when the DC motor is connected to a gear system. These motors typically have a finite rotation of 90 to 180 or even 360 degrees. Servos, however, are limited in their movement to certain angles and do not rotate continually (Amrita et al. 2015).
B. Electric linear actuators: Sometimes linear actuator is also useful for controlling motion. It is made up of a gearbox and a DC motor. A linear actuator transforms rotating movement into linear motion, is easy to integrate into automated processes, has heavy-duty performance, and has slow speed.
C. Hydraulic actuators and pneumatic actuators: Hydraulic actuators are used for heavier work. The system comprises an oil tank, solenoid valves, and an electric pump. Besides of these pneumatic actuators are used for lighter work. It has a simple design, is less accurate, and has high speed. To create a linear motion from compressed air, pneumatic linear actuators are designed.

23.4.5 Furrow Opening System

The soil must be open to a certain desired depth for good germination of seeds. After that, the seed is placed and covered with the soil. For this, the furrow openers are automatically raised and lowered by the mean of electric linear or hydraulic actuators. A feedback system for desired depth is also incorporated into this system using an ultrasonic sensor, LiDAR sensor, or rotary encoder. Once the seeds are dispensed in the furrows, seeds are covered by soil with the help of pressed wheels, planks, etc.

23.4.6 Processing Unit

The electronic control unit serves the function of collecting signals from the sensors and controlling the motion of motors and actuators. Arduino microcontroller and Raspberry Pi are the major advances in this field. They are compact in size, easy to use, and required less power.

Arduino microcontroller: The open-source Arduino platform is used to create electrical projects. Integrated Development Environment (IDE) software and a physical programmable circuit board (often referred to as a microcontroller) compose the Arduino platform. Additionally, the Arduino IDE employs a condensed form of C++ that makes learning to programme simpler. It is a key component of the whole system, interconnected with sensors and other parts to carry out a task. Arduino performs according to the given inputs by sensors. It is used to control stepper motors, actuators, and servo motors. Arduino Mega and Arduino UNO have been widely used microcontrollers recently.

Raspberry Pi: A computer that may be utilized for a variety of purposes is the Raspberry Pi processor. It runs on the Linux operating system. The processor is dumped using Python. RAM is offered in capacities ranging from 1 to 8 GB and the processor runs at 3.3 V. Real-time and computer vision applications can all benefit from the Raspberry Pi CPU. It has Wi-Fi and Bluetooth connectivity. So, it is easy to use in IoT applications.

23.4.7 Power Supply System

The electric power requirement of the robots is fulfilled by batteries and solar panels. Nowadays Lithium Polymer (Li-Po) and Lithium Ion (Li-Ion) batteries are preferred over lead acid batteries because the high energy density of Lithium batteries results in a compact size. They charge at a faster rate and have a higher service life. Solar panels are also used to charge the battery.

23.5 Economic Perspective of the Seeding Robots

Seeding robots can increase agricultural productivity, reduce environmental impact, and stimulate economic growth. However, careful consideration of the impact on employment and appropriate policies are necessary for a smooth transition to a more automated agriculture sector. Farmers and agribusiness owners frequently seek economic research studies before investing in new seeding or planting technology. They recognize that such studies seldom reproduce actual on-farm or company conditions exactly, but they can indicate important impediments to the efficient use of seeding robots and characteristics of organizations that could employ them effectively (Lowenberg-DeBoer et al. 2020).

i. Most studies on the economics of seeding robots employ partial financial arrangements to focus on changes in variable cost. As a result, a more systematic assessment of the potential impact is desperately needed (Pedersen et al. 2017).
ii. On-farm testing will become viable robotic seeding become widely accessible. On-farm classic trial methods might be updated, or few modern on-farm trial designs that incorporate computerized data collecting could be adopted (Bullock et al. 2009). For the trials to be unbiased and credible, they should be overseen by a non-commercial organization (university, research institute, NGO).
iii. Majority of the main farm equipment companies are concentrating on their largest equipment (autonomous versions), none of the publicly accessible research papers on the economics of robotic seeding focus on relatively small robots.
iv. Agriculture machinery market structure and size, environmental factors, crop and robot safety aspects, prime preference for the agriculture data, studies and proper information on crop materials to be used with the robotic seeding, and co-competitors in the present developing world will put their impact on the economic benefit.
v. The ideal size of the seeding robot for a particular farm type and set of duties has to be proven. Numerous elements would need to be considered in this study, such as the cost of the seeding robot, maintenance costs, work rates under various farm circumstances, energy consumption, the effects of compaction on yield, legal and regulatory constraints, and transportation between fields from farmstead to the field.

23.6 Conclusions

To address the need of the growing population, the smart agriculture is playing a major role. Robotic seeding is the most important component of this smart and precision agriculture. Seed sowing robots help in sowing of seeds in the desired position hence assisting the farmers in saving time and money. The robots localize the path and can navigate itself without human action. The remote-control-based seed sowing robots can precisely determine the relative position using dead reckoning and absolute location using GPS. The robots localize the path with the help of sensors (ultrasonic, infrared, Lidar, Radar, etc.), GPS/DGPS/RTK-GPS, electronic compass, and gyroscope. However, GNSS is being used as the most popular method to navigate the robots. Statistical based algorithms, FLC, NN, GA, and SLAM are the most popular algorithms used to localize the path and navigate the robots. The integration of the Internet of things technologies in such robots promises sustainability in future agriculture. The continuous development of robotics and improvement in the field of electronics leads to more reliable, compact, and cost-effective sensors and components.

References

Agarwal S, Snavely N, Seitz SM, Szeliski R (2010) Bundle adjustment in the large. In: Daniilidis K, Maragos P, Paragios N (eds) Computer vision—ECCV 2010. Springer, Berlin/Heidelberg, Germany, pp 29–42

Aguiar AS, Dos Santos FN, Cunha JB, Sobreira H, Sousa AJ (2020) Localization and mapping for robots in agriculture and forestry: a survey. Robotics 9(4):1–23

Anderson IP, Carlson AC, Edwards DB, Anderson MJ, Feeley JJ (2005) An autonomous forest robot that uses a hierarchical, fuzzy logic controller. Trans ASAE 48(4):1603–1617

Aqel MO, Marhaban MH, Saripan MI, Ismail NB (2016) Review of visual odometry: types, approaches, challenges, and applications. Springerplus 5:1–26

Arif A, Butt KM (2014) Computer vision based navigation module for sustainable broad-acre agriculture robots. Sci Int 26(5)

Bailey T, Durrant-Whyte H (2006) Simultaneous localization and mapping (SLAM): part II. IEEE Robot Autom Mag 13:108–117

Bullock DS, Ruffo ML, Bullock DG, Bollero GA (2009) The value of variable rate technology: an information-theoretic approach. Am J Agr Econ 91(1):209–223

Chen SY (2011) Kalman filter for robot vision: a survey. IEEE Trans Industr Electron 59(11):4409–4420

Crist E, Mora C, Engelman R (2017) The interaction of human population, food production, and biodiversity protection. Science 356:260–264

Durrant-Whyte H (1988) Uncertain geometry in robotics. IEEE J Robot Autom 4:23–31

Durrant-Whyte H, Bailey T (2006) Simultaneous localization and mapping: part I. IEEE Robot Autom Mag 13:99–110

Fraundorfer F, Scaramuzza D (2012) Visual odometry: part II: matching, robustness, optimization, and applications. IEEE Robot Autom Mag 19:78–90

Guo J, Li X, Li Z, Hu L, Yang G, Zhao C, Fairbairn D, Watson D, Ge M (2018) Multi-GNSS precise point positioning for precision agriculture. Precision Agric 19(5):895–911

Guo LS, Zhang Q, Han S (2002) Position estimate of off-road vehicles using a low cost GPS and IMU. In: ASAE annual international meeting/CIGR XVth world congress, Chicago, Illinois, USA, July 28–July 31

Haibo L, Dong S, Zunmin L, Chuijie Y (2015) Study and experiment on a wheat precision seeding robot. J Robot 1:1–9

Hassan MU, Ullah M, Iqbal J (2016) Towards autonomy in agriculture: design and prototyping of a robotic vehicle with seed selector. In: Proceedings of the 2016 2nd international conference on robotics and artificial intelligence (ICRAI), Los Angeles, CA, USA, pp 37–44

Iacomi C, Popescu O (2015) A new concept for seed precision planting. Agric Agric Sci Procedia 6:38–43

Inoue K (2008) Steering control system for autonomous tractor. In: Proceedings of the JFPS international symposium on fluid power. The Japan Fluid Power System Society, 7, 53–58

Kala R, Shukla A, Tiwari R, Rungta S, Janghel RR (2009) Mobile robot navigation control in moving obstacle environment using genetic algorithm, artificial neural networks and A* algorithm. In: 2009 WRI world congress on computer science and information engineering, 4, 705–713

Khairuddin AR, Talib MS, Haron H (2015) Review on simultaneous localization and mapping (SLAM). In: 2015 IEEE international conference on control system, computing and engineering, pp 85–90

Kumar GVP, Raheman H (2008) Vegetable transplanters for use in developing countries a review. Int J Veg Sci 14(3):232–255

Kumar GVP, Raheman H (2011) Development of a walk-behind type hand tractor powered vegetable transplanter for paper pot seedlings. Bio Syst Eng 110(2):189–197

Leonard JJ, Durrant-Whyte HF (1991) Simultaneous map building and localization for an autonomous mobile robot, vol 3. In: Proceedings of the IEEE/RSJ international workshop on intelligent robots and systems (IROS'91), Osaka, Japan, pp 1442–1447

Lowenberg-DeBoer J, Huang IY, Grigoriadis V, Blackmore S (2020) Economics of robots and automation in field crop production. Precision Agric 21(2):278–299

Luo X, Zhang Z (2007) DGPS navigation control system for rice transplanter. In: ASABE annual international meeting, Minnesota

Mattos PG (1994) Integrated GPS and dead reckoning for low-cost vehicle navigation and tracking. In: Proceedings of VNIS'94-1994 vehicle navigation and information systems conference, pp 569–574

Mizushima A, Noguchi N, Ishii K (2004) Automatic navigation of agricultural vehicle using a low cost attitude sensor. In: ASAE proceedings conference, Kyoto, Japan

Mousazadeh H (2013) A technical review on navigation systems of agricultural autonomous off-road vehicles. J Terramech 50(3):211–232

Nattharith P, Güzel MS (2016) Machine vision and fuzzy logic-based navigation control of a goal-oriented mobile robot. Adapt Behav 24(3):168–180

Nehme SA, Abirami E, Ankita A, Praveena R, Srimeena R (2015) Agricultural robot for automatic ploughing and seeding. In: 2015 IEEE technological innovation in ICT for agriculture and rural development (TIAR), pp 17–23

Nehme H, Aubry C, Solatges T, Savatier X, Rossi R, Boutteau R (2021) LiDAR-based structure tracking for agricultural robots: application to autonomous navigation in vineyards. J Intell Rob Syst 103(4):1–1

Nister D, Naroditsky O, Bergen J (2004) Visual odometry. In: Proceedings of the 2004 IEEE computer society conference on computer vision and pattern recognition, 2004, CVPR 2004, Washington, DC, USA, 27 June–2 July 2004

Omrane H, Masmoudi MS, Masmoudi M (2016) Fuzzy logic based control for autonomous mobile robot navigation. Computational intelligence and neuroscience

Parish RL (2005) Current developments in seeders and transplanters for vegetable crops. HortTechnology 15(2):346–351

Pedersen SM, Fountas S, Sørensen CG, van Evert FK, Blackmore BS (2017) Robotic seeding: economic perspectives, 167–179

Perez-Ruiz M, Upadhaya KS (2012) GNSS in precision agricultural operations. New approach of indoor and outdoor localization systems, 3–26

Pinto de Aguiar AS, Neves dos Santos FB, Feliz dos Santos LC, de Jesus Filipe VM, Miranda de Sousa AJ (2020) Vineyard trunk detection using deep learning—an experimental device benchmark. Comput Electron Agric 175:105535

Pradhan NC, Sahoo PK, Kushwaha DK, Mani I, Srivastava A, Sagar A, Kumari N, Sarkar SK, Makwana Y (2021) A novel approach for development and evaluation of LiDAR navigated electronic maize seeding system using check row quality index. Sensors 21(17):5934

Reina G, Vargas A, Nagatani K, Yoshida K (2007) Adaptive kalman filtering for gps-based mobile robot localization. In: 2007 IEEE international workshop on safety, security and rescue robotics, pp 1–6

Sakaue O (1996) Development of seeding production robot and automated transplanter system. Jpn Agric Res Q 30:221–226

Savage J, Munoz S, Matamoros M, Osorio R (2013) Obstacle avoidance behaviors for mobile robots using genetic algorithms and recurrent neural networks. IFAC Proc Vol 46(24):141–146

Sirigool W, Kesvarakul R (2020) Particle filter for hector slam to improve the performance of robot positioning by image processing based. Int J Mach Learn Comput 10(3):490–494

Smith R, Self M, Cheeseman P (1990) Estimating uncertain spatial relationships in robotics. In: Autonomous robot vehicles. Springer, New York, NY, USA, pp 167–193

Smith RC, Cheeseman P (1986) On the representation and estimation of spatial uncertainty. Int J Robot Res 5:56–68

Subramanian V, Burks TF, Arroyo AA (2006) Development of machine vision and laser radar based autonomous vehicle guidance systems for citrus grove navigation. Comput Electron Agric 53(2):130–143

Subramanian V, Burks TF (2005) Autonomous path navigation in citrus groves using machine vision and laser radar. In: ASAE annual international meeting, Tampa, Florida, July 17–20

Sukkarieh S (2017) Mobile on-farm digital technology for smallholder farmers. In: Proceedings of the 2017 crawford fund annual conference on transforming lives and livelihoods: the digital revolution in agriculture, Canberra, Australia, p 9

Sylvester G (2018) E-agriculture in action: drones for agriculture. FAO, Bangkok, Thailand

Tavakoli M, Lopes P, Sgrigna L, Viegas C (2015) Motion control of an omnidirectional climbing robot based on dead reckoning method. Mechatronics 30:94–106

Terrier P, Schutz Y (2005) How useful is satellite positioning system (GPS) to track gait parameters? A review. J Neuroeng Rehabil 2(1):1–11

Thrun S (2008) Simultaneous localization and mapping. In: Robotics and cognitive approaches to spatial mapping. Springer, Berlin/Heidelberg, Germany, pp 13–41

Trautman P, Ma J, Murray RM, Krause A (2015) Robot navigation in dense human crowds: statistical models and experimental studies of human–robot cooperation. Int J Robot Res 34(3):335–356

Tu X, Gai J, Tang L (2019) Robust navigation control of a 4WD/4WS agricultural robotic vehicle. Comput Electron Agric 164:104892

Xing H, Guo S, Shi L, Pan S, He Y, Tang K, Su S, Chen Z (2017) Kalman filter-based navigation system for the amphibious spherical robot. In: 2017 IEEE international conference on mechatronics and automation (ICMA), pp 638–643

Xue J, Zhang L, Grift TE (2012) Variable field-of-view machine vision based row guidance of an agricultural robot. Comput Electron Agric 84:85–91

Yousif K, Bab-Hadiashar A, Hoseinnezhad R (2015) An overview to visual odometry and visual SLAM: applications to mobile robotics. Intell Industr Syst 1:289–311

Chapter 24
Revolutionising Crop Yield Prediction: The Synergy of Remote Sensing and Artificial Intelligence Technologies

Sarabjot Kaur Sandhu and Shubham Anand

Abstract Crop yield prediction is becoming progressively essential due to increasing concern about food security. For anticipating the amount of food that will be available for the expanding global population, early crop yield prediction performs a significant role in avoiding food crisis. Yield prediction is an extremely significant agricultural dilemma. But, nowadays advanced digital technologies help us understand things better by constantly observing and measuring various physical events and generating a substantial amount of data, called big data. Agricultural big data is used for some important activities like weed recognition, pests and diseases detection and forecasting and risk management. It is evident that satellite remote sensing has capacity to generate a sufficient volume of data. The multispectral satellite's geographical and temporal resolution allows them to provide a wealth of data on agricultural productivity. The raw bands from remote sensing technologies are utilised to develop vegetation indices, and features are retrieved and then combined with machine learning to forecast agricultural yield. Machine learning and artificial intelligence (AI) approaches have been started recently to study big data. India's agriculture contributes significantly to its GDP. The current statistics show that 17.32% of India's GDP comes from agriculture. Therefore, the government urgently needs an appropriate method for predicting crop yields before harvest. Utilising artificial intelligence technologies in agriculture has potential to completely change the way how policies are made. This chapter includes the work of several researchers so that the present state of remote sensing and AI approaches can be comprehended to know the applications of AI and remote sensing technologies in crop yield prediction.

Keywords Artificial intelligence · Big data · Crop yield · Remote sensing · Yield prediction

S. K. Sandhu · S. Anand (✉)
Department of Climate Change and Agricultural Meteorology, Punjab Agricultural University, Ludhiana, Punjab 141004, India
e-mail: shubham-1904001@pau.edu

K. Pandey et al. (eds.), *Artificial Intelligence and Smart Agriculture*,
Advances in Geographical and Environmental Sciences,
https://doi.org/10.1007/978-981-97-0341-8_24

24.1 Introduction

Crop yield prediction (CYP) is one of the most challenging tasks in agriculture due to weather aberrations and changing climatic scenarios. On global, regional and local levels, decision-making depends on it. At the beginning of crop season, agricultural policymakers must get estimates of productivity of each crop (Frausto-Solis et al. 2009). The decisions taken by farmers regarding crop operations like rotating crops, soil preparation, irrigation, pesticides and fertiliser application, and uncontrollable variables like climatic conditions, market and subsidies may have a significant impact on production. According to Ruß (2009), historically, the accuracy of yield forecasts has depended on growers' in-depth knowledge of specific areas, crops and climatic conditions. Basic estimators are also used, such as the most recent yield or the average of several past yields. However, agricultural productivity exhibits nonlinear regional and temporal variability, which results in significant year-to-year variations (Drummond et al. 2003; Schlenker and Roberts 2006). Since advance prediction about crop production as well as the condition has a significant impact on a country's economy, remote sensing has been utilised to accomplish this goal ever since it was created. Three significant agricultural assessment-related experiments have been conducted on a global scale. These were Corn Blight Watch Experiment (CBWE) (1971), the Crop Identification Technology Assessment for Remote Sensing (1973) and the Large Area Crop Inventory Experiment (LACIE) (1974–1978). The National Aeronautics and Space Administration, the National Oceanic and Atmospheric Administration and the USDA collaborated to create LACIE, the first global project to reveal practical viability of remote sensing technology for agricultural production forecasting (United States Department of Agriculture). Since 1970s, crop acreage estimates and evaluations have been conducted using remote sensing. It has become a more important player in nations' endeavours to estimate agricultural acreage and production as a result of technological advancements in sensor quality, availability and processing. However, it is not a miracle cure. Its use still has problems, particularly with timing, effectiveness and guarantees of ongoing supply (Ray 2000).

More effective methods have been developed as a result of technological improvements, including data-driven models and crop growth models. By merging the dynamics of physiological processes of plant into a mathematical model, crop models abstract such dynamics (Safa et al. 2004). Many mechanistic models, such as AFRCWHEAT2 (Porter 1993), Sirius (Jamieson et al. 1998b) and CERES-Maize (Jones and Kiniry 1986) for wheat, maize and legumes, respectively, are unique to individual crops. Additional crop types can be accessed via parameter fitting, including STICS, SUCROS and SUCROS2 (Spitters et al. 1988; Goudriaan and Van Laar 1994). A review of these models for the wheat crop by Jamieson et al. (1998a) revealed that several of them have reasonable accuracy, with variances falling within 10%. Varcoe (1990) and Drummond et al. (2003) stated that this type of model is impracticable for widespread application and agricultural planning due to its high time and money costs. A complete understanding of physical processes that produced data is not mandatory when creating data-driven models, which are instead developed empirically. These methods don't require any specified structure of a model and are generally cheap and simple to use. Data-driven models incorporating traditional statistics and machine learning were frequently utilised in previous years (Dixon et al. 1994; Sudduth et al. 1996). The parametric structures updated with sum-of-squares residuals, statistical models are represented, and validity of these models is assessed using confidence intervals (Breiman 2001). The primary regression implementation for CYP was primarily linear and had generally subpar results (Drummond et al. 2003). The validation of machine learning (ML) approaches depends on prediction accuracy and is based on semi-parametric as well as nonparametric structures (Breiman 2001). ML approaches used for CYP frequently include regression trees, artificial neural networks and support vector regression as revealed by different researchers, viz. Drummond et al. (2003), Roel and Plant (2004), Irmak et al. (2006) and Fortin et al. (2011). Due to their user-friendly application and performance, earlier investigations suggest that data-based models are more flexible for crop planning than crop growth models (Safa et al. 2004; Irmak et al. 2006). Utilisation of artificial intelligence technologies in agriculture has potential to entirely alter the way to handle difficult tasks in relatively easier way. This chapter presents the work of several researchers so that the existing state of remote sensing and AI approaches can be comprehended to know the execution of AI and remote sensing technologies in crop yield prediction.

Crop area and yield estimation using remote sensing

Remote sensing is a method of collecting information about any object without actually touching it. The majority of the remote sensing observations are derived from conventional platforms like planes and satellites. Aerial photography is the first and maximum used form of remote sensing. It is regularly used for topography mapping, agricultural estimating, environmental and engineering research, military observations, crop disease information and mineral extraction. Pasted or taped paper picture index coverage serves as a primary source of stratified data for area frames. Later, as paper prints and subsequently as digital photos, images of earth's surface taken from space provided a complete overview of enormous areas. Throughout the crop season, area estimation is often done using ground surveys data or surveys data improved by using satellite data. Remote sensing images are typically utilised to carry out stratification, and estimate is commonly done this way as well. As the last few decades, technology and methodology have evolved, remote sensing procedures have become widely used in area estimation (Ray 2000).

24.2 Historical Background

In order to estimate crop acreage in the districts of Anantpur in Andhra Pradesh and Patiala in Punjab, the Indian Council of Agricultural Research (ICAR) and Indian Space Research Organization (ISRO) used aerial colour infrared (CIR) images as part of the joint Agricultural Resources Inventory and Survey Experiment (ARISE). This was first organised attempt at agricultural inventorying with utilisation of remote sensing in India. Three schemes: crop yield modelling, crop yield forecasting and crop stress detection were established in 1984 under the Indian Remote Sensing Utilisation Project (IRSUP) of the Department of Space to investigate various facets of crop inventory using RS data. Then, in 1986, the Remote Sensing Applications Mission established the Crop Acreage and Production Estimate (CAPE) initiative. For country's main crop-growing districts, the CAPE project generated production forecasts at district level for all important crops, including rice, *rabi* sorghum, groundnut, mustard, wheat, rapeseed and cotton. Using stratified sample segment methodology and remote sensing data, crop acreage is estimated. Empirical formulas using crop production and remote sensing parameters, the district-level remote sensing-based yield models were formulated. Some yield models also include other agrometeorological factors and/or a temporal trend variable in addition to remote sensing features.

The CAPE project evolved into the Forecasting Agricultural Output Using Space, Agrometeorology and Land-Based Observations (FASAL) programme. FASAL's major objective is to integrate many strategies and organisations in order to form a classified information system that aids in delivering details on crop conditions and productivity at any point in the growing season, from sowing through harvesting. In accordance with FASAL, numerous production predictions are provided at the national and state levels for commodities such as *kharif* and *rabi* rice, mustard, rapeseed, winter potatoes, jute (using microwave data) and wheat (Parihar and Oza 2006). For the estimation of yield, remote sensing data may be incorporated with weather-based models or crop simulation models like WOFOST. In order to implement the FASAL programme, a new institute called the Mahalanobis National Crop Forecast Centre (MNCFC), New Delhi, was established under the Ministry of Agriculture in 2012. Area under crop and yield of crop are the two elements of crop production statistics. Both these elements are crucial for managing the agricultural market and planning the economy (Gallego et al. 2014). Estimating the area is generally thought to be the simpler and easier task to perform as compared to predicting the crop output. Crop area estimation, however, has a variety of challenges and complications that may not be immediately perceived (Craig and Atkinson 2013).

24.3 Traditional Methods of Estimating Crop Area via Remote Sensing

Crop area was previously determined either through a thorough inventory of all farms or through sampling. Area frame sampling (AFS), a farm list sample, or a combination of both may be used as the sampling technique; the latter is known as multiple frame sampling. In other circumstances, it also considers the skilled judgement of volunteer crop reporters. Markets, agricultural processing facilities (such as cotton or jute mills) and administrative surveys could be additional sources of crop area information. The ultimate estimate is generated either directly from survey data or by a panel of experts who assess statistics from various sources and make the ultimate estimations.

24.4 Satellite Remote Sensing

The traditional approach to estimate crop area is laborious, expensive, time-consuming and vulnerable to human bias. Additionally, it is very challenging on a variety of terrains, including mountainous areas. In order to solve these issues, crop area estimation using satellite remote sensing has been used, either directly or as a support for area sampling systems. Different crops may be categorised using satellite remote sensing, which also gives multispectral, synoptic, temporal and multi-resolution perspectives of land cover as well as land use. The satellite-based remote sensing was started to be used for crop area assessment in early 1970s, when the corn blight watch project was conjointly conducted in 1971 by USDA, NASA and many institutions (Sharples 1973). The ERTS-A was successfully launched in 1972. NASA and the USDA research looked at the feasibility of using multispectral remote sensing to evaluate the main crop kinds from space (Bryan 1974). To highlight the potential of remote sensing in agricultural inventory, experiments like the Large Area Crop Inventory Experiment (LACIE) and Crop Identification Technology Assessment for Remote Sensing (CITARS) experiment have been done (MacDonald 1984).

For maize and soybean crops, the CITARS experiment assessed classification methods and alternative analytic methodologies. The first project funded by the US government, called LACIE, was tested regarding its possibility to predict wheat production across wide geographic areas using data from remotely sensed satellites, notably Landsat data (Nellis et al. 2009). The LACIE effort was launched in the US during 1974, and it eventually moved to Soviet Union and Canada (MacDonald 1984). The Agricultural and Resources Inventory Surveys using Aerospace Remote Sensing (AgRISTARS) project was launched during 1980 as a result of the LACIE findings. By incorporating crop monitoring for more crops like barley, wheat, cotton, corn, soybeans, rice and soybeans, this new initiative was attempted to build on LACIE (Holmes et al. 1979). AgRISTARS was successful project for the use of timely data.

Methods of crop area estimation through satellite remote sensing

As crops differ in structure, physiology, cultural practices and phenology, they appear differently in multispectral data and have distinctive spectral fingerprints. This is the fundamental idea behind using remote sensing for crop identification and categorization. Crops may be identified with the aid of carefully chosen ground data, or ground truth. Different methods used for assessing crop area using remotely sensed data are mentioned in Fig. 24.1.

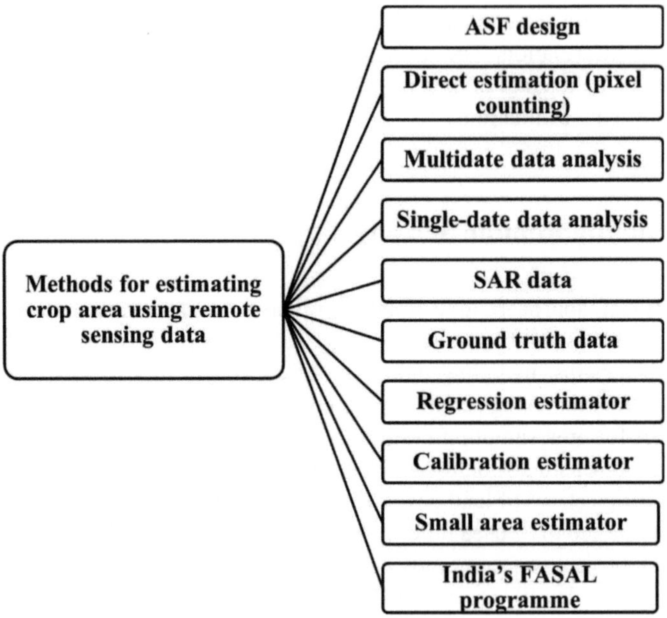

Fig. 24.1 Different methods of remote sensing used for crop area estimation

24.4.1 ASF Design

By dividing areas of interest into enumeration areas (EAs), where list frames of holdings are established, it aids to create area frames made up of primary and secondary sampling units (PSUs and SSUs, respectively), which can be easily differentiated in the digital images (Delince 2015). The satellite photography is essential to frame areas.

Area frames are developed by using a variety of methods in addition to square segments. The FAO used irregular segments with physical boundaries as bounds to determine the amount of rice that can be cultivated in Afghanistan (FAO 2017). The ASF was constructed on many levels, including PSUs (of 500–700 ha), SSUs (from 200 to 300 ha) and terminal sampling units (between 25 and 35 ha) due to the complex landscape of region. Crop intensity (> 75%, 50–75%, 25–50% and < 25%) was the basis for stratification. Based on images from Pleiades to MODIS, frame design definition and survey optimization were carried out. Each portion of 5 × 5 km has its own identifying number. Following stratification, roughly 15% of each type's sample segments are chosen for the final analysis.

Efficiency of the MARS Project's pilot sites reportedly varied normally from 1.1 to 1.6%, according to Carfagna (2013). The efficiency was in the range of 1.7–2.2% for Spain's principal crops, according to Gallego et al. (1999). The efficiency of stratification for calculating the rice production in India ranged from 1.0 to 2.68% using microwave data (special aperture radar, or SAR). Another component that describes the level of stratification is the coefficient of variation (CV). According to samples with sizes of 1229 and 450 ha, the CVs in India for paddy fields between one million and three million ha ranged from 1.15 to 3% (on a state-level basis).

24.4.2 Pixel Counting or Direct Estimation

This method uses ground truth data acquired from common sites to classify satellite images. The crop's area is calculated by multiplying the number of pixels observed in each crop by the pixel size. During photo analysis, either entire scene or a sample of a scene is employed (complete enumeration). To establish the overall area for sample segments, the estimated areas beneath each component are statistically averaged. When employing complete enumeration, the administrative boundary (district, state, county or province) is overlaid on top of the image, and the crop area is determined by adding the cropped pixels and dividing the result by the pixel size.

A crop map is also available under complete enumeration, and it can be used for a variety of other things, like yield sampling. In this method, supervised or unsupervised classification can be used, depending on how the classes are determined. The most popular supervised classification methods are Parallelepiped, Maximum Likelihood and Minimum-Distance-to-Means; unsupervised classification algorithms include ISODATA and K-Means. Other more recent methods for classifying crops include support vector machines, fuzzy-set classifiers, hierarchical classifiers and artificial neural network. A survey of the techniques for categorising images was presented by Prasad et al. (2015). Satellite data from MODIS, Resourcesat AWiFS are used for categorization. Single-date high-resolution data and multitemporal moderate-resolution satellite data are also used (Landsat OLI, Resourcesat LISS III or Sentinel-2 MSI).

24.4.3 Multidate Data Analysis

The variations in phenological stages of different crops grown in one region serve as the foundation for multidate data analysis. The most common data sources used are Resourcesat-2 AWiFS, SPOT VGT and MODIS (250 m, daily or eight-day products, or 1 km, daily or ten-day products), which have intermediate spatial resolution and high temporal frequency (56 m, five-day products). The majority of the crop-growing season's 7–10 dates of co-registered data serve as the foundation for crop classification.

24.4.4 Single-Date Data Analysis

It is established on the presumption that target may be distinguished from other crop and land use/land cover classes at its peak vegetative development when there is sufficient ground data available. Higher-resolution satellite data are utilised in this situation, such as Sentinel-2 MSI, Landsat 8 OLI, or Resourcesat-2 LISS III data. In order to create crop signatures, ground truth data is used. The pixels in a particular crop are classified using supervised classification techniques (like the ML classifier method). The crop area inside an administrative boundary can be calculated by dividing the pixel area with total number of pixels.

24.4.5 SAR Data

In most South Asian countries, where paddy is cultivated during the rainy season, which is also the primary rice-growing season, makes it difficult to gather optical data. The cloud cover issue can be solved using microwave SAR data. Surface roughness has an impact on SAR sensitivity. Rice is typically transplanted in flooded fields. The newly transplanted rice plants have a very low backscatter value because of the reflection of water from flooded field (Suga and Konishi 2008; Choudhury and Chakraborty 2006). As the plant grows, produces tillers and gets closer to physiological maturity, the radar backscatter increases. Data of different dates with an interval of roughly 20–25 days between each date are frequently desired for rice area estimation using SAR data. The typical procedures for processing SAR data for the rice crop are shown in Fig. 24.2.

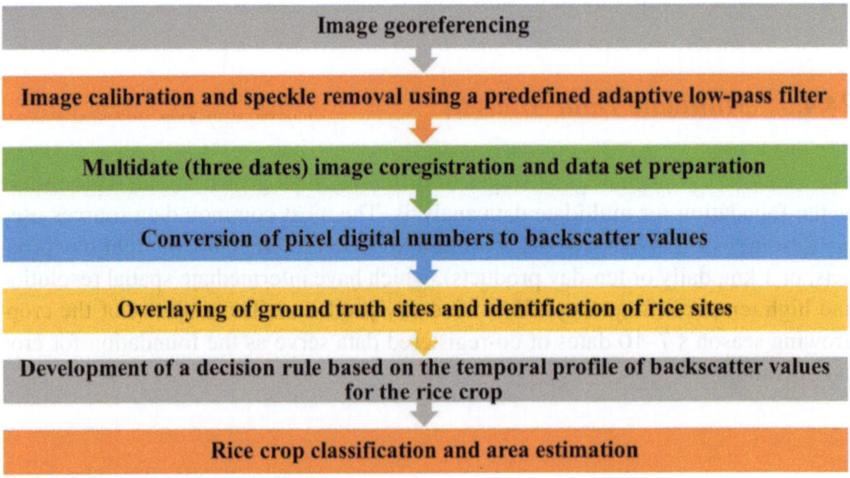

Fig. 24.2 Steps involved in processing SAR data

24.4.6 Ground Truth Data

Ground truth data is a crucial part of crop classification, whether it is used as a building block for a classification method or as validation for classification. Regarding land usage and land cover, the raw data is collected. The typical ground truth data for crop classification consists of two images (close and wide view) of the field, name of the crop, condition, stage, coverage, location, village, district, or state, estimated yield, irrigated or rainfed, sowing and harvesting dates. This type of information is assembled from scheduled sample loci from whole study region. To gather data on the go, many Android smartphone apps have been developed. When data from mobile devices is sent to a central server, analysts working on image analysis can access it instantly. In India, field data is collected by representatives of each state's agriculture department and submitted to the Bhuvan server, an ISRO geoportal.

24.4.7 Regression Estimator

Many countries, most notably the USA, China and Brazil, use regression estimators to calculate crop acreage using remote sensing data. To upsurge the precision of crop data, FAO (2017) suggested a hybrid strategy based on the incorporation of the area frame and photo classification. The Afghan rice acreage was calculated using this method. Images from Sentinel-2 and SPOT-5 satellites were used to build the region frame (having a spatial resolution of 10 m). In order to estimate the crop statistics based on the area frame, the agricultural land inside the pilot project area was stratified, and random parts were visually examined in routine alongside the ground data. The pixel-based crop estimations were obtained using the satellite picture visually inspected as a training sample in the supervised image classification algorithm. The segment variables' rice pixels and rice area's R^2 value in the linear regression was 0.96. This demonstrated that the precision between these two systems was extremely good. Similar methods were used to estimate crop area in China as part of the CAERSS project. An area frame is first built and updated for crop sample design using multisource and multitemporal remote sensing. A sample selection approach is also created in order to carry out a fair stratification and choose samples for a ground truth survey. Finally, a linear model is used to provide a crop acreage estimation with sufficient accuracy. This is accomplished by combining the outcomes of the field investigation with the quantity of cropland that has been identified utilising supplementary information from remote sensing photography. As a result, the estimated planted acreage for major crops is produced at the local and national levels.

24.4.8 Calibration Estimator

By utilising remotely sensed data, calibration estimators offer more information to the estimation process (Benedetti et al. 2015). Stehman (2013) proclaims that by using the reference data from the sample which was utilised to develop the confusion matrix, it is possible to directly or indirectly determine the target's area. The confusion matrix is used by both direct and inverse estimators to change the pixel count area. These two estimators are dissimilar from one another since the former makes use of user accuracy while the later makes use of producer accuracy. The confusion matrix must be calculated by utilising ground data on statistical sample of points or segments, and extrapolation must be done appropriately while taking the sampling technique into consideration (area elements).

To estimate the areas of various land cover classes using simple random sampling without replacement, the researchers utilised the confusion matrix calibration inverse estimator, confusion matrix calibration direct estimator, simple regression and ratio estimator. They revealed that compared to the fundamental random sampling estimator, confusion matrix calibration estimators, ratio and simple regression estimators may produce more accurate and dependable estimates.

24.4.9 Small Area Estimator

When the domain (subpopulation) sample sizes are too small to offer sufficient accuracy for direct domain estimators, little area estimating is crucial in survey analysis. Due to the higher sample size, remote sensing-based estimations may have good accuracy for big regions (countries, states or provinces), but this may not be the case for small regions or areas like districts or counties. The small area estimation (SAE) technique is used to increase estimation accuracy for smaller areas. There are numerous alternate methods that can be used to complete the SAE. In general, they can be divided into statistical and spatial microsimulation approaches, each of which includes a range of distinctive methodologies (Whitworth 2013). The statistical technique is based on regression modelling, which permits a formal evaluation of the relationship between an interest characteristic and the explanatory variable (s)@@. To obtain county- and town-level estimates of rice and maize.

24.4.10 India's FASAL Programme

The ICAR and the ISRO carried out the first comprehensive crop inventory survey using CIR aerial data in India in the late 1970s as part of the ARISE project (Sahai et al. 1977). Following this, frequent research was carried out using data from Landsat-1 and later, data from airborne sources. Bhavsar (1980), Navalgund and Sahai (1985) and Sahai and Dadhwal (1990) documented the pioneering studies based on aerial and Landsat data. A substantial national-level programme known as CAGREP was established after launching IRS 1A, enabling the production and area estimation of important crops at the district and state levels. An optimum sample approach, sensor specifications, accuracy metrics, an ideal analysis technique and internal crop evaluation software were all developed with the help of research studies carried out as part of the CAPE programme (Dadhwal et al. 2002).

To solve the issue of cloud coverage during *kharif* (rainy season), strategies were devised for using SAR data (first from ENVISAT, later from Radarsat) (Patel et al. 1995; Panigrahy et al. 2000). An extensive crop inventory project, the FASAL programme, was established using the knowledge collected via the CAPE project and numerous other pilot studies that were performed, as well as the requirements of the Indian Ministry of Agriculture.

The FASAL project, which was formally established in 2007, aims to provide abundant yield estimates before harvesting of crop at district, state and national levels for the country's 11 key crops using a variety of approaches (econometric, remote sensing and agrometeorological) (Parihar and Oza 2006). Following the creation and advancement of the agricultural output forecasting methodology by the ISRO's Space Applications Centre, the project was functional at MNCFC, i.e. under the Ministry of Agriculture (Ray et al. 2015).

In order to create the sample plan and estimate the size of area, the FASAL approach for area estimation uses remote sensing data. This data might be single-date high-resolution data or multidate medium-resolution data. It employed RISAT-1 SAR data for rice and jute, and Resourcesat-2 AWiFS, LISS III, Landsat OLI or Sentinel-2 MSI data for other crops (rapeseed, wheat, cotton, mustard, sorghum and sugarcane). The rice and jute crops are analysed using a sample segment technique, with a sample size of 5×5 km, a sampling percentage of 15% and four strata (on the basis of crop cover). The comprehensive enumeration method is used for other crops. With smartphone Android apps, the real-world data is captured.

Models based on samples from crop cutting experiments (CCE), crop simulation models and remote sensing data are utilised to forecast the yield. The CCE uses a sampling strategy based on remote sensing. The estimations produced under the FASAL programme are used as an important input to formulate the ultimate estimates for the Ministry of Agriculture and Farmers Welfare. In the light of the success of crop production using remote sensing, the Ministry of Agriculture and Farmers Welfare has launched a new programme called Coordinated Horticulture Assessment and Management using Geoinformatics (CHAMAN) for the inventorying of horticultural crops (banana, potato, chilli, tomato, mango, citrus and onion) and horticultural development planning (Ray et al. 2015).

24.4.11 Hyperspectral Remote Sensing Data Analysis for Agricultural Crops

Hyperspectral data are far larger and more complicated than multispectral data, making data processing and interpretation much more difficult (Navalgund and Ray 2011). They are acquired over a large number of tiny bands in continuous spectral coverage. To manage high-dimensional datasets, it necessitates the development of new methodologies and procedures. The most frequent problem is information redundancy caused by inter-band correlation, which calls for knowledge of application-specific "optimal bands" that can effectively collect the majority of crop trait data (Thenkabail 2002; Thenkabail et al. 2004).

It has also been demonstrated that four principal components calculated from 76 bands can explain 96% of the data's variability. In addition, hyperspectral spectra are typically noisier than those obtained under controlled laboratory conditions. This is due to the fact that the sensors' limited bandwidth can only effectively record a small amount of energy before being overwhelmed by internal noise. The incoming signal is also significantly diminished by the sun's fluctuating light. Prior to further analysis, it is necessary to smooth the reflectance spectra obtained from remotely sensed images or from spectroradiometer data taken in the field. The best bandwidth choice must be made while maintaining absorption characteristics, local minima or maxima and inflection points. The signal-to-noise ratio may be decreased for very narrow bandwidth (Ray et al. 2006).

With hundreds of bands, hyperspectral pictures (such as those produced by Hyperion, CASI and Headwall Micro-Hyperspec) may record more precise spectral responses, making them better equipped to identify minute differences in ground coverings and changes in those variations over time. In order to overcome the aforementioned difficulties and enable more precise and fast identification of crop physiological state, hyperspectral photography can be employed (Lucieer et al. 2014; Gonzalez et al. 2015). The best data on a particular feature will be acquired if the bandwidth is kept between 5 and 10 nm as opposed to the average conditions recorded in broadbands. Ray et al. (2006) suggested optimal bandwidths of 5–10 nm in the red edge and 25 nm in the 500–700 and 800–900 nm sectors in the early near-infrared (NIR) area for agricultural stress experiments. The results of numerous researchers' investigations indicate that the nominal bandwidth for all wavebands is between 5 and 10 nm.

Although hyperspectral data's high spectral resolution allows for the recording and identification of even the smallest differences among crop kinds, it also contains redundant information at the band level, which complicates processing. The most popular technique for reducing the number of wavebands is band/feature selection, which selects the best discriminating bands using discriminant statistics like principal component analysis (PCA) and stepwise discriminant analysis (SDA). These techniques have been effectively applied by many researchers (Thenkabail et al. 2004;

Cochrane 2000; Schmidt and Skidmore 2003; Vaiphasa et al. 2005) to choose informative bands in hyperspectral data and differentiate between different plant types or species.

Hyperion hyperspectral satellite data of EO1 were evaluated by Miglani et al. (2008). The crop spectra made it easy to see that the leaf water absorption occurred around 940 and 1104 nm, that there was a noticeable slope in the red edge region (700–750 nm), and that there was evidence of chlorophyll absorption close to 690 nm. The reflectance patterns of many crops grown during the *rabi* season varied, as seen by the spectral curves of potatoes, wheat, sugarcane, mustard, sorghum and other crops. This variance was dependent on the agricultural characteristics and stage of the crop. The NIR (780–870 nm) and SWIR measurements of the wheat stages revealed substantial variations (1000–1080 nm). Wheat's crop radiance ratings were lowest at maturity and highest at the period just before grain filling. This demonstrated that Hyperion may be able to categorise crop phenological groups either by examining the entire 450–2350 nm spectral range or by using a small number of selective, carefully chosen bands.

24.5 Artificial Intelligence Approaches in Crop Area and Yield Estimation

Artificial intelligence (AI) is the study of how the human brain thinks, learns, makes decisions and works to solve problems. AI is used to develop intelligent software and systems. These computer programmes acquire training data, and just like the human brain, they are intelligent computers that respond to each valid input by providing the desired outcome. Two of the many fields that make up the umbrella of AI are machine learning (ML) and deep learning (DL). AI is study of creating intelligent computers and programmes, whereas DL and ML are the learning of deep neural networks and the capacity to learn anything without being explicitly programmed, respectively (Kulkarni and Deshmukh 2013; Kodali and Sahu 2016). The adoption of artificial intelligence (AI) in agriculture can be facilitated by a number of technological advancements, including robotics, the Internet of things, big data analytics, drone technology, the availability of inexpensive cameras, sensors and wide Internet coverage on geographically dispersed areas. Artificial intelligence (AI) systems can forecast which crop should be selected for particular year and optimum time of sowing and harvesting in a particular place by evaluating soil management data like temperature, weather, soil analysis, moisture and prior crop performance. While utilising fewer water, fertiliser and pesticides, this increases crop production. It is possible to lessen the influence on natural ecosystems and increase worker safety by utilising AI technology. This will guarantee that food production will keep up with the ever-increasing population and keep food prices reasonable. There are several ways to predict crop productivity using AI, as shown in Fig. 24.3. These strategies are briefly outlined.

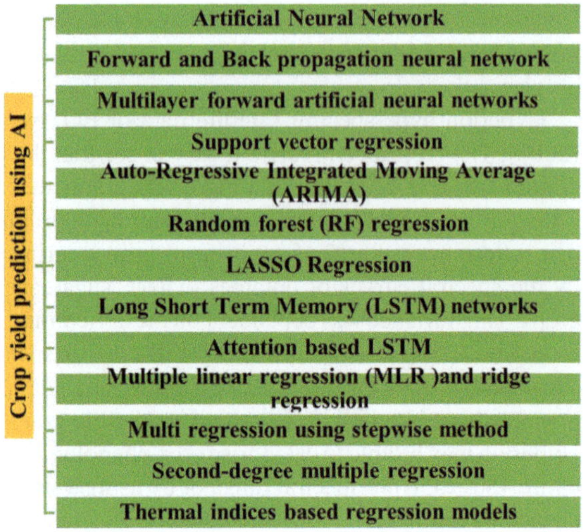

Fig. 24.3 Different artificial intelligence-based yield forecasting approaches

24.5.1 Artificial Neural Network (ANN)

Artificial neural network (ANN) models are primarily made to imitate the capacity for learning and error tolerance of biological brain systems (Stat Soft 2004). The network model has the capacity to remember the knowledge it has learned and to draw on prior experiences (Chang 1993). The accuracy of ANN significantly improves with an increase in the training dataset. Any process can use ANN to establish the relationship between input and output (Mishra et al. 2016). The advantages of using neural networks in agro-ecological modelling were also concisely outlined, including the ANN's capacity to handle both quantitative and qualitative data, merge information and incorporate both linear and nonlinear responses (Schultz et al. 2000). The ANN model has a wide range of uses. Crop yield prediction is one of the major uses of ANNs. ANNs are extremely valuable and desirable for a forecasting assignment because of a number of distinctive properties that set them apart. ANNs are data-based methods that make less prior supposition about models that are appropriate for job at hand than standard based-model approaches. Second, after learning the data (sample), ANNs can generalise. This means that, even if the sample data contains noisy information, ANNs can frequently correct data from the unseen portion of a population. Thirdly, ANNs are universally applicable functional approximators. Conventional statistical techniques cannot handle ANNs' more flexible and general functional forms. Computer simulations based on biological neural networks may provide novel ways to address problems that come up in real-world situations. Artificial neural networks (ANNs) can be useful for modelling nonlinear systems with ambiguous functional relationships (Darbellay and Slama 2000).

Several studies have shown that ANN can accurately estimate agricultural yield (Miao et al. 2006; Baral et al. 2011; Laxmi and Kumar 2011; Dahikar and Rode 2014). A multilayer neural network was employed by Sastny et al. (2011). Liu et al.

(2005) employed ANN to research agricultural yield in response to a soil parameter. The authors of the study analysed data by using a back-propagation neural network. They demonstrated the worth of neural networks as a tool for predicting crop yield. Using artificial neural network (ANN) models, Mistry and Suryanarayana (2014) estimated the cotton production in Vadodara, Gujarat, India. The climatological variables considered to forecast cotton output included maximum and minimum temperatures, relative humidity, wind speed and sunshine hours. They evaluated the model's performance using the coefficient of correlation denoted as r, coefficient of determination (R^2) and discrepancy ratio as performance measures (DR). The outcomes showed that the generalised regression model was the most accurate ANN model, with validation r, R^2 and DR values of 0.93, 0.86 and 0.99. Setiya et al. (2022) used five different techniques, including artificial neural networks (ANNs), Stepwise multiple linear regression (SMLR), elastic net (ELNET), ridge regression and least absolute shrinkage and selection operator (LASSO), in an effort to find the best model for predicting wheat yield for seven different districts in Uttarakhand. For the analysis, 80% of the data were fixed and the remaining dataset was used for validation. The current study discovered that ANN outperformed other models for forecasting crop yields based on the values of RMSE (0.0-5 to 0.474) and nRMSE (0.1-6 to 26.171).

24.5.2 Forward and Back-Propagation Neural Network

ANN is a significant and popular artificial intelligence method. Information is passed through the input layer to the output layer (forward) in a feed-forward neural network (FFNN). The output layer is targeted for additional processing in the back-propagation neural network (BPNN), though. Meena and Singh (2013) suggested a BPNN model for predicting agricultural productivity. Physical elements were used for yield forecasts rather than fuzzy models. When compared, the annual forecast evaluation reports have been decreased from 11.40 to 3.82%. Saad and Ismail (2009) investigated the back-propagation network and the radial basis function (RBF) network as two models for the prediction of rice yield in precision farming. The research observed that in terms of training time, accuracy and hidden layer node count, radial basis network outperformed back-propagation network. Additionally, it was discovered that the multilayer perceptron (MLP) network's training process was frequently too slow, particularly when dealing with complex situations. In general, RBF network was faster than MLP network because it could set its hidden neuron parameters directly from the input data and train the network parameters. Back-propagation neural network structure was used by Uno et al. (2005) to conduct an experiment on relation between maize yield and soil nutrients with the use of ANN. 50 tested soil samples with a 6:9:1 topological structure were employed in the back-propagation neural network, and a model was developed to validate the remaining 13 samples. According to the findings, the crop yield is linearly correlated with soil water content and alkali-hydrolyzable nitrogen, multi-nominally correlated with organic matter, total nitrogen and rapidly available potassium and exponentially correlated with rapidly available phosphorous.

24.5.3 Multilayer Forward Artificial Neural Networks

A study on the prediction of tangerine yield was conducted by Boonprasom and Bumroongitt in 2002 utilising multilayer forward artificial neural networks (MLFANNs) and the back-propagation (BP) learning technique of ANN. Data on yield and weather variables were collected over a nine-year period in this study, which included weather variables as influencing factors. Despite the limited amount of accessible data, the study showed that the ANN had a high potential and capacity to anticipate tangerine yield properly. According to the reports, average temperature had less impact on tangerine production than rainfall which had a stronger impact.

24.5.4 Support Vector Regression

A non-probabilistic algorithm, support vector machine (SVM) is technically described by separating hyper-plane. It comes in both linear and nonlinear SVM varieties. When the data can be split into linear groups, linear SVM is utilised. In essence, it divides the data into two categories. When there are just two output classes, such as a healthy and diseased leaf, linear SVM uses a hyper-plane, which is a straight line. Nonlinear SVM is used for data that cannot be classified using a straight line but can be transformed into high-dimensional, linearly separable data. Data having more than two classes are classified using multi-class SVM (Prakash et al. 2012). Using SVM, Kumar et al. (2019) predicted the paddy yield in India. The dataset covers the yield of rice in India from 1950 to 2014. They developed SVM and demonstrated its efficiency in forecasting the production of the rice crop. SVM classification models were evaluated using threefold, fourfold and fivefold cross-validation techniques, as well as a one-versus-one multiple classification strategy. The best forecast accuracy for the four-year relative mean growth was found to be 75.06% after applying the fourfold cross-validation process.

24.5.5 Auto-regressive Integrated Moving Average (ARIMA)

The ARIMA methodology was first made popular by Box and Jenkins (1976), and Box–Jenkins models are now regularly used to refer to ARIMA models. These models use time-series data to predict the output. Kandala (2003) used weekly weather data on a variety of variables, including temperature, relative humidity and sunshine hours to forecast wheat productivity in the Junagadh District, neural networks are an effective tool for solving problems that call for knowledge that is hard to define but for which there are sufficient data points. He also compared an ARIMA and RBF neural network-based synergistic technique with ARIMA and feed-forward networks, which had an R^2 of 0.77 and an RMSE of 0.50. Five models for predicting wheat yield were created by Paul et al. (2013) utilising an ARIMA model with an exogenous variable (ARIMAX). The weekly maximum temperature during the tillering, anthesis, milk and dough stages as well as evapotranspiration at the crown root initiation (CRI) stage were used to create the models. ARIMAX models were used to anticipate wheat production in the Kanpur District of Uttar Pradesh. The fitted models were compared from the perspective of the relative mean absolute prediction error (RMAPE). During different phases of wheat crop starting from initial to reproductive stage, it was shown that the ARIMAX approach is capable of providing pre-harvest projections based on meteorological factors (126 days after sowing). By utilising relevant statistical models including regression and time-series models, Dharmaraja et al. (2020) did an empirical analysis for agricultural yield estimating with the goal of concentrating on predicting the production of pearl millet crop. Models like ARIMA and the ARIMAX were utilised for estimation and ARIMAX model generated the best outcome.

24.5.6 Random Forest (RF) Regression

As a supervised learning model, the RF algorithm employs several decision trees that share the same nodes. To reach a conclusion that reflects the average of all these decision trees, it produces numerous decision trees and aggregates output of numerous further decision trees. The decision tree algorithm integrates the Iterative Dichotomiser (ID3), Classification and Regression Tree (CART) and other well-known techniques. The performance of an algorithm can be evaluated using the mean square error (MSE) measure (Anbananthen et al. 2021). To predict the agricultural yield, Almahdi (2020) used a random forest classifier. They developed a graphical web-based interface to forecast crop yields prior to planting in Maharashtra.

24.5.7 LASSO Regression

A type of linear regression that benefits from shrinkage is called least absolute shrinkage and selection operator (LASSO) regression. In order to improve accuracy, it makes use of both variable selection and regularisation. Simple, sparse models are encouraged by the LASSO model. This specific type of regression is appropriate when one needs to automate some model selection processes, such as parameter removal and variable selection, or when a model exhibits significant multicollinearity (Anbananthen et al. 2021). A statistical model was utilised by Singh et al. (2019) to forecast crop yields for wheat crops at various growth stages. During the crop-growing season, this model took into account the highest and lowest temperatures, precipitation and morning and evening relative humidity. The forecasting model was built using regressors made from created weather indicators. Stepwise and LASSO regression both were utilised to identify the most important climatic factors affecting agricultural productivity. LASSO outperformed stepwise regression in terms of results. Stepwise regression and LASSO have R^2 values of 0.84 and 0.85, respectively. In 2020, Saeed et al. introduced a convolutional neural network-recurrent neural network (CNN-RNN) framework for crop yield prediction. The performance of CNN-RNN was compared to that of RF, DFNN and LASSO algorithms, as well as other models for forecasting maize and soybean yields. The US Corn Belt was forecasted for three years, i.e. 2016–2018. The accuracy for maize and soybean was 87.82% and 87.09%, respectively, based on three categories with soil, weather and management as the parameters.

24.5.8 Long Short-Term Memory (LSTM) Networks

Long short-term memory (LSTM) networks are utilised for time-series modelling, and they can find long-term temporal correlations in complex multivariate sequences (Malhotra et al. 2015). LSTMs have achieved cutting-edge results in a variety of areas, including natural language processing (Sutskever et al. 2014), engineering systems (Gangopadhyay et al. 2020) and offline handwriting recognition. LSTMs have also been effective in multivariate time-series prediction problems (Gangopadhyay et al. 2018; Shook et al. 2018). LSTM-based models have been used to estimate maize yield (Jiang et al. 2019); however, these models don't have interpretability based on spatial data alone and temporal resolution in the absence of daily meteorological data.

24.5.9 Attention Based LSTM

During previous years, number of strategies have been used to address the interpretability issue (Zhang et al. 2018; Wu et al. 2017). One such method is the employment of an attention mechanism, where the objective is to localise significant volumes of incoming data using attention weights. The attention-based approach was initially developed (Bahdanau et al. 2014) to outperform the encoder-decoder paradigm for neural machine translation (Cho et al. 2014). Despite the fact that the phrase has been applied frequently in the field of computer science, the notion of attention has lately been profoundly utilised in time-series modelling (Kim et al. 2017; Qin et al. 2017). The attention mechanism in time-series modelling can be leveraged to improve interpretability, which is another advantage that is considered as more apparent to ML community (Gangopadhyay et al. 2020).

24.5.10 Multiple Linear Regression (MLR) and Ridge Regression

By applying a single prediction approach throughout the whole sample space, MLR generates global models. MLR divides the sample space into smaller regions that are each represented by a simple model, repeatedly (Quinlan 1992). When models have several parameters, ridge regression has a specific use. The multi-collinearity issue in linear regression is diminished by RIDGE. When linear regression models contain strongly correlated independent variables, ridge regression offers a ridge regression estimate that helps in solving the problem (Wang et al. 2011). The crop output and input climatic variables influencing agricultural productivity were modelled by Rahman et al. in 2005. Multiple regression and ridge regression techniques were used to analyse the model. For the sensitivity analysis, ridge regression simulation was performed. The findings demonstrated how crop growth phases impacted yields. The crop's yield and degree of production were significantly influenced by the climate. According to this study, global warming may also have an impact on the distribution and species makeup of grasslands along with impact on production of food grains.

24.5.11 Multiple Regression Using Stepwise Method

Stepwise multiple regression analysis involves methodology to add independent variables stepwise. It is very common technique to forecast crop yield. A significant correlation was recorded (Ghodrati et al. 2013) between seed yield and 100 seed weight (*r*: 0.59), pod number per plant (*r*: 0.50), first pod (*r*: 0.56), days to maturity (*r*: − 0.51; days to flowering initiation; *r*: − 0.47); total dry matter (*r*: 0.82; harvest index; *r*: 0.64; *r*: − 0.60; *r*: 0.60; and days to seed filling). Multiple regression studies using a stepwise methodology showed that factors such as total dry matter, harvest index, nod plant^{-1} count and days to maturity ($R^2 = 0.88$) had a significant effect on soybean seed output. Total dry matter (0.72), followed by harvest index, had the largest direct beneficial impact on seed yield, according to route coefficient analysis (0.59). Days to maturity and the number of nods per main stem had direct effects on seed yield of 0.25 and 0.31, respectively.

24.5.12 Integration of Remote Sensing and Artificial Intelligence

Big data, which can be categorised as an enormous amount of information provided by remote sensing, is useful for crop modelling and decision-making. Big data is described as "huge amount of data with a wide variety that may be obtained, analysed and used for decision-making" by Wolfert et al. (2017). Machine learning has evolved into a viable tool for finding patterns and rules in datasets so that this data can be used more effectively. (Zhang 2006), as well as addressing nonlinear issues on their own. Many studies have shown how well machine learning algorithms anticipate agricultural production for different kinds of crops, enabling policymakers and farmers to put successful marketing and storage plans into place. The possibility of ANNs for yield mapping and formulating the forecasting systems was scrutinised by Uno et al. in 2005. They took hyperspectral pictures of maize plots in eastern Canada with different weed control and fertilisation rates using a tiny, aerial spectrum imager. To develop models for predicting yield, statistical and ANN methods were combined with different vegetation indices. They verified that ANNs were very helpful in forecasting corn production and producing yield maps for precision farming techniques. The artificial neural network (ANN) method was used by using research data from a previous experiment. This data was then utilised to train and test the network. They build network by using the back-propagation technique. The prediction accuracy of the neural network model outperformed the statistics model by a wide margin. Using neural network approaches, Panda et al. (2010) investigated the efficacy of major spectral vegetation signals for predicting crop productivity in agriculture. At the Oakes Irrigation Test Area research site in North Dakota, USA, a study of the yields of irrigated maize crops was conducted. BPNN models were created using four widely used spectral indices: normalised difference vegetation index (NDVI) based on red and NIR, green and NIR-based green vegetation index (GVI), soil adjusted vegetation index (SAVI) based on red and NIR, and red and NIR. The best accurate maize yield predictions came from BPNN models which were based on means and standard deviations of PVI grid images. The most precise model, which had a standard error of prediction of 0.05 t/ha and an average testing accuracy of 93% ($R^2 = 0.72$), was created using the pool PVI converted values. Pantazi et al. (2016) combined online multilayer soil data with crop growth parameters from satellite imagery using neural networks to estimate wheat yield. They compared the effectiveness of XY-fused networks (XY-Fs), supervised Kohonen networks (SKNs) and counter-propagation artificial neural networks for estimating wheat yield on a 22-ha field in Bedfordshire, UK, for a single cropping season (CP-ANNs). Given the complex interactions between limiting factors and yield, the results demonstrated that cross-validation-based yield prediction using the SKN model recognised 91% variability, which may be deemed to be quite accurate. The study found that the SKN model performed the best overall, with an average overall accuracy of 81.65% for SKN, 78.3% for CP-ANN and 80.92% for XY-F. In order to assess and anticipate maize and soybean yields on a county-by-county basis in the corn belt area of the Midwestern and Great Plains regions of the USA, Li et al. (2007) developed a new methodology employing an artificial neural network (ANN). The models were developed using historical yield data as well as long time-series NDVI data from AVHRR and MODIS. The ANN model was trained in a novel manner by utilising the SCE-UA optimisation technique. The effectiveness of ANN and MLR models was evaluated, and the models' stability and predictability were confirmed. The ANN models were successfully trained using the novel techniques, and prediction accuracy was 85%. Performance analysis included three statistical variables: *r*, RMSE and mean difference (AVDIF).

24.6 Future Scope

Farmers can manage their crops more effectively if they have reliable estimates regarding crop yield. They can maximise the use of irrigation, fertilisers and pesticides to increase agricultural yields while reducing their negative effects on the environment. By facilitating improved agricultural planning, distribution and storage, the capacity to accurately anticipate crop yields can contribute to an increase in food security. This is crucial in areas where food insecurity is a serious problem.

24.7 Conclusion

Crop yield prediction is crucial for decision-making at global, regional and local levels as well as in the field. In the modern era, crop status can be intelligently considered using remote sensing and artificial intelligence approaches to further anticipate production based on crop photos and meteorological data. From the review, it can be inferred that using remote sensing and AI techniques, relatively adaptable and extendable weather-based yield prediction models may be created for many crops. By estimating how much food would be available for the expanding global population, early crop yield prediction plays a crucial role in reducing hunger. Crop yield prediction may benefit from machine learning, which enables the model to automatically collect attributes and learn from the datasets. Hence, using AI and remote sensing techniques separately or together is the best option for forecasting agricultural productivity in future.

References

Almahdi H (2020) Machine learning nano-degree capstone project data-science-exercise

Anbananthen KSM, Subbiah S, Chelliah D, Sivakumar P, Somasundaram V, Velshankar KH, Khan MKAA (2021) An intelligent decision support system for crop yield prediction using hybrid machine learning algorithms. F1000 Res 10:1–18. https://doi.org/10.12688/f1000research.73009.1

Bahdanau D, Cho K, Bengio Y (2014) Neural machine translation by jointly learning to align and translate. arXiv preprint arXiv: 14090473

Baral S, Tripathy AK, Bijayasingh P (2011) Yield prediction using artificial neural networks. In: Das VV, Stephen J, Chaba Y (eds) Computer networks and information technologies, vol 142. Springer, Berlin, pp 315–317

Benedetti R, Piersimoni F, Postiglione P (2015) Sampling spatial units for agricultural surveys. Springer, Berlin
Bhavsar PD (1980) Demonstrated applications in India of earth resources survey by remote sensing. Proc Natl Acad Sci India 46(3):275–285
Boonprasom P, Bumroongitt G (2002) Prediction of tangerine yield using artificial neural network. Chiang Mai University, Chiang Mai, p 50200
Box GEP, Jenkins GM (1976) Time series analysis: forecasting and control. Holden-Day, San Francisco
Breiman L (2001) Statistical modeling: the two cultures (with discussion). Statist Sci 16:199–231
Bryan ER (1974) The ERTS-1 investigation (ER-600). Vol I: ERTS-l agricultural analysis. NASA Technical Memorandum X-58117
Carfagna E (2013) Evaluating the cost-efficiency of remote sensing in developing countries. Presentation prepared for the first Scientific Advisory Committee of the Global Strategy—improving AG-statistics, 18–19 July 2013, Rome, FAO
Chang KC (1993) A neural network approach to geographical analysis of population pattern change. Ph.D. dissertation, University of Minnesota, Minneapolis, MN, USA
Cho K, Van Merrienboer B, Gulcehre C, Bahdanau D, Bougares F, Schwenk H (2014) Learning phrase representations using RNN encoder-decoder for statistical machine translation. arXiv preprint arXiv: 14061078
Choudhury I, Chakraborty M (2006) SAR signature investigation of rice crop using RADARSAT data. Int J Remote Sens 27(3):519–534
Cochrane MA (2000) Using vegetation reflectance variability for species level classification of hyperspectral data. Int J Remote Sens 21(10):2075–2087
Craig M, Atkinson D (2013) A literature review of crop area estimation. FAO Publication, Rome
Dadhwal VK, Singh RP, Dutta S, Parihar JS (2002) Remote sensing based crop inventory: a review of Indian experience. Trop Ecol 43(1):107–122
Dahikar SS, Rode SV (2014) Agricultural crop yield prediction using artificial neural network approach. Int J Innov Res Electr Electron Inst Control Eng 2(1):683–686
Darbellay GA, Slama M (2000) Forecasting the short-term demand for electricity: do neural networks stand a better chance? Int J Forecast 16:71–83
Delince J (2015) Technical report on cost-effectiveness of remote sensing for agricultural statistics in developing and emerging economies. Technical report series global strategy. Technical report, Rome
Dharmaraja S, Jain V, Anjoy P (2020) Empirical analysis for crop yield forecasting in India. Agric Res 9:132–138
Dixon BL, Hollinger SE, Garcia P, Tirupattur V (1994) Estimating corn yield response models to predict impacts of climate change. J Agric Resour Econ 19(1):58–68
Drummond ST, Sudduth KA, Joshi A, Birrel SJ, Kitchen NR (2003) Statistical and neural methods for site-specific yield prediction. Trans ASABE 46(1):5–14
FAO (2017) Afghanistan: Monitoring of rice crop using satellite remote sensing and GIS technologies. FAO Publication, Rome
Fortin JG, Anctil F, Parent L, Bolinder MA (2011) Site-specific early season potato yield forecast by neural network in Eastern Canada. Precision Agric 12(6):905–923
Frausto-Solis J, Gonzalez-Sanchez A, Larre M (2009) A new method for optimal cropping pattern. In: Proceedings of the 8th Mexican international conference on artificial intelligence, pp 566–577
Gallego FJ, Peedell S, Carfagna E (1999) The use of CORINE land cover to improve area frame survey estimates in Spain. Res Off Stat 2(2):99–122
Gallego FJ, Kussul N, Skakunb S, Kravchenko O, Shelestov A, Kussul O (2014) Efficiency assessment of using satellite data for crop area estimation in Ukraine. Int J Appl Earth Obs Geoinf 29:22–30
Gangopadhyay T, Tan SY, Huang G, Sarkar S (2018) Temporal attention and stacked LSTMs for multivariate time series prediction. NeurIPS

Gangopadhyay T, Tan SY, Jiang Z, Meng R, Sarkar S (2020) Spatiotemporal attention for multivariate time series prediction and interpretation. arXiv preprint arXiv: 200804882
Ghodrati R, Sekhavatm S, Mahmood H, Gholam A (2013) Evaluation of correlations and path analysis of components seed yield in soybean. Int J Agric Res Rev 3(4):795–800
Gonzalez-Dugo V, Hernandez P, Solis I, Zarco-Tejada P (2015) Using High-resolution hyperspectral and thermal airborne imagery to assess physiological condition in the context of wheat phenotyping. Remote Sens 7:13586–13605
Goudriaan J, VanLaar H (1994) Modelling potential crop growth processes. Kluwer Acad Publishers, Dordrecht, The Netherlands
Holmes QA, Horvath R, Cicone RC, Kauth RJ, Malila WA (1979) Development of landsat-based technology for crop inventories. NASA technical report SR-E9-00404. NASA Publication
Irmak A, Jones JW, Batchelor WD, Irmak S, Boote KJ, Paz JO (2006) Artificial neural network model as a data analysis tool in precision farming. Trans ASABE 49(6):2027–2037
Jamieson PD, Porter JR, Goudriaan J, Ritchie JT, Van Keulen H, Stol W (1998a) A comparison of the models AFRCWHEAT2, CERES-Wheat, Sirius, SUCROS2 and SWHEAT with measurements from wheat grown under drought. Field Crops Res 55:23–44
Jamieson PD, Semenov MA, Brooking IR, Francis GS (1998b) Sirius: a mechanistic model of wheat response to environmental variation. Eur J Agron 8:161–179
Jiang H, Hu H, Zhong R, Xu J, Xu J, Huang J (2019) A deep learning approach to conflating heterogeneous geospatial data for corn yield estimation: a case study of the US Corn Belt at the county level. Glob Change Biol
Jones CA, Kiniry JR (1986) CERES-maize: a simulation model of maize growth and development. Texas A&M Univ Press, College Station, Texas, USA, p 194
Kandala V (2003) Study of statistical modeling techniques in agriculture
Kim J, Canny J (2017) Interpretable learning for self-driving cars by visualizing causal attention. In: Proceedings of the IEEE international conference on computer vision, pp 2942–2950
Kodali RK, Sahu A (2016) An IoT based soil moisture monitoring on Losant platform. In: 2nd International conference on contemporary computing and informatics. IEEE, pp 764–768
Kulkarni VA, Deshmukh AG (2013) Advanced agriculture robotic weed control system. Int J Adv Res Electr Electron Instrum Eng 2(10):5073–5081
Kumar S, Kumar V, Sharma RK (2019) Rice yield forecasting using support vector machine. Int J Recent Technol Eng 8(4):2588–2593
Laxmi RR, Kumar A (2011) Weather based forecasting for crops yield using neural network approach. Stat Appl 9(1 and 2):55–59
Li A, Liang S, Wang A, Qin J (2007) Estimating crop yield from multi-temporal satellite data using multivariate regression and neural network techniques. Photogramm Eng Remote Sens 73(10):1149–1157
Liu G, Yang X, Li M (2005) An artificial neural network model for crop yield responding to soil parameters, vol 3498. Springer, Berlin, pp 1017–1021
Lucieer A, Malenovsky Z, Veness T, Wallace L (2014) HyperUAS-imaging spectroscopy from a multirotor unmanned aircraft system. J Field Robot 31:571–590
MacDonald RB (1984) A summary of the history of the development of automated remote sensing for agricultural applications. IEEE Trans Geosci Remote Sens 22:473–481
Malhotra P, Vig L, Shroff G, Agarwal P (2015) Long short term memory networks for anomaly detection in time series. In: European symposium on artificial neural networks, computational intelligence and machine learning, Bruges, Belgium, 22–24 Apr 2015. Presses universitaires de Louvain, p 89
Meena M, Singh PK (2013) Crop yield forecasting using neural networks. Swarm, evolutionary, and memetic computing, SEMCCO. Lecture notes in computer science, 80. Springer, Cham, pp 319–331
Miao Y, Mulla DJ, Robert PC (2006) Identifying important factors influencing corn yield and grain quality variability using artificial neural networks. Precision Agric 7:117–135

Miglani A, Ray SS, Pandey R, Parihar JS (2008) Evaluation of EO-1 Hyperion data for agricultural applications. J Indian Soc Remote Sens 36:255–266

Mishra S, Mishra D, Santra GH (2016) Applications of machine learning techniques in agricultural crop production: a review paper. Indian J Sci Technol 9(38):1–14

Mistry P, Suryanarayana TMV (2014) Development of artificial neural network models for estimation of yield of cotton. Int J Sci Res 3(7):290–293

Navalgund RR, Ray SS (2011) Hyperspectral data, analysis techniques and applications. J Indian Soc Remote Sens 156

Navalgund RR, Sahai B (1985) Remote sensing applications in agriculture—Indian experience and plans. In: Proceedings of the fourth Asian agricultural symposium, 28 Feb–1 Mar 1985. Kyushu Tokai University, Kumamoto, Japan, pp 329–343

Nellis MD, Price KP, Rundquist D (2009) Remote sensing of cropland agriculture. Papers in natural resources. Paper 217. University of Nebraska–Lincoln Publication, Lincoln, NE, USA

Panda SS, Ames DP, Panigrahi S (2010) Application of vegetation indices for agricultural crop yield. Remote Sens 2:673–696

Panigrahy S, Chakraborty M, Manjunath KR, Kundu N, Parihar JS (2000) Evaluation of Radarsat ScanSAR synthetic aperture radar data for rice crop inventory and modeling. J Indian Soc Remote Sens 28:59–65

Pantazi XE, Moshou D, Alexandridis T, Whetton RL, Mouazen AM (2016) Wheat yield prediction using machine learning and advanced sensing techniques. Comput Electron Agric 121:57–65

Parihar JS and Oza SR (2006) FASAL: an integrated approach for crop assessment and production forecasting. In: Kuligowski RJ, Parihar JS, Saito G (eds) Proceedings of SPIE. Vol 6411: Agriculture and hydrology applications of remote sensing. SPIE Publication, Goa, India

Patel NK, Medhavv TT, Patnaik C, Hussain A (1995) Multi-temporal ERS-1 SAR data for identification of rice crop. J Indian Soc Remote Sens 23:33–39

Paul RK, Prajneshu, & Ghosh, H. (2013) Statistical modelling for forecasting of wheat yield based on weather variables. Indian J Agric Sci 83(2):180–183

Porter JR (1993) AFRCWHEAT2: a model of the growth and development of wheat incorporating responses to water and nitrogen. Eur J Agron 2:69–82

Prakash JS, Vignesh KA, Ashok C, Adithyan R (2012) Multi class support vector machines classifier for machine vision application. In: International conference on machine vision and image processing (MVIP). IEEE, pp 197–199

Prasad SVS, Satya Savithri T, Murali Krishna IV (2015) Techniques in image classification; a survey. Glob J Res Eng Electr Electron Eng 15(6):17–32

Qin Y, Song D, Chen H, Cheng W, Jiang G, Cottrell G (2017) A dual-stage attention-based recurrent neural network for time series prediction. arXiv preprint arXiv: 170402971

Quinlan JR (1992) Learning with continuous classes. In: Adams A, Sterling L (eds) Proceedings of the AI'92, 5th Australian joint conference on artificial intelligence. World Scientific, Singapore, pp 343–348

Rahman MS, Huq MM, Sumi A, Mostafa MG, Azad MR (2005) Statistical analysis of crop-weather regression model for forecasting production impact of Aus rice in Bangladesh. Int J Stat Sci 4(1):57–77

Ray SS (2000) Remote sensing applications. Indian Exp 251–264

Ray SS, Das G, Singh JP, Panigrahy S (2006) Evaluation of hyperspectral indices for LAI estimation and discrimination of potato crop under different irrigation treatments. Int J Remote Sens 27:5373–5387

Ray SS, Neetu Mamatha S, Gupta S (2015) Use of remote sensing in crop forecasting & assessment of impact of natural disasters: operational approaches in India. In: Srivastava MK (ed) Crop monitoring for improved food security. Proceedings of the expert meeting, Vientiane, Lao People's Democratic Republic, 17 Feb 2014. RAP Publication 2014/28. Joint publication of the Food and Agriculture Organization of the United Nations (FAO) and the Asian Development Bank (ADB), Rome; Manila, pp 111–122

Roel A, Plant RE (2004) Factors underlying yield variability in two California rice fields. Agron J 96:1481–1494

Ruß G (2009) Data mining of agricultural yield data: a comparison of regression models. In: Proceedings of the 9th industrial conference on advances in data mining-applications and theoretical aspects, 20–22 July, Leipzig, Germany

Saad P, Ismail N (2009) Artificial neural network modelling of rice yield prediction in precision farming. Artificial Intelligence and Software Engineering Research Lab, School of Computer & Communication Engineering, Northern University College of Engineering (KUKUM), Jejawi, Perlis

Saeed K, Lizhi W, Archontoulis Sotirios V (2020) A CNN-RNN framework for crop yield prediction. Front Plant Sci 10:1750–1755

Safa B, Khalili A, Teshnehlab M, Liaghat A (2004) Artificial neural networks application to predict wheat yield using climatic data. In: Proceedings of the 20th international conference on IIPS, 10–15 Jan, Iranian Meteorological Organization, pp 1–39

Sahai B, Dadhwal VK (1990) Remote sensing in agriculture. In: Verma JP, Verma A (eds) Technology blending and Agrarian prosperity. Malhotra Publishing House, New Delhi, pp 83–98

Sahai B, Chandrasekhar S, Barde NK, Nag Bhushna SR (1977) Agricultural resources inventory and surveys experiment. In: Rycroft MJ, Stickland AC (eds) COSPAR space research. Pergamon Press, Oxford, UK, pp 3–8

Sastny J, Konecny V, Trenz O (2011) Agricultural data prediction by means of neural network. Agricco N-Czech 57(7):356–361

Schlenker W, Roberts MJ (2006) Estimating the impact of climate change on crop yields: the importance of non–linear temperature effects. Discussion papers 0607-01, Department of Economics, Columbia University

Schmidt KS, Skidmore AK (2003) Spectral discrimination of vegetation types in a coastal wetland. Remote Sens Environ 85(1):92–108

Schultz A, Wickland R, Lutze G (2000) Neural network in agroecological modeling-stylish application or helpful tool? Comput Electron Agric 29:73–97

Setiya P, Satpathi A, Nain AS, Dass B (2022) Comparison of weather-based wheat yield forecasting models for different districts of Uttarakhand using statistical and machine learning techniques. J Agrometeorol 24(3):255–261

Sharples JA (1973) The corn blight watch experiment: economic implications for use of remote sensing for collecting data on major crops. LARS technical report, Paper 121. Purdue University Publication, West Lafayette, IN, USA

Shook JM, Wu L, Gangopadhyay T, Ganapathysubramanian B, Sarkar S, Singh AK (2018) Integrating genotype and weather variables for soybean yield prediction using deep learning. Bio Rxiv 331561

Singh KN, Singh KK, Sudheer K, Sanjeev P, Bishal G (2019) Forecasting crop yield through weather indices through LASSO. Indian J Agric Sci 89(3):540–544

Spitters CJT, Van Keulen H, Van Kraailingen DWG (1988) A simple but universal crop growth simulation model, SUCROS87. In: Rabbinge R, Van Laar H, Ward S (eds) Simulation and systems management in crop protection. PUDOC, Wageningen, pp 87–98

Stat Soft Inc. (2004) Electronic statistics textbook. Tulsa, OK, USA. http://www.statsoft.com/textbook/stathome.html

Stehman SV (2013) Estimating area from an accuracy assessment error matrix. Remote Sens Environ 132:202–211

Sudduth KA, Drummond ST, Birrell SJ, Kitchen NR (1996) Analysis of spatial factors influencing crop yield. In: Robert PC, Rust RH, Larson WE (eds) Proceedings of the 3rd international conference on precision agriculture. ASA-CSSA-SSSA, Madison, WI, USA, pp 129–140

Suga Y, Konishi T (2008) Rice crop monitoring using X-, C- and L-band SAR data. In: Neale CMU, Owe M, d'Urso G (eds) Proceedings of SPIE. Vol 7104: Remote sensing for agriculture, ecosystems, and hydrology X. SPIE Publication, Cardiff, Wales, UK

Sutskever I, Vinyals O, Le QV (2014) Sequence to sequence learning with neural networks. Adv Neural Inf Process Syst 3:104–3112

Thenkabail PS (2002) Optimal hyperspectral narrow bands for discriminating agricultural crops. Remote Sens Rev 20:257–291

Thenkabail PS, Enclona EA, Ashton MS, Van Der Meer B (2004) Accuracy assessments of hyperspectral waveband performance for vegetation analysis applications. Remote Sens Environ 91(3–4):354–376

Uno Y, Prasher SO, Laeroix R, Goel PK, Karimi Y, Viau A (2005) Artificial neural network to predict corn yield from compact airborne spectrographic imager data. Comput Electron Agric 47:149–161

Vaiphasa C, Ongsomwangc S, Vaiphasa T, Skidmore AK (2005) Tropical mangrove species discrimination using hyperspectral data: a laboratory study. Estuar Coast Shelf Sci 65:371–379

Varcoe VJ (1990) A note on the computer simulation of crop growth in agricultural land evaluation. Soil Use Manag 6(3):157–160

Wang Q, Zhang T, Cui J, Wang X, Zhou H, Han J, Gislum R (2011) Path and ridge regression analysis of seed yield and seed yield components of Russian wildrye (*Psathyrostachys juncea* Nevski) under field conditions. PLoS ONE 6(4):1–10. https://doi.org/10.1371/journal.pone.0018245

Whitworth A (2013) Evaluations and improvements in small area estimation methodologies. Discussion paper, National Centre for Research Methods (NCRM), University of Sheffield Publication, Sheffield, UK

Wolfert S, Ge L, Verdouw C, Bogaardt MJ (2017) Big data in smart farming—a review. Agric Syst 153:69–80

Wu T, Li X, Song X, Sun W, Dong L, Li B (2017) Interpretable R-CNN. arXiv preprint arXiv: 171105226

Zhang D (2006) Advances in machine learning applications in software engineering. IGI Global

Zhang Q, Wu YN, Zhu SC (2018) Interpretable convolutional neural networks. In: Proceedings of the IEEE conference on computer vision and pattern recognition, pp 8827–8836

Chapter 25
Harvesting the Future: Navigating the Challenges and Limitations of AI in Agriculture

Pooja Purushotham, Gaurav Rakhonde, Namburi Karunakar Reddy, Anusha M. Nayak, K. M. Pravalika, and G. Sahana

Abstract Artificial intelligence (AI) is among perhaps the most paramount disciplines of computer science and has impacted a wide range of domains, including the field of education medical care, finance, and industry, due to its aptitude for problem-solving that people are unable to solve. Employing AI in agriculture is a remarkably promising alternative for addressing food insecurity and adapting to the requirements of a rising population which examines the Internet of Things (IoT), an innovation with enormous opportunities for the future. Despite all of AI's trump cards in agriculture, there are manifold serious hurdles in its use. Predominantly, AI issues are that these technologies are not affordable to many farmers in underdeveloped countries. These technologies can only be employed on large farms. AI systems' distribution in agriculture may be uneven due to limited internet access and expertise in handling operations. This could result in delayed and uneven adoption, potentially impacting food production beyond natural limits. Factors like insects, soil, and inert materials also influence grain characteristics. Since many physical devices, including the IoT, may be left unattended for extended periods of time, they are initially susceptible to hardware attacks. This chapter provides: (a) tutorial on IoT and AI advancements in smart agriculture; (b) analyzing two technologies' adoption challenges and obstacles; and (c) a thorough discussion of the technological and social future trends that will be present after smart agriculture systems are widely used by farmers around the world.

Keywords Artificial intelligence · Internet of Things · Unarmed vehicle · Smart agriculture · Challenges

---

P. Purushotham (✉) · G. Rakhonde · N. K. Reddy · A. M. Nayak
Department of Plant Pathology, University of Agricultural Sciences, GKVK, Bangalore 560065, India
e-mail: poojapurushotham17@gmail.com

K. M. Pravalika · G. Sahana
Department of Agronomy, College of Agriculture, V C Farm, Mandya, University of Agricultural Sciences, GKVK, Bangalore 560065, India

K. Pandey et al. (eds.), *Artificial Intelligence and Smart Agriculture*,
Advances in Geographical and Environmental Sciences,
https://doi.org/10.1007/978-981-97-0341-8_25

25.1 Introduction

Agricultural production with high productivity and quality is a challenging task. There is an imperative need to practice innovative approaches in order to continue producing nutritious food sustainably at a minimal cost. To confront all challenges in food safety production, novel techniques must be cutting-edge and environmentally friendly (Lioutas and Charatsari 2020). The employing intelligent automation and novel innovations has transformed agriculture, allowing it to experience a revolution that has led to a significant rise in productivity and profitability for the industry (Himesh et al. 2018).

Mechanization and the genetically modified "green revolution" made up the contemporary agricultural revolution's initial and subsequent phases. The third wave, known as precision agriculture, is characterized by the ability to apply the necessary inputs by defining the three variables that determine the matter, the location, and at what point the inputs are needed. Now that larger aggregates of data are more readily available, precision agriculture is being updated with improved knowledge of farm systems using artificial intelligence (Mohr and Kuhl 2021). From Agriculture 1.0 to Agriculture 4.0, a parallel agricultural revolution can also be explained (Fig. 25.1). Agriculture 1.0 started with the use of local tools; Agriculture 2.0 involved the utilization of tractors and fertilizers; Agriculture 3.0 involved systems for the formulation of decisions and evaluation; and Agriculture 4.0 involved intelligent farming or innovative agriculture (Liu et al. 2020). Implementation of cutting-edge technology like the Internet of Things (IoT), artificial intelligence (AI), blockchain, unmanned aerial vehicles (UAVs), nanotechnology, robotics, etc., is what is known as Agriculture 4.0 (Fig. 25.2). Traditional agriculture has transformed into sustainable, intelligent, efficient, environmentally friendly featuring technology-based tools like IoT, AI, blockchain, UAV, nanotechnology, and robotics. "Smart" agriculture is displacing long-established old-world agriculture. The terms "smart farming," "precision farming," and "digital farming" are becoming more common.

The emphasis of "Smart Farming" is on data retrieval and using those analytics to determine a complex system with the goal of increasing produce quality and yield while reducing human labor (Mitra et al. 2022). The United States Department of Agriculture (USDA) stated in October 2016 that employing AI in precision agriculture has increased operational profits and net returns (Schimmelpfennig 2016).

Artificial intelligence is intelligence displayed by computers that is similar to human intelligence. Sensor technology advancements in the 1980s permitted flawless signal extraction, real-time data communication, and physical environment adaptation. The 1990s witnessed additional promising results in this discipline. The advancement of computational technologies was responsible for the evolution of intelligence technology. These intelligent devices have three main functions: (i) signal information extraction; (ii) signal processing; and (iii) instruction execution (Sharma et al. 2022).

AI and machine learning (ML) advancements have shown a great potential in a diversified of fields, including e-commerce and marketing (Soni 2020), human

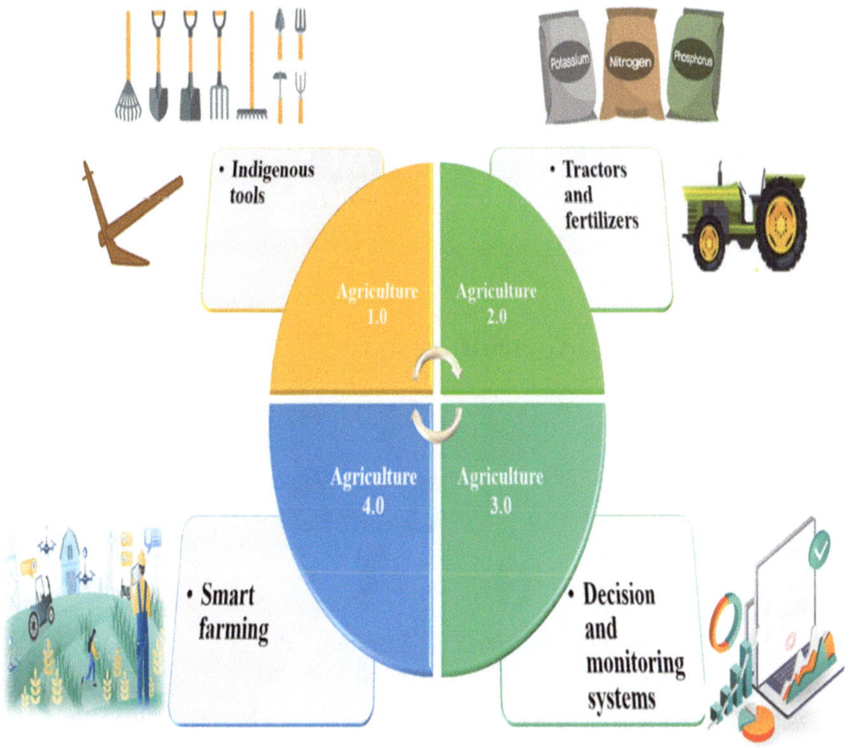

Fig. 25.1 Parallel agricultural revolution

resources (Strohmeier and Piazza 2015), computer vision (Mitra et al. 2021a), multimedia forensics (Mitra et al. 2021b, c), health care (Jiang et al. 2017), social media (Mitra et al. 2020, 2021d), gaming (Palaus et al. 2017; Skinner and Walmsley 2019), vehicles, and agriculture. AI has made substantial strides in disciplines such as machine learning (ML), processing of natural language, and vision of computer, among others. AI-based methods entitle a computer to comprehend and evaluate activities in real time by recognizing the source of information. The amalgamation of these two applied science and technologies presents an intriguing solution in the disciple of intelligent sensing (Sharma et al. 2022). AI is utilized in agriculture to improve efficiency, crop yield, and profitability. Techniques like SVM, ANN, DNN, CNN, and LSTM are used to monitor crop health, weather, supply chain optimization, irrigation systems, pesticide and fertilizer use, weed control, smart sensing, livestock tracking, and geofencing (Liakos et al. 2018) (Fig. 25.3). Recently, the agricultural division has had applications of AI that were noticeable in its line of work. Saving money and labor, improving production and cost reduction with less effort, and yielding high-quality food through environmentally friendly methods have all been demonstrated by technologically advanced farms (Diez 2017). Farmers must have the drive to adopt innovative technology in their areas, as well as comprehending

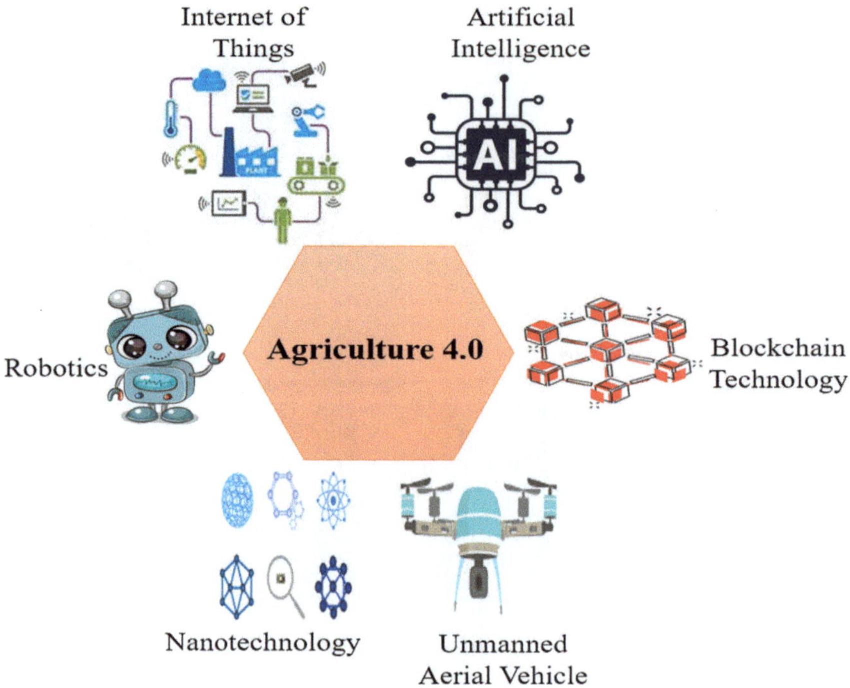

Fig. 25.2 Technologies involved in Agriculture 4.0

the prospective of each particular farm in the context of economic growth, as larger farms have higher profit margins (Schimmelpfennig 2016).

There have been difficulties in achieving the full advantage and boosting crop yields, which included numerous impediments, like paucity of technological knowledge on the part of farmers, pest infestations, and diseases they spread, low output, the prerequisite for extensive information encompassing all the information and ineffective soil treatments. Numerous methods have been put into place within farms to manage and sustain their production in light of the surrounding environment. Regrettably, the operation of such methods has led to a number of trade-offs and uncertainties. According to market research, elements that make acceptance easier of environmentally sound agronomic inventions consist of properly educating and workshops and knowledge exchange of farmers, easier access to financial resources, and an increase in the number of consumers demanding organic food (Saiz-Rubio and Rovira-Mas 2020). When new techniques are used, the task of collecting data from crops should be valuable and relevant, as data is usually just numbers, pictures, or pictorial representations. In the section that follows, a detailed discussion on the drawbacks of AI applications as well as the techniques for using expert systems to produce more outputs has been represented.

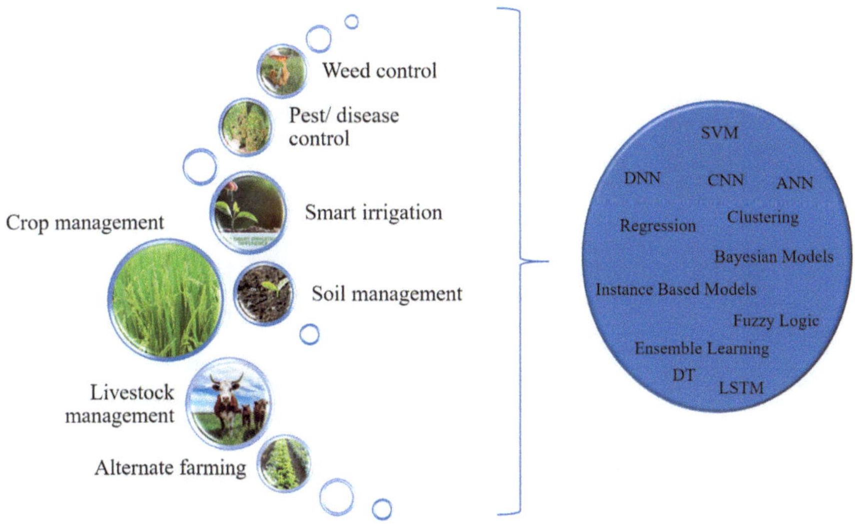

Fig. 25.3 Artificial intelligence approaches used in smart agriculture

25.2 Challenges of Artificial Intelligence in Agriculture

Smart agriculture techniques have standardized and modernized traditional agriculture. Although AI is an obvious progression toward smart agriculture for sustainable, cost-effective, and efficient farming, the following constraints represent significant hurdles to deploying AI in agriculture. In order to adapt technologies, there are still many challenges that require to be resolved. These problems pertain to a multiplicity of issues discussed in the present chapter. Several of the main obstacles to smart agriculture are illustrated in Fig. 25.4.

25.2.1 Disparity in Agricultural Research and AI Innovations

presence of gap between the fields of AI advancement and the agricultural sector. As a result, AI researchers are unaware of farmers' challenges, and farmers are similarly unaware of available AI technology. To overcome this two-pronged dilemma, more interdisciplinary collaboration is required.

25.2.2 Lack of Well-Established Policies and Regulations

AI applications in agriculture are yet in their early development stages, insufficiency of well-established regulations and policies. As a result, many legal questions

Fig. 25.4 Challenges of artificial intelligence in agriculture

about smart farming remain unsolved. To maintain data integrity, standard policies and practices must be followed. Information presentation in machines must go through a number of stages. The majority of the machines can function both as standalone computational units and connected resource-constrained devices. Because the majority of AI-IoT solutions were cloud-based until recently, farmers were unable to use AI techniques due to concerns about cyberattacks, data security, and privacy. A new IoT setting called "Edge AI" has developed in response to this problem. Edge AI handles sensor data locally, providing greater data protection and privacy while reducing latency and cost.

25.2.3 *Lack of Data*

Lack of data is another hurdle for AI in agriculture because it is an endeavor that depends on information; hence, a major obstacle is the lack of suitable information for many people using AI algorithms. Gathering the sensor data and combining it with data from other sources to identify and process in accordance with the results are the first step in intelligent sensing. If data can be examined locally, more actions are required. The majority of devices, however, proceed to send the data to database storage or the cloud as their next step. These types of data are then gathered and processed using user-specific queries for data analysis and presentation. There are several different security risks that must be addressed throughout the data collection and presentation processes. Security must be considered at every stage of the intelligent sensing process, which includes sensing as well as information analysis (Sharma et al. 2022).

25.2.4 Threat of Unemployment

The most serious public issue is the threat of unemployment. In effect, the majority of repetitive labor and activities may be replaced by intelligent machinery and robots; however, there is no significant human participation, which will present serious problems with employment standards. Machines, for example, can only do duties that have been designed or developed to deal with other technological hurdles. Anything that causes them to crash or produces little outcomes could be a background for them.

25.2.5 High Cost

Smart farming will increase the income and productivity of farmers. As an illustration, farming in enclosed spaces or earlier crop disease detection through sophisticated agricultural advising platforms powered by AI may provide precise predictions and security against crop diseases that are airborne and carried by the soil. Similar to this, crops may grow up to three times more quickly with intelligent irrigation techniques like aeroponics (Rahman et al. 2018). But the high cost of developing and maintaining smart machines and smart computers may also be considered a technological restriction of AI technologies, particularly when AI evolves on a daily basis, necessitating the regular updating of hardware and software to meet the most recent requirements. Machines necessitate pricey repairs and maintenance.

25.2.6 Concerns of Power

The majority of technologies in AI and automated large machine systems used in smart agriculture consume a lot of power to operate. Farms frequently have very high-power requirements because they typically have large surface areas and need numerous electronic components. The widespread adoption of these automation processes in big farms has been hampered by this. Some of the solutions call for the machinery to be continuously and uninterruptedly powered by sustainable energy produced by environmentally friendly sources such as hydroelectric power, wind power, and the sun. Many researchers have been interested in this area, and work is still being done for implementation and improvement of such renewable energy sources for smart farming (Ram et al. 2020; Huang et al. 2020). These alternative power sources have some drawbacks, including uneven energy needs across the farm and problems with power transmission and storage. To resolve these problems, an effective microgrid design is needed, and considerable study has been conducted in this field to offer recommendations for a smart microgrid that is efficient and compatible with renewable energy sources (Desai et al. 2021; Ebrahimi et al. 2019).

25.2.7 Consumption of Power

As the Internet of Agricultural Things (IoAT) devices must be powered by alternative power sources for the smart agriculture farm to operate smoothly, dependably, and sustainably, the deployed models must use less power. They ought to be capable of working in a space with limited resources. Wearable flexible sensors are now receiving a prime attention in the disciple of medicine (Son et al. 2014; Xu et al. 2014; Kim et al. 2012). In medical science, in order to assess physiological parameters including temperature, ECG, EMG, muscular activity, and cardiovascular issues, these sensors come into touch with the wearer's clothing. These devices' power usage is a crucial problem that demands attention. In addition, a difficult problem that needs to be solved is the price of manufacturing a versatile sensor (Nag et al. 2017). To lower the wearable flexible sensors' power consumption, low-power sensors like Telos and Shimmer should be used for health monitoring.

25.2.8 Hardware Accessibility

Different sensors and gadgets are needed for smart agriculture in order to sense various system and environmental parameters. The devices act on those signals after collecting the data to produce a better, more predictable yield. A pinch point in this situation is the availability of particular hardware.

25.2.9 Hardware Installation

Having undergone the perks of AI, creating hardware installation strategies demands a lot of computational power and resources and is a difficult task (Sze et al. 2017). Therefore, serious efforts must be made to improve intelligent communication between hardware and AI components. Large memory footprint and sensor data impact hardware training accuracy and processing speed; lack of hardware-specific libraries leads to improper deployment on low-power devices and FPGAs. Researchers aim to improve real-time inference model performance and memory efficiency by reducing hardware complexity in AI and ML algorithms (Talib et al. 2021; Karras et al. 2019).

25.2.10 Hardware Protection

IoT-connected devices are anticipated to total 50 billion by 2020 (Dave 2011). These Internet of Things (IoT) devices must be strong and resilient against different attacks. However, the demand for inexpensive, basic hardware jeopardizes hardware security. The most frequent hardware security risks for IoT devices are hardware Trojan and side-channel attacks, which restrict the widespread utilization of IoT networks' essential applications, but hardware Trojans use adversarial alterations as vulnerabilities to take over systems, launch assaults, and evade detection, and few approaches utilize scanning with an electronic microscope on demetallized chips (Courbon et al. 2015), examining the circuit's power and delays as well as the PUF, the device's distinguishing mark. Another common danger to hardware security is side-channel attacks, which utilize side-channel signals to access confidential information, such as encryption keys. These side-channel information include power assessment, evaluation, and electromagnetic emission (Tang et al. 2018). IoT networks are especially liable to these problems; hence, numerous solutions are put up (Ju et al. 2015; Chakraborty and Bhunia 2009; Bathalapalli et al. 2021).

25.2.11 Networking and Communications

Interaction between machines (M2M) is largely prevalent feature of smart agriculture. In order to share data and cooperate on a project, this uses many network and communication protocols. Most applications use a variety of communication networks, including GPRS, Wi-Fi, LoRA, SigFox, and ZigBee. Due to threats and physical harm, creating and maintaining such vast networks are expensive and impractical in big, open farms. Some solutions for effective communication networks have been put forth and research topics have been investigated (Concepcion et al. 2014; Zhang 2011; Sahitya et al. 2016). Additionally, some studies have combined communication tools with other smart devices, SIL-IoT, a unique agricultural product developed using continuous communication technologies like WSN and solar insecticidal lamps (SILs) (Yang et al. 2020). Applications for smart agriculture have a significant demand for more secure and reliable communication, which calls for additional study and new, reasonably priced technology.

25.2.12 Connection Problems

The lack of reliable high-speed Internet access in many rural locations throughout the world slows the advancement of computing via the cloud and automated farming. The line-of-sight GPS communication can potentially be disrupted by tall trees or hills (Khosla 2010). Remote rural regions with weak cell networks and agriculture as their main sector can be transformed by edge AI. Low-power wide area network (LPWAN) technology transmits compressed data to the fog layer, while convolutional neural networks compress sensor picture data. This strategy strengthens connected agriculture and smart agriculture (Gia et al. 2019).

25.2.13 Privacy and Data Security

Despite the AI and ML models' success, they still have to deal with the extremely difficult problem of data security. By learning patterns that contain information, ML models extract features attacked the real world (Fasano et al. 2015). The quality of datasets and system performance are both impacted by data integrity, which is a legitimate concern in any real-time environment. Hackers might target intelligent, self-driving farming equipment. Self-driving cars are already susceptible to various cyberthreats, such as denial-of-service attacks and their variants (Sybil, Grayhole, Wormhole, and Rushing attacks) (Feng et al. 2020). Jamming/denial-of-service attacks are especially sensitive to IoT sensor node communication, which leverages numerous IEEE wireless communication protocols and is utilized for autonomous cars and, eventually, autonomous farm equipment (Lyamin et al. 2013; Petit and Shladover 2015). For instance, the automated transit system in a smart city is facilitated by UAVs, where malicious entities may leak details about vehicle location and speed (Abbas and khan 2014; Menouar et al. 2017). Only the information needed to complete any operation must be gathered and shared by the sensors. Strong encryption techniques and security measures are required to protect data privacy and security during data transfer. IoAT sensor nodes, however, are not resource-intensive because of their simple design and underlying protocols. With today's technologies, practicing security measures on a gadget with limited resources is challenging. Thus, smart agriculture now has a significant hurdle with regard to data security and privacy. Since agricultural operations are automated, an enemy may control them to wreak havoc on the network. This may result in extremely negative effects on agricultural production quality and yield. These cyber dangers require strong counter-attack solutions at both the core and the periphery in order to protect the interests of smart farmers.

25.2.14 Scalability and Durability

Smaller private farms and larger commercial farms are both common sizes for agricultural farms. They require various numbers sensors to be deployed in the field. Any agriculture technology must be scalable because these sensors generate a wide variety of data. Devices must be dependable in order to minimize the several of redundant devices needed to support fault tolerance. The cost will be much decreased.

25.2.15 Big Data Challenge

The need to manage the vast amount of data, including music, videos, photos, gadget data, and online community data, is urgent.. Poor data management will make it challenging to evaluate the accuracy of the information gathered by sensors and will further pose implications on decision-making phases (Kang et al. 2014). Because there is an enormous amount of data available, we recognize the use of ML and AI approaches to increase the overall performance of the sensor-based system. Hence, to extract meaningful data and avoid overlap, sophisticated AI is needed. Smart agriculture uses cameras or sensor nodes to collect vast amounts of varied data. Since traditional techniques fall short, BD analysis must be used to analyze this enormous amount of data. Big data can be used to explore enormous databases. It offers real-time decision-making and predictive analysis by introducing new business models and discussing issues with food security that boosts the efficiency of the throughout its entirety chain of supply in intelligent farming operations (Poppe et al. 2015; Wolfert et al. 2017). BD platforms for milk chain of supply security have been integrated using support vector machines (SVMs) and ANN (Kempenaar et al. 2016). The very first phase is collecting data via multiple sensor network nodes via multiple nodes in the sensor network, followed by a range of evaluation methods, including traditional and big data evaluation.

25.2.16 Technical Failure

Technology disruption can result from technical issues, such as sensor damage. Devices' poor decisions might cause a significant amount of loss and multi-domain damage. If the sensors in a paddy field are damaged by hail, they will be unable to accurately predict the amount of water in the soil, which can harm crops, have an effect on the food supply chain, and lead to an imbalance in the price of rice.

25.2.17 Lack of First Capital Investment

The initial investment for cutting-edge technologies is not usually available in rural areas of underdeveloped nations where farmers struggle with extremely small profit margins. It can slow down the widespread adoption of smart technologies.

25.2.18 Lack of Standardized Practices

Different nations employ varying technological and unit standards, necessitating specialized solutions. The cost is raised as a result. The issue will be resolved by a global standard (Khosla 2010).

25.3 Practical Applicability of AI in Agriculture

AI revolutionizes agriculture by increasing yields and reducing waste and enhancing the efficiency of farming operations. Here are some practical applications of AI in agriculture:

a. Precision Farming: Al can be used to monitor crop growth and health, soil moisture, and weather patterns to maximize the usage of resources. This can result in increased crop yields and reduced environmental impact.
b. Crop Monitoring and Prediction: Al can be used to analyze data from sensors, drones, and satellite imagery to monitor crop health, detect disease outbreaks, and predict crop yields. This information can be used to make more informed decisions about when to harvest, how much to plant, and which crops to grow.
c. Farming Automation: Al can be used to automate tasks such as planting, harvesting, and irrigation. This can reduce labor costs and improve efficiency, allowing farmers to focus on more strategic tasks.
d. Livestock Monitoring: Al can be used to monitor the health and behavior of livestock, detect disease outbreaks, and optimize feeding schedules. This can improve animal welfare and reduce the risk of disease transmission
e. Supply Chain Optimization: Al can be used to optimize the supply chain, from farm to consumer, by predicting demand, improving logistics, and reducing waste. This can result in more efficient distribution, lower costs, and higher profits for farmers and retailers.

Overall, AI enhances agriculture efficiency, productivity, and sustainability. By using AI-powered tools, farmers can optimize their operations, reduce waste, and produce more food with less resources.

25.4 Future Prospects

AI has the potential to revolutionize agriculture streamlined data-driven decision-making for farmers, optimizing crop yields, reducing resource waste. Through machine learning algorithms, computer vision, and other AI technologies, farmers can analyze vast amounts of data on soil health, weather patterns, pest infestations, and other variables to make informed decisions that improve crop production and overall sustainability. As AI continues to evolve, it will become more accessible and affordable for small-scale farmers, allowing them to compete with larger farms. AI-powered drones and robots are automating tasks like planting, harvesting, and monitoring, reducing labor costs and improving accuracy. .

Moreover, AI can assist with mitigating several of the most gravest problems confronting the agriculture business, including as environmental degradation, security of food, and water shortages. By using AI to analyze weather patterns and predict crop yields, farmers can adapt their practices to changing environmental conditions and avoid crop failures. Additionally, AI-powered precision irrigation systems can help to conserve water and reduce water waste.

AI-based smart agriculture system (SAS) applications will contribute a paradigm shift similar to other applications, in which the edge devices like wireless sensors will develop the intelligence to make independent decisions on their own without the assistance of potent central servers running AI algorithms. Because to recent developments in electronics, embedded systems with more processing and memory capacity, also known as System on Chip (SoC) (Gopalakrishnan et al. 2020; Merchant et al. 2018), are now capable of providing a comprehensive solution without depending on other external entities. Utilizing portable architectures, smart embedded systems powered by artificial intelligence and vision of computer algorithms have been established that helped in building real-time pest detection (Brunelli et al. 2019). Gia et al. (2019) have developed an experimental framework where CNN-based image compression is integrated inside the sensor node because of the low data transmission rates of current WSN networks based on LoRa and NB-IoT. As a result, solutions based on edge computing will replace cloud computing solutions. Additionally, real-time threat detection and prompt threat prevention are features of these solutions. According to Alsamhi et al. (2021), drones equipped with the B5G physical layer wireless technologies could be used to process edge AI for federated learning in smart environments. Due to B5G networks' high data rates with low latency, intensive ML is possible.

As the cloud evolves, farmers shift from proprietary to open-source solutions. It is feasible that farmers have little of the technological skills necessary to create AI-based algorithms from scratch (Zhu et al. 2018). In future edge AI-based solutions will heavily rely on publicly accessible software frameworks like OpenCV libraries for computer vision and other open-source AI-based algorithms. The price of SAS will decrease as a result. Consultancy businesses supply farmers with plug-and-play software and hardware solutions, lowering design time and costs. Farmers might also benefit from open-source software solutions to freely share their measurements with the other farmers, allowing them to develop a solid knowledge.

The majority of farming equipment will be self-driving in the future, similar to smart cars (Qazi et al. 2020). These include computer vision-enabled smart farming tractors and unmanned aerial vehicles (UAVs), which are utilized to control a diverse of tasks such as planting, irrigation, weed control, fertilizer application, herbicide spray, and detection of pest. Farmers that are tech-savvy rely on AI and big data analytics to analyze agricultural conditions, leveraging edge AI apps for smartphone software-based services like fruit quality and insect monitoring.

25.5 Conclusion

This investigation looks at AI-based strategies in smart agriculture systems, providing tips and tricks for IoT and AI developments. It also criticizes their extensive acceptance and anticipates future technical and social developments when SASs are widely used by farmers throughout the world.

The relevance of sustainable farming practises in resolving global food supply and demand imbalances is highlighted in this chapter. Traditional approaches are insufficient, and the present hardware components of smart agriculture systems include IoT nodes and sensors. These sensors remotely monitor agricultural conditions, allowing farmers to take corrective action manually or automatically by transmitting pre-programmed signals to actuators. The paper covers the constraints of smart agricultural systems' hardware, software, and wireless communications. It also addresses the essential components and their possible influence on future innovations, such as unmanned aerial vehicles (UAVs) and drones, in order to assist the paradigm shift of incorporating technology breakthroughs into conventional systems.

The social impact of smart agricultural systems on conventional farming practises is one of the issues. These improvements may render old practises outdated, necessitating the use of technology by farmers in order to stay current. Additional technical obstacles include component standardization, manufacturer compatibility, and cost reduction. Smart agricultural system implementation will need a mix of technological and societal problems. Overall, smart agricultural system development poses major technological and societal hurdles. Future directions will mainly concentrate on enhancing the accessibility of open-source, adaptable solutions for farmers, putting cybersecurity measures in place, and moving more AI-based solutions away from the core and toward edge devices for real-time threat detection and prompt corrective actions.

Abbreviations

AI	Artificial intelligence
ANN	Artificial neural networks
CNN	Convolutional neural networks
DNN	Deep neural networks
GPS	Global positioning system
GPRS	Ground penetrating radar service
IoAT	Internet of agricultural things
IoT	Internet of Things
LPWAN	Low-power wide-area network
LSTM	Long short-term memory
M2M	Machine-to-machine
ML	Machine learning

PUF	Physical unclonable functions
SAS	Smart agriculture system
SoC	System on chip
SIL	Solar insecticidal lamps
SVM	Support vector machine
UAV	Unmanned aerial vehicle
USDA	United States Department of Agriculture
WSN	Wireless sensor network

References

Abbas A, Khan SU (2014) A review on the state-of-the-art privacy-preserving approaches in the e-health clouds. IEEE J Biomed Health Inform 18:1431–1441

Alsamhi SH, Almalki FA, Afgha F, Hawbani A, Shvetsov AV, Lee B, Song H (2021) Drones' edge intelligence over smart environments in b5g: blockchain and federated learning synergy. IEEE Trans Green Commun Netw 6(1). https://doi.org/10.1109/TGCN.2021.3132561

Bathalapalli VK, Mohanty SP, Kougianos E, Yanambaka VP, Baniya BK, Rout B (2021) A PUF-based approach for sustainable cybersecurity in smart agriculture. In: 2021 19th OITS international conference on information technology (OCIT). IEEE, pp 375–380

Brunelli D, Albanese A, d'Acunto D, Nardello M, (2019) Energy neutral machine learning based IoT device for pest detection in precision agriculture. IEEE Internet Things Mag 2(4):10–13. https://doi.org/10.1109/IOTM.0001.1900037

Chakraborty RS, Bhunia S (2009) Security against hardware Trojan through a novel application of design obfuscation. In: 2009 IEEE/ACM international conference on computer-aided design-digest of technical papers. IEEE, pp 113–116

Courbon F, Loubet-Moundi P, Fournier JJ, Tria A (2015) A high efficiency hardware trojan detection technique based on fast sem imaging. In: 2015 design, automation & test in europe conference & exhibition (DATE), pp 788–793

Dave E (2011) The internet of things. How the next evolution of the internet is changing everything. cisco internet business solutions group. http://www.cisco.com/c/dam/en_us/about/ac79/docs/innov/IoT_IBSG_0411FINAL

De La Concepcion AR, Stefanelli R Trinchero D (2014) Adaptive wireless sensor networks for high-definition monitoring in sustainable agriculture. In: 2014 IEEE topical conference on wireless sensors and sensor networks (WiSNet). IEEE, pp 67–69

Desai A, Mukhopadhyay I, Ray A (2021) Techno-economic-environment analysis of solar PV smart microgrid for sustainable rural electrification in agriculture community. In: 2021 IEEE 48th photovoltaic specialists conference (PVSC). IEEE, pp 2281–2285

Diez C (2017) Hacia una agricultura inteligente. Cuaderno De Campo 60:4–11

Ebrahimi M, Kelati A, Nkonoki E, Kondoro A, Rwegasira D, Dhaou IB, Tenhunen H (2019) Creation of CERID: challenge, education, research, innovation, and deployment "in the context of smart MicroGrid". In: 2019 IST-Africa week conference (IST-Africa). IEEE, pp 1–8

Fasano G, Accardo D, Tirri AE, Moccia A, Lellis ED (2015) Radar/electro-optical data fusion for non-cooperative UAS sense and avoid. Aerosp Sci Technol 46:436–450

Feng X, Zhan Y, Wang Q, Yang X, Yu C, Wang H, He Y (2020) Hyperspectral imaging combined with machine learning as a tool to obtain high-throughput plant salt-stress phenotyping. Plant J 101(6):1448–1461. https://doi.org/10.1111/tpj.14597

Gia TN, Qingqing L, Queralta JP, Zou Z, Tenhunen H, Westerlund T (2019) Edge AI in smart farming IoT: CNNs at the edge and fog computing with LoRa. In: 2019 IEEE AFRICON, pp 1–6. https://doi.org/10.1109/AFRICON46755.2019.9134049

Gopalakrishnan N, Pradeep K, Raman CJ, Ali LJ, Gopinath MP (2020) A systematic review on image processing and machine learning techniques for detecting plant diseases. In: 2020 international conference on communication and signal processing (ICCSP), 0465-0468

Himesh S, Prakasa Rao EVS, Gouda KC, Ramesh KV, Rakesh V, Mohapatra GN, Ajilesh P (2018) Digital revolution and big data: a new revolution in agriculture. CAB Rev Perspect Agric Veter Sci Nutr Nat Resour 13(21):1–7

Huang K, Shu L, Li K, Yang F, Han G, Wang X, Pearson S (2020) Photovoltaic agricultural internet of things towards realizing the next generation of smart farming. IEEE Access 8:76300–76312

Jiang F, Jiang Y, Zhi H, Dong Y, Li H, Ma S, Wang Y (2017) Artificial intelligence in healthcare: past, present and future. Stroke Vascular Neurol 2(4):230–243

Ju H, Jeon Y, Kim J (2015) A study on the hardware-based security solutions for smart devices. In: 2015 international conference on computational science and computational intelligence (CSCI). IEEE, pp 833–834

Kang J, Yin S, Meng W (2014) An intelligent storage management system based on cloud computing and internet of things. In: Proceedings of international conference on computer science and information technology. Springer, New Delhi, pp 499–505

Karras K, Pallis E, Mastorakis G, Nikoloudakis Y, Batalla JM, Mavromoustakis CX, Markakis E (2019) A hardware acceleration platform for AI-based inference at the edge. Circuits Syst Signal Process 39:1059–1070

Kempenaar C, Lokhorst C, Bleumer EJB, Veerkamp RF, Been T, Van Evert FK, Noorbergen H (2016) Big data analysis for smart farming: results of TO_2 project in theme food securit, Wageningen University & Research, p 655

Khosla R (2010) Precision agriculture: challenges and opportunities in a flat world. In: 19th world congress of soil science, soil solutions for a changing world, Brisbane, Australia

Kim DH, Ghaffari R, Lu N, Rogers JA (2012) Flexible and stretchable electronics for biointegrated devices. Annu Rev Biomed Eng 14:113–128

Liakos KG, Busato P, Moshou D, Pearson S, Bochtis D (2018) Machine learning in agriculture: a review. Sensors 18(8):2674

Lioutas ED, Charatsari C (2020) Smart farming and short food supply chains: are they compatible? Land Use Policy 94:104541

Liu Y, Ma X, Shu L, Hancke GP, Abu-Mahfouz AM (2020) From Industry 4.0 to Agriculture 4.0: current status, enabling technologies, and research challenges. IEEE Trans Indus Inf 17(6):4322–4334

Lyamin N, Vinel A, Jonsson M, Loo J (2013) Real-time detection of denial-of-service attacks in IEEE 802.11 p vehicular networks. IEEE Commun Lett 18(1):110–113

Menouar H, Guvenc I, Akkaya K, Uluagac AS, Kadri A, Tuncer A (2017) UAV-enabled intelligent transportation systems for thesmart city: applications and challenges. IEEE Commun Mag 55:22–28

Merchant M, Paradkar V, Khanna M, Gokhale S (2018) Mango leaf deficiency detection using digital image processing and machine learning. In: 2018 3rd international conference for convergence in technology (I2CT), pp 1–3. https://doi.org/10.1109/I2CT.2018.8529755

Mitra A, Mohanty SP, Corcoran P, Kougianos E (2020) A novel machine learning based method for deepfake video detection in social media. In: 2020 IEEE international symposium on smart electronic systems (iSES) (Formerly iNiS). IEEE, pp 91–96

Mitra A, Mohanty SP, Corcoran P, Kougianos E (2021a) Detection of deep-morphed deepfake images to make robust automatic facial recognition systems. In: 2021 19th OITS international conference on information technology (OCIT). IEEE, pp 149–154

Mitra A, Mohanty SP, Corcoran P, Kougianos E (2021b) EasyDeep: an IoT friendly robust detection method for GAN generated deepfake images in social media. In: IFIP international internet of things conference. Springer, Cham, pp 217–236

Mitra A, Mohanty SP, Corcoran P, Kougianos E (2021c) iFace: a deepfake resilient digital identification framework for smart cities. In: 2021 IEEE international symposium on smart electronic systems (iSES) (Formerly iNiS). IEEE, pp 361–366

Mitra A, Mohanty SP, Corcoran P, Kougianos E (2021d) A machine learning based approach for deepfake detection in social media through key video frame extraction. SN Comput Sci 2(2):1–18

Mitra A, Vangipuram SL, Bapatla AK, Bathalapalli VK, Mohanty SP, Kougianos E, Ray C (2022) Everything you wanted to know about smart agriculture. arXiv preprint arXiv:2201.04754

Mohr S, Kuhl R (2021) Acceptance of artificial intelligence in German agriculture: an application of the technology acceptance model and the theory of planned behavior. Precision Agric 22(6):1816–1844

Nag A, Mukhopadhyay SC, Kosel J (2017) Wearable flexible sensors: a review. IEEE Sens J 17:3949–3960

Palaus M, Marron EM, Viejo-Sobera R, Redolar-Ripoll D (2017) Neural basis of video gaming: a systematic review. Front human Neurosci 248–288

Petit J, Shladover SE (2015) Potential cyberattacks on automated vehicles. IEEE Trans Intell Transp Syst 16(2):546–556. https://doi.org/10.1109/TITS.2014.2342271

Poppe K, Wolfert J, Verdouw C, Renwick A (2015) A European perspective on the economics of big data. Farm Policy J 12(1):11–19

Qazi S, Sabir F, Khawaja BA, Atif SM, Mustaqim M (2020) Why is internet of autonomous vehicles not as plug and play as we think? Lessons to be learnt from present internet and future directions. IEEE Access 8:133015–133033. https://doi.org/10.1109/ACCESS.2020.300933

Rahman F, Ritun IJ, Biplob MRA, Farhin N, Uddin J (2018) Automated aeroponics system for indoor farming using Arduino. In 2018 Joint 7th International Conference on Informatics, Electronics & Vision (ICIEV) and 2018 2nd International Conference on Imaging, Vision & Pattern Recognition (icIVPR), pp 137–141

Ram SK, Sahoo SR, Das BB, Mahapatra K, Mohanty SP (2020) Eternal-thing: a secure aging-aware solar-energy harvester thing for sustainable IoT. IEEE Trans Sustain Comput 6(2):320–333

Saiz-Rubio V, Rovira-Mas F (2020) From smart farming towards agriculture 5.0: a review on crop data management. Agronomy 10(2):207

Sahitya G, Balaji N, Naidu CD (2016) Wireless sensor network for smart agriculture. In: 2016 2nd international conference on applied and theoretical computing and communication technology (iCATccT). IEEE, pp 488–493

Schimmelpfennig D (2016) Farm profits and adoption of precision agriculture (No. 1477-2016-121190)

Sharma A, Sharma V, Jaiswal M, Wang HC, Jayakody DNK, Basnayaka CMW, Muthanna A (2022) Recent trends in AI-based intelligent sensing. Electronics 11(10):1661

Skinner G, Walmsley T (2019) Artificial intelligence and deep learning in video games a brief review. In: 2019 IEEE 4th International conference on computer and communication systems (ICCCS). IEEE, pp 404–408

Son D, Lee J, Qiao S, Ghaffari R, Kim J, Lee JE, Song C, Kim SJ, Lee DJ, Jun SW (2014) Multifunctional wearable devices for diagnosis and therapy of movement disorders. Nat Nanotechnol 9:397–404

Soni VD (2020) Emerging roles of artificial intelligence in ecommerce. Int J Trend Sci Res Dev 4(5):223–225

Strohmeier S, Piazza F (2015) Artificial intelligence techniques in human resource management—a conceptual exploration. In: Intelligent techniques in engineering management. Springer, Cham, pp 149–172

Sze V, Chen YH, Emer J, Suleiman A, Zhang Z (2017) Hardware for machine learning: challenges and opportunities. In: Proceedings of the IEEE custom integrated circuits conference (CICC), Austin, TX, USA, pp 1–8

Talib MA, Majzoub S, Nasir Q, Jamal D (2021) A systematic literature review on hardware implementation of artificial intelligence algorithms. J Supercomput 77:1897–1938

Tang M, Luo M, Zhou J, Yang Z, Guo Z, Yan F, Liu L (2018) Side-channel attacks in a real scenario. Tsinghua Sci Technol 23(5):586–598

Wolfert S, Ge L, Verdouw C, Bogaardt MJ (2017) Big data in smart farming—a review. Agric Syst 153:69–80

Xu S, Zhang Y, Jia L, Mathewson KE, Jang KI, Kim J, Fu H, Huang X, Chava P, Wang R (2014) Soft microfluidic assemblies of sensors, circuits, and radios for the skin. Science 344:70–74

Yang F, Shu L, Huang K, Li K, Han G, Liu Y (2020) A partition-based node deployment strategy in solar insecticidal lamps Internet of Things. IEEE Internet Things J 7(11):11223–11237

Zhang Y (2011) Design of the node system of wireless sensor network and its application in digital agriculture. In: 2011 international conference on computer distributed control and intelligent environmental monitoring. IEEE, pp 29–35

Zhu N, Liu X, Liu Z, Hu K, Wang Y, Tan J, Guo Y (2018) Deep learning for smart agriculture: concepts, tools, applications, and opportunities. Int J Agric Biol Eng 11(4):32–44. https://doi.org/10.25165/j.ijabe.20181104.4475

Index

A
Accountability, 15, 18
Adaptability, 122, 123, 131, 133, 139, 243, 264, 403
Architecture, 7, 8, 55, 56, 65, 152, 153, 177, 205, 212, 220, 231–234, 288, 317, 318, 323, 331, 401, 405, 452, 464, 465, 468–477, 480, 482, 483, 487, 488, 507, 559
Artificial Intelligence (AI), 3–7, 9, 11–18, 48, 103, 109, 114, 121, 122, 125, 127, 138, 139, 149, 163, 194, 195, 200, 204, 206, 318, 327, 332, 397, 400, 402, 403, 406, 408, 449, 507, 521, 533, 547, 548
Ashraf, 384
Automation, 4, 5, 13, 15, 19, 48, 54, 123–125, 139, 151, 211, 212, 390, 393, 398, 402, 407, 503, 507, 548, 553, 558

C
Census, 26
China, 47, 262, 319, 450, 529
Collaboration, 19, 47, 185, 311, 430, 551
Communication, 35, 51, 53, 123, 150, 154, 155, 179, 184, 196, 199, 205, 212, 232, 234, 235, 240, 242, 252, 285, 286, 293, 298, 326, 327, 402, 405, 423, 425, 427, 437, 453, 454, 497, 512, 548, 554–556, 560
Consumption, 6, 14, 15, 47, 83, 95, 127, 150, 195, 226, 241, 249, 252, 254, 275, 288, 294, 298, 307, 369, 375, 377, 381, 383, 392, 451, 517, 554
Culture, 353, 392, 393
  historical, 407, 453
  traditional, 342, 353, 389

D
Data analysis, 57, 109, 123, 124, 150, 154, 190, 244, 266, 300, 309, 316, 413, 453, 527, 528, 532, 552
Data collection, 4, 18, 53, 93, 109, 149, 151, 152, 155–157, 187, 199, 204, 209, 211, 226, 227, 288, 295, 299, 301, 306, 321, 398, 452, 453, 476, 477, 488, 553
Decision making, 110, 111, 135, 136, 178, 235, 281, 454, 468
Design, 29, 33, 34, 47, 48, 62, 65, 90, 97, 132, 199, 217, 221, 243, 254, 275–277, 297, 311, 318, 327, 330, 342, 365, 413, 463, 468, 477, 480, 482, 487, 488, 495, 500, 515, 517, 526, 529, 554, 556, 559
Digital technologies, 43–45, 48, 55, 56, 445, 449, 521
Diversification, 94

E
Education
  higher, 140
  system, 53, 98, 122, 180, 200
Empowerment, 47
Environment

K. Pandey et al. (eds.), *Artificial Intelligence and Smart Agriculture*, Advances in Geographical and Environmental Sciences, https://doi.org/10.1007/978-981-97-0341-8

economic, 99, 181, 251, 285, 369, 495, 498, 550
social, 15, 43, 99, 285, 549
Exchange, 29, 54, 160, 297, 362, 398, 399, 550
Expectations, 135, 204

F
Family, 80, 417
Fossil fuels, 98

G
GDP, 124, 231, 521
Globalisation, 252, 414, 415
Gulf Cooperation Council (GCC), 297
Gulf, the, 297

H
Hail, 49, 557
Heuristics, 7
History, 122, 126, 197, 233, 388
Housing benefits, 13

I
Identity
social, 48
India, 40, 43–47, 49, 51, 53, 54, 57, 62, 112, 114, 180, 189, 194, 232, 283, 294, 315, 521, 524, 526, 529, 530, 535, 536
Information Technology (IT), 24, 95, 98, 317
Infrastructure
skill, 186, 188
systemic, 183
Innovation, 18, 19, 40, 44, 45, 51, 79, 96, 97, 252, 310, 311, 344, 547, 548, 551, 560
Interactive learning, 483
Internet, 16, 44, 47, 54, 55, 127, 130, 131, 138, 151–154, 159, 180, 193–197, 204, 205, 217, 225, 227, 231–234, 237, 238, 240, 241, 247, 294, 326, 397, 398, 401, 403, 408, 425, 427, 431, 437, 441, 452, 477, 518, 533, 547, 548, 555
Islam, 128, 152, 154, 251
Israel, 342

L
Labour productivity, 125, 274, 284, 288, 416, 496, 500
Lifestyle, 415
Literature, 112, 165, 166, 188, 228, 452, 462, 463, 465, 468, 488

M
Massachusetts Institute of Technology (MIT), 196, 398
Ministry of Civil Service (MCS), 321
Music, traditional, 557

P
Participation, 97, 138, 418, 553
Policy
implementation, 52, 185
Poverty, 23, 79, 81, 179
Private sector, 288
Production, 4, 6, 24, 40, 44, 47–49, 54, 62, 79, 82, 91, 93, 95, 98, 107, 108, 110, 112, 122–125, 127, 128, 135–139, 148, 178, 180, 182, 193, 194, 198, 206, 231, 232, 245, 246, 249, 251–253, 258, 264–266, 275, 279, 282–284, 293–295, 298, 301, 303, 311, 315–317, 332, 337, 338, 342, 344, 355, 358, 364, 365, 370, 377, 383, 384, 387–390, 392, 396, 400–402, 406, 408, 414, 418, 422, 425, 427, 430, 438, 439, 441, 453, 496, 497, 522, 524–526, 531, 533, 535–537, 539–541, 547–550, 556, 558
Productivity, 4, 8, 9, 11, 13, 14, 17, 19, 43, 44, 48, 54–57, 62, 79, 83, 87–89, 94, 98, 121, 122, 131, 135–137, 139, 148, 149, 169, 179, 180, 188, 196, 199, 206, 210, 211, 232, 235, 238, 252, 254, 264, 266, 271, 273, 294, 295, 300, 303, 306, 311, 315, 316, 338, 383, 388, 391, 392, 397, 398, 401, 414, 416, 444, 496, 498, 510, 517, 521, 522, 524, 533, 535, 536, 538–541, 548, 553, 558

Q
Quantitative data, 11

R
Resilience, 17, 46, 55, 80, 97, 183, 264

Responsibility, 294

S
Satellite television, 155
Science, 23, 32, 33, 45, 46, 50, 109, 124, 148, 158, 163, 188, 189, 197, 228, 316, 338, 366, 399, 414, 422, 441, 538, 547
Security, 15, 16, 24, 43, 47, 57, 80–82, 103, 111, 123, 148, 169, 181, 185, 205, 206, 211, 231, 235, 237, 249, 251, 252, 264, 265, 274, 278, 285, 315, 318, 387, 401, 414, 428, 521, 541, 552, 553, 555–557, 560
Sharma, 84, 85, 243, 245, 341, 350, 364, 394, 400, 423, 548, 549, 553
Sport, 182
Stability, 277
Sustainable Development Goals (SDGs), 286, 293

T
Technology, cyber, 556, 560
Transformation
  socio-economic, 44
Transparency, 15, 16, 18, 265

U
Unemployment, 124, 553
United Kingdom (UK), 182, 540
United States (US), 112, 184, 194, 287, 468, 522, 525, 538, 548
Urbanisation, 80

W
Wages, 288
World bank, 123
World economic forum, 54

MIX
Papier aus verantwortungsvollen Quellen
Paper from responsible sources
FSC® C105338

FSC
www.fsc.org

If you have any concerns about our products,
you can contact us on
ProductSafety@springernature.com

In case Publisher is established outside the EU,
the EU authorized representative is:
Springer Nature Customer Service Center GmbH
Europaplatz 3, 69115 Heidelberg, Germany

Printed by Libri Plureos GmbH
in Hamburg, Germany